ORTHOPOXVIRUSES
PATHOGENIC FOR HUMANS

Shchelkunov   Marennikova   Moyer

Orthopoxviruses Pathogenic for Humans

With 112 Figures

 Springer

S. N. Shchelkunov
Biotechnology Vector Head
Dept. of Molecular Biology
    of Genomes
Center of Virology and
    Biotechnology Vector
Koltsovo 33159, Novosibirsk
Russia
snshchel@vector.nsc.ru

S. S. Marennikova
Biotechnology Vector
    Chief Researcher
Dept. of Molecular Biology
    of Genomes
Center of Virology and
    Biotechnology Vector
Koltsovo 33159, Novosibirsk
Russia

R. W. Moyer
Dept. of Molecular Genetics
    and Microbiology
University of Florida
College of Medicine
Gainesville, FL
USA
rmoyer@ufl.edu

Library of Congress Control Number: 2005923401

ISBN-10: 0-387-25300-9   eISBN: 0-387-25306-8   Printed on acid-free paper.
ISBN-13: 978-0387-25300-8

Printed in the United States of America.   (EB)

9   8   7   6   5   4   3   2   1

springeronline.com

Preface

The viruses belonging to the genus *Orthopoxvirus* of the family Poxviridae are among the pathogens heading the list of microorganisms that have had an important role in the interactions between the humankind and infectious agents. Until recently, smallpox, caused by variola virus, was the most dangerous epidemic disease of humans, spreading as a conflagration. The toll of this infection was a tremendous number of human lives. Only in the previous century, smallpox killed about 300 million people. The variola virus is unique in that the only sensitive host of this pathogen is the man; moreover, the case-fatality rate of smallpox may exceed 30%. Variola virus is a strict anthroponosis unable to be retained in wild nature in animal organisms.

Another orthopoxvirus–cowpox virus–occupies one of the most honorable places in the history of medicine. In 1796, already one hundred years before the kingdom of viruses was discovered by Dmitri Ivanovsky in 1892, the famous experiments of Edward Jenner commenced use of cowpox virus for infecting people in order to protect them from smallpox, thereby opening the era of vaccine prevention of communicable diseases.

The origin of vaccinia virus, which substituted cowpox virus during massive vaccinations of humans against smallpox about one hundred years ago under vague circumstances, is a great mystery for the modern science. It is yet unclear whether vaccinia virus is a result of long-term artificial selection of a highly immunogenic against smallpox and lowly reactogenic virus. This virus so far has not been discovered in nature. Nonetheless, the availability of effective live vaccine against smallpox involving vaccinia virus as the major constituent, the lack of natural reservoir of variola virus, and joint efforts of the world medical community under the auspice of the World Health Organization (WHO) allowed the most hazardous infectious disease, smallpox, to be defeated by 1977 for the first time in the history of humankind. Hitherto, this is a unique example of successful campaign that eventuated in eradication of epidemically dangerous human disease.

As the massive immunization against smallpox was accompanied by side effects and complications, WHO in 1980 recommended to stop the routine vaccination. This resulted in an ever vanishing protection of the majority of the world population against not only smallpox, but also other infections caused by cowpox and monkeypox viruses. Smallpox is beaten; however, the circulation of monkeypox virus, a pathogen capable of infecting a wide range of animals and humans, in the zone of tropical rainforests in Africa arouses the concern of the scientific and medical communities. Moreover, monkeypox virus causes a human disease similar to smallpox in its clinical manifestation and course. Monkeypox virus is less efficient in person-to-person transmission compared with variola virus. Nevertheless, if this virus acquires the capability of highly efficient transmission in the human population as a result of evolutionary changes, the humankind will face the problem more complex than that when it encountered variola virus, as monkeypox virus is zoonotic, making it virtually impossible to eradicate the pathogen and the corresponding disease. Potential penetration of monkeypox virus to other continents also presents a considerable menace. The human monkeypox outbreak in the USA in 2003 was the first recorded outside the African continent. This disease was imported into the USA with Western African animals intended as pets. This was the first alarm for the public health services worldwide.

Thus, the orthopoxviruses pathogenic for humans are still attracting a rapt attention of scientists as well as medical researchers and practitioners. A large amount of information about these viruses has been accumulated recently. This made us consider it timely and necessary to summarize the data, obtained in many laboratories of the world as well as in our laboratories, on biological, ecological, and molecular genetic features of these unique viruses, which have played and continue to play an important role in the history of humankind.

Acknowledgments

It is a pleasant duty for us to express a sincere gratitude to Galina B. Chirikova for her tremendous work on technical aspects of preparation of the monograph and translation of its major part as well as to Natalie S. Krylova and Viktor V. Gulevich for assistance in translation.

We are very grateful to Richard C. Condit for his fruitful participation in preparing Section 3.4.

Contents

List of Figures

List of Tables

Chapter 1

SMALLPOX IN HUMAN HISTORY

1.1 Introduction

The role of smallpox in human history has been discussed in detail (Fenner *et al.*, 1988; 1989). These references, while excellent, are out of print, but they remain available at many libraries. In addition, one of these, the superb reference *Smallpox and its Eradication* (Fenner *et al.*, 1988) is available from the World Health Organization on their WEB site (http://www.who.int/emc/diseases/smallpox/Smallpoxeradication.html).

There is also an excellent historical treatise, which dwells in depth on the role of smallpox in human events entitled *Princess and Peasants: Smallpox in History*, originally published in 1983 by Donald R. Hopkins. This book was recently reissued under the title of *The Greatest Killer—Smallpox in History* (Hopkins, 2002). A perusal of any of these sources will readily convince the reader that smallpox has had an enormous impact on human history for over 2000 years. We have relied heavily on these references, as the purpose of this Chapter is to provide the reader with an appreciation of the role of smallpox (variola major) in human history. It is difficult today, when the world has been freed of this dreadful disease for nearly thirty years, to appreciate the past terror and apprehension concerning smallpox, a disease ever present, prior to its eradication, throughout the world. Prior to the third quarter of the 20th century, when eradication succeeded, it was widely known and appreciated that epidemics had occurred and would again, the only variables being when and how severe. Parenthetically, thirty years after eradication, it is tragic that serious worldwide concerns have been rekindled for fear this virus might be intentionally released by terrorists as a bioweapon. The possibility that smallpox, or smallpox-like viruses, such as monkeypox or engineered recombinant poxviruses, might be deliberately released would knowingly undo what is arguably the single most significant medical achievement of mankind, namely, the eradication of smallpox from the face of the earth. This possibility poses a potential tragedy of the first order.

1.2   The Origins of Smallpox

Smallpox, a uniquely human virus, has no known animal reservoirs and therefore must rely on human-to-human transmission to be maintained in the population. Therefore, while considered an ancient disease, it should be appreciated that a requirement for the virus to become endemic was that sufficient numbers of susceptible individuals within a large enough population must exist to allow the virus to be sustained. The true origins of the virus are murky, but it is clear that in addition to a critical population density, dissemination from a point of origin depended on the emergence of commerce between nations and groups and armed conflicts both of which facilitated contact within different populations.

The virus as we know it probably originated in either Egypt or India no later than roughly 1000 B.C. In Egypt, the mummified remains of the pharaoh Ramses V (Ruffer, 1921; Figure 1.1), which could date as early as 1157 B.C., show pustular eruptions consistent with smallpox. Examination of several other mummies dating from this period or even earlier also suggested the presence of pustular lesions again consistent with smallpox (Ruffer & Ferguson, 1910). These mummies date to roughly 3000 years ago, which precedes reliable descriptions of the virus anywhere else by approximately 1000 years and makes a strong argument that the virus originated in Egypt and was then carried to India and Asia by caravans of commerce, ocean going vessels, or traveling armies. Consistent with this notion are reports dating even earlier from the 14th century B.C. describing Hittite attacks that ultimately defeated Egypt. A "pestilence" is described as having broken out among the Hittites, contracted from the Egyptians, which persisted for some 20 years, killing large numbers of people including at least two Hittite leaders. There is some but not-conclusive evidence that this epidemic could have been due at least in part to smallpox.

Figure 1-1. The mummified head of Ramses V, pharaoh of Egypt who ruled from 1150–1145 B.C., showing facial pustules believed to be consistent with smallpox.

In India, there are ancient Sanskrit writings "Charka Samhita" and "Susruta Samhita", which could be as old as 1500 B.C., that also describe a disease consistent with smallpox. It is interesting to note that Hopkins describes writings of an Indian scholar, Dhanwantari, who some 2000 years ago described a preventative procedure astonishingly similar to the procedure described in the last years of the 18th century by Jenner. Dhanwantari writes, "Take the fluid of the pock on the udder of the cow or on the arm between the shoulder and elbow of a human subject on the point of a lancet and lance with it the arms between the shoulders and elbows until the blood appears. Then, mixing this fluid with the blood, the fever of the smallpox will be produced" (Hopkins, 2002). There are also descriptions by Brahmin priests, who describe rituals and prayers directed toward the "Goddess of Smallpox" from 1000 B.C. to the birth of Christ. What is abundantly clear from the writing and descriptions is that smallpox existed in India as well as Egypt well before the birth of Christ.

Therefore, the most reliable evidence would suggest that the disease originated from either Egypt or India, but anecdotal evidence suggests that the virus was present in ancient Greece as well. Hippocrates (400 B.C.) has written references to a disease that could have been smallpox. Better evidence is provided by Thucydides, a resident of Athens, who described the "plague of Athens" that occurred during the Peloponnesian war. This pestilence lasted for a number of years, killed roughly 25% of the Athenian army as well as private citizens, and ultimately resulted in the introduction of the virus into Persia. This again would be consistent with virus that originated in Egypt, but entered Greece through seafarers through the port of Piraeus roughly 30–50 years before this war. Whether this was smallpox is not known with certainty, but the end result was a serious erosion of Athenian strength, which diminished their capacity for later conflict with the Spartans and their ultimate decline (Hopkins, 2002).

China in ancient times would have had the population to allow both epidemics and the virus to become endemic. Scholars estimate that the virus was introduced into China from the North about 250 B.C. An epidemic is described about 243 B.C., which, from descriptions, could have been smallpox. However, the first clinical descriptions date from Ko Hung in 340 A.D. It was roughly 200 years later, before Chinese writings describe the disease in either Korea or Japan with introductions into Korea likely in 583 A.D. and Japan in 585 A.D.

It is clear that while the virus was well established in North Africa, India, China, and Persia, there is no evidence of smallpox in Europe until far later, or roughly the 6th century A.D. Very likely, the virus was introduced during the Islamic invasions, which originated from North Africa and entered Europe via Iberia in the 7th and 8th centuries. In the first millennium, several notable writings were produced. Al-Razi, an outstanding Persian physician and philosopher who lived from 850–925 A.D., has been credited as the first to use animal gut for sutures and plaster of paris for casts. He produced many medically related texts including his most famous *A Treatise on the*

Figure 1-2. A translation in 1776 of the Abu Bakr Mohammad Ibn Zakariya al-Razi (864–930 A.D.) treatise on smallpox and measles.

Smallpox and Measles (Figure 1.2). In Japan, Ishinho described smallpox hospitals and the "red treatment", which was to completely cover rooms with red cloth, similarly outfit patients in red clothing, and then expose them to red light. By 1000 A.D., smallpox was endemic in the more densely populated regions of Europe and Asia encompassing North Africa and Spain in the west to Japan in the east. In some ancient cultures, smallpox was so devastating, that infants were not named until it was clear they had caught the disease and survived.

The establishment of an endemic infection in Europe was aided in great measure by the Crusades, which took place between European countries and those of South West Asia. At the same time, in Africa, trade caravans transversed the Sahara Desert to spread the disease into those countries of West Africa that had sufficient population to sustain the infection. The interior of Africa was largely spared, even though Arabs likely introduced the virus sporadically during these years, because these regions lacked the population necessary to sustain the infection. By the 16th century, smallpox was common throughout Europe but did not become a major problem until the 17th century (Carmichael & Silverstein, 1987). During the 17th and 18th centuries, the London Bills of Mortality provided accurate documentation of the nature and effect of smallpox on Europe. During the 18th century alone, smallpox killed five reigning European monarchs. The spread of smallpox from Egypt and India into Asia and Europe is summarized in (Figure 1.3). Well into the 20th century, epidemics had huge social, economical, and clinical impacts. In 1962, a Pakistani traveler initiated a smallpox epidemic in Rhondda, Wales, in the UK. Ultimately, 25 people contracted smallpox and 6 died. The public demanded vaccination and ultimately 900,000 people in South Wales were vaccinated. One of the last of the major European outbreaks occurred in 1972 in Yugoslavia. The Yugoslavian epidemic was apparently initiated by a Muslim pilgrim named Latin Muzza, who had return from Mecca via Iraq to his home in Kosovo. He likely contracted the disease while in Iraq where smallpox was active. Muzza, upon falling ill in Yugoslavia, visited a number of hospitals before being admitted to a Belgrade hospital. Unfortunately, the epidemic had progressed extensively before definitive diagnosis was made. Strict government measures were implemented in order to control this outbreak, yet there were 175 cases

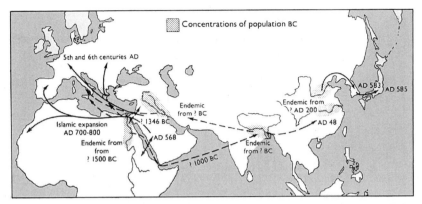

Figure 1-3. Proposed spread of smallpox within the ancient world (from Fenner *et al.*, 1988; reprinted with permission of the World Health Organization).

reported with 35 deaths before the epidemic ended (World Health Organization, 2004). Officials were so concerned that mass vaccination was employed as a control measure in Belgrade and Serbia with the result that 8,160,000 people were vaccinated out of a total population of 8,437,000. A similar vaccine strategy ensued in Kosovo, which also reported some cases, and 1,200,000 persons were vaccinated out of a total of 1,244,000.

1.3   The Spread and Effect of Smallpox on Naïve Populations

Extensive documentation details the effects of smallpox when introduced into naïve, previously unexposed populations. Up to the 1500s, the disease probably was not present to any significant degree in Southern Africa, the Americas, or Australia. In the Americas, it was probably the Spanish who brought the disease into the Caribbean in 1507 (Figure 1.4). The effect was to completely decimate the native island population, which encouraged the importation of African slaves to fill the population void. There is little doubt that the conquest of both Mexico and Peru by the Spanish was influenced by smallpox. An introduction of the virus in 1520 by Cortes on the mainland was a major factor in the devastation of the Aztecs by the Spanish invading armies. Similarly, Pizarro brought the disease to the Incas, which was a major factor in the downfall of that empire as well. In Brazil, missionaries carried the disease far into the interior of the continent.

In North America, smallpox was a major factor in the pattern of settlement of both the English and French and was a major factor in the political evolution of Canada and the United States (Stearn & Stearn, 1945). Initially, the effects of the disease, while devastating on individual introductions, had less overall effect because of the lesser density of native

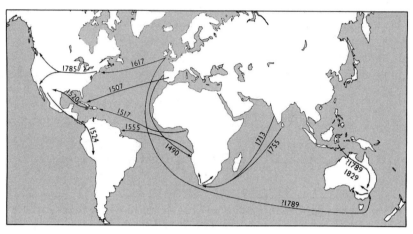

Figure 1-4. The spread of smallpox by Europeans into the Americas, Australia, and South Africa (from Fenner *et al.*, 1988; reprinted with permission of the World Health Organization).

populations. The virus was repeatedly introduced by seamen into North America with the result of devastation of the native population mostly at the site of introduction, which involved initially coastal tribes. As the virus spread inland, the Iroquois Nation in upstate New York suffered no less than five separate epidemics during the 17th and 18th centuries. In the 18th and 19th centuries, the Tripara and Mandan Tribes further west were similarly exposed and essentially destroyed with the result that the Sioux tribes annexed their territory as a consequence of a lessened ability to resist.

Smallpox was a major issue that factored into the British, French, and American strategies during the American Revolutionary war. For example, during the American siege of Boston from June 1775–March 1776, the length of the siege was protracted because of George Washington's reluctance to enter Boston, which was known at the time to have smallpox, and fear that the virus would be introduced into his American Army. Indeed, when the British finally departed, the city was initially occupied by troops who had survived smallpox. In Canada, smallpox was a major factor in determining that Canada would eventually come under British rule. In the winter of 1775–1776, the Americans were attempting to liberate Quebec province from British rule. The Americans captured Montreal and moved to attack Quebec City. Although the situation for the Americans looked promising, the British commander had citizens recently recovered from smallpox fraternize with the Americans. Smallpox broke out among the American troops and about half of the 10,000 soldiers ultimately fell ill followed by a retreat because the forces were too weakened to continue the battle.

Historically, one of the greatest of American Presidents was a victim of smallpox. Lincoln, who gave his famous Gettysburg address on April 19, 1865, fell acutely ill two days later. It is likely in retrospect that he was symptomatic the day he gave the address. While Lincoln survived, his death could have altered the course of American history as the Nation was in the midst of a civil war.

Introduction of the disease into Southern Africa was into Angola by the Portuguese. The virus was introduced into South Africa in 1713 via contaminated bedding and resulted in the decimation of the native Hottentots. Despite these introductions, the disease did not become endemic in Central Africa until the late 19th century.

Introduction of the virus into Australia, despite being an isolated continent, occurred within one year of European arrival in the city of Sydney. From 1829–1831, the disease broke out among the aborigines of Southeast Australia, which clearly facilitated European settlement of this region.

Smallpox raged throughout areas of the world from the 17th to mid-20th centuries despite the monumental discovery of vaccination by Jenner in 1796, which had long been recognized as being effective. It was only through the concerted efforts of the world community through the United Nations that smallpox was effectively eradicated with the last natural case occurring in October 1977, the patient being Ali Maow Maalin, a hospital cook in the town of Merak, Somalia.

1.4   Smallpox as a Historical Bioweapon

Given the influence of smallpox on world events and the common knowledge that once infected, a survivor was immune to the disease, it is not surprising that smallpox has been considered as a possible weapon. A couple of illustrations suffice. During the French and Indian wars, the British commander, Lord Jeffrey Amherst, for whom the City and University in the US State of Massachusetts are named, deliberately introduced the virus into the warring Indian factions. Amherst, in a letter to one of his officers, Colonel Henry Bouquet, in 1763 stated, "Could it not be contrived to send the smallpox among these disaffected tribes of Indians? We must on this occasion use every stratagem in our power to reduce them". Bouquet replied, "I will try to inoculate the Indians with some blankets that may fall into their hands and take care not to get the disease myself" (Duffy, 1951).

During World War II, both the British and Americans considered using smallpox as a deliberate weapon. One factor mitigating this was the fact that there was available a good vaccine, which was widely distributed. Therefore, further consideration was abandoned. In 1969, President of the United States, Richard M. Nixon officially banned development of any biological weapon. In 1972, the UK, US, and Soviet Union all signed the Biological Weapons Convention, superficially ending consideration of smallpox, by this time near eradication, as a weapon of war. Over the years, unfortunately, it appears that efforts to weaponize smallpox were not universally abandoned. In 1972, a Soviet defector, Dr. Kanatjan Alibekov (Ken Alibek) claimed to have been in charge of a bioweapons program designed to develop smallpox into an offensive weapon. Another soviet scientist, Dr. Vladimir Pasechnik, who died in 2001, also supported the thesis that the former Soviet Union had intensified development and perfection of an aerosolized form of smallpox through a company Biopreparat, which had been established in 1973 and which was reportedly active until the end of the Gorbachov era (Alibek, 1999). These activities have since been abandoned. While there is no official acknowledgment of efforts to weaponize smallpox, heightened terrorist activities and concern about rogue Nations have led to renewed efforts to develop better vaccines and ameliorative measures against smallpox infections.

1.5   Monkeypox Virus and Engineered Viruses: The Future of Smallpox and Smallpox-like Infections

Formally, the world is certified as "smallpox-free", as the last natural case occurred in 1977. However, there are two aspects of smallpox that merit vigilance for the future. First, there is monkeypox virus, which can cause serious disease in both monkeys and man. Then, there is the possibility of genetically engineering of monkeypox virus, clandestine strains of smallpox, or even the more attenuated orthopoxviruses to exacerbate or create a smallpox-like disease.

Monkeypox virus, an indigenous disease of rodents in equatorial Africa, causes a disease in monkeys and humans indistinguishable from smallpox. Fatality rates of humans in central Africa infected with monkeypox virus are similar to those of smallpox caused by variola major virus. However, monkeypox virus is less transmissible from human to human. Nevertheless, unlike variola strains, which are present in closely guarded deposits only in Russia and the United States, monkeypox virus is ever present in zoonotic reservoirs and poses a continuing threat to human populations, as the territory allocated to wildlife continues to shrink, thereby increasing the likelihood of contact between humans and wild animals.

Monkeypox was first recognized in 1958 at the State Serum Institute in Copenhagen (von Magnus *et al.*, 1959) in a colony of monkeys. The disease was noted in 1970 in humans living in tropical rainforest areas in western and Central Africa (Lourie *et al.*, 1972; Marennikova *et al.*, 1972a). The first case of human monkeypox was reported in Zaire in 1972 (Ladnyi *et al.*,

1972). It is now recognized that monkeypox is a zoonotic disease of rodents with transmission to both monkeys and man. Human infections are generally seen when humans increase their contact with forests where the natural animal reservoirs of the virus are located. Between February 1996 and October 1997, there were some 511 suspected cases in the Democratic Republic of the Congo (formerly Zaire). The increased frequency of the disease, while related in part to population movements resulting from the political instability of the region, also generated concern that the virus had somehow changed or mutated into a more virulent form. There were also some concerns that the virus might be a progenitor of smallpox virus; however, molecular studies have clearly indicated that this is not the case. The two viruses are distinct, and it is virtually impossible that monkeypox virus could naturally evolve into smallpox virus (Douglass & Dumbell, 1992; Shchelkunov *et al.,* 2001).

The aspects of human monkeypox that have garnered the most attention are the clinical similarities to smallpox and the similar fatality rates. The virus is classified as a possible emerging pathogen by the CDC in Atlanta, USA, and has received considerable attention. A perusal of the map of Africa showing the case distribution of monkeypox virus indicates two "pockets" of the virus, one in Central and the other in West Africa (see Figure 5.11; Jezek & Fenner, 1988). Evidence now suggests that fatalities are much more prevalent in Central than West Africa. Indeed, it now appears that there are two variants of the virus in circulation, and that the strain circulating in West Africa is considerably attenuated as compared to the Central African (Zairian) strain.

Further appreciation of the concerns about monkeypox by Public Health Officials is highlighted by a recent outbreak of the disease in the United States. In early June 2003, monkeypox was reported among several people in the United States. Ultimately, 72 cases were reported but no deaths. It is likely that humans became infected after coming in contact with wild prairie dogs. The prairie dogs in turn in all likelihood became infected after coming in contact with the Gambian giant pouched rat imported as part of a shipment of animals into the US on April 9 for use as pets. Since that time, importation of the Gambian rat has ceased as has importation of tree squirrels, rope squirrels, dormice, brush-tailed porcupines, and striped mice. Interestingly, unlike what was reported for monkeypox infections in the Congo, there were no deaths in the United States. A possible explanation for this dichotomy is that the strain imported into the US was the more attenuated West African strain of monkeypox virus. A second concern highlighted by the American outbreak, was the fear that the virus might become zoonotic in North American rodents. At the present time, at least for this outbreak, this does not appear to be the case, although the zoonotic potential of the virus cannot be ignored.

Another current concern stems from data in the scientific literature that suggest that poxviruses can be engineered in the laboratory to partially overcome the immune response. Ectromelia virus is an orthopoxvirus of mice and hence a member of the same virus family as smallpox. In certain strains of mice (BALB/c), the virus is virtually uniformly lethal, whereas in other strains (C57BL/6) the animals are relatively resistant to the virus. However, when a recombinant ectromelia virus engineered to express the cytokine IL-4 is used to infect mice, all animals including those of the normally resistant C57BL/6 strain develop systemic disease with uniformly high mortalities. A second equally sobering finding was the report that ectromelia expressing IL-4 was able to overcome the effects of vaccination (Jackson *et al.,* 2001). While more recent data suggest that vaccination can protect animals against most of the effects of the ectromelia IL-4 virus (R.M. Buller, personal communication), the concept of engineering poxviruses of increased virulence is a significant issue that argues for continued worldwide vigilance.

Chapter 2

CLASSIFICATION OF POXVIRUSES AND BRIEF CHARACTERIZATION OF THE GENUS *ORTHOPOXVIRUS*

Representatives of the family of poxviruses (Poxviridae) infect diverse members of the animal kingdom including insects, amphibians, and mammals. Taxonomically, the Poxviridae family is divided into two subfamilies of viruses: the poxviruses of vertebrates (Chordopoxvirinae) and the poxviruses of insects (Entomopoxvirinae). The first subfamily contains eight genera; the second, three. A number of poxviruses, particularly those of insects, remain unclassified.

Classification of the family of poxviruses and their general characterization are summarized in Table 2.1. Poxviruses that cause disease in humans belong to four different genera of the vertebrate poxvirus subfamily. The majority these viruses belong to the genus *Orthopoxvirus*. In recent years, representatives of this genus have expanded because of new data, modifications in the classification criteria, and discovery of new viruses. For example, one of the earliest *Orthopoxvirus* classifications assigned only four species (variola, vaccinia, cowpox, and mousepox viruses) to the vaccinia virus group. The most recent classification lists nine orthopoxvirus species and two subspecies of one of them (vaccinia virus) designated as buffalopox and rabbitpox viruses (Virus Taxonomy, 2000).

Current representatives of the *Orthopoxvirus* genus were originally isolated from humans and various animal species of the Old and New World. Among these viruses are both the pathogens displaying clinical infections in a wide range of animals (cowpox and monkeypox viruses) and the agents with a very selective host range (camelpox and variola virus). All the orthopoxviruses cause distinctive visible lesions—pocks on chorioallantoic membranes (CAMs) of developing chick embryos. In the infected cell, these viruses form cytoplasmic inclusions. The main characteristics of orthopoxviruses are listed in Table 2.2 (Fenner *et al.,* 1989; Virus Taxonomy, 2000). Two viruses (skunkpox and Uasin Gishu viruses) are omitted from Table 2.2, because they are yet only tentatively designated as orthopoxviruses (Virus Taxonomy, 2000) due to incomplete data. Skunkpox virus was isolated from a skunk in North America, and Uasin Gishu disease virus, causing a disease of horses, was isolated in Central Africa. While some believe (Fenner *et al.,* 1989) these two viruses are orthopoxviruses, official designation remains for the future.

In addition to the official prototypic virus for each species, based on international taxonomic considerations (vaccinia virus for orthopoxviruses), there are also "reference strains" for the majority of species. For example, the reference strain Elstree (Lister) represents vaccinia virus; strain Brighton, cowpox virus; strain Harvey, variola virus; and strain Copenhagen, monkeypox virus.

The orthopoxviruses are closely related with respect to their antigenic and immunologic characteristics as well as serologically related and provide immunological cross-protection. The virions of orthopoxviruses are also morphologically indistinguishable. Buller and Palumbo (1991) consider the

Table 2-1. Brief characterization and classification of the poxvirus family (Poxviridae)*
Family

Typically, virions are brick-shaped or ovoid with a size of 220–450 × 140–260 × 140–260 nm or 250–300 × 160–190 nm, respectively. The outer lipoprotein membrane of brick-shaped virions consists of tubular or spherical elements; of ovoid virions, is formed of helical filamentous structures (10–20 nm in diameter). The virion contains a biconcave or cylindrical core—the nucleoprotein complex comprising viral genomic DNA and proteins enclosed in membrane. One or two lateral bodies are located between the outer membrane and core membrane. Intracellular mature virions (IMV) are covered with membrane. Some of them have additional Golgi-derived envelope containing virus-specific proteins. Certain vertebrate viruses, for example, cowpox virus, may also be sequestered within inclusion bodies. The weight of the particle amounts to 5×10^{-15} g. Virions are usually sensitive to detergents, formaldehyde, oxidizing agents, and temperature over 40°C and are relatively stable at a room temperature when dried.

The nucleic acid—one double-stranded linear DNA molecule with a length of 130–375 kbp and covalently bound ends—comprises about 3 wt % of the virus particle.

The proteins represent 90 wt % of the virus particle. Depending on particular species, the genome encodes 150 to 300 proteins; the virion itself contains about 100 proteins. Virus particles contain numerous enzymes involved in DNA transcription and modification of proteins and nucleic acids. The virus within enveloped virions encodes the polypeptides of encompassing double lipid envelope.

The lipids compose 4 wt % of the particle. The virions with external envelope contain lipids, including glycolipids.

The carbohydrates amount to 3 wt % of the particle. Certain viral proteins, for example, hemagglutinin in the envelope of orthopoxviruses, may have *N*- and *C*-bound glucans.

All the poxviruses communicable to humans are zoonotic, except for variola (now eradicated) and molluscum contagiosum viruses.

continued

Subfamilies				
Vertebrate subfamily (Chordopoxvirinae)			**Insect subfamily (Entomopoxvirinae)**	
The shape is brick-like or ovoid; DNA G + C content is low (30–40%), except for parapoxviruses and molluscum contagiosum virus. Individual genera display a very pronounced (except for avian poxviruses) serological and immunological cross-reactivity. The neutralizing antibodies are genus-specific. Representatives of orthopoxviruses contain hemagglutinin antigens, which are rare in the other genera. Certain viruses produce pocks on developing chick embryo CAMs. Intraspecies DNA cross-hybridization is common; in certain cases, interspecies DNA cross-hybridization is reported.			The shape is brick-like or ovoid; DNA G + C content is 20%. The virions contain at least four enzymes identical to those found in vaccinia virus. Entomopoxviruses and chordopoxviruses are serologically unrelated.	
Genus	**Prototype**	**Species pathogenic for humans**	**Genus**	**Prototype**
Orthopoxvirus	Vaccinia virus	Cowpox, monkeypox, variola, vaccinia (and its subspecies, buffalopox and rabbitpox) viruses	*Entomopoxvirus A*	*Melolontha melolontha* entomopoxvirus (MMEV)
Parapoxvirus	Orf virus (contagious pustular dermatitis or contagious ecthyma virus)	Orf, paravaccinia (pseudocowpox), and bovine papular stomatitis viruses	*Entomopoxvirus B*	*Amsacta moorei* entomopoxvirus (AMEV)
Avipoxvirus	Fowlpox virus	–	*Entomopoxvirus C*	*Chironomus luridus* entomo-poxvirus (CLEV)
Capripoxvirus	Sheeppox virus	–		
Leporipoxvirus	Myxoma virus	–		
Suipoxvirus	Swinepox virus	–		
Molluscipoxvirus	Molluscum contagiosum virus	Molluscum contagiosum virus		
Yatapoxvirus	Yaba monkey tumor virus	Tanapox virus and Yaba monkey tumor virus (in case of accidental inoculation)		

*Data of *Virus Taxonomy* (2000).

Table 2-2. Genus Orthopoxvirus: species, natural hosts, regions of circulation, and main properties

Species	Natural host		Accidental host[b]	Properties					
	Species[a]	Geographic range		Pocks on CAMs[c]	Rabbit skin test		Ceiling temperature of pock formation on CAM	Type of intracellular inclusions	Pathogenicity for laboratory animals
					Intracutaneous	Cutaneous			
Cowpox virus	At least 14 species of wild terrestrial rodents and insectivores	Asia and Europe	Humans, cattle, domestic cats, white rats, etc., and several exotic species in zoos and circuses	Predominant—large, white flat hemorrhagic; single—white dense	Large dense hemorrhagic infiltrate with central necrosis	Confluent papulopustular rash with hemorrhages	Varies from 38.6 to 40°C depending on the strain	B and A	Varies from high to medium depending on the strain
Monkeypox virus	Tropical squirrels, several species of terrestrial rodents, and primates	Equatorial zone of Central and West Africa	Humans and several animal species	Predominant—small with central hemorrhages; single—larger and white dense	Large dense slightly hemorrhagic infiltrate with necrosis	Confluent papulopustular rash	39–39.5°C	B	High
Variola virus	Humans (the disease is eradicated)	Previously, all over the globe	Orangutans	Monomorphic white dome-shaped with distinct border	Insignificant infiltrate	No reaction	37.5–38.5°C	B	Low
Vaccinia virus	Not found		Upon inoculation, humans, cattle, horses, and several other animal species	Vary depending on the strain (white large and/or grayish flat with ulceration)	Dense infiltrate (with central necrosis in the case of certain strains)	Papulopustular rash	41°C	B	Varies from medium to high depending on the strain

Virus	Reservoir/host	Geography	Humans/cows	Lesion description	Infiltrate	Reaction	Temperature	Type	Pathogenicity
Buffalopox virus (subspecies of vaccinia virus)	Buffalos (?)	Southeast Asia, Egypt	Humans and cows (?)	Two types: white raised and grayish flat (similar to those of vaccinia virus)	Varies depending on the strain from small nodule to dense infiltrate	Scarce data	Varies from 38.5 to 41°C depending on the strain	B	Medium
Rabbitpox virus (subspecies of vaccinia virus)	Colonized rabbits	North America and Europe	Unknown	Grayish with hemorrhages	Extensive lesions with necrosis	Papular rash with hemorrhages	<40°C	B	High
Camelpox virus	Dromedaries and Bactrian camels	Asia and Africa	Unknown	White small monomorphic with hardly noticeable central hemorrhages	Insignificant infiltrate	Weak hyperemia	38.5°C	B	Low
Ectromelia virus	Colonized mice; natural reservoir is unknown	Europe	Unknown	Tiny (dot) white	Insignificant reddening	No reaction	39°C	B and A	Low
Raccoonpox virus	*Procion lotor*	North America	Unknown	Tiny white dense	Small nodule	No data	No data	B and A	No data
Taterapox virus	*Tatera kimpi*	West Africa	Unknown	Small white dense	Insignificant nodule	No reaction	38°C	No data	Low
Volepox virus	*Microtus californicus* and *Peromyscus truei*	North America	Unknown	Tiny white dense	Small nodule	No data	40°C	No data	Scarce data

Notes:
[a]For details, see the corresponding sections on ecology of these viruses;
[b]Incubation temperature was 34.5–35°C; incubation time, 72 h, except for vaccinia virus (48 h); and
[c]Onto scarified skin.

ability of one orthopoxvirus to non-genetically reactivate another virus (Fenner & Woodroofe, 1960) as a property by which to group orthopoxviruses. Molecular weights of orthopoxvirus DNA vary from 120 MDa (186 kb) for variola virus to 145 MDa (224 kb) for cowpox virus.

More details on the structure of orthopoxvirus virions, their morphogenesis, replication, DNA structure, and polypeptides are reported in Chapter 3.

According to restriction enzyme analysis (Mackett & Archard, 1979; Esposito & Knight, 1985), the DNA sequences in the central region of virtually all orthopoxvirus genomes display very similar patterns of restriction sites. This, now commonly accepted property, reflects a central core of conserved genes common to all orthopoxviruses (see Chapter 3). This conserved core region encodes the structural polypeptides and essential enzymes involved in the metabolism of viral nucleic acids, particularly, primary transcription (Wittek *et al.*, 1984a, b; Traktman *et al.*, 1984; etc.). The defining regions of orthopoxvirus genomes that distinguish one species from another are the left and right most terminal regions, which are unique for each species (Esposito & Knight, 1985; Ichihashi, 1990; etc.; for details, see Chapters 3–6).

As early as in 1930s and 40s, Craigie & Wishart (1936a, b) and Shedlovsky & Smadel (1942) defined the so-called LS antigen, composed of two components, thermolabile L and thermostable S, which could be isolated from vaccinia virus particles and considered to be a common component and feature of all orthopoxviruses. Introduction of this antigen into animals induced the formation of antibodies that reacted in complement-binding and precipitation tests with all the orthopoxviruses. However, these antisera, while useful in classifying unknown poxviruses as orthopoxviruses, were incapable of neutralizing these viruses. Recently, this antigen was better defined, and additional polypeptides have been detected, including virus-encoded nonstructural cell-associated polypeptides (Ichihashi, 1990).

Concurrently with discovery of LS antigen, NP antigen was isolated from purified virions of vaccinia virus following treatment with alkali (Smadel *et al.*, 1942). This nucleoprotein structure again contains antigens common to all the vertebrate poxviruses (Woodroofe & Fenner, 1962). The NP antigen induces formation of precipitating, and possibly, neutralizing antibodies. This antigen, again a mixture of components, comprises a number of polypeptides; some of these polypeptides are 40 and 64–65 kDa in size (Ikuta *et al.*, 1979; Kitamoto *et al.*, 1987).

In addition to NP and LS antigens, orthopoxviruses produce a hemagglutinin, also an antigen. Initially, hemagglutinin was discovered in vaccinia virus (Nagler, 1942), Subsequently, it was shown that the hemagglutinin is a glycoprotein with a molecular weight of 85 kDa localized to the lipoprotein envelope of EEV virions as well as to the surface of infected cells (Weintraub & Dales, 1974; Payne & Norrby, 1976; Ichihashi, 1977; Shida & Dales, 1981). The hemagglutinin, as the name implies, determines the phenomenon of hemadsorption and the hemagglutination of certain avian erythrocytes.

Stern and Dales (1976) identified a virion protein with a molecular weight of 58 kDa within surface tubular structures (as their major component) of vaccinia virus. This protein, ascribed to the group of protective antigens, induces antibodies that neutralize intracellular virus (see below) and block fusion of infected cells. During infection, two major forms of virus are formed. The first, intracellular mature virus (IMV), is found within infected cells. A small portion of the IMV undergoes further envelopment, is transported to the cell surface and to the outside of the cell, and released. This extracellular enveloped form of the virus (EEV) is required in animal models for production of serious clinical infections. These two forms of virus are equally infectious in cell culture but antigenically distinct. Some protective antigens are embedded within the lipoprotein membrane of EEV (see Section 3.4). These antigens are absent in the intracellular mature virions (IMV), which lack the outer EEV envelope (see Section 3.4; Ichihashi, 1990). IMV virions represent an overwhelming majority of infectious virus in homogenates of infected cells. Thus, these two morphologically distinct types of virions differ in the composition of their surface epitopes, i.e., are antigenically dissimilar. While EEV is a relatively minor component of the total infectious virions, it has been shown that EEV is required in animals for viremic dissemination of the infection to secondary tissues and organs (Appleyard *et al.,* 1971; Boulter & Appleyard, 1973; Payne & Kristensson, 1985).

As was mentioned above, progress in development of the methods for isolation and analysis of viral polypeptides allowed for discovery of a considerably larger number of previously unknown antigens in both the virion itself and extracts of infected cells. Undoubtedly, this collection of antigens plays a role in both the humoral and cell-mediated immunological responses of the host. Despite the identification of a variety of antigens, it is not completely clear which antigens are the most critical as determinants of serological identification and differentiation of orthopoxviruses.

A great deal of effort has been expended on identifying antigens that are species-specific. Many approaches using different monoclonal and polyclonal antibody preparations to proteins of individual members of this genus have been tested. The existence of species-specific antigenic distinctions was successfully demonstrated by monoclonal antibodies to monkeypox virus (Marennikova *et al.,* 1988a) and variola virus (Ichihashi, 1990) as well as by preliminary adsorption of sera with reference to the antigens common for orthopoxviruses. Using this approach, Gispen and Brand-Saathof (1974) succeeded in detecting species-specific antigens produced by vaccinia, variola, and monkeypox viruses. Maltseva and Marennikova (1976) used adsorption directly during the gel precipitation assay and demonstrated that this allowed for species-level differentiation of closely related orthopoxviruses. Esposito *et al.* (1977) also reported detection of species-specific antigens of variola, monkeypox, and vaccinia viruses.

Preliminary pre-adsorption of sera also was essential in detecting species-specific antibodies in sera of wild animals and convalescent patients recovering from orthopoxvirus infections. Pre-adsorption with heterologous and homologous viral antigens has been successfully used to detect species-specific antibodies using radioimmunoassay (RIA–adsorption; Hutchinson *et al.*, 1977), enzyme immunoassay (EIA–adsorption; Marennikova *et al.*, 1981; Mal'tseva *et al.*, 1984), and immunofluorescence (Gispen *et al.*, 1976).

Chapter 3

VACCINIA VIRUS

3.1 Origin of Virus

Surprising as it is, the origin of the virus used for billions of vaccinations and being the major contributor to the global victory over smallpox is yet vague. It has preserved the name *vaccinia* (derived from *variola vaccinae*, cowpox) since Jenner's time. Nevertheless, it is an independent orthopoxvirus, distinct from cowpox virus, used originally for vaccination. Probably, we will never know when and how the substitution occurred. There is indirect evidence that at least by the mid-19th century vaccinia virus rather than with cowpox virus was used for vaccination. The natural reservoir of vaccinia virus is unknown, making the issue even more intricate. Many scientists believe that the natural reservoir does not exist at all, and the virus is maintained only as laboratory strains.

Several hypotheses were put forward on the origin of vaccinia virus. Two of them, later discarded, assumed transformation of smallpox virus during passages in animals or during arm-to-arm vaccination, or variolation. Arguments against one of them are presented in Chapter 4. Concerning the other hypothesis, we refer to Baxby's (1981) remark that the long practice of variolation before Jenner's experiments did not yield a virus with vaccinia properties in any case. Sequencing of smallpox and vaccinia viruses also argues against such transformation (see Section 4.5.3 for details). Another hypothesis assumed transformation of cowpox virus during passages on skin of calves and other animals. This was advocated by Downie (1970), a prominent expert in poxviruses. One of the mechanisms of such transformation might have involved deletions and/or transpositions in the cowpox virus genome. Cowpox virus has the largest genome of all orthopoxviruses and frequently produces deletion mutants. Many of these mutants displayed certain properties of vaccinia virus yet preserving the ability to form type A inclusions, characteristic of cowpox virus and exhibited similar restriction endonuclease profiles (Archard & Mackett, 1979; Fenner *et al.*, 1989).

Bedson and Dumbell (1964) provided experimental evidence for the origin of vaccinia virus from recombination of cowpox and variola viruses. Such recombination could occur at the beginning of vaccination practice, when some vaccinees with the material containing cowpox virus were at the smallpox incubation period. The authors obtained manifold recombinants, some of which were close to vaccinia virus in their properties. Unfortunately, their paper, lacked DNA restriction assay data, preventing from an unambiguous conclusion.

One more hypothesis should be mentioned—the origin of vaccinia virus from a virus that has disappeared from the environment (at least, in Europe). In the 18th and 19th centuries, the causative agent of horsepox was considered such a virus. Even now, this hypothesis has its proponents (Baxby, 1981). Possibly, this hypothesis will be further developed while studying a vaccinia-like isolate recovered from a horse as well as other vaccinia-like isolates recovered from animals and humans. Find below a briefed relevant data.

With the exception of many human and animal cases apparently related to the vaccinia virus used for immunization, such relation was not evident for several isolates reported. For example, isolates having the properties of vaccinia virus were recovered from kidney cell culture of the monkey *Macaca cynomolgus* (Gispen & Kapsenberg, 1966), kidneys of a wild Zaire monkey (Shelukhina *et al.*, 1975), and two humans with a pox-like disease in Nigeria and South Africa (Bourke & Dumbell, 1972; Fenner *et al.*, 1989; etc.). In the last two cases, despite some specific phenotypic features of the isolates, DNA restriction assay showed that they were identical to anti-smallpox vaccines used in those countries (Carra & Dumbell, 1987). As for the isolates recovered from monkeys, they were considered as accidental contaminations (Fenner *et al.*, 1989), although received no sufficient investigation.

Data of recent decades obtained while studying agents of various pox-like diseases of cattle melted the skepticism that surrounded the existence of vaccinia virus natural reservoir. One of the diseases in question is buffalopox. Initially, emergence of this disease was considered related to infection transmission from recent vaccinees. This opinion was confirmed by the similarity between the viruses isolated from ill buffaloes in India and vaccinia virus (Mathew, 1967; Singh & Singh, 1967; etc.). However, buffalopox outbreaks upon cessation of anti-smallpox vaccination both in India and all over the world suggested that even if the mechanism mentioned above actually took place before, it was not the only one. Presumably, the buffalopox outbreaks during the post-eradication era were caused by a virus independently circulating in nature. Interestingly, a thorough study of the group of Indian isolates recovered from buffaloes, performed, on the one hand, as early as at the late 1960s–early 1970s (Baxby & Hill, 1969, 1971) and, on the other, decades later, after cessation of anti-smallpox vaccination (Fenner *et al.*, 1989; Dumbell & Richardson, 1993), detected two varieties of this pathogen. One is virtually indistinguishable from vaccinia virus, whereas the other displays several specific biological features, although the genomic structure of the latter is very similar this that of vaccinia virus too. The latter variant is considered the genuine buffalopox virus. This assumption might be more preferable by the analogy with cowpox virus. Evidently, circulation of buffalopox virus or vaccinia-like viruses in nature requires the natural carriers. It is yet unclear whether buffaloes are the carrier in question or they are just an indicator of infection, while the pathogen persists in other animal species.

A similar situation is observed in Brazil. Recently, a sufficiently large outbreak affecting dairy cattle and humans in Rio de Janeiro State attracted attention in this country (Damaso *et al.*, 2000). Comparative investigation of the isolate recovered during this outbreak (Contagalo) and several vaccine strains demonstrated that Contagalo was genetically related to one of these vaccine strains, namely, JOC, used previously in Brazil for anti-smallpox vaccination. However, unlike India, several other vaccinia virus-like isolates were recovered from rodents in Brazil during the last four decades. The first isolate, named Cotia virus (CV) was obtained as far back as during vaccination against smallpox (1961) from mice in southeastern Brazil (Lopes *et al.*, 1965). In 1979, six years after the vaccination was stopped, another virus close to vaccinia virus, named SPAnv, was isolated in the same region from the same rodent species; however, SPAnv differed from CV in a number of molecular biological characteristics (da Fonseca *et al.*, 2002). Interestingly, the ATI gene of SPAnv virus appeared identical to that of the vaccinia virus vaccine strain WR, which was among other strains used for immunizing the population in Brazil. In addition to the aforementioned isolates, one more vaccinia-like virus, BAV, was recovered from a wild rodent trapped at the edge of Amazonian forest (da Fonseca *et al.*, 1998). These data suggest that although the isolates mentioned are not a uniform group, they are closely related and sometimes identical to vaccinia virus. Two possibilities could be considered to explain these findings. The first explanation implies accidental introduction of vaccine strains into the environment during smallpox eradication and massive vaccinations, resulting in infection of animal populations of susceptible to vaccinia virus (rodents) and eventual emergence of the corresponding natural reservoir. In this process, the viruses adapted to new natural carriers, and certain changes in their properties could ensue. The other explanation is evidently an assumption that the vaccinia-like viruses initially existed in nature.

In addition to buffalopox virus and the vaccinia-like viruses isolated in India and Brazil, study of a virus isolated from an ill horse in Mongolia was recently reported (Tulman *et al.*, 2002). In the late 1960s–early 1970s, outbreaks of a horse benign disease, characteristic of which were pock-like eruptions on teats and other parts of the body, were repeatedly recorded in this country. Similarly to the situation in India, these outbreaks were frequently considered connected with infection transmitted from recent vaccinees (P.F. Kurchenko, personal communication). We have no information on whether there were any pox outbreaks among horses after the vaccination against smallpox was ceased. According to the data of Tulman *et al.* (2002), the isolate recovered from the horse and named horsepox virus is rather close to several vaccine strains in its genomic structure. However, as we see it, an insufficiency of the available data yet prevent from speaking definitely about the nature and origin of this virus.

The results reported by Bryant *et al.* (2000), who studied three strains isolated during outbreaks in a colony of white mice in Poland in 1986 and 1988, were quite unexpected. Initially, these strains were erroneously attributed to ectromelia virus. However, further study demonstrated that they were closely related to vaccinia virus. The authors believe that their finding suggests a possible existence of the natural vaccinia virus reservoir among rodents of central Europe. However, drawing this inference, the authors do not take into account that these data were actually obtained only with laboratory colonized mice.

Summing up the data related to the origin of vaccinia virus, it is evident that the main and basic question yet remains open. Probably, investigation of the structure–function genomic organization of the closely related orthopoxviruses will unravel the evolutionary relationships among the viruses and thereby shed light upon the question. In addition, broad-scale field studies directed to discover natural carriers of vaccinia virus might be most promising. So far, we can only state that at least a part of the isolated vaccinia-like viruses circulating in nature is derived from vaccine strains used for inoculation.

3.2   Subspecies of Vaccinia Virus

As mentioned in Chapter 2, the modern classification of orthopoxviruses includes, in addition to vaccinia virus itself, two naturally occurring subspecies—agents of buffalopox and rabbitpox.

3.2.1   Buffalopox

Buffalopox is recorded in several Asian countries and Egypt (Lal & Singh, 1973; Tantawi *et al.*, 1977; etc.). Outbreaks of this infection were most frequently recorded in India since the 1930s of the last century until nowadays. The majority of outbreaks were related to contacts with recent vaccinees, especially during the period of massive vaccination of the population against smallpox. However, it appeared that buffalopox outbreaks continued even after the cessation of anti-smallpox vaccination (Dumbell &Richardson, 1993; Kolhapure *et al.,* 1997).

Characteristic of this buffalo disease itself is development of pox lesions on teats, udder, and rarer, on other parts of the body (perineum, hips, ears, etc.). In the case of suckling buffaloes, lesions are mainly located in mouth cavity, tongue, and lips. During the outbreaks, dairy and animal care staff usually contracts the disease. As a rule, ill people develop skin eruptions; although, disseminated lesions were also reported (Kolhapure *et al.,* 1997). In 1992–1996, the authors observed children who had no contacts with ill buffaloes yet got infected with buffalopox, which suggested the possibility of infection transmission from their ill parents. Cases of positive seroconversion in persons without any clinical manifestation of the disease were also recorded. Usually the course of human buffalopox is benign. There were also some cases when besides buffaloes and humans, other animal species, such as white cattle, horses, and goats contracted the disease (Mathew & Mathew, 1986).

As for the buffalopox agent itself, its isolation and studies commenced in the late 1960s (Mathew, 1967; Singh & Singh, 1967). The authors, who studied the virus isolates recovered from ill animals during outbreaks in various regions of India, obtained similar results. According to their data, the pathogen in question in its biological properties appeared very close to vaccinia virus. This similarity was also confirmed by serological studies (Lal &Singh, 1973).

A comprehensive study of the isolated strains of buffalopox virus, however, demonstrated that they were heterogeneous in their properties. Baxby and Hill (1969; 1971) were first to discover the heterogeneity in question as early as in 1971. These scientists examined four isolates and found that three of them were indistinguishable in their biotype from vaccinia virus, whereas the fourth (named virus A) differed in the ceiling temperature of pock formation on CAM, pock size, and some other features. These data formed the background to regard the virus in question as buffalopox virus. However, subsequent DNA restriction mapping showed that all the four strains were identical to vaccinia virus (Fenner *et al.,* 1989). This was confirmed by the electrophoretic patterns of intracellular polypeptides induced by virus A (Harper *et al.,* 1979).

Virtually the same situation was observed during buffalopox outbreaks in 1985–1987, six–eight years after the discontinuation of routine vaccination. Along with typical vaccinia virus isolates, strains similar to the aforementioned virus A were recovered. All of them had vaccinia-like

*Hin*dIII restriction profiles. However, digestion with *Pst*I restriction endonuclease revealed distinctions from both vaccinia virus strains used for vaccination and virus A (Fenner *et al.*, 1989; Dumbell & Richardson, 1993).

It is assumed that natural carriers are involved in circulation of buffalopox virus and other vaccinia-like viruses in nature (see Section 3.1).

3.2.2   Rabbitpox

Rabbitpox in its main biological features resembles vaccinia virus. However, on the other hand, unlike vaccinia virus, it can cause a highly contagious disease of colonized rabbits with a high mortality rate. Such outbreaks were recorded in several rabbit breeding facilities in the USA and Europe (Greene, 1933; Rosahn & Hu, 1935; Jansen, 1941; E. Shelukhina, unpublished data; etc.). Fenner investigated the virus isolated during one of such outbreaks at the Rockefeller Institute and suggested that the source of the rabbit epizootics was neurovaccinia virus, obtained as early as in 1920–1930s at the Pasteur Institute in Paris and then widely used in virological studies (Fenner, 1958; Fenner *et al.*, 1989). However, only some rabbitpox outbreaks were related to contamination with neurovaccinia virus; causes of the rest episodes remain vague.

As has been demonstrated, transmission of rabbitpox is airborne (Bedson & Duckworth, 1963). The clinical course of the disease varies from latent, manifesting itself only after various impacts on the rabbits, to acute and lethal, with lesions on the skin and mucous membranes. A rabbitpox variant causing animal death before the appearance of skin lesions was also described (Jansen, 1941; Westwood *et al.*, 1966).

3.3   Biological Features

The virion morphology, genomic organization, and encoded proteins are described in Section 3.4. Here, we describe the behavior of the virus in various biological systems and some other virus properties.

3.3.1   Pathogenicity for Animals

Generally, characteristic of vaccinia virus is a relatively high pathogenicity and a wide range of susceptible animal hosts. Numerous studies have demonstrated that under experimental conditions that vaccinia virus host range includes species of many mammalian orders: ungulates, primates, lagomorphs, rodents, etc. Birds (Levaditi & Nicolau, 1923) and reptiles (Sergiev & Svet-Moldavsky, 1951; Svet-Moldavsky, 1954) were also reported to be susceptible to this virus. The susceptibility of larger animals (cattle, buffaloes, sheep, goats, etc.) and the course of disease after scarification are well known from the experience of production of smallpox dermovaccine accumulated for over a century. It was pioneered by Negri in Italy, who introduced continuous passages of the virus on calf skin in 1840. This provided a constant source of the material for inoculation in a form of scrapings from the infected skin (detritus or calf lymph). In the long run, this innovation replaced everywhere the previously used vaccination method of arm-to-arm administration. In addition to cattle and calves, used for this purpose in Russia as well (Morozov & Solov'ev, 1948), buffaloes, sheep, and other animals were infected in several countries to obtain the detritus. As it was believed that continuous passages in one animal species (in particular, calves) reduced the activity of the virus, intermediate passages using other animal species or humans were practiced (Gamaleya, 1913). At first, it was retrovaccine, i.e., inoculation of the contents of child pustules to calves. Later, donkeys, rabbits, and other animals were used for this purpose. Intermediate passages in rabbits became most widespread (Morozov & Solov'ev, 1948).

Rabbits. This animal species is most frequently used in vaccinia virus studies because of its sensitivity to the pathogen in question. The first close experimental study of rabbit infection with vaccinia virus showed that application of the virus to epilated skin caused local lesions resembling those developed upon vaccination of humans (Calmette & Guerin, 1901). This data was repeatedly confirmed by other researchers; some scientists reported differences in the patterns of lesions caused by dermo-, neuro-, and testiculovaccines (Levaditi & Nicolau, 1923; Fenner, 1958; Herrlich, 1960; Solov'ev & Mastyukova, 1961). Unlike dermovaccine, neurovaccine applied onto the skin caused an extensive hemorrhagic necrotic lesion, generalization of the infection, and death of the animal. A similar course of the infection was observed with testicular vaccine.

Still more pronounced differences between these variants of vaccinia virus as well as other orthopoxviruses were observed after subcutaneous infection of rabbits. In these experiments, the administration of vaccinia virus dermal strains induced a dense infiltration of a doze-dependent size. Note also that this method was of wide laboratory use until the early 1960s for quantitative estimation of the virus content (Groth & Münsterer, 1935). Intracutaneous infection with neurovaccinia and testicular vaccinia strain caused extensive necrotic response with acute edema of underlying tissues and hemorrhages (Levaditi & Nicolau, 1923; Fenner, 1958; Solov'ev & Mastyukova, 1961).

Unlike smallpox virus, vaccinia virus can be maintained by passaging in subcutaneously infected rabbits (Ledingham & McClean, 1928; etc.). Interestingly, the adaptation of vaccinia virus was accompanied by a loss of the ability to reproduce on rabbit skin and increase in the virus reproduction upon subcutaneous inoculation.

Mal'tseva (1965) compared various dermal vaccinia virus strains using infection on scarified skin and subcutaneous inoculation of rabbits. At scarification, all the strains developed similar responses: confluent papular–pustular rashes on an edematous hyperemic area developing from papules through vesicles and pustules to scabs. In contrast, subcutaneous infection revealed clear differences between the strains. A common feature was the development of distinct dense infiltrates. Most strains caused necrotic lesions in the centers of the infiltrates. Two strains, Elstree and EM-63, produced smaller and less dense infiltrates without necroses, which involuted more rapidly. Mal'tseva also confirmed the formerly reported ability of some dermal vaccinia strains to cause encephalitis and the death of rabbits upon intracerebral administration. Only four of the twenty strains studied by Mal'tseva did not cause the death of rabbits or any significant changes in their general state, not mentioning a small weight loss and mild fever during the first 3 days post infection. Infection of rabbits with the rest strains induced a severe disease involving paralyses of legs and sphincters, convulsions, and death. The brains of dead rabbits contained high concentrations of the virus (up to 10^8 PFU/g). Similar experiments on isolation of the virus from the brains of rabbits infected with non-lethal strains showed that the virus disappeared from the brain tissue 3 days post inoculation. In earlier Fenner's experiments, seven of eight dermal vaccinia virus strains did not cause the death if rabbits (Fenner, 1958). All the variants of dermal vaccinia virus strains passaged in chick embryos strains studied by Marennikova (1958) were clearly neurovirulent for rabbits.

Administration of the virus into rabbits' testes induced orchitis, which led in some cases to testis atrophy. The degree of orchitis development (edema, hyperemia, and sclerosis) after administration of equal doses of various dermal vaccine strains varied from a short inflammation to grave orchitis with necroses and hemorrhagic lesions followed by generalization of the infection and death. The specificity of the process was confirmed in all cases by isolation of the virus from testicular tissue (Mal'tseva, 1965).

The results of intravenous infection of rabbits are determined mainly by strain-specific features. According to Herrlich and Mayr (1954), dermal strains and their derivatives cultivated on chick embryo CAMs and in tissue cultures did not cause apparent infection generalization when administered intravenously. Neural vaccine and testiculovaccine caused a lethal generalized process with rashes on the mucous membranes of the nose, mouth, esophagus, and conjunctiva and internal lesions.

Inoculation on the scarified cornea or anterior chamber caused specific keratitis in rabbits (Marennikova, 1962).

White mice. Like rabbits, white mice are susceptible to vaccinia virus. This specific process was reported to be caused by cutaneous, intranasal, intracerebral, and other ways of infection. Nevertheless, opposite results were also reported. This contradiction in early data on the results of infection in these animals is largely due to the pronounced age difference in susceptibility. This was convincingly confirmed by Duran-Reynals and Phyllis (1963), who showed that, as the age of mice increased from one to three weeks, increasing virus doses were required to cause a lethal generalized infection (in particular, with subcutaneous neural vaccine infection), whereas four-week-old mice were totally resistant to the infection. Therefore, we present only selected data on the susceptibility of these animals to different methods of infection, obtained by comparison of various vaccinia strains.

Intranasal infection of 2–5-day-old suckling mice showed that most vaccinia strains caused specific pneumonia ending fatally 5–8 days after inoculation. Only one (EM-63) of seven strains studied was comparatively low virulent. Infection with 7–8 log PFU/dose caused pneumonia ending lethally in only 25% of the sample. With this infection method, LD_{50} was significantly strain-dependent: It varied from <1 to 4.75 log PFU/dose (Mal'tseva, 1965). Fenner (1958) and Mal'tseva (1965) studied the neurovirulence of several dermal vaccine strains in laboratory mice. In Fenner's experiments with five-week-old mice, six strains were virulent with intracerebral infection, and two were moderately virulent. Mal'tseva intracerebrally infected mice of two age groups: 2–2.5-week-old and five-day-old sucklings. Both age groups outperformed rabbits as sensitive indicators of the ability of the virus to cause encephalitis: the number of strains causing illness and death was greater in both the 2–2.5 week-old group and, particularly, the suckling group. However, even in the latter case no neurovirulence was found in two of the strains studied (Elstree and EM-63).

Intravenous infection of 2–2.5-week-old mice with large doses led to animal death because of the toxic effect of the virus as early as in 24–48 h independent of the strain. Autopsy showed spleen enlargement, hemorrhagic lesions in the liver, hyperemia of intestine vessels, and blood effusion in the heart apex and other organs. Smaller doses caused neither death nor macroscopically visible changes in organs.

Guinea pigs. Formerly, this species was widely used for evaluating the specific properties of smallpox vaccines. Being administered to the scarified cornea or into the anterior chamber, the virus induced vaccine-related keratitis, which was the marker of activity of the preparation administered at various dilutions.

Owing to specific features of guinea pig skin, the response of these animals to cutaneous or subcutaneous virus administration and the resulting infiltration are significantly more massive than in rabbits. The vaccinia infection is accompanied by virus dissemination throughout internal organs: the liver, kidneys, lungs, and bone marrow (Herrlich, 1960).

White rats. It is generally accepted that these animals are little, if at all, susceptible to vaccinia virus. According to Svet-Moldavskaya (1970), they show no notable response even to intravenous administration of large virus doses. Intranasal infection of 5-day-old sucklings did not yield any disease either (Shelukhina, 1980). However, as reported by Herrlich (1960), application of the virus to the scarified rat cornea caused keratitis, and replication of the virus in testis tissue was observed after intratesticular infection. Total gamma irradiation of laboratory rats with sublethal doses (400 R) rendered them susceptible to vaccinia virus (Svet-Moldavskaya, 1968).

Cotton rats (*Sigmodon hispidus*). The susceptibility of these animals to vaccinia virus with various infection routes was closely investigated by Rytik *et al.* (1976), Gudkov *et al.* (1976), and Gudkov (1980). They found that cotton rats were most sensitive to intracerebral infection. The disease was characterized by adynamia, impaired coordination, and tonic and clonic spasms of back and limb muscles. Doses of 6 log PFU caused death of 80% of animals. The death rate after intranasal infection with the same dose was 66.6%; after intraperitoneal infection, 20%.

Concurrent experiments with white mice of 5–6 g in weight showed that they were significantly less susceptible than cotton rats. Virological examination of cotton rats at the acme of the disease confirmed the generalized mode of the infection. High concentrations of the virus were found in the brain, lungs, liver, kidneys, and blood. The pathomorphological patterns of the infection were clearly dystrophic and necrobiotic. The infection was accompanied by intense antibody production and development of a high resistance of immune individuals to subsequent administration of certainly lethal virus doses.

Investigation of close virus strains in this animal model revealed slight differences in their virulence and antigenic activity. As in mice, the susceptibility of cotton rats was inversely dependent on their age: the LD_{50} values for animals at the ages of 2, 4, and 20 weeks were 4.8, 6.2, and 7.0 log PFU/ml, respectively.

Primates. Monkeys are highly sensitive to vaccinia virus. The vaccine process following cutaneous infection resembles that in humans. Vaccinated monkeys show significant amounts of the virus at the inoculation site and in regional lymph nodes. The virus appears in blood on day 4 to 8 after infection. After infection with some strains, the virus was also isolated from nasopharynx and internal organs: spleen and liver (Shenkman, 1972; Marennikova, 1973). Earlier, Solov'ev *et al.* (1962) observed viremia in cutaneously infected monkeys before day 6. Differences in the response of monkeys to intravenous and intracerebral infection with various vaccinia strains were reported by Gendon and Chernos (1964). Of three strains tested by them, one was not virulent, whereas the other two caused 100% death of animals independently of infection routes. All three strains caused extensive lesions at the site of subcutaneous administration.

Infection of monkeys with the virus causes intense antibody production and resistance to subsequent infection with both vaccinia and smallpox viruses (Brinckerhoff & Tyzzer, 1906; Rao, 1952; Hahon, 1961; Solov'ev *et al.*, 1962). Similarly, vaccinia virus produces immunity to monkeypox virus, virulent for primates (McConnell *et al.*, 1968).

3.3.2   Behavior in Chick Embryos

As early as 24 h after the virus inoculation onto CAM, weakly notable surface lesions with indistinct borders are observed mainly at the inoculation site. After 48 h, the lesions grow larger and acquire specific morphologic features—pocks, either slightly convex, round, rather dense, and white or smaller flat grayish. White pocks can have central ulcers or dot hemorrhagic lesions. The latter are also observable in grayish pocks. The pock diameter varies among strains from 1.3 to 2.1 mm. After 72 h, the number of pocks increases dramatically doe to development of secondary pocks, similar in density but smaller in size, around the primary pocks. Note that some strains produce only white pocks (e.g., Elstree), whereas others develop both white and grayish flat (Figure 3.1). The differences are related to the presence of two different subpopulations of the virus within the population in the latter case. The death of the embryos caused by generalized infection occurs after 56–120 h. Pock-like lesions are observed on the embryo surface; necrotic foci, in the liver, spleen, and kidneys.

In addition to pock morphology, considered below in more detail, individual strains of vaccinia virus differ only in the when secondary pocks develop and embryos die. Some strains cause death by 56–72 h; others, by 96–120 h. Lesions on CAM also develop, although slower, after infection into the allantoic sac. However, their pattern is somewhat changed: The lesions are located closer to the surface and are more diffuse in shape. Pocks are fewer than after the infection with the same dose on CAM. Embryo death occurs at the same time that after inoculation onto CAM (Mal'tseva, 1965). She also showed that intravenous administration of large doses of the virus caused rapid (after 18–24 h) embryo death with severe hemorrhages on the body surface and in internal organs. Except for hemorrhages, no apparent lesions were observed on CAMs. Smaller virus doses caused embryo death by 72–144 h. In this case, pocks typical of this virus developed on CAMs. Pock-like lesions were detectable on the skin and mucous membranes of embryos; hemorrhagic and necrotic foci, in the liver.

Individual strains showed no differences in the pathogenicity for chick embryos—the developing pattern depended on the virus dose and infection route.

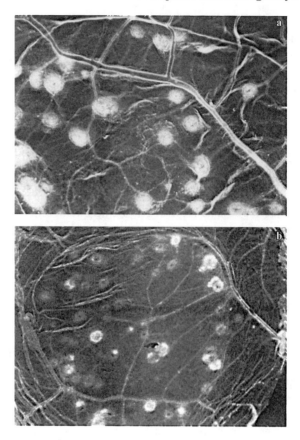

Figure 3-1. Vaccinia pocks on CAMs: (a) white pocks (strain
L-IVP) and (b) pocks of two types—white and flat grayish
(strain Tashkent).

3.3.3   Growth in Cell Cultures

By now, a great body of information has been accumulated on the
behavior of vaccinia virus in cell cultures. It has been shown that the range
of cells sensitive to this virus is very wide, and cytopathic effect (CPE) is
observed in all the cultures tested. Here are the main differences of the CPE
of vaccinia virus from those of variola and alastrim viruses: a faster
development, diffuse destructive mode, and early and ample formation of
large symplasts connected with each other and individual cells with
cytoplasmic bands (Figures 3.2 and 3.3). The cells are granulated, and their
boundaries are poorly distinguishable. Total degeneration of the cell layer

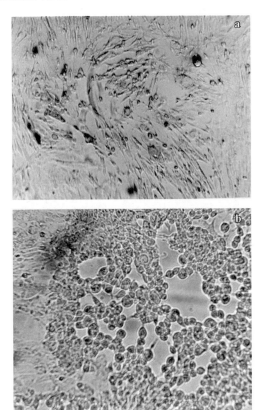

Figure 3-2. Initial stage of the CPE caused by vaccinia virus (24 h after inoculation) in (a) human embryo fibroblasts and (b) continuous cell line of human amnion.

occurs 2–3 days after the first signs of cytopathic transformations, independent of the infective dose (Yumasheva, 1959; 1968; Solov'ev & Mastyukova, 1961; Marennikova *et al.,* 1961a; Al'shtein *et al.,* 1961; Gurvich, 1964; etc.). Symplasts containing hundreds of nuclei were also observed in cell cultures infected with vaccinia virus. Many authors reported in cultures of various origins these features of the CPE of vaccinia virus. Comparison of several primary and continuous cell cultures showed that, although most of them are highly sensitive, some cultures (primary culture of monkey kidneys) were more susceptible, whereas others (HeLa) less susceptible to the virus (Gurvich, 1964).

Experiments with cell cultures of various organs of one species (rabbit embryo) revealed tissue-dependent differences in the rate of virus

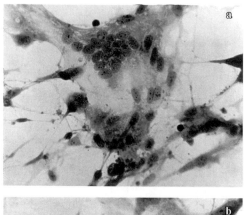

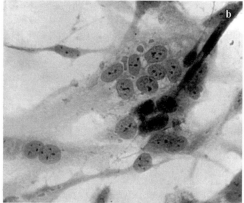

Figure 3-3. A preparation of human embryo fibroblast
culture 48 h after inoculation with vaccinia virus: (a) a
large symplast (magnification 7 × 20) and (b) a symplast
with clearly visible B-type inclusion bodies (×500).

accumulation and its yield (Solov'ev & Mastyukova, 1961). The maximal
virus titers were observed in cultures of skin–muscular and lung tissues.
These authors observed the same patterns in cultures from other animal
species and humans.

Cytoplasmic B-type inclusion bodies (Guarnieri bodies) are formed in
cells infected with vaccinia virus. Their number increases with time after the
infection and reaches its maximum at 48–72 h. The number of the bodies is
much less than in cultures infected with variola virus. Most cells contain 2–3
bodies, sometimes 8–11. Their shapes are diverse (Figure 3.3).

Like other orthopoxviruses, vaccinia virus causes hemadsorption in
infected cell cultures (Figure 3.4). It can be observed even before CPE
manifestation, sometimes, even without it, for instance, at very small virus
doses (Marennikova *et al.*, 1964).

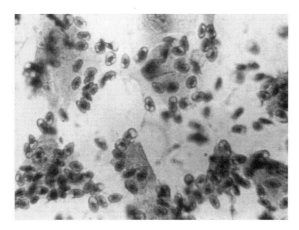

Figure 3-4. Hemadsorption phenomenon in a preparation of cell culture infected with vaccinia virus (magnification 8 × 40).

Probably, Noyes (1953) was the first to report the plaque-forming ability of vaccinia virus. Later, Mika and Pirsch (1960) showed that this ability also manifested itself in cell cultures under agar overlay. They also, like other authors (Porterfield & Allison, 1960; Solov'ev & Bektemirov, 1962; Gurvich & Marennikova, 1964; Mal'tseva, 1965) paid attention to the differences between vaccinia and variola viruses in plaque sizes and times of their development. The latter appear by 48 h, which is 48–72 h earlier than variola virus plaques and they are larger: 2.96–3.95 mm. The times when plaques and their sizes for vaccinia, variola, alastrim, and cowpox viruses are shown in Table 4.2. As with CAM inoculation (formation of white and grayish superficial pocks), various vaccinia virus strains form two plaque types in cell cultures: transparent and reticular (Gendon & Chernos, 1963; Gurvich, 1964; Mal'tseva, 1965). Transparent plaques have a lytic center and even or slightly diffuse edges. Microscopy examination reveals two zones in these plaques: the central lytic zone with few round faintly colored cells and the peripheral zone, consisting of adjoined granular cells with lost processes.

The reticular plaques are smaller, with irregular shape, and a fine reticular structure formed by degenerated cells. The plaques are bordered by a deeply colored zone of partly degenerating cells, intensely absorbing the stain (Figure 3.5). In the culture without agar overlay, the plaques form earlier, by 24 h, and increase in number by 48 h accompanied by formation of smaller secondary (daughter) plaques.

The ability to form plaques was used (Gendon & Chernos, 1963; Mal'tseva, 1965; etc.) for cloning vaccinia virus and, in particular, to isolate genetically identical clones from isolates containing mixed populations of the virus according to the Dulbecco method (1952).

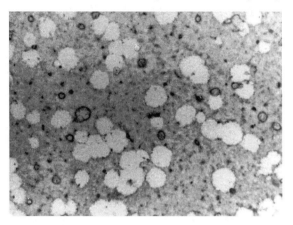

Figure 3-5. Transparent and reticular plaques formed by vaccinia virus in chick embryo fibroblast culture (strain Tashkent, day 6 post infection).

3.3.4   Hemagglutinin

As was shown, the hemagglutinin of vaccinia virus is produced at the late stage of the viral infection and accumulated in the cell membrane. This constituent of the lipoprotein envelope EEV is reported to be 85 kDa (Shida, 1986) to 89 kDa (Smith, 1993). One of the intriguing biologic functions of vaccinia virus hemagglutinin is its inhibiting effect on the virus-induced aggregation of infected cells, which eventually moderates the infection spreading (Oie *et al.*, 1990; Ortiz & Paez, 1994).

The hemagglutinating effect of the virus was first studied by Nagler (1942) and Clark and Nagler (1943), who found that vaccinia virus grown on CAM agglutinated erythrocytes of about 50% chicks. Later, it was found that the virus had an agglutinating effect, although less pronounced, on erythrocytes of chickens, doves, turkeys, laboratory mice, laboratory rats, and cats (Marennikova & Akatova, 1958; McCarthy & Helbert, 1960; Marennikova, 1962; Shafikova, 1970; etc.). The substrate used for virus growth influences hemagglutinin production to a great degree. Some authors noted, for instance, that the virus grown on animal skin (dermovaccinia) had no hemagglutinating effect, which was attributed to the presence of hemagglutination inhibitors in the scrapings or hemagglutinin degradation caused by necrotic processes in the skin (Nagler, 1942; Stone & Burnet, 1946; Marennikova, unpublished results; etc.). The effect may be weak in other substrates (suspensions of rabbit brains and testes) despite high titers of the virus (Solov'ev & Mastyukova, 1961). The relation of this phenomenon to the presence of hemagglutinin inhibitors in tissues was confirmed by Cassel *et al.* (1962). They found that vaccinia virus passaged in Ehrlich ascitic carcinoma and L cells cultivated in a medium with ascitic fluid, containing an inhibitor of vaccinia hemagglutinin, lost the hemagglutinating effect (Cassel & Fater, 1959; Cassel *et al.*, 1962). The presence of vaccinia hemagglutinin inhibitor in various body fluids was also reported by Szathmary (1961).

The highest rate of hemagglutinin production by the virus cultivated on CAM was observed by all authors. In addition to the substrate for virus growing, the hemagglutinating effect depended on the individual susceptibility of erythrocytes, their content in suspension, reaction temperature, pH, etc. Differences in the hemagglutinating effects of various vaccinia strains were reported by Mayr (1958) and Fenner (1958).

3.3.5 Virus Stability

The virus stability and, in particular, heat tolerance, are determined by its state and the ambient medium. This problem was comprehensively studied by Li Khe Min' (1955), who demonstrated that the virus withstood heating to 50°C in a liquid medium (CAM suspension), although a 3-h exposure decreased its titer significantly. At 60°C, the virus is completely inactivated after 2 h; at 70°, after 30 min; and at 80° or more, within 5 min. The virus freeze-dried under vacuum in the presence of serum retained its activity after three-month storage at 41° or after 2 h heating at 100°. As reported by Cross *et al.* (1957), smallpox vaccines prepared with 5% peptone as a stabilizer were used for vaccination of naive children with 100% efficiency after two years of storage at 45° and 37°.

Fenner (1958) compared the heat resistance of various dermovaccinia and neurovaccinia strains and discovered a strain whose activity was maximally decreased by a 4-min heating at 55°C. Similar results were obtained by Bedson and Dumbell (1961) using the same heating mode. Mal'tseva (1965) studied the heat stability of seven production vaccinia virus strains at five temperatures (37, 41, 50, 60, and 100°C) and found two strains with higher stability at all the temperatures except 100°C. Of interest is the observation by Kaplan (1958), who found that some strains contained virions differing in their heat resistances.

Bedson and Dumbell (1961) also determined the ceiling temperature of pock induction on chick embryo CAMs by various orthopoxviruses and found no correlation with heat resistance. However, the degree of virus pathogenicity for chick embryos did show such a correlation.

The pH optimum for vaccinia virus ranges within 7–7.8 (Akatova-Shelukhina, 1962). Both pH increase and decrease reduce the virus stability, whereas at pH 3.0, it is inactivated within 1 h. The virus is very

UV-sensitive. It is inactivated by UV irradiation faster than variola and alastrim viruses under identical conditions (Marennikova *et al.*, 1965).

Irradiation with radioactive cobalt eliminates the virus activity at 2×10^6 R. However, this dose retained partially the hemagglutinating and antigenic activities of the virus (Solov'ev & Mastyukova, 1961). This prompted to use γ-irradiation as a method for producing inactivated smallpox vaccine (Matsevich, 1970).

Many dyes degrade vaccinia virus even at low concentrations. A review of such dyes and data on the effect of other chemicals are given by Solov'ev & Mastyukova (1961).

3.3.6   Variability and Intraspecies Variation

Vaccinia virus is probably the most flexible of all orthopoxviruses pathogenic for humans. Its properties vary significantly depending on the number of passages and the biological species where it is passaged. A classic example of such variation is neurovaccinia, the virus variant obtained by multiple passages on rabbit brain (Levaditi & Nicolau, 1923; Solov'ev & Mastyukova, 1961) and testiculovaccinia, the variant adapted to rabbit testes. Both these variants became extremely virulent, first of all, for rabbits. They cause a grave lethal generalized disease of rabbits after inoculation into the brain, testes, or veins and extensive necrotic skin lesions after subcutaneous inoculation. Certain changes were observed in while passaging the virus on chick embryo CAMs. Mal'tseva (1965) reported that after 10 passages on chick embryos, some virus strains became pathogenic for rabbits and laboratory mice upon inoculation into the brain and for suckling mice with intranasal administration, which had not been the case before the passaging.

Multiple passages can also cause other changes. At the Bavarian State Institute of Vaccination, 500 passages of the Ankara strain in a chick embryo fibroblast culture yielded a variant with much lower pathogenicity for chick embryos, laboratory animals (including monkeys), and humans. The antigenic activity of this variant also decreased significantly (Stickl *et al.*, 1973). This variant of vaccinia virus, named MVA, changed not only its phenotypic features. As shown by Mayr *et al.* (1978), the molecular weight of the MVA strain DNA decreased by 9%. Other examples of changes in vaccinia virus features have been reported. The practice of preparation of smallpox vaccine using calves shows that addition of one to three intermediate passages in donkeys increases reactogenicity of the preparation drastically (Gamaleya, 1913; M.A. Morozov, personal communication).

Obviously, this property of the virus taken together with a wide diversity of passage schemes have determined the wide range of interstrain differences that we observe during the investigation, in particular, of commercial strains of the virus (see below).

3.4   Vaccinia Virus Molecular Biology

3.4.1   Overview

Vaccinia virus is a large (about 200 genes) double stranded DNA-containing virus with a complex enveloped virion. The virus replicates entirely in the cytoplasm of infected cells; thus, it must encode all of the enzymes and structural proteins necessary to transcribe the DNA genome, replicate the viral DNA, and assemble a complex membraned particle, all independently of a significant active contribution from the cell nucleus. The virus also encodes an array of functions designed to combat the host immune response. A summary of the virus replication cycle is presented in Figure 3.6. The following subsections address the details of the molecular biology of the virus structure, replication, and assembly, and molecular mechanisms of immune evasion.

3.4.2   Vaccinia Gene Nomenclature

Genetic nomenclature in poxvirus is an evolving process. The first poxvirus genome to be sequenced in its entirety was vaccinia Copenhagen (Goebel *et al.,* 1990). The open reading frame (ORF) designations adopted for the Copenhagen sequence were based on the *Hin*dIII restriction map of the genome and the direction of transcription. For example, the viral DNA topoisomerase was designated as H6R, the sixth ORF from the left end of the *Hin*dIII H DNA fragment, transcribed in a rightward direction. Because it was the first complete vaccinia sequence available, the Copenhagen nomenclature became the standard in the field. However, the vaccinia strain WR has been the *de facto* strain used for experimentation for over 30 years, even though the complete WR genome sequence only recently became available. Recent nomenclature rules adopted by the ICTV and applied to the more recently sequenced poxvirus genomes uses a linear designation beginning at the left hand of the genome and continuing to the right end of the genome. Using the newer nomenclature, the WR ortholog of the vaccinia Copenhagen H6R gene is designated as VACWR104. Therefore, while acknowledging the history of the field and the use of Copenhagen nomenclature in much of the literature to date, yet mindful that vaccinia WR is the prototypic vaccinia strain and that nomenclature procedures have been changed, we have designated ORFs by both the original Copenhagen and the newer vaccinia WR designations where possible.

3.4.3   Virion Structure

Mature vaccinia virions exist in four forms, which differ in the number of membranes and hence surface antigens surrounding the particle (reviewed in

Smith G.L. *et al.,* 2002). The four forms are designated intracellular mature virus (IMV), intracellular enveloped virus (IEV), cell associated enveloped virus (CEV), and extracellular enveloped virus (EEV) (Figures 3.6 and 3.7). IMV, the first assembled and simplest infectious form of virus, is a membraned particle, which as the name implies remains inside cells following virus maturation (Boulter & Appleyard, 1973). IEV is essentially IMV that has acquired two additional membranes via wrapping in Golgi derived cisternae (Ichihashi *et al.,* 1971). CEV is derived from IEV by fusion of the outermost IEV membrane with the plasma membrane; CEV remains attached to the outer surface of the cell (Blasco & Moss, 1992). EEV is CEV that has been released from the surface of the cell (Boulter & Appleyard, 1973). EEV and CEV are primarily responsible for spread of a poxvirus infection in culture and in infected animals.

The structure of intact IMV, the simplest and predominant form of the virus, has been extensively studied by electron microscopy (EM). Analysis

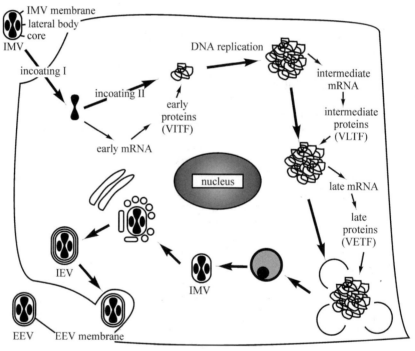

Figure 3-6. The vaccinia replication cycle. Features of growth shared by all poxviruses include a temporal pattern of gene expression, a cytoplasmic site of development. See text for details. IMV, intracellular mature virus; IEV, intracellular enveloped virus; EEV, extracellular enveloped virus; VETF, viral early transcription factors; VITF, viral intermediate transcription factors; and VLTF, viral late transcription factors.

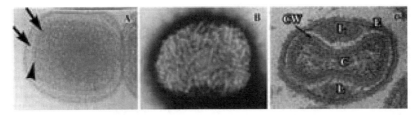

Figure 3-7. Vaccinia virion structure. A: Cryoelectron microscopy of positively stained isolated particles. Two layers are evident (arrows) that are separated by a spike or spicule layer (arrowhead). B: Negatively stained isolated particles showing surface tubule elements. C: Thin section of a vaccinia virion showing envelope (E), lateral bodies (L), and the core (C) surrounded by the core wall (CW). A is from (Griffiths *et al.*, 2001b); B, from (Wilton *et al.*, 1995); and C, from (Pogo & Dales, 1969). (Reprinted with permission of the National Academy of Sciences.)

of whole virions by cryoelectron microscopy, which is thought to preserve best the natural surface features of the virion, reveals a rectangular structure with dimensions of 350×270 nm, including a 30 nm thick "surface domain", which is less electron dense relative to the inner domain of the particle (Figure 3.7; Dubochet *et al.*, 1994; Griffiths *et al.*, 2001b). By contrast, conventional EM of negatively stained, dehydrated samples of purified IMV reveals numerous rod-like protrusions, called "surface tubule elements" (STE), on or near the surface of the virion (Figure 3.7; Wilton *et al.*, 1995). Cross sections of IMV reveal four distinct virion substructures: the envelope, the lateral bodies, the core wall, and the core (Figure 3.7; Dales & Pogo, 1981; Ichihashi *et al.*, 1984; Griffiths *et al.*, 2001a; 2001b). In many sections, the core and its wall possess a biconcave dumbbell or peanut shape, however in many sections, the core appears round, rectangular, or even somewhat irregularly shaped, ultimately suggesting a three dimensional structure similar to an elongated biconcave disk. The lateral bodies fill the space between the concavities in the core and the envelope. The envelope, which contains the lipid outer membrane, is also a bilaminar structure (Risco *et al.*, 2002).

The biochemical makeup of IMV has been investigated by controlled degradation of purified virions coupled with deductions made from studies of virus assembly. Virions contain as many as 100 proteins, most if not all of which are virus coded (Essani &Dales, 1979; Jensen *et al.*, 1996). The precise substructure of the virion envelope is controversial: while some studies conclude that the envelope contains a single lipid bilayer, others conclude that it contains two tightly apposed bilayers (Hollinshead *et al.*, 1999; Sodeik & Krijnse-Locker, 2002) (see Subsection 3.4.7, IV and IMV membrane formation, below). Treatment of virus with neutral detergent and a reducing agent solubilizes the envelope and as many as a dozen envelope associated proteins (Easterbrook, 1966; Ichihashi *et al.*, 1984; Jensen *et al.*, 1996).

Several of the IMV envelope proteins exist in a disulfide bonded complex (Rodriguez *et al.,* 1997). Upon solubilization of the outer envelope, the virion core, which now assumes a more rectangular or brick shaped structure, can be isolated with the lateral bodies still attached (Easterbrook, 1966). The lateral bodies can be removed with controlled trypsin treatment and therefore are proteinaceous in nature. Within the core, resides the viral DNA complexed with several virus-coded DNA-binding proteins (Ichihashi *et al.,* 1984).

Because poxviruses initiate infection and thereafter express information from a DNA molecule in the cell cytoplasm, the virions contain the full complement of enzymes required for synthesis of early viral mRNA. Virion cores are thus a rich source of the viral enzymes necessary for viral transcription and modification (Kates & McAuslan, 1967; Kates & Beeson, 1970a; Shuman & Moss, 1990b; Table 3.1). The subviral core is capable of synthesizing authentic early viral mRNAs *in vitro,* which are capped, methylated, and polyadenylated (Kates & Beeson,, 1970b; Wei & Moss, 1975; Bossart *et al.,* 1978; Pelham *et al.,* 1978). The enzymology of viral mRNA metabolism is discussed in detail below (see Subsection 3.4.5). The precise localization and organization of viral enzymes within the core is not known. In addition to the enzymes and factors listed in Table 3.1, which have an obvious direct role in RNA metabolism, the virion contains other enzymes with activities suggesting roles in nucleic acid or protein modification (Table 3.2).

Table 3-1. Vaccinia encoded mRNA metabolism enzymes and factors

Function Enzyme/factor[a] Encoding gene(s) (Common/Copenhagen/WR)[b]	MW (kDa)	Virion[c]	Homologies; alternate functions	Reference[d]
RNA polymerization				
RNA polymerase				
RPO 147/J6R/VACWR098	147	Yes	Homology to eukaryotic RNA polymerase	(Broyles & Moss, 1986)
RPO 132/A24R/VACWR144	133	Yes	Homology to eukaryotic RNA polymerase	(Amegadzie *et al.,* 1991a)
RAP 94/H4L/VACWR102	94	Yes	Early promoter specificity; early termination	(Ahn *et al.,* 1994)
RPO 35/A29L/VACWR152	35	Yes		(Amegadzie *et al.,* 1991b)
RPO 30/E4L/VACWR060	30	Yes	TFIIS homology; VITF-1	(Ahn *et al.,* 1990a)
RPO 22/J4R/VACWR096	21	Yes		(Broyles & Moss, 1986)

continued

Function Enzyme/factor[a] Encoding gene(s) (Common/Copenhagen/WR)[b]	MW (kDa)	Virion[c]	Homologies; alternate functions	Reference[d]
RPO 19/A5R/VACWR124	19	Yes		(Ahn *et al.*, 1992)
RPO 18/D7R/VACWR112	18	Yes		(Ahn *et al.*, 1990b)
RPO 7/G5.5R/VACWR083	7	Yes	Homology to eukaryotic RNA polymerase	(Amegadzie *et al.*, 1992a)
mRNA capping Capping enzyme D1R/VACWR106	97	Yes	Early termination, intermediate transcription	(Morgan *et al.*, 1984)
D12L/VACWR117	33	Yes	Early termination, intermediate transcription	(Niles *et al.*, 1989)
(Nucleoside-2'-O-)- methyltransferase VP39/J3R/VACWR095	39	Yes	Poly(A) polymerase subunit; intermediate/ late elongation	(Schnierle *et al.*, 1992)
mRNA polyadenylation Poly(A) polymerase VP55/VACWR057	55	Yes		(Gershon *et al.*, 1991)
VP39/J3R/VACWR095	39	Yes	mRNA cap methylation; intermediate/ late elongation	(Gershon *et al.*, 1991)
Early transcription factors VETF A7L/VACWR126	82	Yes		(Gershon & Moss, 1990)
D6R/VACWR111	73	Yes		(Gershon & Moss, 1990)
Early termination factors VTF D1R/VACWR106	97	Yes	mRNA capping, intermediate transcription	(Shuman *et al.*, 1987)
D12L/VACWR117	33	Yes	mRNA capping, intermediate transcription	(Shuman *et al.*, 1987)
Nucleoside phosphohydrolase I D11L/VACWR116	72	Yes		(Deng & Shuman, 1998)
H4L RAP 94/H4L/VACWR102	94	Yes	RNA polymerase subunit; early promoter specificity	(Mohamed & Niles, 2001)

continued

Function Enzyme/factor[a] Encoding gene(s) (Common/Copenhagen/WR)[b]	MW (kDa)	Virion[c]	Homologies; alternate functions	Reference[d]
Intermediate transcription factors				
Capping enzyme				
D1R/ VACWR106	97	Yes	Early termination, mRNA capping	(Vos *et al.,* 1991)
D12L/ VACWR117	33	Yes	Early termination, mRNA capping	(Vos *et al.,* 1991)
VITF-1				
RPO 30/E4L/VACWR060	30	Yes	RNA polymerase subunit; TFIIS homology	(Rosales *et al.,* 1994a)
VITF-3				
A23R/VACWR143	44	No		(Sanz & Moss, 1999)
A8R/VACWR127	33	No		(Sanz & Moss, 1999)
Late transcription factors				
VLTF-1				
G8R/VACWR086	30	No		(Keck *et al.,* 1990)
VLTF-2				
A1L/VACWR119	17	No		(Keck *et al.,* 1990)
VLTF-3				
A2L/VACWR120	26	No		(Keck *et al.,* 1990)
VLTF-4				
H5R/VACWR103	22	No		
Intermediate/late elongation factors				
Negative elongation/RNA release factor				
A18R/VACWR138	57	Yes		(Lackner & Condit, 2000)
Positive elongation factor				
G2R/VACWR080	26	No		(Black & Condit, 1996)
Positive elongation factor				
VP39/J3R/VACWR095	39	Yes	mRNA cap methyla- tion; poly(A) polymerase	(Latner *et al.,* 2000)

[a]Multiple genes indented under a single enzyme/factor designate a multisubunit enzyme/factor.
[b]See text for description of nomenclature. Many genes do not have a "common" name; thus, only the Copenhagen and WR designations are used.
[c]Designates whether or not the gene product is packaged in virions.
[d]A single reference is given that identifies the gene or assigns an alternate function. See text for additional references.

Table 3-2. Vaccinia virion enzymes[a]

Function Enzyme/factor[b] Encoding gene(s) (Common/Copenhagen/WR)[c]	MW (kDa)	Comment	Reference[d]
Protein modification			
F10L ser/thr kinase			
F10L/VACWR049	52	Required for virion morphogenesis	(Traktman *et al.*, 1995)
B1R ser/thr kinase			
B1R/VACWR183	34	Required for DNA replication	(Rempel & Traktman, 1992)
Tyr/ser protein phosphatase			
VH1/H1L/VACWR099	20	Required for early transcription	(Guan *et al.*, 1991)
Thiol oxidoreductase			
E10R/VACV-COP_082	11	Required for virion morphogenesis	(Senkevich *et al.*, 2002a)
G4L/VACV-COP_099	14	Required for virion morphogenesis	(Senkevich *et al.*, 2002a)
A2.5L/VACWR121	9	Required for virion morphogenesis	(Senkevich *et al.*, 2002a)
Glutaredoxin			
O2L/VACWR069	12	Non-essential	(Rajagopal *et al.*, 1995)
Nucleic acid topology			
Topoisomerase			
H6R/VACWR104	37	Function unknown	(Cheng *et al.*, 1998)
RNA helicase			
I8R/VACWR077	78	Required for early transcription	(Gross & Shuman, 1996)

[a]Enzymes in this table are those that have not been shown to play a direct role in transcription. See Table 3.1 for virion transcription enzymes.
[b]Multiple genes indented under a single enzyme/factor designate a multisubunit enzyme/factor.
[c]See text for description of nomenclature. Many genes do not have a "common" name; thus, only the Copenhagen and WR designations are used.
[d]A single reference is given that identifies the gene or assigns a function. See text for additional references.

3.4.4   Genome Organization

The vaccinia genome is a linear, double stranded DNA molecule with covalently closed hairpin ends, approximately 190 kbp in length. The genome contains relatively large (approximately 10 kbp) inverted terminal repeats. Near the ends of the terminal repeats are numerous short tandem repeated sequences of unknown function. The hairpin termini play a critical role in DNA replication (see Subsection 3.4.6). The inverted terminal repeats also contain genes; thus, the virus is diploid for these genes. The genetic organization of vaccinia genes parallels that for virtually all vertebrate poxviruses. While not absolute, essential genes required for virus growth cluster around the centralmost portion of the virus chromosome, which comprises about 60% of the total genome. Genes that reside outside this region, which vary considerably from virus to virus (Mackett & Archard, 1979), are generally not required for virus growth, generally associated with immune modulation and deflection of host defenses, and serve to provide each poxvirus unique and specific characteristics relating to host range in cell culture as well as virulence and patterns of disease in animals. The genetic map of vaccinia virus is shown in Figure 3.8 in conjunction with the *Hin*dIII restriction map. The variable regions at the both termini, the centralmost core genes, and the small terminal inverted repetitions (TIRs) at the terminal extremes of the variable regions of the genome are also shown (Figure 3.8 and Subsection 3.4.6).

Genes are very closely packed in the vaccinia genome; gaps of larger than 50 nt between coding sequences are rare, and in fact, genes frequently overlap by several nucleotides. Vaccinia genes do not contain introns, probably because of the cytoplasmic site of replication of the virus. Most genes within each terminal third of the genome are oriented such that transcription proceeds outward towards the termini of the genome, i.e., leftward on the left end and rightward on the right end. In the middle third of the genome, genes are interspersed in either transcriptional orientation. With the broad exception of the conserved and variable regions of the genome described below, there is no obvious logic to the distribution of genes on the genome with respect to function; however, genes often seem to be arranged in a fashion that would minimize transcriptional interference. Several additional features of the general organization of individual poxvirus genes are noteworthy.

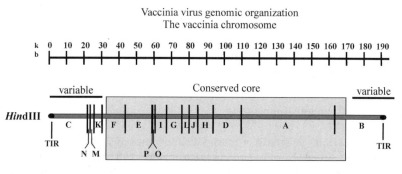

Figure 3-8. The organization of the vaccinia virus chromosome. The conserved orthopoxvirus virus core genes are indicated as are the variable gene regions located at both termini. The *Hin*dIII restriction pattern is shown. The small (~1–2 kb) terminal inverted repetitions (TIRs) at the both extremes of the chromosome are shown.

3.4.5 Vaccinia Transcription and Regulation of Viral Gene Expression

Gene expression during poxvirus infection occurs in a tightly regulated temporal cascade featuring sequential synthesis of early, intermediate, and late gene (the last two termed postreplicative) products. Mechanistically, gene expression is regulated at the level of transcription (Condit & Niles, 2002; reviewed in Broyles, 2003). The early, intermediate, and late gene classes are distinguished by class-specific transcriptional promoters and cognate trans-acting factors. In general, each gene class encodes factors required for transcription of the succeeding gene class, resulting in the observed temporal cascade of gene expression (Kovacs *et al.*, 1994; see Figure 3.6). Thus, early mRNAs, transcribed by enzymes packaged within infecting virions, encode intermediate gene transcription factors; intermediate genes encode late gene transcription factors; late genes encode early gene transcription factors, which are packaged into virions together with RNA polymerase, and other RNA metabolism enzymes for the next round of infection. Both intermediate and late transcription are coupled to DNA replication, i.e., inhibitors of DNA replication prevent synthesis of intermediate and late RNAs (Oda & Joklik, 1967; Vos & Stunnenberg, 1988; Baldick & Moss, 1993). Transcription of intermediate and late genes also requires synthesis of new RNA polymerase, which is synthesized early during infection (Hooda-Dhingra *et al.*, 1989).

All poxvirus genes apparently contain their own promoters. Regardless of gene class, poxvirus promoters are relatively simple and contained entirely within approximately 30 nt upstream of the transcription initiation site (Davison & Moss, 1989a; 1989b; Baldick *et al.*, 1992). Importantly, several genes have been described that contain compound promoters and are, therefore, expressed at multiple stages of infection (Wittek *et al.*, 1980; Broyles & Pennington, 1990).

As stated previously, because poxviruses replicate in the cytoplasm of infected cells, the virus encodes virtually all the enzymes required for regulated synthesis and modification of mRNAs suitable for translation in a eukaryotic environment. Thus, at least 23 virus genes have been described that play a direct role in viral mRNA synthesis and modification (Table 3.1). Interestingly, viral transcription may not be completely independent of the host; a limited number of host proteins have also been identified that may play a role in virus transcription (Rosales *et al.*, 1994b; Broyles *et al.*, 1999; Wright *et al.*, 2001).

In the paragraphs below, we first describe the enzymes and reactions common to synthesis and modification of all classes of poxviral mRNA and then focus in detail on differences in the expression of each gene class.

RNA polymerase. The virus-encoded RNA polymerase is a multisubunit enzyme that exists in two forms, one specific for early genes and one specific for intermediate and late genes (Table 3.1). The two forms of the enzyme contain eight subunits in common; the only difference between the two forms of the enzyme is that the early gene-specific enzyme contains one additional subunit, RAP94 (H4L/VACWR102; Nevins & Joklik, 1977; Baroudy & Moss, 1980; Ahn et al., 1994). Among the eight common subunits, the two largest, RPO147 (J6R/VACWR098) and RPO132 (A24R/VACWR144), share extensive homology with the largest subunits of both eukaryotic and prokaryotic multisubunit RNA polymerases (Broyles & Moss, 1986; Patel & Pickup, 1989; Amegadzie et al., 1991a). Another subunit, RPO30 (E4L/VACWR060), also shows homology with the eukaryotic transcription elongation factor TFIIS (Ahn et al., 1990a). Interestingly, this subunit also serves as an intermediate gene transcription factor (Rosales et al., 1994a). The smallest vaccinia RNA polymerase subunit, RPO7 (G5.5R/VACWR083), shows homology to the smallest subunit of eukaryotic RNA polymerase II (Amegadzie et al., 1992a). Neither the remaining four common subunits nor RAP94 show significant homology to other eukaryotic or prokaryotic proteins.

Both forms of the viral RNA polymerase, RAP94+ and RAP94-, are found in virions as well as in the cytoplasm of infected cells. The activities of these enzymes are mutually exclusive, i.e., the RAP94+ enzyme transcribes only early viral genes and not postreplicative genes, whereas the RAP94- enzyme transcribes only postreplicative genes and not early genes (Wright & Coroneos, 1995). Importantly, this means that the RNA polymerase that transcribes either early or postreplicative genes retains a memory of the class of promoter at which transcription was initiated. This impacts significantly on termination of transcription, which occurs differently on early and postreplicative genes, as described in more detail below.

mRNA capping and polyadenylation. All classes of poxviral mRNAs contain a 5' cap-1 structure, in which the first transcribed nucleotide is modified by addition of a 7 methylguanine in a 5'–5' triphosphate linkage, and is also methylated on the 2' hydroxyl group of the ribose moiety (Wei & Moss, 1975). The entire process is carried out by two virus-coded enzymes (Table 3.1). The cap 0 structure (5'–5' 7-methylguanine only) is added by the heterodimeric mRNA capping enzyme (D1R/VACWR106; D12L/VACWR117) (Shuman & Morham, 1990; Higman et al., 1994; Mao & Shuman, 1994). Methylation of the ribose on the first transcribed nucleotide is catalyzed by the J3R (nucleoside-2'-O-) methyltransferase (J3R/VACWR095; Barbosa & Moss, 1978a; 1978b; Schnierle et al., 1992). Interestingly, both the capping enzyme and the methyltransferase are multifunctional enzymes. The capping enzyme also participates in early gene transcription termination, intermediate gene transcription initiation, and may also play a role in telomere resolution during processing of replicated viral

DNA concatemers to monomeric genomes (Shuman *et al.*, 1987; Vos *et al.*, 1991; Carpenter & DeLange, 1991; Hassett *et al.*, 1997). In addition to its role in mRNA capping, the J3R (nucleoside-2'-O-) methyltransferase serves as a stimulatory subunit for the viral poly(A) polymerase and also as a transcription elongation factor for postreplicative genes (Gershon *et al.*, 1991; Latner *et al.*, 2002). These three activities of the J3R protein are independent of one another (Latner *et al.*, 2002).

All classes of poxviral mRNAs are polyadenylated by a virus-coded heterodimeric poly(A) polymerase (E1L/VACWR057; J3R/VACWR095) (reviewed in Gershon, 1998; Table 3.1). The poly(A) polymerase is relatively promiscuous and will polyadenylate virtually any available RNA 3'-end providing it contains uridylate residues positioned at specific sites near the RNA 3'-end (Gershon & Moss, 1993).

Early gene transcription. Early vaccinia genes are transcribed in the cell cytoplasm from partially uncoated virion cores. Early transcripts are detectable within 20 min following infection, they reach a peak of synthesis by 100 min following infection, and decline thereafter with a halflife of approximately 30 min (Baldick & Moss, 1993). Early gene products number at least 50, representing at least 25% of the coding capacity of the vaccinia genome, and include enzymes required for DNA replication, new viral RNA polymerase and capping enzyme, host defense functions, plus intermediate gene-specific transcription factors (VITFs), required for transcription from intermediate transcriptional promoters.

Transcription of early genes requires the viral early transcription factor, VETF (A7L/VACWR126; D6R/VACWR111; Gershon & Moss, 1990; Broyles & Fesler, 1990; Table 3.1). VETF is the only member of the poxvirus transcription factors that has been shown to bind directly and specifically to its cognate promoter (Broyles *et al.*, 1991; Cassetti & Moss, 1996). Upon binding to the promoter, VETF recruits RNA polymerase to the promoter and then dissociates, hydrolyzing ATP in the process (Broyles, 1991). As noted above, the RAP94 subunit of the RNA polymerase is also required for initiation at early promoters.

Termination of early gene transcription is an energy-dependent, sequence-specific event requiring virus-coded trans-acting factors (reviewed in Condit & Niles, 2002). The sequence that triggers early gene transcription termination is UUUUUNU, occurring in the nascent RNA (Yuen & Moss, 1987; Shuman & Moss, 1988). Termination occurs 30–50 nt downstream of this sequence. Three viral proteins are required for early termination: the capping enzyme (called the viral termination factor or VTF in the context of termination), RAP94 early RNA polymerase subunit, and DNA-dependent ATPase, called NPH I (D11L/VACWR116; Shuman *et al.*, 1987; Christen *et al.*, 1998; Deng & Shuman, 1998; Mohamed & Niles, 2001; Table 3.1). NPH I provides the energy required for termination and directly contacts

RAP94 (Mohamed & Niles, 2000). The activity of the capping enzyme in termination is independent of its mRNA capping activity (Luo *et al.,* 1995; Yu & Shuman, 1996). Otherwise, the precise roles of these three proteins in early gene termination and the mechanism of sequence recognition remain to be determined.

Intermediate gene transcription. Intermediate transcription is triggered by the synthesis of intermediate gene transcription factors coupled with the initiation of viral DNA replication (Vos & Stunnenberg, 1988). Intermediate mRNAs are detectable by 100 min post infection and peak by 120 min post infection, declining thereafter with a 30 min halflife (Baldick & Moss, 1993). The identification of intermediate genes from among postreplicative genes is not straightforward and requires one or more specific *in vivo* or *in vitro* tests. Only seven intermediate genes have so far been positively identified, and these encode a variety of functions including some host defense functions, DNA and RNA metabolism functions, and, importantly, late transcription factors (VLTF), which are required for recognition of late viral transcriptional promoters (Keck *et al.,* 1990; Zhang *et al.,* 1992; Xiang *et al.,* 1998). Many of the postreplicative genes, which are now classified as late, will probably ultimately prove to be intermediate when subjected to appropriate tests for expression. Intermediate (and late) vaccinia mRNAs contain an unusual 30-nt non-templated 5'-"poly(A) head" structure, which is added during initiation by slippage of the RNA polymerase at an AAA sequence, common to both intermediate and late promoters. No function for the poly(A) head has been demonstrated; however, it would provide a 5'-untranslated region for mRNAs transcribed from genes in which the promoter is directly abutted to the translation initiation ATG.

Four factors have been identified that are required for maximal expression of intermediate genes *in vitro:* the mRNA capping enzyme and three intermediate transcription factors—VITF 1, VITF 2, and VITF 3 (Vos *et al.,* 1991; Rosales *et al.,* 1994a; 1994b; Sanz & Moss, 1999; Table 3.1). VITF 1 (E4L/ VACWR060) and 3 (A8R/ VACWR127; A23R/ VACWR143) are virus-coded, while VITF 2 is a cellular protein. VITF 1 is identical to the 30 kDa RNA polymerase subunit RPO30. The cellular protein remains to be identified. The precise mechanism of action of these proteins in intermediate gene transcription remains to be determined.

Termination of intermediate and late gene transcription occurs by a similar mechanism and is fundamentally different than termination of early gene transcription (reviewed in Condit & Niles, 2002). During transcription of intermediate genes, the early transcription termination signal, UUUUUNU, is ignored, presumably because the RNA polymerase, which transcribes intermediate genes. lacks RAP94 (Condit *et al.,* 1996). Instead, RNA polymerase transcribing intermediate genes apparently terminates inefficiently at multiple sites, such that each intermediate gene is transcribed to yield an extremely 3'-end heterogeneous family of transcripts ranging in size from 1 to 4 kb, regardless of the size of the gene being transcribed (Mahr & Roberts, 1984). Genetic and biochemical experiments have identified three virus-coded factors—A18R (A18R/VACWR138), G2R (G2R/VACWR080), and J3R(J3R/VACWR095), which influence the length of intermediate transcripts, and therefore, play a role in intermediate gene transcription elongation or termination (Table 3.1). The A18R protein is likely to be a transcription termination factor, while G2R and J3R behave *in vivo* like transcription elongation factors (Black & Condit, 1996; Lackner & Condit, 2000; Xiang *et al.*, 2000).

Late gene transcription. Late viral mRNAs are detectable by 140 minutes post infection, and continue to be synthesized throughout infection (Baldick & Moss, 1993). Genes currently classified as late genes may comprise as much as 75% of the vaccinia genome; however, some of these are likely to be intermediate genes, as discussed above. Late mRNAs encode the majority of virus structural and assembly functions and virion enzymes, including the viral early transcription factor (VETF), which is packaged into newly assembled virions along with RNA polymerase, and RNA modification enzymes for use in the next round of infection. Like intermediate transcripts, late gene mRNAs contain a 5'-poly(A) head, synthesized via RNA polymerase slippage at a TAAAT initiator motif (Patel & Pickup, 1987; Ahn & Moss, 1989).

Genetic and biochemical experiments have identified at least five late transcription factors (VLTF), which are required for maximal transcription from late promoters (Table 3.1). VLTF 1–4 (G8R/VACWR086, A1L/VACWR119, 17 kDa, A2L/VACWR120, 26 kDa, H5R/VACWR103) are virus-coded factors. A fifth cellular factor, VLTF-X, which stimulates transcription from late vaccinia promoters *in vitro,* has recently been identified as consisting of the heterogeneous nuclear ribonucleoproteins A2/B1 and RBM3 (Wright *et al.*, 2001). The specific roles of each of the late transcription factors have yet to be determined.

Elongation and termination of late mRNAs appears to occur by mechanisms identical to intermediate mRNAs. Specifically, late mRNAs are 3'-end heterogeneous, and their lengths are regulated by the products of genes A18R, G2R, and J3R (Condit & Niles, 2002).

Additional regulation: host shutoff and RNA turnover. Sequential synthesis of transcription initiation factors accounts only for the onset of synthesis of successive classes of poxvirus gene products; additional mechanisms must regulate the shutoff of host gene expression and the decay in expression of early and intermediate genes as infection progresses. One factor that clearly plays a role in the decay of cellular and viral gene expression is a global reduction in stability of mRNA induced by poxvirus infection. For example, β-actin mRNA, which normally has a halflife of 10 h in uninfected cells, is rapidly degraded upon vaccinia infection (Rice & Roberts, 1983). In general, late viral mRNAs have a halflife of less than 60 min (Oda & Joklik, 1967). In addition, host gene expression may be altered by virus infection by inhibition of both cellular RNA polymerase and translation of cellular mRNA (Puckett & Moss, 1983; Cacoullos & Bablanian, 1993).

3.4.6   DNA Replication

Structural considerations. DNA replication not only serves to provide DNA for progeny virus, but also provides a functional template for the temporal expression of intermediate and late viral genes. Synthesis of DNA occurs within the cytoplasm and is readily visualized by electron microscopy as electron dense bodies or "factories", which serve as the site of initiation for the assembly of mature virions. The mechanism of DNA synthesis takes into consideration the following features of viral DNA already discussed: the TIRs, the terminal DNA hairpin loops, and the generation of replicative concatemeric DNA.

The onset of replication is heralded by introduction of a nick within one or both of TIRs, most likely within 200 bp of the end of the molecule (Du & Traktman, 1996). The nicking can occur at either or both termini, but for simplicity, is only shown at one end (Figure 3.9). Strand extension at the 3'-terminus of the nick with concomitant strand displacement allows the terminal nucleotides to be copied. Hybridization of the newly synthesized terminalmost nucleotides to the parental strand then allows polymerization to continue, eventually giving rise to concatemers comprised of complete genomic units of fused head-to-head or tail-to-tail genome length monomers (Moyer & Graves, 1981). Additional complexity can be conferred to replicating DNA through either strand invasion by other molecules, by initiating a second replication cycle on molecules actively replicating, or through recombination. There are a number of biochemical studies consistent with the proposed mechanism.

Concatemeric DNA must be resolved and unit length genomes excised in order to assemble mature virus. A graphic illustration of a possible mechanism for this process is shown in Figure 3.10, in which resolution is initiated by a sequence-specific nick. The detailed requirements for concatemer resolution include a small sequence (Merchlinsky & Moss, 1986; 1989; Delange *et al.*, 1986) present as an inverted repetition on both sides of the fused RR or LL hairpin loop contained within the L-RR-LL-R head-to-head or tail-to-tail concatemers that results from the replication process. It has been proposed that resolution may also occur via site-specific recombination and oriented branch migration (Merchlinsky, 1990), or by nicking and sealing of extruded cruciform, holiday like structures (McFadden *et al.*, 2003).

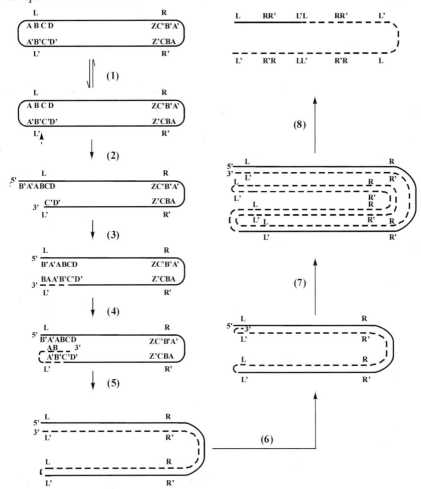

Figure 3-9. The replication of vaccinia virus DNA. The viral genome is indicated at the top left with "L" and "R" orientation of the molecule as indicated. Replication begins with introduction of a nick near one or both termini of the genome (indicated schematically by arrow between B' and C') followed by strand extension. Parental DNA is indicated by solid lines; newly synthesized DNA, by dashed lines. Following strand extension, self-primed replication creates a new hairpin, replication of the genome and creation of a concatemeric genomic dimer. See text for details (from Moyer & Graves, 1981; reprinted with permission of Elsevier).

It is possible for concatemers to exist as large cruciform structures and deletions can readily occur within such structures. During genome maturation, such deletion events allow for expansion and contraction of the terminal regions of the chromosome and creation of TIRs of varying length as described (Figure 3.11). The terminal variability is theoretically limited

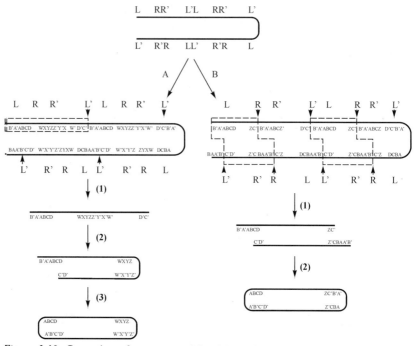

Figure 3-10. Generation of mature vaccinia virion DNA from replicative concatemers. Generation of monomeric genomic DNA is initiated by single-stranded nicking (B'C'), as indicated by the arrow. Genomic DNA can be derived in either a cis- (left) or trans- (right) fashion from the concatemer, as indicated by the red boxes. The designations L and R serve to orient the left and right termini of the molecules and where L' and R' sequences complementary to those of L and R respectively (from Moyer & Graves, 1981 reprinted with permission of Elsevier).

only by the need to retain essential genes in or at the conserved core and the need to retain cis-sequences required for replication (Moyer *et al.*, 1980b; Moyer & Graves, 1981) and provides a mechanism for rapid evolution of genes within this region. The documented ability of terminal regions of the virus to expand and contract implies that the amount of DNA packaged within a virion is not fixed and is determined by location of the concatemer resolution sequences rather than the amount of DNA within a given chromosome.

The enzymology of DNA replication. Although poxviruses encode many enzymes associated with aspects of DNA synthesis, those that are conserved among many members are likely to constitute the "basic" or core replication machinery (Table 3.3). A subset of these conserved proteins has been shown through the study of mutants to exhibit defects in DNA synthesis (Table 3.3, top). The role of other conserved genes is in post-synthetic steps of DNA maturation or unknown. The subset of proteins

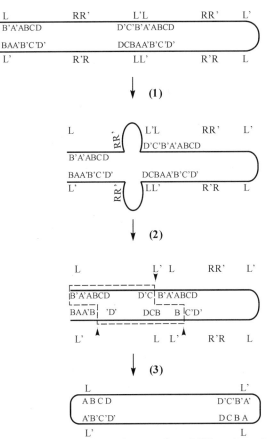

Figure 3-11. Mechanism for expansion and contraction of TIR regions. A replicative, concatemeric DNA intermediate is shown. Deletion from within the array (step 1) followed by molecular excision is governed by the hypothetical concatemeric resolution signal (B'C') and indicated by the red box. Limits of the deletion are theoretically determined by viability of the deleted genomic molecule. In the example shown, the TIR is expanded by addition of sequences from the left end of wild type molecule (from Moyer & Graves, 1981; reprinted with permission of Elsevier).

known to be involved in DNA synthesis include a DNA polymerase (ORF E9L, VACWR 065; McDonald & Traktman, 1994); a uracil DNA glycosylase (ORF D4R, VACWR 109; Upton *et al.,* 1993; Millns *et al.,* 1994; Stuart *et al.,* 1993); a serine/threonine protein kinase (ORF B1R, VACWR 183; Banham & Smith, 1992); an NTPase (ORF D5R, VACWR 110; Evans & Traktman, 1992; Evans *et al.,* 1995); and ORF A20, VACWR 141, a processivity factor (Ishii & Moss, 2001; Punjabi *et al.,* 2001), which acts in concert with the E9L DNA polymerase. Inactivation of any of these proteins leads to DNA negative phenotype and is lethal to the virus.

Table 3-3. Vaccinia genes involved in DNA replication. Poxvirus genes involved in DNA replication are divided into two groups: viral genes directly involved in the synthesis of DNA for which genetic evidence (ts mutants) exist (top) or those genes involved in replication and/or postreplicative events for which genetic evidence needed to conclude these genes are essential for replication is generally lacking.

VACCOP ORF	VACWR ORF	Size (AA)	Description
A20R	141	49.1 kDa (426)	DNA polymerase processivity factor
B1R	183	34 kDa (300)	Serine/threonine phosphokinase
D4R	109	25 kDa (218)	Uracil DNA glycosylase
D5R	110	90 kDa (785)	Nucleoside triphosphatase
E9L	065	116 kDa (1006)	DNA polymerase

Orthopoxvirus proteins required for other aspects of DNA replication

VACCOP ORF	VACWR ORF	Size (AA)	Description
A22R	142	22 kDa (187)	Holliday junction resolvase
H6R	104	36 kDa (314)	DNA topoisomerase
I3L	072	24 kDa (269)	SS DNA binding protein
A50R	176	63 kDa (552)	DNA ligase
D5R	110	90 kDa (785)	Nucleotide phosphatase

A number of the other conserved proteins are assumed to be required for later steps in the replicative process but direct genetic evidence is lacking for most of them. Mutations in these genes would be predicted to not lead to defects in DNA synthesis *per se*, but rather to defects in maturation of the newly synthesized DNA, which in turn would lead to inhibition of progeny virus formation. Examples of such genes would include a resolvase encoded by A22R (VACWR 142; Garcia *et al.*, 2000; Garcia & Moss, 2001); H6R (VACWR104), a type 1 topoisomerase (Shuman & Moss, 1987; Shuman *et al.*, 1989); and a single-stranded DNA-binding protein encoded by I3L (VACWR 072; Rochester & Traktman, 1998; Tseng *et al.*, 1999). The I3L protein localizes to virosomes or factories during DNA replication (Rochester & Traktman, 1998). However, elucidation of the precise role of both the H6R and I3L genes in the replicative process awaits isolation of conditional lethal mutants.

Recombination is an essential feature of DNA replication. Indeed, recombination between genetically marked viruses or transfected plasmids and viral DNA is rampant and in practice, facilitates generation of genetically engineered viruses. The processes of recombination and replication are intimately linked, and mutations that impair recombination are also impaired in DNA replication (Merchlinsky, 1989; Colinas *et al.*, 1990; Evans & Traktman, 1992).

Finally, there are a number of poxvirus genes devoted to the purpose of maintaining intracellular nucleotide pools, which serve as substrates for DNA synthesis. These enzymes, many of which are not absolutely essential for growth include (1) a thymidine kinase encoded by the J2R gene (VACWR 094; Hruby & Ball, 1982; Black & Hruby, 1990); (2) a thymidylate kinase encoded by the A48R gene (VACWR 178; Hughes *et al.,* 1991); (3) a dUTPase encoded by gene F2L (VACWR 041; Broyles, 1993); and (4) a heterodimeric ribonucleotide reductase comprised of the products of genes I4L (VACWR 073) and F4L (VACWR 043; Slabaugh & Mathews, 1986; Slabaugh *et al.,* 1988; Tengelsen *et al.,* 1988; Rajagopal *et al.,* 1995).

3.4.7 Vaccinia Morphogenesis

Electron microscopic examination of poxvirus-infected cells reveals discrete stages of virus morphogenesis with characteristic intermediate structures associated with each stage (Dales & Siminovitch, 1961; Dales & Pogo, 1981; Griffiths *et al.,* 2001a; 2001b; Sodeik & Krijnse-Locker, 2002; Risco *et al.,* 2002). The earliest evidence of poxvirus infection is the appearance in the cytoplasm of discrete granular foci, which are largely devoid of normal cellular organelles. These foci, called "viroplasm" or "viral factories", contain replicating viral DNA. As the infection progresses, viral factories increase in size. The first evidence of virion morphogenesis is the appearance within the viroplasm of rigid, crescent shaped membrane structures (cupules in three dimensions), which are precursors to the IMV membrane (Figure 3.12). The origin of these viral membranes is discussed in more detail below. Crescents evolve into circles (spheres in three dimensions), called immature virions (IV), which enclose relatively dense viroplasm (Figure 3.12). Some IVs contain internal "nucleoids", which are unbounded, extremely electron dense, DNA-containing structures presumed to be precursors to the virion core (Figure 3.12). Formation of these

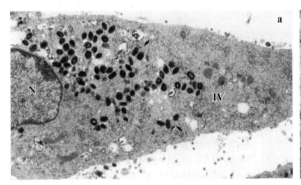
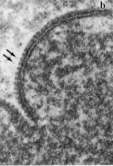

Figure 3-12. Electron microscopy of vaccinia virus infected cells: (a) low magnification field of infected HeLa cells 24 h post infection showing the characteristic accumulation of spherical IV and IVN (marked IV) and dense brick-shaped mature viruses (arrowheads) and (b) high-magnification field showing viral crescents (from Risco *et al.,* 2002; reprinted with permission of the American Society for Microbiology).

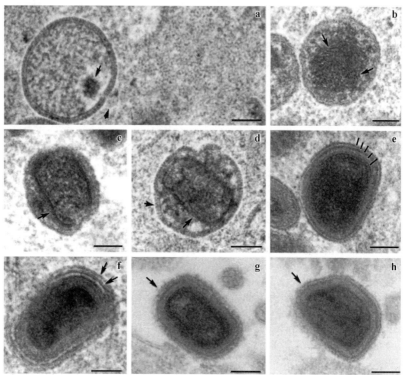

Figure 3-13. Intermediates in vaccinia virus assembly: (a) IV particle packing DNA (arrow); (b) spherical dense particles with fibrous, DNA-like, internal material (arrows); (c and d) potential intermediate maturation stages in the construction of the internal viral core (arrows) and the IMV; (e) IMV; (f) IEV with the additional double membrane (arrows); (g and h) two different section planes of EEV, which have an external fuzzy coat (arrows); bars, 100 nm (from Risco *et al.*, 2002; reprinted with permission of the American Society for Microbiology).

nucleoid-containing IV, called here IVN, is associated with viral DNA packaging, described in more detail below. IVN then mature to IMV, containing the classical biconcave core, core envelope, lateral bodies, and outer membrane (Figure 3.7). Maturation of IV to IMV is essentially a metamorphosis, which is presumed to take place via rearrangement of the contents of IVN; while structures that are thought to be intermediates in the IVN to IMV transition have been observed; they are relatively rare and, therefore, the process of IMV maturation is probably rapid (Figure 3.13). Maturation from IVN to IMV is accompanied by proteolysis of several major virion structural proteins; mutation of one of the viral proteases involved in virion protein cleavage interrupts morphogenesis at the IVN to IMV transition (Ericsson *et al.*, 1995; Byrd *et al.*, 2002). Maturation from IVN to IMV is also accompanied by microtubule-mediated migration of the particles from viroplasm toward the periphery of the cell (Sanderson *et al.*, 2000). Greater than 90% of virus does not mature past the IMV stage; a small percentage matures further to eventually be exported from the cell as EEV (Boulter & Appleyard, 1973). As described earlier, EEV formation involves first the formation of intracellular enveloped virus (IEV). IEV consists of IMV surrounded by two additional membranes, acquired via wrapping of IMV in trans Golgi network-derived cisternae that have been modified by the incorporation of specific virus proteins (Figure 3.6; Schmelz *et al.*, 1994). IEV are mobilized to the undersurface of the plasma membrane on microtubules, where the outermost IEV membrane fuses with the plasma membrane to form cell-associated enveloped virus (CEV; Blasco & Moss, 1992; Hollinshead *et al.*, 2001). CEV thus consists of IMV wrapped in one additional membrane, and remains attached to the external surface of the plasma membrane. Actin filament bundles then form beneath CEV pushing outward, resulting eventually in the formation of long, specialized actin-containing microvilli, each of which contains a single CEV perched at the tip (Stokes, 1976; Cudmore *et al.*, 1996). CEV is eventually released from the surface of the cell to become extracellular enveloped virus (EEV), which like CEV, consists of IMV surrounded by one additional membrane. Both CEV and EEV mediate short range and long range cell to cell spread of virus, respectively (Blasco & Moss, 1992; Law *et al.*, 2002).

Over the past decade, information regarding the contribution of individual virus genes to virus morphogenesis has been gathered by studying virus assembly following infection with mutants in at least 32 different virion proteins. The mutants that have been characterized can be roughly staged into six categories corresponding to the major intermediates in virus assembly (one recent representative reference is provided for each class): (1) mutants defective in viral membrane synthesis, which that accumulate viroplasm devoid of crescents (Demasi & Traktman, 2000); (2) mutants defective in IV synthesis, which accumulate viroplasm containing crescent membranes but no complete IV (Chiu & Chang, 2002); (3) mutants defective in IVN synthesis, which accumulate viroplasm containing IV (Yeh *et al.*, 2000); (4) mutants defective in IMV synthesis, which accumulate viroplasm containing IV and IVN but no IMV (Garcia & Moss, 2001); (5) mutants that accumulate normal looking IMV which are nevertheless not infectious (Liu *et al.*, 1995); and (6) mutants defective in EEV synthesis, which accumulate IMV (Sanderson *et al.*, 2000). Mutants in several of these categories accumulate structures that are not normally seen during wild type virus infections; such structures may be either normal intermediates in virus assembly or dead end defective products. Logically, mutants in several IMV membrane proteins are defective in crescent or IV formation (Rodriguez *et al.*, 1995; Wolffe *et al.*, 1996; Rodriguez *et al.*, 1998; Traktman *et al.*, 2000), mutants in several core proteins are defective in maturation of IV to

IMV (Klemperer *et al.,* 1997; Williams *et al.,* 1999), and mutants in several EEV membrane proteins are defective in maturation of IMV to EEV (Blasco & Moss, 1991; Engelstad & Smith, 1993; Wolffe *et al.,* 1993). However, it is also true that mutants in several core proteins and enzymes affect viral membrane formation (Wang & Shuman, 1995; Demasi & Traktman, 2000), and conversely that mutants in some membrane proteins affect the transition from IV to IMV (Ravanello & Hruby, 1994). These genetic studies have also revealed that poxviruses encode a complete pathway for disulfide bond formation, which is required for formation of at least some of the disulfide bonds critical for virus structure and assembly (Senkevich *et al.,* 2002b). Mutation of any of the genes in the pathway interrupts morphogenesis at the IV to IVN stage.

IV and IMV membrane formation. Two distinct models have been proposed to describe the origin and structure of viral crescents and the IMV membrane. One model proposes that the crescent membranes comprise a single lipid bilayer which arises *de novo,* i.e., without continuity with any other intracellular membrane structure (Dales & Mosbach, 1968; Hollinshead *et al.,* 1999). This *de novo* model is based on the observation that, using conventional electron microscopic techniques, the crescents and the IMV membrane have dimensions consistent with a single lipid bilayer, and that in most standard electron microscopic preparations of virus-infected cells, no convincing connection can be observed between the crescents and any other intracellular membrane system. The notion of *de novo* biogenesis of a lipid bilayer membrane with free ends and two cytoplasmic surfaces is unprecedented in cell biology. An alternate model, which is more consistent with established cell biology principles, is that the crescents are formed from membrane cisternae derived from the endoplasmic reticulum-Golgi intermediate compartment (ERGIC), which through incorporation of viral proteins collapse into rigid crescents, such that the crescents and IMV membranes actually comprise two tightly apposed lipid bilayers (Sodeik & Krijnse-Locker, 2002). This "double membrane" model is supported by electron microscopic images of IV prepared by freeze substitution and by electron microscopic examination of thin cryosections of infected cells, which under some circumstances reveal apparent localized separation of two membranes within a crescent. In addition, cryosections stained with antibodies to both cellular and viral proteins show labeling of ERGIC membranes with viral proteins and continuity between crescents and cell derived ERGIC membrane cisternae (Sodeik *et al.,* 1993; Salmons *et al.,* 1997; Rodriguez *et al.,* 1997; Risco *et al.,* 2002). The double membrane model is further supported by the clear demonstration of a remarkably similar mechanism of virion membrane biogenesis during maturation of another cytoplasmic DNA virus, African swine fever virus (Andres G. *et al.,* 1998). Interestingly, infection with mutants in either of the poxvirus IMV membrane proteins A17L (A17L/VACWR137) or A14L (A14L/VACWR133) results in aberrant or deficient synthesis of crescents, and accumulation of numerous small vesicles containing viral membrane proteins (Rodriguez *et al.*, 1995; Wolffe *et al.*, 1996; Rodriguez *et al.*, 1998; Traktman *et al.*, 2000). It has been proposed that these vesicles may normally be intermediates in crescent formation (Traktman *et al.*, 2000). The vesicles could be shed from the ERGIC and would normally be short lived. Crescents formed by fusion and collapse of these vesicles would comprise tightly apposed double membranes, but consistent continuity with the intermediate compartment would be rare, thus explaining the apparent *"de novo"* evolution of viral membrane crescents. A similar model of viral membrane formation proposes that the IV membrane is formed by the fusion of ERGIC derived, viral protein modified tubules (Risco *et al.*, 2002). Phosphorylation may play a critical role in regulating viral membrane formation, since both the A17L and A14L membrane proteins are phosphorylated by the virus-coded F10L kinase, and mutational inactivation of the F10L (F10L/VACWR049) kinase abrogates crescent formation (Traktman *et al.*, 1995; Betakova *et al.*, 1999; Derrien *et al.*, 1999).

Viral DNA packaging. The mechanism by which viral DNA is packaged into virions is incompletely understood. Viral DNA concatemers are resolved to unit length genomes during infection with virus mutants that are incapable of forming even the earliest crescent membranes; therefore, the final stages of viral DNA processing do not seem to be coupled to morphogenesis (Demasi & Traktman, 2000). In theory, unit length viral DNA genomes could be incorporated into IV during membrane synthesis and subsequently condense into nucleoids. However, several investigators have observed IV that are incompletely sealed and possess a chromatin- or nucleoid-like material positioned across a small opening or pore in the IV, sometimes apparently continuous with the internal nucleoid (Morgan, 1976; Ericsson *et al.*, 1995; Risco *et al.*, 2002). These images are interpreted to mean that DNA is specifically packaged late during IV assembly to form IVN. Interestingly, mutation of the viral gene A32L (A32L/VACWR155), which contains a predicted ATP/GTP binding site "P loop" motif, interrupts virus assembly at the IV to IVN transition, resulting in the accumulation of reduced amounts of IVN and large amounts of novel, electron dense spherical particles that do not contain DNA (Cassetti *et al.*, 1998). These observations are consistent with a late DNA packaging model and suggest that the A32L gene product may mediate DNA packaging.

3.4.8   Vaccinia Genetics

Genetic manipulation of poxviruses has taken liberal advantage of both classical "forward" genetic techniques and modern "reverse" genetic techniques. Classical forward genetics consists of chemical mutagenesis and random isolation of temperature sensitive, host range, drug resistant, and drug dependent mutants, followed by genetic mapping procedures to determine which genes have been affected. Modern reverse genetics consists of site-specific engineering of the viral genome, from alteration of existing poxvirus genes *in situ* to introduction of foreign genes and heterologous transcriptional operons into the poxvirus genome.

Regardless of the application, reverse genetic engineering of poxvirus genomes follows the same fundamental "*in vivo* recombination" protocol (Nakano *et al.*, 1982). Specifically, if cells are first infected with a poxvirus and then transfected with a DNA molecule that has homology with the infecting poxvirus genome, the transfecting DNA molecule will undergo homologous recombination with the infecting genome. A simple example of this principle is marker rescue mapping of temperature sensitive mutants of vaccinia virus (Thompson & Condit, 1986). In this example, cells are infected with mutant virus and transfected with candidate DNA molecules derived from wild type virus. Homologous recombination between the mutant genome and a transfected DNA molecule, which contains the wild type allele of the gene, will generate wild type virus, thus mapping the temperature sensitive lesion to a specific rescuing DNA fragment. Several variations on the basic *in vivo* recombination scheme have been derived that improve the efficiency of the procedure and/or enhance selection for recombinant genomes (Chakrabarti *et al.*, 1985; Franke *et al.*, 1985; Fathi *et al.*, 1986; Falkner & Moss, 1990; Merchlinsky & Moss, 1992; Scheiflinger *et al.*, 1992). In general, poxvirus recombination is efficient; the genome contains numerous non-essential insertion sites and can accommodate large amounts of foreign DNA, so that almost any recombinant that can be imagined can be constructed. Recently, a novel engineering scheme has been developed in which the entire vaccinia genome is contained in a bacterial artificial chromosome, where it can be engineered and subsequently reactivated by infection of cells infected with a non-replicating helper fowlpox virus (Domi & Moss, 2002).

Several laboratories have used classical forward genetic methods to assemble collections of conditional lethal mutants to be used for the study of the function of individual poxvirus genes (reviewed in Condit & Niles, 1990). Recently, a complete complementation analysis combining two independent mutant collections has been conducted, resulting in a combined collection of 138 temperature sensitive mutants comprising 53 complementation groups (Lackner *et al.*, 2003). These temperature sensitive mutants, along with numerous drug resistant, drug dependent, and host range mutants, have been and will continue to be a powerful resource for determining poxvirus gene function. However, given that the poxvirus genome contains approximately 200 genes and assuming that approximately

150 of these genes are essential for virus growth in culture, even this combined collection provides only about one third coverage of the genome. With this in mind, two alternative methods for reverse genetic engineering of mutants in specific poxviral genes have been developed.

One reverse genetic method for creating conditional lethal poxvirus mutants involves regulating transcription of the poxvirus gene of interest using elements from a bacterial operon (Fuerst *et al.*, 1989; Rodriguez & Smith, 1990; Traktman *et al.*, 2000). In its simplest form, this involves construction of a recombinant poxvirus that constitutively expresses a bacterial repressor protein, for example the *lac* repressor, and also contains a cognate operator (i.e., a repressor-binding site) between the transcriptional promoter and the coding sequence of the target gene to be regulated. In such a recombinant, the expression of the gene can be controlled by use of an inducer molecule during virus infection, for example, IPTG in the case of *lac* repressor control. Appropriate regulation of the target gene sometimes requires constructions that are more baroque. For example, recombinants can be constructed that contain bacteriophage T7 RNA polymerase under *lac* repressor control, and in which the target gene is regulated by both a T7 RNA polymerase promoter and a *lac* operator (Ward *et al.*, 1995). Numerous inducible recombinant poxviruses have been constructed that have proven invaluable for study of individual poxvirus genes. One drawback of this approach is that some poxviral genes, most notably early genes, cannot be cleanly regulated.

An alternative approach to creating conditional lethal poxvirus mutants is the construction of temperature sensitive mutants using clustered charge to alanine scanning mutagenesis (Hassett & Condit, 1994). Clusters of charged residues are identified within the linear amino acid sequence of a given target gene, and all of the charged residues in a given cluster are changed to alanine in a cloned copy of the gene. Individual clustered charge to alanine mutants are then recombined into the viral genome and recombinant viruses are tested for temperature sensitivity. Empirically, approximately 30% of clustered charge to alanine mutants prove to be temperature sensitive; theoretically, because the charge clusters normally reside on the surface of a protein, and the genetic neutralization of the charge weakens protein–protein interactions. Useful clustered charge to alanine temperature sensitive mutants have been created in several vaccinia genes that were inaccessible via bacterial operon regulation (Hassett *et al.*, 1997; Demasi & Traktman, 2000; Ishii & Moss, 2001; Punjabi *et al.*, 2001).

3.4.9 Vaccinia Virus Genes Responsible for Immune Evasion, Host Range, and Control of Apoptosis

Introduction. Poxviruses adopt a wide variety of strategies to ensure survival (see additionally Section 7.5). These strategies include the shut-off of host transcription, synthesis of different forms of infectious virus particles with differing surface components that facilitate cell spread and viremia, and assembly of virus particles into environmentally resistant inclusion bodies. In addition, as the virus family evolved and became established in different vertebrate hosts or sites within a host, poxviruses developed the means to capture essential host genes, which then evolved to target critical aspects of host defenses, such as host range, control of apoptosis, inflammation, and other immunomodulatory mechanisms serving to enhance virus survival. As described previously, many of these modulatory genes are located within the variable region of the genome external to the core of conserved genes and are generally non-essential for virus growth in cell culture. There has been much speculation as to how poxviruses, which develop in the cytoplasm, might capture cellular genes. One clue is suggested by the finding that strains of fowlpox virus have incorporated a reticuloendotheliosis virus (a retrovirus) within the fowlpox virus genome (Hertig *et al.*, 1997). The reverse transcriptase encoded by a retrovirus might reverse transcribe cellular transcripts, which might subsequently be incorporated into the poxvirus genome.

Many of the vaccinia-encoded immune modulatory, host range, and apoptosis-regulating genes and the resulting strategies they employ for deflecting the host responses to infection are shared by a number of poxviruses, particularly those that cause similar disease even though, like ectromelia and variola, they are very host (mouse and human. respectively) restricted viruses. There is little doubt that these genes play a major role in the disease process, and their function is to overcome the immune and other host responses directed against an overt generalized infection. While cautioning against oversimplification, this is a useful model to consider. These genes are summarized in Table 3.4.

Lastly, we need to consider the disease caused by orthopoxviruses, such as vaccinia. All orthopoxviruses cause a general, disseminated infection. Much of what we have learned about vaccinia deflection of host responses can be extrapolated to other poxviruses, not only to viruses within the same genera, such as variola and ectromelia, but to poxviruses of other genera that cause a similar type of disease, most notably the leporipoxviruses exemplified by myxoma virus. While the overall strategy of these viruses is clear, it is important to take note of the fact that individual viruses have adapted to their specific hosts. For example, there is data to suggest that variola orthologs of vaccinia virus genes may be particularly attuned to human hosts (Rosengard *et al.*, 2002) and the myxoma virus orthologs tend to be particularly attuned to the rabbit. There are at least five excellent, comprehensive reviews of poxvirus immunomodulatory and host regulatory proteins (Moss *et al.*, 2000; Moss & Shisler, 2001; Shisler & Moss, 2001; Alcami, 2003; Seet *et al.*, 2003).

Table 3-4. Vaccinia virus immunomodulatory, anti-apoptotic, and host range genes

VVCOP Orf	VACWR Orf	Size (AA)	Description
C3L	025	263	Complement Control Protein
K3L	034	88	eIf2-α homolog
E3L	059	190	dsRNA binding protein
B8R	190	272	IFN-α receptor decoy
B19R	200	353	IFN-α/β binding protein
Absent	013	126	IL-18 binding protein
B13/14R	195	345	Serpin (SPI-/crmA)
A46R	172	240	T0LL/IL-1 receptor-like
A52R	178	190	T0LL/IL-1 receptor-like
A53R	179	103	TNF-like receptor
C22L/B28R	004, 215	122	Soluble TNF receptor
A41L	166	219	Chemokine binding protein, unknown ligand, antiflammatory
A38L	162	277	CD47-like protein
K2L	033	369	Serpin (SPI-3)
A39R	163	295	Semaphorin
A44L	170	346	Hydroxysteroid dehydrogenase
C23L (B29R)	218	244	Type II chemokine binding protein
B16R	197	326	IL-1β binding protein
C12L/B22R	205	353	Serpin (SPI-1)
K1L	032	284	Host range, ankyrin repeat protein
C7L	021	150	Host range, ankyrin repeat protein
C11R	009	142	Growth factor

Immunomodulation. Generally, the initial exposure of a given host to a poxvirus infection involves macrophage and lymphocytes and leads to inflammation in response to tissue damage and the production of interferon. Interferon in turn stimulates natural killer (NK) cells, which can attack virus-infected cells as a first line of defense in a non-specific fashion independent of the more sophisticated, adaptive immune response. Similarly, the activation of the complement cascade also functions as a first line of defense. Vaccinia virus encodes proteins designed to interfere with both complement- and interferon-mediated signaling. Syntheses of antigenic virus proteins are processed by antigen-producing cells, such as dendritic cells and macrophages. This processing generates peptides that then complex with the major histocompatibility complex (MHC) and are presented by these antigen-presenting cells to lymphocytes. There are a number of lymphocytes including CD8 T-lymphocytes (CTLs), which expand to clear infected cells that comprise the basis of cell-mediated immunity (CMI). There are two types of T helper cells, which contribute to CTL activity. Type 1 CD4+ helper T lymphocytes are involved in inducing long-lasting CTL activity, and the type 2 CT4+ helper T lymphocytes interact with B cells to produce neutralizing antibodies, which prevent newly released virus from initiating subsequent infection (humoral immunity). CTLs recognize antigen in conjunction with class I MHC, whereas CD4+ helper cells recognize antigen presented in conjunction with MHC class II (Esposito & Fenner, 2001).

Following the initial innate response, in a naive host, the CTL response occurs, and it is these aspects of the immune response that are primarily responsible for clearing an infection (Buller & Palumbo, 1991; Palumbo *et al.,* 1994; Ada & Blanden, 1994; Barry & Bleackley, 2002). However, antibodies formed soon thereafter may be instrumental in preventing newly synthesized particles from invading sites not readily accessible to adaptive CTLs, such as the skin. Therefore, it is reasonable to assume that virus-encoded genes that deflect host responses would be designed not only to interfere with antigen presentation in context of class I and II MHC, but also to interfere with consequences of that antigen presentation, namely, virus clearance and ultimately immunological memory. There is little evidence to suggest that orthopoxviruses, such as vaccinia, do so directly. However, a significant and general effort is mounted by all poxviruses against cytokines and chemokines, which are involved in both innate and adaptive immunity. The virus-mediated deflection of these responses serves to mask the infection by various means and "buy time" during which the infection becomes established. In addition to encoding proteins that interfere with complement and interferons, vaccinia encodes a variety of proteins that interfere with immunoregulatory chemokine and cytokine signaling, which impact the immune response, the virus also encodes various other assorted immunomodulators as well as proteins that effect cellular processes, such as apoptosis. The vaccinia gene products involved in these processes are summarized in Table 3.4.

Control of apoptosis and host range. Apoptosis is a programmatic induction of cell death designed to eliminate unwanted, defective cells. Hosts frequently respond to virus infection by invoking apoptosis to eliminate the infected cells (Hay & Kannourakis, 2002). There are two methods by which these viral proteins function to control these processes. The first is by antagonizing the intracellular signaling pathways that serve to induce apoptosis and the second is by secretion of proteins designed to complex with pro-apoptotic ligands and prevent engagement of such ligands with cellular receptors and subsequent signaling. Although not strictly immunomodulatory, there are certainly links and parallels with the innate immune responses, and some gene products, such as crmA/SPI-2, have roles in both the innate response as well as control of apoptosis. In general, these proteins are synthesized early in the infectious cycle and include both secreted and intracellular proteins (see Appendix 2).

3.5 Smallpox Vaccines and Production Strains of the Virus

3.5.1 Dermal Smallpox Vaccine

Dermal smallpox vaccine is the oldest preparation used for specific prevention of infectious diseases. Year 1996 was the beginning of the third century since the vaccination of the first patient. We have already touched upon some stages of the complicated history of smallpox vaccine (see also Section 6.2). Here we just remind its most important events.

As commonly known, during the first decades after Jenner's discovery, most of the material for vaccination was either taken from fresh cowpox cases in cattle, horses, or dairymaids or maintained by passaging from one vaccinee to another. The shortage of material (humanized lymph) obtained from humans and the rare occurrence of cowpox outbreaks limited the expansion of vaccination. The invention of virus passages on calves by Negri in 1840 solved at least two important problems: constant reproduction of sufficient quantities of inoculation material and elimination of the risk of infection of vaccinees with syphilis, hepatitis, etc. These infections were often disseminated with vaccination substrate of human origin.

In spite of the obvious advantages of virus passages on calves by cutaneous inoculation proposed by Negri, its introduction demanded considerable time and lasted virtually until the end of the 19th century. Note that this apparently obsolete method for producing dermovaccine by passages on the calf skin or skin of other animals (sheep and buffaloes) still exists at present as the main stage in vaccine manufacture. Nevertheless, the smallpox vaccine produced by this method has been by right considered the most efficient of all currently available vaccines. Another important step to improve smallpox vaccine was the use of glycerol since the 1860s, which assisted preservation of the inoculation material and decrease in its bacterial contamination. Although liquid glycerol-stabilized vaccine was widely used in immunization against smallpox for almost one century, including the very beginning of the global smallpox eradication program, it had grave disadvantages. The most important of disadvantages were a low thermal stability and short shelf life. These shortcomings became even less tolerable during massive vaccination in countries with a hot climate, where, as a rule, there were no appropriate conditions for low-temperature storage of the vaccine. As early as at the beginning of the 20th century, an increase in the stability of smallpox vaccine by desiccation was attempted. In the 1920s, Otten obtained highly thermostable dried "lymph" at the Pasteur Institute (Bandung, Indonesia). It allowed for eradication of smallpox in Indonesia even before World War II (Spaander, 1973). However, the crucial success was reached later. In Russia, Morozov *et al.* (Morozov *et al.*, 1943; Morozov

& Solov'ev, 1948) developed a vacuum-dried thermostable smallpox vaccine with protein and sucrose as stabilizers. An even more efficient technology by Collier (1955) involved use of 5% peptone as a stabilizer and Freon treatment of pox detritus for elimination of contaminant proteins. Smallpox vaccines obtained by this method were immeasurably more thermostable. As was mentioned in Section 3.2, these vaccines were 100% efficient after two years of storage at 37 and 45°C. The Collier's method was then universally accepted.

For a long time, the quality of smallpox vaccines was the manufacturers' concern. The reasons for this were the absence of both the legal acts determining the requirements for vaccines and the corresponding supervisory bodies. One more reason was the absence of satisfactory methods for quantitative assessment of the specific activity of preparations. In the course of time, some countries developed national requirements for smallpox vaccine and commenced producing it according to them. At the beginning of the worldwide smallpox eradication program, a research team convened by WHO in 1958 developed International Requirements for these preparations (WHO, Requirements for Biological Substances, 1959). In 1965, these requirements were revised and supplemented (WHO, Requirements for Biological Substances, 1966). In particular, these requirements strictly limited infective activity, thermostability, bacterial contamination, etc. The main method for evaluating the infective activity of a preparation involved highly precise titration on chick embryo CAMs by pock counting. Prior to this, WHO supervised international comparative tests of smallpox vaccines manufactured in various countries and established a reference preparation of smallpox vaccine (Krag & Bentzon, 1963). This reference preparation was the smallpox dermovaccine produced at the Lister Institute of Preventive Medicine, UK. It was developed involving the Elstree strain (alias Lister) by cultivation on sheep skin.

The result of the tremendous scientific, managing, and practical efforts by WHO aimed at the improvement of smallpox vaccines was that since 1972, 95% of preparations used in the global program met the international requirements (Arita, 1973).

Prior to reviewing other smallpox vaccine types, let us dwell on the important issue of the properties of so-called production vaccinia virus strains.

3.5.2   Virus Strains Used for Production of Smallpox Vaccines

One of the specific features of smallpox vaccine distinguishing it from other live vaccines was a wide diversity of strains used for its manufacture. It happened so that in the late 1960s—early 1970s, virtually every laboratory manufacturing the vaccine had a production strain and/or a passage scheme of its own. The actual origin of the strains was unknown. It was believed that some of the strains originated from variola virus (e.g., Elstree, Tashkent, and Dairen) and others were progeny of strains used since Jenner's times.

Studies initiated by Fenner (1958) and continued by Marennikova and Mal'tseva (1964a; 1964b) and Gendon and Chernos (1964) showed that vaccinia virus strains differed in a variety of biologic markers. Moreover, Polak *et al.* (1963) and Marennikova *et al.* revealed differences in the reactogenicity of smallpox vaccines made involving different strains.

Our study of a large set of strains collected from various production facilities throughout the world showed that they differed most in their pathogenicity for laboratory animals. This manifests, first of all, in neurovirulence, ability to cause necroses after subcutaneous inoculation to rabbits, rabbit response to intratesticular inoculation, etc. We classified the strains studied into three groups according to their pathogenicity: high, moderate, and low (Table 3.5).

Note that tests with intracerebral and intranasal inoculations of 2–5-day-old suckling mice reveal differences between the strains belonging to one moderate or low pathogenicity group. According to these tests, EM-63 appeared the least pathogenic in the latter group. Later, our results were confirmed by Hashizume *et al.* (1973), who evaluated strain pathogenicity by intracerebral inoculation of monkeys.

Further studies at our laboratory demonstrated a correlation between strain pathogenicity for animals and the severity of their vaccination response. For example, the strains highly pathogenic for laboratory animals (in particular, Tashkent) also produced the most severe response in humans. In contrast, strains low pathogenic for animals produced significantly milder vaccination reactions with temperatures not exceeding 37.5°C in most cases (Marennikova & Tashpulatov, 1966; Unanov *et al.*, 1967; Marennikova, 1968; Marennikova *et al.*, 1969; Shneiderman, 1978; etc.). Fedorov (1977) reported that vaccination of children with vaccines of low pathogenic strains (EM-63, Elstree, and NYCBH) using a jet injector caused no fever in 80–85% of cases.

The following facts were discovered during a closer study of the virus behavior in vaccinated animals (rabbits and monkeys). Highly pathogenic strains accumulated more intensely in the skin (at the inoculation site) and regional lymph nodes (Figures 3.14 and 3.15). In the postvaccination period, they were also isolated from blood and internal organs (in monkeys, also from pharynx). In contrast, replication of low virulent strains in rabbits was confined to the inoculation site and regional lymph nodes. In monkeys (*M. rhesus*), these strains less intensely (as well as in rabbits) accumulated in the skin and regional lymph nodes. The virus was not detected in the pharynx or internal organs, although viremia was observed in individual animals (Shenkman, 1972; Marennikova, 1973). Correspondingly, these

Table 3-5. Classification of vaccinia virus strains according to their pathogenicity for laboratory animals

Major assessment criteria	Degree of pathogenicity	Strain name	Country of use
Death of rabbits and laboratory mice after intracerebral inoculation; formation of necroses after intracutaneous inoculation to rabbits; pronounced response to intratesticular inoculation to rabbits; etc.	High	Tashkent	USSR
		White Tashkent clone (cloned from a white pock)	USSR
		Gamaleya institute of Epidemiology and Microbiology (IEM), Tomsk, Perm	USSR
		Chinese variant of Gamaleya IEM*	China
		Copenhagen	Denmark
		Budapest	Hungary
		Paris	France
		Dairen, Ikeda	Japan
No or low death rate of rabbit after intracerebral inoculation; death of young laboratory mice after intracerebral inoculation; death of suckling mice after intranasal inoculation: formation of necroses after intracutaneous inoculation of rabbits	Moderate	Bern	Germany
		Patwadangar	India
		Bandung	Indonesia
		B-51	USSR
		L-IVP (variant of Elstree)	USSR
No virulence after intracerebral inoculation to rabbits or laboratory mice; no necroses after intracutaneous inoculation: no or low death rate of suckling mice** after intracerebral or intranasal inoculation	Low	EM-63	USSR
		Elstree (Lister)	UK and other countries
		New York City Board of Health (NYCBH)	USA

*Maintained by passaging in China for 10 years (since 1954);
**Age 2–5 days.

strains had lower antigenic activity (Chimishkyan, 1971; Marennikova *et al.*, 1972c). It was also found that low pathogenic strains were more sensitive to interferon (Bektemirov *et al.*, 1971; Marennikova, 1973).

These data shed some light upon the reasons for different reactogenicities of smallpox vaccines prepared from different strains. Later, a similar correlation was found with reference to the frequency of postvaccination complications (Marennikova & Matsevich, 1971; Marennikova, 1973; Gurvich, 1983; Marennikova *et al.*, 2003; Table 3.6).

These data on the effect of the properties of production vaccine strains on the reactogenicities of smallpox vaccines and the frequency of postvaccination complications, reported at numerous international

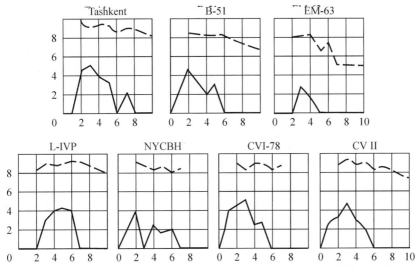

Figure 3-15. Time courses of the virus accumulation on scarified skin (dash line) and in regional lymph nodes (solid line): X axis, days post inoculation; Y axis: virus titer, log PFU/ml.

conferences, justified the substitution of low and moderately reactogenic strains for highly reactogenic ones. This was realized in the USSR, Germany, Austria, Japan, and several other countries. At the majority of production facilities, old strains were replaced with Elstree (Lister) or its variant passaged on heifers (L-IVP) or, in the USSR, with EM-63 and B-51 strains. Later data demonstrated conclusively that the replacement of the strains resulted in a significant reduction of the severity of postvaccination reactions and the rate of complications, including postvaccination encephalitis (Berger & Heinrich, 1973; Marennikova, 1975; Gurvich, 1983; etc.).

Many authors studying various strains of the virus noted the heterogeneity of their populations. Fenner (1958) detected such nonuniformity in the Lederle strain, and King-Andersen (1959, personal communication), in the Gamaleya Institute of Epidemiology and Microbiology strain. Later, it was found that the genetic heterogeneity of populations was characteristic of many other production strains. Only 5 of 21 strains studied at our laboratory were

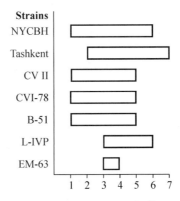

Figure 3-14. Duration of virus release from regional lymph nodes, days post inoculation.

Table 3-6. Effects of properties of production strains on reactogenicities of smallpox vaccines and the rate of postvaccination complications, PVC (data of vaccination during 1975–1979)

Production vaccine strain	Laboratory assessment (pathogenicity for animals)	Reactogenicity	PVC number per 10^7 vaccine doses produced				Notes
			Total	Postvaccination encephalitis*	Encephalitic reactions	PVC with skin manifestations	
					Including		
Tashkent	High	Extremely high	Administered before commencement of these studies; according to available data, displayed the highest PVC rate				Withdrawn from production in the USSR since 1967–1968
B-51	Moderate	Moderate	28.4	11.5 (2.2)	7.4 (0.22)	9.6	Since 1975, substituted with strain L-IVP
EM-63	Low	The least of the strains compared	16.1	5.7 (1.4)	3.4 (0)	6.9	Is not used since 1976 according to economical reasons; substituted with L-IVP
L-IVP (Elstree variant adapted to the calf skin)	Moderate	Moderate	34.9	6.8 (1.5)	14.2 (0.42)	13.9	Used until cessation of anti-smallpox vaccination

*In addition to postvaccination encephalitis, cases of meningoencephalitis and encephalomyelitis are included.

genetically uniform (Marennikova & Mal'tseva, 1964a; Mal'tseva, 1965; Marennikova *et al.*, 1969). This heterogeneity manifested itself as development of two pock types on chick CAMs: dense white and smaller indistinct grayish pocks. Gendon and Chernos (1963) revealed the same phenomenon in a cell culture with agar overlay: genetically heterogeneous strains formed two plaque types, transparent and opaque (or reticular).

All genetically homogeneous strains (Elstree, EM-63, Patwadangar, Copenhagen, and Budapest) formed dense white pocks on CAM and transparent plaques in cell cultures.

Counts of pocks of both types, which were formed by genetically heterogeneous strains, showed that their ratio was fairly constant for each strain. In some strains (Tashkent, B-51, NYCBH, Bern, and Paris) the population forming white pocks was dominating; in other (Gamaleya IEM, Dairen, Ikeda, and the strain applied in Poland in the 1950-60s), indistinct grayish ones. For example, the ratio of white to grayish pocks in NYCBH was 87.3 : 12.7%, and in Ikeda, 18.3 : 81.7, respectively. Interestingly, the predominance of the population forming flat grayish pocks was characteristic mainly of strains obtained from China, Indonesia, and Japan, i.e., within a certain geographic area. The experience of cloning genetically heterogeneous strains accumulated at our laboratory showed that both the subpopulations forming white and grayish pocks could be cloned as stable lines. A clone could easily be isolated from a predominant population on both CAMs and in cell cultures. It was much more difficult to obtain clones from the subordinate population.

Subsequent experiments showed that clones isolated from the same strain differed not only in the pattern of pocks on CAMs and plaques in cell cultures, but also in other features: pathogenicity for animals, reactogenicity after inoculation to humans, etc. The most pathogenic and reactogenic clones were isolated from the subpopulation forming flat grayish pocks. It was also found that the subpopulation predominant in a certain virus strain determined the properties of the whole strain (Mal'tseva, 1965). Closer studies showed that purification of genetically heterogeneous strains with domination of by the white-pock-forming subpopulation from the admixture of the population forming the other pock type could alter the strain properties. For example, cloning of some of such strains led to a decrease in virus accumulation, antigenic activity, reactogenicity, and vaccination efficiency.

3.5.3 Ovovaccine and Culture Smallpox Vaccine

In addition to dermovaccine, which is the main vaccine used for controlling smallpox, vaccines made of virus cultivated on other substrates have been developed. These are ovovaccine and culture vaccine. The virus-bearing substrate of the former is chick embryo CAM infected with vaccinia virus. The virus grown in cell cultures was used for preparing the latter type of vaccine. In the 1950-70s, ovovaccine was manufactured in the USSR (Marennikova, 1958) and some other countries (New Zealand, Sweden, etc.), but it gained no wide acceptance. This vaccine type was most widely used in Brazil, where it constituted 90% of the overall vaccine administered for the smallpox eradication program (Voegeli, 1973). The advantages of this vaccine were its sterility and simple production. Nevertheless, it was thought that this substrate did not allow for producing a vaccine meeting the International Requirements, in particular, concerning thermostability (Fenner *et al.*, 1988). Its manufacture demanded use of eggs from farms free of diseases, including avian leukemia.

Few of numerous attempts to obtain smallpox vaccines using various cell cultures were entirely successful. In the Netherlands, a culture smallpox vaccine perfectly meeting the International Requirements was obtained from a primary culture of rabbit kidney cells with the Elstree strain. This vaccine successfully passed comprehensive trials, including sufficiently wide clinical administration (Hekker *et al.*, 1973; Fenner *et al.*, 1988). In 1994, this method was recommended by WHO for manufacturing preparations in case of emergency (threat of smallpox reemergence).

Another culture vaccine completely meeting the International Requirements was produced in Japan (Hashizume *et al.*, 1973) in a rabbit kidney cell culture from the attenuated Lc16m8 strain, described below in more detail. Note also the Yugoslavian experience of manufacturing the smallpox vaccine in the human diploid cells WI-38 using the strain Bern–Zagreb (Ikič *et al.*, 1973). The reactogenicity of the strain was reduced significantly after 11 passages in the culture. Field trials of the culture vaccine involving primary vaccination of 724 persons showed its 100% take moreover, 72.1% of the vaccinees displayed no fever. However, note the infectious titers of this vaccine were less than those required by WHO by nearly 1 log.

To produce a culture vaccine, Chernos *et al.* (1977) used suspension culture of Japanese quail embryos for growing the strain L-IVP (the strain Elstree adapted to calf skin). Laboratory trials of this vaccine demonstrated that it met completely the International Requirements. Clinical trials in a cohort of revaccinees (about 1000 persons) demonstrated that the vaccine take-up rate and antigenic activity was equal to the dermal vaccine. In addition, its reactogenicity was less pronounced (Andzhaparidze *et al.*, 1980). End of vaccination against smallpox prevented testing this vaccine for primary vaccination of children and organizing its commercial manufacture. Concurrently, a highly active and stable smallpox vaccine was produced in Germany (Hochstein-Mintzel, 1977) using virtually the same technology (roller cultivation) but another culture (primary culture of chick embryo fibroblasts). The last vaccine completely substituted the dermal vaccine in 2 years.

Recently, the menace of bioterrorism, including potential actions with variola virus, revived the interest in development of new smallpox vaccines (LeDuc & Jahrling, 2001; WHO Advisory Committee on Variola Virus Research, 2003). An example is designing of new variant of culture vaccine using cloned strain NYCBH grown in human diploid cells MRC-5, which successfully passed experimental and clinical trials (Weltzin *et al.,* 2003). However, note that despite evident advantages of culture vaccines from both biological and economical–production standpoints, several scientists call for their careful administration due to the absence of proved epidemiological efficacy and safety (Mortimer, 2003).

The culture smallpox vaccines are only one of the directions in the field of development of smallpox vaccine prophylaxis. The research now follows various courses—development of DNA vaccines, subunit vaccines involving antigens of vaccinia or variola viruses, genetic modifications of vaccine strains, etc. (WHO Advisory Committee on Variola Virus Research, 2003).

3.5.4 Inactivated Smallpox Vaccine and Vaccines from Attenuated Strains

In addition to the above-described vaccines, the vaccines mentioned in the heading were developed, first of all, for vaccination of the persons with high risks of postvaccination complications. These vaccines belong to two types: inactivated vaccines and vaccines involving attenuated strains.

Inactivated vaccines. Development of inactivated vaccines attracted attention for a long time. This attention was enhanced by the problem of postvaccination complications.

Various methods were used for inactivating the virus: heating, treatment with formalin, UV-irradiation, photodynamic inactivation, and γ-irradiation (Janson, 1891; Herrlich, 1959; Kaplan, 1962; Turner *et al.,* 1970; Svet-Moldavskaya, 1970; Marennikova & Macevič, 1975).

The history of study and application of inactivated smallpox vaccines has shown that they are able to induce production of antibodies, including the virus-neutralizing antibodies. However, their immunogenicity is considerably less than that of the live vaccines. This is appreciably due to both the inability of antibodies resulting from inoculation of inactivated vaccine to neutralize extracellular virus (Appleyard *et al.,* 1971; Boulter & Appleyard, 1973) and their lower avidity for extracellular virus (Matsevich, 1983). The cellular mechanisms of immunity formation also proved to be inadequate after immunization with inactivated vaccine (Turner & Squires, 1971; Matsevich *et al.,* 1978). Thus, inactivated vaccine cannot be an adequate substitute for the live vaccine. Nevertheless, taking into account the fact that inactivated vaccine induces a certain immunological transformation of the organism, subsequent immunization with live vaccine can be regarded as a sort of revaccination. The latter is known to cause much fewer postvaccination complications than the primary vaccination. For these reasons, inactivated vaccine was used at the first stage of so-called two-stage vaccination, proposed by Herrlich (1959) for preventing postvaccination complications in high-risk groups.

The experimental and clinical studies of two-stage vaccination with a vaccine inactivated with γ-irradiation (^{60}Co) carried out at our laboratory (Marennikova & Macevič, 1975; Marennikova *et al.*, 1977a; 1978c; Matsevich, 1983) demonstrated that this vaccination method provided a quicker and more intense antibody formation than in the case of conventional vaccination. The antibodies were highly active against both intra- and extracellular virus. The two-stage vaccination also caused a quicker (by 3–3.5 days) formation and development of skin reaction development. Virological examination of such vaccinees showed a reduced virus circulation after vaccination with the live vaccine (Matsevich, 1983). The two-stage vaccination was used in the USSR and Bulgaria for the persons with high risk of postvaccination complications (naïve children of older cohorts and adults). The vaccinations were usually performed seven days apart, which had been shown an optimum interval. The response to the live vaccine was moderate and milder than in vaccinees immunized with this vaccine conventionally. Of nearly 23,000 vaccinees, only one child (with inborn macrocephaly) had a postvaccination encephalitis (Marennikova *et al.*, 1978c; Matsevich, 1983). This rate of such complications was much lower than the rate recorded in vaccinations by the conventional method. In the study of Gurvich (1992), this rate amounted to 1 : 9000 for 3–5 year-old children and 1 : 3200 for children older than 5 years. The decrease in the frequency of postvaccination encephalitides after a two-stage vaccination was recorded in another series of experiments, where formalin-inactivated vaccine was used as the killed vaccine (Ehrengut & Ehrengut-Lange, 1973).

Vaccines involving attenuated strains. Such vaccines were produced of strains treated using a special attenuation procedure. These strains are CVI-78, CVII, MVA, and Lc16m8. Most attention was paid to studying the vaccines involving CVI-78 and CVII strains, obtained by Rivers through passaging the NYCBH strain in explants and on chick embryo CAMs. Their lower reactogenicity was recorded by all scientists. Kempe *et al.* (1968) and Tint (1973) reported a satisfactory result of administering CVI-78 vaccine in children with eczema. However, later experiments showed its insufficient antigenic activity. According to Galasso *et al.* (1973) and Neff (1973), an increase in the titer of virus-neutralizing antibodies was observed only in 16–30% of vaccinees showing a positive response to vaccination. These results favored the opinion of Rivers *et al.* (1939), who had obtained these strains, that the protective effect should be enhanced by subsequent revaccination with a conventional vaccine. Moreover, our laboratory study of CVI-78 and CVII strains (S.S. Marennikova and L.S. Shenkman, unpublished data) showed that their pathogenicity for rabbits and laboratory mice was higher than that of EM-63, with various administration methods (Table 3.7). The CVI-78 strain was also more neuropathogenic for monkeys than EM-63 and Elstree, as shown by Japanese scientists (Hashizume *et al.*, 1973). A significant invasivity of CVI-78 and CVII was also confirmed by their behavior in the bodies of vaccinated rabbits (Figures 3.14 and 3.15), which was similar to that of highly reactogenic strains (Shenkman, 1972; Marennikova, 1973; Hashizume *et al.*, 1973). Taken together, these facts cast doubt on the classification of these strains as truly attenuated.

Another attenuated vaccine was developed in Germany involving MVA strain. This strain was obtained by a long-term passaging (over 500 passages) of Ankara vaccine strain in a culture of chick embryo fibroblasts. The biological characterization of the strain is shown in Table 3.7.

Table 3-7. Biological characterization of the strains classified as attenuated in comparison with the "mildest" commercial EM-63 strain

Markers\Strain		CVI-78	CVII	MVA	Lc16m8[g]	EM-63
Pock-type uniformity of a population		Nonuniform	Nonuniform	Uniform	Uniform	Uniform
Pocks on CAM	Size, mm	–	–	0.6	<1	2.1
	Pock type	Two types: white with central ulcers and flat grayish with hemorrhage at the center	Two types: white with central ulcers and flat grayish with hemorrhage at the center	Raised distinct white	Dense distinct white	Dense distinct white
	Formation at 40.5°C	+[h]	+[h]	+	+	+
Helbert's test[a] (log PFU/g)		6.4–6.6	6.3–6.5	7.2–7.4	–	8.0
Hemagglutinating activity		64	128	10–80	–	20–80
Pathogenicity for rabbits[b]	Cutaneous inoculation	Confluent papular–pustular rash with hyperemia and edema	Confluent papular–pustular rash with hyperemia and edema	No response	–	Confluent papular–pustular rashes
	Intracutaneous inoculation[c]	15 mm (+++), necrosis 4 mm	25 mm (+++), necrosis 5 mm	Erythema 5 mm in diameter	Infiltrate is less pronounced than in the initial Lister strain	12 mm (++)

continued

Markers\Strain		CVI-78	CVII	MVA	Lc16m8[g]	EM-63
LD$_{50}$ for laboratory mice (log PFU/dose)	Intracerebral[d]	2.5	1.9	Not pathogenic	Not pathogenic	Not pathogenic
	Intracutaneous into the paw[e]	1.6	1.6	Virtually no response (slight edema 3 days post inoculation of 6 log PFU/dose	–	Pronounced local response; a 50% mortality rate
	Intranasal[f]	1.9	1.9	No apparent response	–	25% mortality rate after inoculation of 7–8 log PFU/dose

Notes: (a) accumulation in the liver of chick embryos; (b) upon inoculation of 6 log PFU/0.1 ml; (c) diameter of infiltrate (density scored on a four-plus scale); (d) 10-day-old mice; (e) 3-day-old mice; (f) 1–2-day-old mice; (g) according to Hashizume (1975); (h) of two subpopulations of the strain, only the white-pock-forming variants accumulated at 45.5°C; and (–) not studied.

Both our experiments and the studies of the authors of the vaccine (Stickl _et al._, 1973) showed that this strain had an insignificant pathogenicity for laboratory animals and formed considerably smaller pocks on CAMs compared with the conventional strains (Figure 3.16). The limited trials of MVA vaccine performed by its authors showed no positive results in humans after cutaneous administration in contrast to the conventional vaccine. Only subcutaneous administration was accompanied by a weakly pronounced local response (small infiltrate or node appearing by day 5 post vaccination and persisting until day 20). However, this response induced virtually no humoral antibodies in vaccinees, but subsequent revaccination with conventional vaccine caused a modified or accelerated vaccination response. Therefore, the authors proposed inoculation of the MVA vaccine for preimmunization followed by revaccination with conventional vaccine after 6–12 months and started to apply this approach (Stickl _et al._ 1973; 1974). As far as we know, there was no comprehensive analysis of the results of the practical implementation of this protocol. Nevertheless, it is known that more than 100,000 people were vaccinated by this method without any serious complications.

In the experiments of Matsevich (1983) on estimation of the two-stage vaccination method, a vaccine inactivated by γ-irradiation and MVA vaccine were administered at the first stage. According to the study, the immunological efficiency of the second variant of two-stage vaccination was somewhat lower than that of the first one and was significantly lower than in the case of conventional vaccination. Earlier, this fact was also observed by the authors of the vaccine themselves.

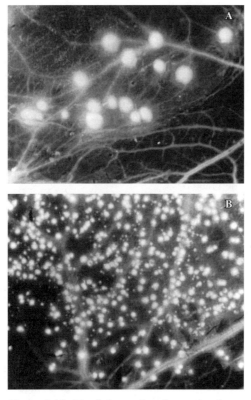

Figure 3-16. Morphology of pocks produced on CAMs under the same conditions by vaccinia virus (A) EM-63 strain and (B) attenuated MVA strain.

The latest of attenuated vaccines developed was the vaccine involving strain Lc16m8. This strain was obtained from the Lister (Elstree) strain passaged at 30°C in rabbit kidney cell culture by cloning from a single pock on CAM (Hashizume *et al.*, 1973). Laboratory studies showed that it was not neuropathogenic for laboratory mice (Table 3.6) or rabbits. It was also less neuropathogenicity for monkeys than any other strain including CVI-78. In clinical observations covering over 10,000 primarily vaccinated children, the consolidated data on temperature response, diameter of infiltration at the inoculation site, encephalogram deviations, etc., confirmed a considerable attenuation of Lc16m8 strain as compared with the initial strain. Unlike other attenuated strains, it preserved a high take after cutaneous inoculation and satisfactory antigenic and immunogenic activities (Hashizume, 1975; Fenner *et al.*, 1988). However, noted that despite the obvious advantages of this vaccine, postvaccination complications were recorded, including eight cases of generalized vaccinia and one *eczema vaccinatum* in 10,578 vaccinees. The protocol for manufacture of this vaccine in a rabbit kidney cell culture developed for this vaccine gives the preparation meeting perfectly the International Requirements. Although the vaccine was not epidemiologically tested because of smallpox eradication, the available data allowed public health authorities in Japan to approve its use as a reserve smallpox vaccine in case of emergency.

3.6   Vaccination Response and Vaccination Immunity

Prior to presenting the data on vaccination response, let us dwell on consider vaccination methods.

3.6.1   Vaccination Methods

Various methods have been used for administering the vaccination material: skin cuts and incisions, scarification, injections, subcutaneous inoculation, etc. Number of inoculation sites, scarification lengths, and number of injections also varied. However, the overall diversity of vaccination methods pursued the single goal—introduction of the virus into deep epidermis layers. The outer surface of the shoulder in the area of the deltoid muscles was chosen almost always as the vaccination site.

During the smallpox eradication program, many obsolete or traumatic vaccination methods and tools (e.g., round lancet) were abandoned. Owing to the improved quality of smallpox vaccine and high standard of its infection activity, the number of inoculation sites was reduced to one or two. The most important improvement in vaccination methods was multiple pricking with a bifurcated needle invented in the USA by Dr. B. Rubin and adapted for mass-scale vaccination by WHO (Fenner *et al.*, 1988). Vaccination with bifurcated needles (15 pricks in a 3–5 mm skin area) was virtually ubiquitous, although in some countries, including the USSR, the old scarification method (two incisions for vaccination and three incisions for revaccination) was applied until the cessation of smallpox vaccination.

Another method used for massive vaccination during the smallpox eradication program involved jet injectors. Inoculation by the needle-free method into one site on a shoulder skin area was successfully used in Brazil, Afghanistan, and Western and Central Africa. Administration of this method reveals certain drawbacks—need in specially trained staff for maintenance of injectors, a high price, the necessity of gathering large population cohorts in one place at once, etc. For these reasons, the method could not compete with the vaccination with bifurcated needles.

In addition to vaccine administration through the skin, other methods were developed. One of their goals was to eliminate the side effects of conventional vaccination method. These alternative methods were aerosol and oral. After experimental studies and limited clinical trials, the former was abandoned because of serious pathological responses in lungs (Vorob'ev *et al.*, 1996). The oral method had its partisans, whose long-term and comprehensive studies proved the safety and efficiency of the method and allowed a pill vaccine against smallpox to be developed (Vorob'ev & Lebedinsky, 1977; Vorob'ev *et al.*, 1996).

3.6.2   Vaccination Response

Cutaneous application of smallpox vaccine is accompanied by both local reaction and general response. The manifestation of the latter varies depending of the reactogenicity of the vaccine, individual sensitivity, and several other factors.

Early signs of reaction to vaccine appear days 3–4 after vaccination and include development of erythema, swelling, and a papule at the site of inoculation. On the next day, the papule increases in size, and a narrow deep-red zone appears around it. On day 6, the papule transforms into a vesicle filled with transparent liquid. The margin of the vesicle is raised, and the center is sunken. The vesicle consists of multiple chambers. On day 7 or 8, another, wider red ring, an areola, appears around the vesicle. Starting from day 8, the contents of the vesicle (pock) becomes turbid and pussy (Figure 3.17). The pock reaches its maximal size by days 8–10 and then undergoes involution— it becomes dry and forms a scab. The areola turns pale, and the scab falls off (usually, during week 3), forming a specific scar. The development of the pock is accompanied by a general response of the organism. A pronounced fever is concurrent with the appearance of areola (days 7–10). At this time,

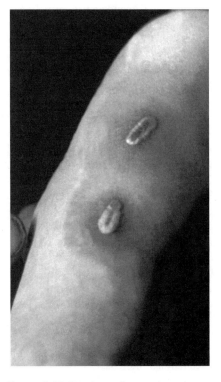

Figure 3-17. Local reaction to the primary vaccination by scarification technique (day 10).

malaise, headache, and anorexia may occur. Lymph nodes at the vaccination side are sometimes swollen and painful. The described response type is referred to as primary vaccination reaction. It is observed primary vaccinees.

Note that the replacement of highly reactogenic vaccine strains with low and moderately reactogenic variants significantly reduced the severity of vaccination reactions. For example, the vaccination reaction was usually accompanied by a temperature elevation to 38–39.5°C (Morozov & Solov'ev, 1948; Solov'ev & Mastyukova, 1961). After the replacement of strains and reduction in the number of inoculation sites to 1–2, a temperature elevation above 38°C was observed in a small part of the vaccinees. According to data obtained at our laboratory, vaccination with a vaccine prepared from EM-63 strain by the scarification technique with two incisions, the vaccination reaction occurred at normal temperature in 38.6% of the vaccinees; and only 19.4% had a temperature above 38°C. Similar results were obtained by Polak *et al.* (1963) and Noordaa (1964), who used a vaccine involving Elstree strain. The number of vaccinees with fever above 38.7°C was 5.7-fold less than in case of the vaccine from a highly reactogenic Copenhagen strain. Local reactions was also notably milder; areolae were smaller, additional pustules at the vaccination sites developed much less frequently, and the rate regional lymphadenitis decreased from 67–85 to 30–32%.

The reaction to revaccination is more diverse. It is determined primarily by the level of anti-smallpox immunity at the instant of the revaccination. In addition to the primary-type response, in some cases developed by revaccinees long after the first vaccination, accelerated and immediate responses are recognized. According to Ogorodnikova (1969), primary-type response was observed in as few as 0.17% of revaccinees. In all cases, the interval between the primary vaccination and revaccination was not less than 20 years. Accelerated reactions occur in vaccinees with a partially preserved immunity. The reactions follow the pattern of the primary response but develop faster. The manifestation of the response is, as a rule, less pronounced than after the primary vaccination. Immediate reactions to revaccination are characteristic of cases with high levels of humoral and cell-mediated immunities against smallpox. Actually, this is a hypersensitivity reaction of a delayed type. The reaction appear after 24–48 h in a form of erythema, swelling, or, sometimes, a papule at the inoculation site. In some cases, the papule is transformed to a vesicle, whose development reaches its acme after 48–72 h. Immediate responses were most frequent in the cohorts of vaccinees who received the first immunization 1–3 years before the revaccination (61.4–91.7%). In groups with the interval between vaccinations increased to 6–7 years, the rate of those with immediate reactions was less—33.4–61%. In contrast, the rate of accelerated responses increased with the interval and reached 70% revaccinees first immunized

10 years before (Ogorodnikova, 1969). Similar patters were observed by other researchers (Benenson, 1950; etc.).

Studies specially performed to find a correlation between the antibody level in blood serum and type of skin vaccination response yielded no positive results. A certain correlation was found only when average antibody levels in cohorts with different response types were compared. The highest average antibody level was found in the cohort of revaccinees with the immediate response type. A lower initial titer was recorded in the cohort displaying the accelerated reactions. However, the cohort with the immediate reaction to revaccination showed a lower seroconversion rate and a lesser multiplicity of antibody titer increase (Ogorodnikova, 1969).

3.6.3   Vaccination Immunity

A successful vaccination induces an immunity state in vaccinees. Provided that a high-quality vaccine and proper vaccination method were used, the degree of immunity stress and duration are determined primarily by the immunological reactivity of an individual. Antibodies to smallpox appear in vaccinees not earlier than on days 10–14 (Downie, 1951; McCarthy *et al.*, 1958; Ehrengut, 1966), and their titer gradually rises. The titer of virus-neutralizing antibodies reaches its maximum after 3–4 weeks; however, it is lower than in the case of genuine smallpox (Downie, 1951; S.S. Marennikova, unpublished data). Later on, the titer of antibodies gradually decreases. Complement-binding antibodies disappear after 6 months. After 8 months, AHAs are also undetectable in the majority of vaccinees. Virus-neutralizing antibodies are retained for the longest period— they can be found 5 and more years after the vaccination (Figure 3.18).

The protective effect of vaccination is induced rather rapidly. This is proven by the results of epidemiological studies of the vaccination efficiency after a contact with a smallpox case. As mentioned above, the duration of vaccination immunity varies individually. It is believed that vaccination ensures a sufficient protection, as a rule, for the first three years. Later, the protective effect gradually decreases. According to Ladnyi (1985), contacts of vaccinees with smallpox cases occurring within 1–3 years after vaccination resulted in a morbidity rate of 0.3%; 3–5 years after vaccination, 0.9%; and over 10 years after vaccination, 10.7%.

Already the first decades of vaccination practice showed that a single vaccination was insufficient. Revaccination was first introduced in Germany in 1829 (Fenner *et al.*, 1988). Later, revaccination was accepted in other countries. The number and terms of revaccination varied over a broad range. In Russia, revaccination schedule changed several times. The last vaccination directions (before the discontinuation of routine vaccination), adopted in 1975, demanded double vaccination at 8 and 16 years old.

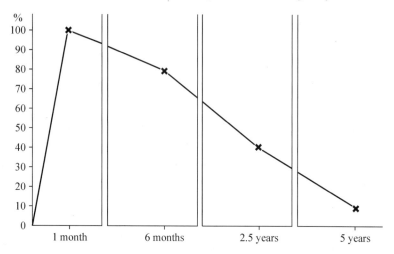

Figure 3-18. Virus-neutralizing antibodies in children after the primary vaccination (strain EM-63). The geometric mean of the titers of virus-neutralizing antibodies 1 month after the vaccination is taken as 100%.

The humoral response to revaccination appears much quicker than the response to primary vaccination. The virus-neutralizing antibodies are detectable on day 7, and their titer is much higher than that after the primary vaccination. The maximal AHA titers are observed from day 10 to 15. Owing to the rapid induction of immune response to revaccination, it can be used as a mean for emergency prevention even during the first days after infection (contact with a smallpox case).

The efficiency of revaccination can be illustrated by the data obtained by Rao (1972). No diseases occurred in a cohort of 212 persons who had family smallpox contacts no later than 2 years after revaccination. Of 30 revaccinees more than two years before, one contracted the disease. Even if the disease took place, the preceding vaccination altered its course drastically. Among the persons with histories of vaccination and revaccination, the modified course (varioloid) was observed in 40% of the cases. This rate was 25% among vaccinated once and as little as 1%, among nonvaccinees.

3.7   Postvaccination Complications and Other Vaccinia Virus-related Pathologies

Along with to the tremendous benefits, the history of vaccination brought about a number of victims related to postvaccination complications. Although such complications were relatively few, they caused sufferings or even death of people healthy before the vaccination. It is commonly known that smallpox eradication was followed by cessation of routine vaccination against smallpox. Nevertheless, the problem of postvaccination complications is of interest not only as a specific pathology caused by vaccinia virus. Attention should be paid to these complications when designing recombinant vaccines involving this virus. Moreover, it should be kept in mind that the vaccination against smallpox may be resumed due to the menace of potential bioterrorism attacks using this most dangerous pathogen (see additionally Chapter 9). Such campaigns of selective vaccination were already realized in several countries (Israel and the USA).

The diverse postvaccination complications may be classified into three groups: neurological disturbances, lesions on the skin and mucous membranes, and impairments of other organs and systems.

Because of the limited volume, we cannot consider this intricate problem in detail. Those interested in the subject can refer to specialized monographs (Ehrengut, 1966; Wilson, 1967; Rohde, 1968; Matsevich, 1970; Ehrengut *et al.*, 1972; Bondarev & Voitinsky, 1972; Gurvich, 1983). We discuss only the topics most important to our opinion.

The bulk of the complications of the first group are constituted by postvaccination encephalitides and encephalopathies. Other diseases (myelitis, serous meningitis, and polyradiculoneuritis) are rare.

Numerous comprehensive studies demonstrate that the rate of postvaccination encephalitides depends on vaccination status and age of the vaccinees. According to studies by Marennikova *et al.* (Marennikova *et al.*, 1968; Marennikova & Matsevich, 1971; 1974) and Gurvich (1983), the rate of this complication type in revaccinees was 40- and 48-fold less, respectively, than in primary vaccinees. Data from several other countries show lesser differences (Dick, 1973). This may be related to longer intervals between the primary vaccination and revaccination than in the USSR, where vaccination immunity was maintained at a high level owing to systematic mandatory vaccinations.

Numerous studies have shown that the rate of neurological complications depends on age, being higher among older persons (Herrlich, 1959; Regamey, 1965; Rohde, 1968; Gancheva & Andreev, 1972; Espmark *et al.*, 1973; Marennikova & Matsevich, 1974; etc.). Of special interest from this standpoint are the data on the rates of such complications in various age cohorts in the USSR within the 5 years preceding the discontinuation of routine vaccination (Table 3.8).

As is evident, the increase in the rate of neurological complications, including postvaccination encephalitides, with the age is apparent without additional comments. Similar data are reported by Espmark *et al.* (1973). There is no explanation why this pattern was not so clear in the studies by US scientists (Lane *et al.*, 1970; Lane, 1973).

Table 3-8. Postvaccination complications in primary vaccinees depending on the age (according to the data of 1975–1979)*

Complication type	Characteristics	Age (years)			
		1–2	2–3	3–5	Over 5
Neurological complications	Rate	1 : 75,200	1 : 24,700	1 : 9000	1 : 3200
	Cases per 1 million	13.3	40.5	111.1	312.5
	Fatal cases per 1 million	1.4	3.6	15.9	26.1
Inclu- ding	Postvaccination encephalitis** Rate	1 : 209,000	1 : 46,700	1 : 13,200	1: 4800
	Cases per 1 million	4.8	21.4	75.7	208.3
	Rate of fatal cases	1 : 714,000	1 : 280,000	1 : 63,000	1 : 38,300
	Encephalitic reactions*** Rate	1 : 118,000	1 : 52,500	Over 3 years old 1 : 46,400	
	Cases per 1 million	8.5	19.05	21.55	
Complications with skin lesions	Rate	1 : 115,800	1 : 27,400	1 : 19,400	1 : 5500
	Cases per 1 million	8.6	36.5	51.5	181.8
	Fatal cases per 1 million	0.23	0	0	0
Number of primary vaccinees	Totally, >19 million, including	17.1 million	2.5 million	252,000	38,300

*According to Gurvich (1983; 1992) and Marennikova *et al.* (2003);
**Including cases of postvaccination encephalitis, meningoencephalitis, and encephalomyelitis; and
***Consolidated cases of febrile and afebrile convulsions; in older children, hallucinatory syndrome.

Postvaccination encephalitis is one of the gravest complications. Its case-fatality rate depends on age and several other factors. According to Polak (1973), the mortality rate was 75% among 1–3-month-old children and 39% among 4–6-month-old ones. Dick (1973) reported that the mortality rate amounted to 39.2% among the children younger than 1 year and 10.5% among children of 5–14 years old. Similar data were reported by Gurvich (1983): 29.2% in a cohort younger than 2 years and 12.5% in a cohort older than five years. During evaluation of mortality, note that this rate decreased significantly after the introduction of treatment with specific immunoglobulin (Marennikova & Matsevich, 1968; Polak, 1973).

Despite the studies and discussion lasting for decades, the nature of this pathology is not understood yet. The distinctive pathomorphological changes detected in the brain during the postvaccination encephalitis are classified as perivenous demyelinating microglial encephalitis. Their specific pattern differs from that observed in animals with experimental meningoencephalitis caused by intracerebral infection with vaccinia virus. Guarnieri bodies, as a direct evidence of the virus replication in the brain, are absent as well. Nevertheless, there are data indicating that vaccinia virus is involved in the development of this pathology. For example, the vaccination immunity decreases its rate manifold. The virus was isolated from the cerebrospinal fluid and brain tissue of some vaccinees with postvaccination encephalitis and the corresponding fatal cases (Matsevich, 1970; Marennikova & Matsevich, 1974; Gurvich, 1983; etc.). This is also confirmed by the efficiency of seroprophylaxis of complications—vaccination under protection with the specific immunoglobulin (Nanning, 1962; Marennikova *et al.*, 1975b) and by a certain effect of specific serotherapy against postvaccination encephalitides (Marennikova *et al.*, 1968; Bondarev & Voitinsky, 1972; Marennikova & Matsevich, 1974; Braginskaya *et al.*, 1978). Encephalopathies occur mainly in younger children. Unlike postvaccination encephalitis, the pathomorphological alterations related to encephalopathy are represented mainly by brain edema and circulation disturbances (stases, hyperemia, etc.).

The gravest complications affecting the skin and mucous membranes are *eczema vaccinatum* and progressive vaccinia. The latter term was introduced by Kempe and Benenson (1955) to designate a postvaccination complication previously known under various names: prolonged generalized vaccinia, *vaccinia necrosum*, or *vaccinia gangrenosa*. The background for the pathogenesis of this complication is immune deficiency, which makes the vaccination response flaccid and accompanied by necrotizing and spreading of lesions from the vaccination site to other skin areas. The disease sometimes lasts for months. Treatment of progressive vaccinia is complex and frequently unsuccessful even with administration of specific immunoglobulin. The disease is highly lethal.

In contrast, specific serotherapy is efficient against another severe dermal complication, *eczema vaccinatum* (Kempe, 1960). It develops in vaccinees suffering from eczema before the vaccination. Until the introduction of treatment with specific immunoglobulin, the lethality of *eczema vaccinatum* was 30–40%.

Generalized vaccinia is less severe and less frequent complication of this group. It manifests itself as rash lesions on the body and mucous membranes and as a rule is benign. Some patients show signs of immunodeficiency. Treatment with specific immunoglobulin yields good results by interrupting further expansion of rash.

The third group of complications, affecting various organs and systems, is very diverse. Postvaccination involvements of the heart (myocarditis and pericarditis), kidneys (nephritis and nephrosis), lungs (pneumonia), skeletal system (osteomyelitis), etc., are reported. Most of these complications develop on the background of preexisting but not diagnosed diseases. Some of such complications had a severe courses and ended fatally.

The complications involving the heart should be considered specially. The recent anti-smallpox vaccination campaign in the USA (2003) demonstrated a rather high rate of such complications. For example, cases of myocarditis, pericarditis, and myopericarditis as well as various ischemic disturbances among the civil population were recorded at a rate of 1 : 1700 vaccinees (CDC, 2003). Among the military cohort, this rate amounted to 1 : 36,000. The majority of such cases resulted in recovery. However, the fatality rate according to Thorpe *et al.* (2004) amounted to 6.1 cases per 1 million, and all the lethal cases had atherosclerosis. The mechanism underlying development of cardiac complications is yet vague; in particular, there are no evidences of the direct effect of vaccinia virus on the heart. Note that this type of complications in Europe was recorded at a considerably lower rate.

Some complications developed due to contamination of highly sensitive areas—genitals or eyes—with the vaccination virus. In the latter case, blepharitis or blepharoconjunctivitis may develop. In the case of cornea contamination, keratitis of various degree of severity may develop (Pepose *et al.,* 2003). More severe eye diseases, for example, keratouveitis (Lee *et al.,* 1994), were also recorded. In certain cases, these complications resulted in a partial or complete blindness.

To conclude, let us mention an unusual form of postvaccination infection we observed. It was caused by airborne ingress of the virus into the body (Marennikova *et al.,* 1993). The disease developed in previously unvaccinated children as a result of inhalation of dry smallpox vaccine from broken ampoules. The vaccine (more than 100 ampoules; titer $>10^8$ PFU) was occasionally found by children; and they broke the ampoules while playing in a small room without ventilation. Four to seven days later, five of the eight children had acute fever, headache, sore throat, and neck ache. They also developed hyperemia of scleras and conjunctivas, ulceration and diphtheria-like patches on the rear mucous surface of the throat, mouth, and nasal passages. The cervical, parotid, and, particularly, submaxillary glands were swollen and painful. Some of the children had skin lesions near nose wings. The disease lasted for 7 to 12 days. Its vaccinia origin was confirmed by the detection of the virus in the pharynx, nasal discharge, tear fluid, and skin lesions. Species-specific antibodies were detected in the blood sera taken after 2.5 weeks. All the children recovered without any complications.

The described course of the vaccinia-caused disease is similar to that after experimental airborne infection of animals and, to a certain extent, to the course of revaccination of people by this method. In animals, lung involvement (interstitial pneumonia) was predominant. Its manifestation depended on the reactogenicity of the vaccine (Vorob'ev *et al.,* 1996). The involvement of the lungs cannot be excluded in the case described above. A situation similar in the mechanism of its development was recorded in summer 2000 in Vladivostok, when eight children at the age of 7 to 13 years old (unvaccinated against smallpox) contracted the disease. As in the case described above, the source of infection was vaccinia virus from broken accidentally found ampoules with a smallpox vaccine after the end of its shelf life. The specific feature of this "outbreak" was predominant development of skin lesions at the sites of previous or existing cuts, abrasions, or scratches. The lesions varied in number, localization, size, and the degree of involvement of the adjacent tissues (edema, hyperemia) as well as regional lymph nodes. The progress of skin lesions resembled the vaccination reaction of the so-called primary type. Several children develop a more pronounced reactions compared with a conventional vaccination response, presumably, due to a larger infective dose. The process was accompanied by a temperature increase. Some children at the height of disease developed a short-term punctate rash. Upon healing, the lesions formed characteristic scars. A comprehensive laboratory investigation (Onishchenko *et al.,* 2003) confirmed the vaccinia virus etiology of the disease and identity of the viruses isolated from children and found in the ampoules.

Chapter 4

VARIOLA (SMALLPOX) VIRUS

4.1   Smallpox

Before the global eradication, smallpox was among highly hazardous infectious diseases of humans. Since ancient times, its disastrous and devastating epidemics filled countries and peoples with consternation over many centuries. However, a mild course of certain smallpox epidemics with virtual absence of deaths attracted attention at the end of 19th–beginning of 20th centuries. Such epidemics were recorded in North and South Americas, Africa, Europe, and other regions. It actually seems that this disease existed and was known considerably earlier. In particular, a vivid description of this disease given by Jenner (1798), who observed an epidemic of such infection in Gloucestershire in 1791, is one of the confirmations. Jenner determined it as "a special type of smallpox". This type of smallpox was referred to in the literature as white pox, milk pox, minor pox, kaffir pox, etc. However, the name *variola minor*, introduced in 1929 (Dixon, 1962) to distinguish it from the classical type—*variola major*, became commonly accepted. Characteristic of *variola minor* disease are a mild course, often even not requiring a bed rest, and a mortality rate of 0.1–2%. The case-fatality rate of *variola major* infection varied from 5 to 40%, amounting on the average in Asia to 20% (Fenner *et al.,* 1988). Before commencement of the smallpox eradication program, *variola major* was spread in countries of Asia and Africa.

Outbreaks of *variola minor* were observed concurrently and independently of *variola major* outbreaks. The main foci of *variola minor* during the smallpox global eradication were South America and some regions of Africa (South Africa and the Horn of Africa). The smallpox in South America is conventionally called *variola alastrim*, unlike Africa, where it retained the name *variola minor*. As was mentioned, this terminological differentiation stems from the discovered biological distinctions between the viruses isolated in these two regions (Dumbell & Huq, 1986). However, no differences in the clinical courses of *variola minor* and *variola alastrim* were found. Note that *variola major* and *variola minor* (*alastrim*) pathogens in the early classifications (Wildy, 1971) were considered as separate species. However, the International Committee on Taxonomy of Viruses later decided that the existing distinctions were insufficient for regarding variola minor virus and variola alastrim virus as separate species (Virus Taxonomy, 2000).

In addition to *variola major* and *variola minor*, note certain specific features of smallpox that was spread on the African continent. Overall, distinguishing of this variant were a milder course (compared to Asian) and lower fatality rate (5–15%). This phenomenon yet lacks a complete explanation.

4.1.1   Classification of Clinical Forms
of Variola Major and Their Courses

As early as at the time of Jenner, it was well-known that classical smallpox had various clinical courses. Jenner himself underlined it referring to two smallpox types—confluent and discrete (Jenner, 1798). Of the smallpox classifications that appeared later, the classification proposed by Rao (1972), based on his first-hand examination of almost 7000 smallpox cases (Table 4.1), became the most generally recognized (including the opinion of the WHO expert committees).

Prevalences of individual *variola major* types and the corresponding mortality rates are listed in Table 4.2.

As is evident, ordinary type of smallpox clinical course is observed in the majority of cases (88.8%), while both hemorrhagic and the so-called flat-type smallpox courses displayed the highest case-fatality rates. During the smallpox outbreak in Yugoslavia in 1974, the last type was unusually prevalent (15.3%; Suvaković *et al.*, 1973).

The latent period of smallpox (from the moment of infection to fever onset) is usually 10–14 days); in rare instances, shorter or longer (from 7 to 19 days; Fenner *et al.*, 1988). From his own observations, Rao determined the latent period as amounting to 12–21 days. In singular cases, it may be as short as 8–9 days or may extend to 24 days (Rao, 1972). The smallpox course comprises the two stages—prodromal (before the rash onset) and eruption. The latter stage consists of papulovesicular, pustular, and scab stages. Below, we are describing the course of ordinary smallpox. The prodromal stage takes from 2 to 4 days (commonly, 3 days); the patients are febrile (to 39.5–40.5°C), have severe headache, and a characteristic sacrum pains. In addition, the symptoms include tachycardia, tachypnoea, nausea, sometimes vomiting, and delirium. On days 2–3, the majority of patients

Table 4-1. Classification of *variola major* clinical types*

Type	Main specific features
Ordinary smallpox	Extensive rash; the lesions appear as macules and evolve through papules, vesicles, and pustules to scabs. Subtypes: *Confluent:* confluent rash on the face and extensor surfaces of the limbs and discrete rash in the rest regions. *Semiconfluent:* confluent rash on the face and discrete rash on the body and limbs. *Discrete:* discrete pocks cover the entire body, while the skin between lesions remains healthy.
Modified smallpox (varioloid)	Similar to common smallpox (may be *confluent, semiconfluent,* and *discrete*); however, differ in an accelerated course and absence of toxicosis.
Variola sine eruptione	Fever and other prodromal symptoms and no rash; occurs in individuals with a high immunity; rarely, in nonvaccinees; diagnosis is confirmed serologically.
Flat smallpox	Rash lesions are flat and soft; usually has fatal outcome. Subtypes: *confluent, semiconfluent,* and *discrete.*
Hemorrhagic smallpox	Numerous hemorrhages into the skin and/or mucosae at one or another stage of the disease. *Early:* hemorrhaging on the skin and mucosae starting from the prodromal stage; always lethal. *Late:* hemorrhaging on the skin and mucosae after the rash onset; usually lethal.

*According to Fenner *et al.* (1988).

develop the so-called prodromal rash in forms of roseolae and petechia in the femoral trigone region and/or on the chest, which disappears in 2–3 days. Temperature subsiding and development of enanthema precede the onset of

Table 4-2. Prevalences of individual variola major clinical types and mortality rates of nonvaccinees and vaccinees (quoted from Fenner *et al.*, 1988)

Clinical type	Nonvaccinees		Vaccinees	
	Prevalence (% of total number of cases)*	Mortality rate (%)	Prevalence (% of total number of cases)**	Mortality rate (%)
Ordinary	88.8	30.2	70.0	3.2
Confluent	22.8	62.0	4.6	26.3
Semiconfluent	23.9	37.0	7.0	8.4
Discrete	42.1	9.3	58.4	0.7
Modified	2.1	0	25.3	0
Flat	6.7	96.5	1.3	66.7
Hemorrhagic	2.4	98.4	3.4	93.9
Early	0.7	100.0	1.4	100
Late	1.7	96.8	2.0	89.8
Total		35.5		6.3

*3544 cases.
**3398 cases.

rash lesions on the skin (several hours or one day prior to eruption). Skin eruptions in the form of small reddish nodules (papules) appear on days 2–4 upon the fever onset first on the face and upper extremities, then spreading to other body areas. Papules grow in size and number to progress during 1–2 days into blisters filled with a clear fluid (vesicles) and a characteristic umbilical depression in the center. By days 6–7 after the rash onset, the fluid in the vesicles gradually changes from clear to pus-like (pustular stage), accompanied by a gradual temperature elevation (the second wave of hyperthermia). The specific pattern of smallpox rash is its centrifugal distribution—increase in the number of lesions per unit area (density) from the center to periphery (face and limbs), including palms and soles. Pustules reach their maximal size by days 10–13 (Figure 4.1) and then gradually flatten, get dry, and start to form scabs. During this stage, amelioration and decrease in body temperature are observed. Dried scabs fall off by days 30–40 of the disease, leaving behind reddish spots. These spots later develop into characteristic smallpox scarring, mainly on the face (pock-marked face). The time course of smallpox rush progress during flat-type confluent smallpox is shown in Figure 4.2.

The main distinctions of the rest smallpox clinical types are listed in Table 4.4. Below are some additional details. Characteristic of the early hemorrhagic smallpox (*purpura variolosa*) are a shorter latent period, longer prodromal stage (to 6 days), and most pronounced manifestation of all the symptoms typical of the latter stage, including toxicosis. Most typical are early and numerous hemorrhages into the skin and mucosa as well as nasal, respiratory, and gastrointestinal bleedings (Morozov & Solov'ev, 1948). Subconjuctival hemorrhages are among the first symptoms of the commencing hemorrhagic diathesis. According to Rao (1972), the patients usually die approximately on day 6 of fever. If the death is delayed by 1–2 days, the surface skin layer rises, and fluid accumulates beneath forming large blisters filled with serous or blood-containing serous fluid. Hematological examination of early hemorrhagic smallpox cases demonstrated severe impairments of the blood clotting system: a pronounced decrease in platelet counts, low thrombin level, increase in circulating antithrombin, etc. Rao (1972) assumed that these hematologic defects might be a result of intensive and long-lasting viremia, observed in such cases (Downie *et al.*, 1953). *Purpura variolosa* is met in adults, more frequently, women, at a rate of 88% (Rao, 1972).

In the late hemorrhagic form (*variolosa pustulosa hemorrhagica*), hemorrhages into pustules are typical, making these lesions dark, almost black (that is why it is called "black smallpox"). However, Rao noted that sometimes hemorrhages developed earlier, starting from the papular stage, and occasionally concurred with the hemorrhages into the skin and mucosae. In certain cases, he observed no hemorrhages into the rash lesions, only into

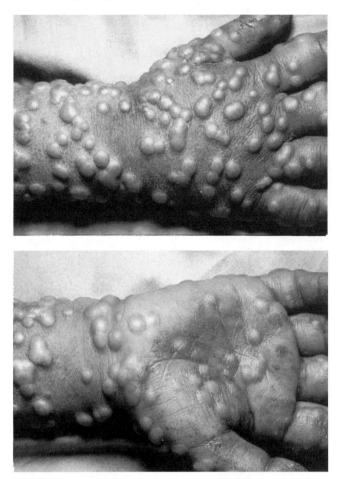

Figure 4-1. Smallpox at the pustular stage; eruptions on hands including palms (courtesy of WHO).

mucosae. Patients die on day 8–10 of the disease. Similarly to early hemorrhagic smallpox, humans over 14 years old are predominantly (80%) afflicted.

In addition to specific features of the lesions themselves (as a rule, not rising above the skin and lacking typical depressions in the center), characteristic of the flat-type smallpox form is a slower evolution of the lesions and in certain cases, halt in development at the stage of vesicles. Pustules are rarely observed. The disease proceeds on the background of high fever and severe toxicosis and is frequently accompanied by lung complications. Skin detachment at the site of lesions is occasionally

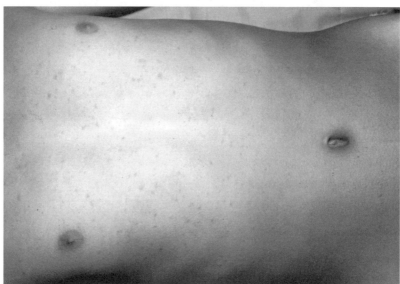

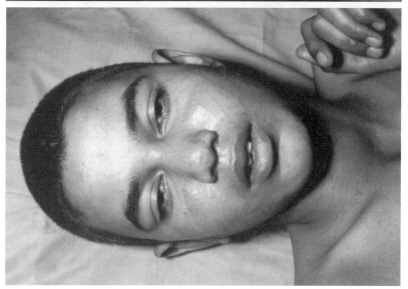

Figure 4-2. (1–12). A patient with a flat-type confluent smallpox with a fatal outcome on day 13. The time course of rash development from day 1 of rash onset to day 10 (courtesy of WHO).

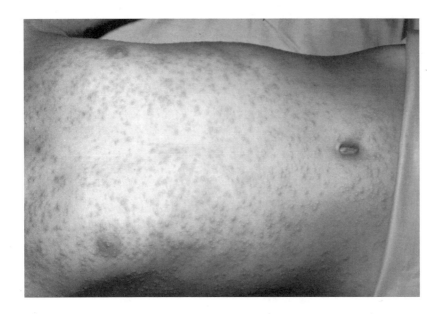

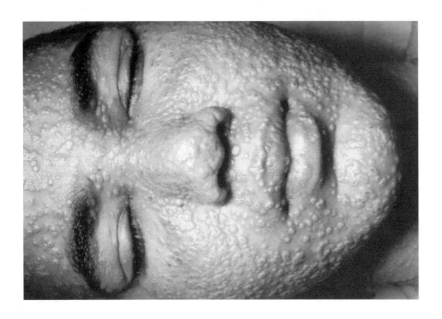

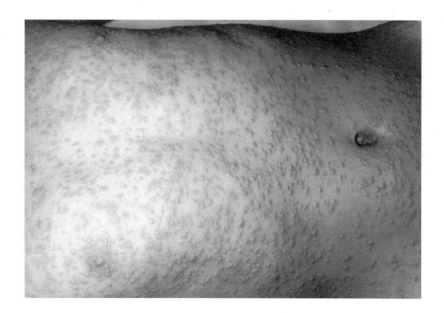

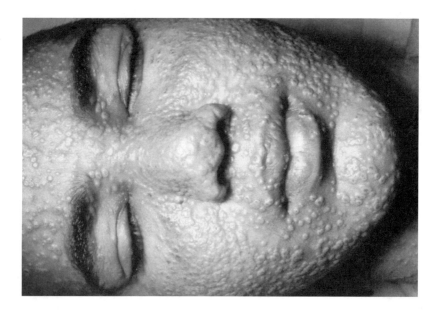

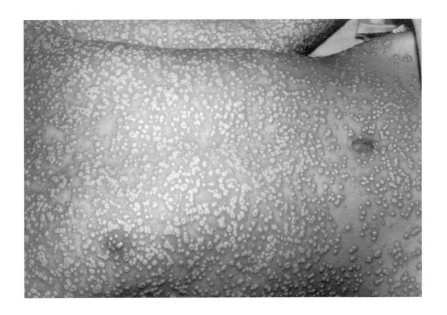

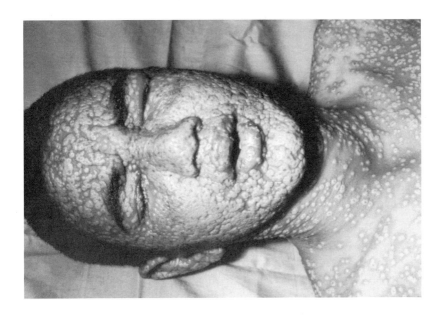

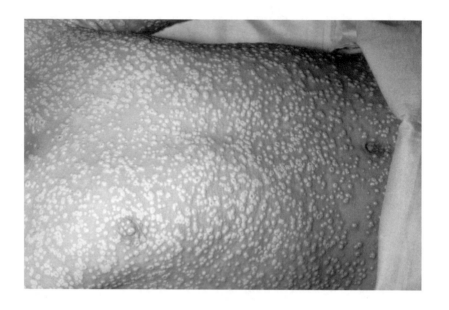

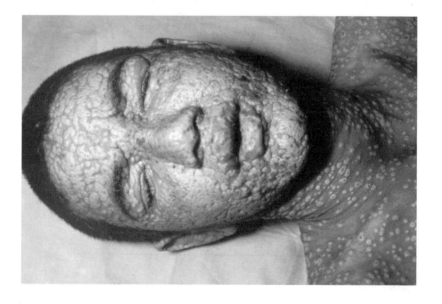

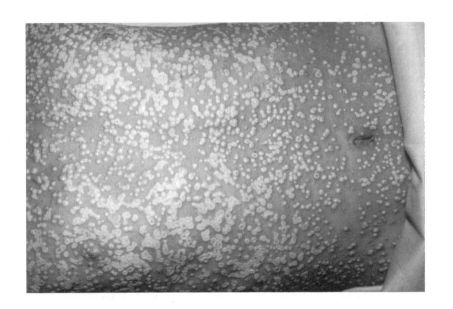

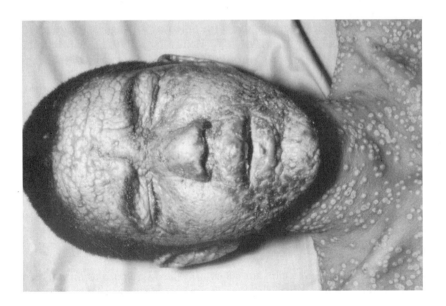

observed. Unlike hemorrhagic forms, the majority of cases are children. The lethality rate is very high, reaching 100% in the case of confluent type.

In the modified smallpox type (varioloid), after a 2–3 days of prodromal stage with a moderate or high body temperature and less pronounced symptoms, skin lesions lacking the typical umbilical depressions appear and evolve quickly. The second wave of hyperthermia is usually absent. Scabs are formed to day 10 of the disease. Fatal outcomes are absent. As a rule, varioloid is observed in vaccinees.

Variola sine eruptione or *variola sine exanthemata* proceeds with the prodrome typical of smallpox, but skin eruptions do not develop. The throat pain observed in some cases suggests enanthema (Rao, 1972). Clinical diagnosis is verified by detecting the specific antibodies in blood serum. Isolation of the virus from pharynx was demonstrated (Verlinde & van Tongeren, 1952; Marennikova *et al.,* 1961a; 1963; Rao, 1972). The so-called pulmonary form of smallpox, when patients develop pneumonia without developing skin lesions, is also attributable to *variola sine eruptione.*

Of interest is the observation of Kempe *et al.* (1969), who reported conjunctivitis as a sole manifestation of variola virus infection in 21 of the 55 persons contacted with smallpox cases. The VARV nature of this conjunctivitis was confirmed by isolation of the virus from conjunctival exudates of 12 examinees. In another set of examinations performed by the same team, the virus was isolated from tears (Dekking *et al.,* 1967).

The so-called subclinical forms of smallpox, when no symptoms are evident and serological examination is the only way of diagnosing, are also described (Downie & McCarthy, 1958; Heiner *et al.,* 1971). Subclinical forms were recorded for both *variola major* and *variola minor* (Salles-Gomes *et al.,* 1965).

Inherited smallpox was known as far back as since the middle of 18th century. It was repeatedly recorded in the cases of both *variola major* and *variola minor* (Marsden & Greenfield, 1934; Dixon, 1962; S.S. Marennikova, unpublished data, 1965; etc.). Dixon divided the inherited smallpox into two variants: genuine inherited smallpox, when the fetus was born with the symptoms of the disease, and the smallpox that developed after delivery (within the incubation period). Births of twins were described when one newborn baby was healthy, whereas the other displayed skin lesions characteristic of smallpox.

4.1.2   Smallpox Complications

Complications of smallpox are numerous. They developed as a result of both the direct effect of variola virus on various organs and body systems and of secondary bacterial infections. The most frequent complications were eye affections (keratites, corneal ulcerations, and as a consequence, partial or complete blindness). According to Morozov and Solov'ev (1948), the majority of convalescents had hearing impairments. Involvement of respiratory organs (bronchitis, bronchopneumonia, and pneumonia) and larynx were also frequent. Rao (1972) described development of encephalitis at a rate of 1 case per 500 smallpox patients. Severe course of the disease were accompanied by serious impairments of the gastrointestinal tract up to mucosal detachment over considerable areas. Some women developed uterine bleeding; a part of pregnant women had miscarriages. Various skin lesions mainly due to secondary infections (abscesses, phlegmons, and even gangrenes) were among frequently observed sequelae. Another type of complications was sepsis.

4.1.3   Smallpox Pathogenesis

The concept of smallpox pathogenesis is based on both studies of variola virus (VARV) itself using sensitive animal (monkeys and white suckling mice) models (Hahon & Wilson, 1960; Westwood *et al.*, 1966; Kaptsova, 1967) and the data on generalized orthopoxvirus infections of animals—mousepox and rabbitpox (Fenner, 1948; Roberts, 1962; Westwood *et al.*, 1966). A part of the materials was obtained while examining patients with various smallpox types at different stages of the disease. Below, we are describing the main stages of smallpox pathogenesis, as it outlines from the sources mentioned.

A stepwise pattern of the virus spreading in the body, comprising the three main stages—infection of cells, virus replication, and its "blowout"—is typical of all the generalized orthopoxvirus infections, smallpox included. In the case of respiratory contagion, the major infection route for smallpox, cells of the upper (olfactory epithelium) and lower (bronchial and alveolar epitheliums, including alveolar macrophages) respiratory tract are infected first. Roberts (1962) demonstrated this in mousepox model using fluorescent antibodies.

Lancaster *et al.* (1966) confirmed these results using three models (rabbitpox, VACV-infected rabbits, and VARV-infected rhesus monkeys) and a combination of immunofluorescent and histological examinations. Already after 24 h, the antigen was detectable in the epithelium of bronchioli and alveolar cells yet without any pathohistological changes. They developed later when the virus titer reached 5 log PFU/g tissue and were represented by the foci consisted of wide central necrotic zone with detectable fragments of bronchioli and external zone formed by clusters of neutrophils and macrophages. The adherent alveoli were edematic, and their cavities were filled with exudate containing neutrophils and free alveolar phagocytes. Larger bronchi also contained the foci of destruction and infiltration of epithelium. The adjacent arteries displayed proliferation of intima and edema of the middle layer and adventitia.

The virus is confined to the respiratory tract, accumulating actively in the lungs until day 3 (Hahon & Wilson, 1960; Westwood *et al.,* 1966; Kaptsova, 1967). From the primary focus, it then penetrates into lymph nodes via lymphatic tract to replicate and invade free macrophages of the lymphoid system. The virus further spreads in the body predominantly through the lymphatic system as well as by a short primary viremia (Hahon & Wilson, 1960). Due to the viremia, it appears, in particular, in the liver and spleen, where its replication presumably occurs initially in phagocytes and then in parenchyma. The accumulated and released virus enters the bloodstream (secondary viremia) and invades with it the skin, kidneys, CNS, and other organs. In addition to this main route the virus uses to enter the bloodstream (via the lymphatic system), Lancaster *et al.* (1966) demonstrated its direct penetration into the blood circulation system through the vessel (arteriolar) wall.

The onset of secondary viremia coincides with the display of the disease clinical symptoms. Penetration of VARV into the skin and mucosa induces development of characteristic smallpox lesions. Note that it is not always possible to detect the viremia in patients with ordinary type smallpox. A possible cause is a transitory type of the viremia and a low concentration of the virus. Positive assays for variola virus in blood or serum are rather rare and confined to the prodromal and early rash stages. On the contrary, intense viremia during hemorrhagic smallpox cases continues until the death, and the virus is readily isolated from the blood in considerable amounts (Downie *et al.,* 1953; Mitra *et al.,* 1966; etc.). In addition, Downie *et al.* (1969) noticed that the viremia during early hemorrhagic smallpox was more acute compared to the viremia of the late hemorrhagic form cases.

In experimental studies, a long-lasting viremia was observed during the fatal smallpox of suckling mice caused by VARV intranasal inoculation (Kaptsova, 1967) and rabbitpox (Westwood *et al.,* 1966). In the latter case, a severe disease course and high viremia continuing to the death suggested that this infection might be used as an experimental model similar in many respects to human hemorrhagic smallpox.

The smallpox lesions developed on oral and nasopharyngeal mucosae (unlike the skin lesions) ulcerate quickly, releasing large amounts of the virus. Variola virus is excreted from the fauces starting from days 3–4, i.e., virtually from the appearance of enanthema, and continuing for 7–9 and even 13 days depending on the severity of disease and its type (Downie *et al.,* 1961; Marennikova *et al.,* 1961a; Sarkar *et al.,* 1973a). One of the *variola sine eruptione* cases, observed by Marennikova *et al.,* displayed the virus excretion from fauces on day 3 of the disease. Shelukhina (1980) reported analogous case.

Presence of the virus in fauces already during the incubation period was demonstrated. Sarkar *et al.* (1973b) described a successful isolation of variola virus from fauces 5–6 days before the disease onset. Fenner *et al.* (1988) mentioned similar cases observed by Huq; they explained this by replication of the virus in fauces and pharynx. Hahon and Wilson (1960) also detected variola virus during the incubation period in fauces of aerogenically infected monkeys. They assumed that the virus came from the lungs, where it was actively replicating (Hahon & Wilson, 1960).

Numerous publications report a high concentration of VARV in skin lesions at all the stages of their evolution. However, data on the presence of VARV at the convalescence stage are sparse. Shelukhina *et al.* (1973) examined 12 smallpox convalescents after scabs fall off and the temperature normalized. The virus was detected in nasopharyngeal discharge of three convalescents on days 24, 25, and 26 of the disease; in urine of two convalescents on days 23 and 25. Moreover, the virus in one case was isolated from both nasopharynx and urine concurrently. Downie *et al.* (1965) reported isolation of variola virus from urine on days 6–16 of the disease onset; Sarkar *et al.* (1973a) on days 3–19.

The pathogenesis of comparatively rare cases of infection through the skin (for example, during variolation, accidental trauma during autopsy of smallpox cases, etc.) is basically similar to the pathogenesis of the disease caused by respiratory contagion. However, the primary skin lesion that appears on days 3–4 accelerates the process. From this skin lesion, the virus with infected macrophages reaches regional lymph nodes already after 24 h to accumulate there and spread in the body. Consequently, the incubation period is shortened by 2–3 days and viremia and rash commence earlier.

4.1.4 Immunity

Smallpox convalescents acquire a long-term and steady immunity. Reinfection cases are very rare—1 per 1000 according to Rao (1972); repeated smallpox infections, exclusively rare (Morozov & Solov'ev, 1948).

Similar to other orthopoxviruses, infection with variola virus induces a complete range of cell-mediated and humoral immune responses. The experimental and clinical observations of the humoral response are numerous, whereas the data on the cell-mediated immunity are yet sparse, although it is known that its role in both the recovery and further protection against reinfection perhaps even more important than the role of antibodies. Despite the insufficiency of the facts on the smallpox *per se,* there are grounds to extend the main patterns of cell-mediated response found for other orthopoxvirus infections to variola virus (Fenner *et al.,* 1989).

Below, we are briefing the data related to humoral response to the infection—formation of antibodies. The development of smallpox is accompanied by appearance of specific virus-neutralizing, hemagglutinating, precipitating, and other types of antibodies in the blood of afflicted or experimentally infected susceptible animals (monkey). The roles of these antibody types in fighting infection are different. Three groups of important antibody types exist: (1) the antibodies neutralizing the infectivity and comprised by those neutralizing extracellular (EEV) and intracellular (IMV) virions; (2) the antibodies that with the help of the complement are involved in lysis of infected cells; and (3) the antibodies eliminating the circulating viral antigens by formation of immune complexes. It is considered that the last type of antibodies is responsible for development of certain toxic symptoms during smallpox (Fenner *et al.*, 1989). In the experimental infection of monkeys, the virus-neutralizing antibodies appeared on day 2 of the disease onset; AHA, on day 4. The virus-neutralizing antibodies were detectable in all the human smallpox cases on day 6, and their titers were higher compared to the titers upon vaccination (Downie & McCarthy, 1958). The virus-neutralizing antibodies remained at a high level for 8 months; in one instance, they were found 4.5 years after recovery. Downie and McCarthy (1958), Mineeva (1961), and other researchers reported detection of virus-neutralizing antibodies in smallpox convalescents many years after the disease.

The complement-binding antibodies in serum of smallpox patients were detected soon after the reaction itself was discovered. They were detectable starting from days 8–10 of the disease and further over several month (Downie & McCarthy, 1958). Further studies demonstrated that mainly short-lived (IgM) antibodies, which remained in circulation for about 1 year, were involved in the complement binding reaction.

According to Herrlich *et al.* (1959), precipitating antibodies were detectable in 65–100% smallpox cases depending on the disease type. These antibodies are also short-lived, so their presence indicates a recent infection.

Collier (1951), Downie and McCarthy (1958), and Marennikova *et al.* (1961a; 1962) demonstrated that smallpox was accompanied by a considerable increase in AHA, that appeared on days 2–3 of the disease onset. High AHA titers were also recorded by days 12–26 with a peak between days 12 and 15. Later, their level decreased.

Comparative study of the time courses of accumulation of antihemagglutinin, complement-binding, and precipitating antibodies during various types of smallpox confirmed that AHA were the earliest and most frequent type of antibodies during smallpox (Herrlich *et al.*, 1959).

4.2   Morphology of Virions

Discovery of the pathogen causing smallpox is commonly associated with Paschen, who in a series of his works starting from 1906 described elementary bodies (variola virions), which got the name Paschen bodies, in the material from pustules of smallpox cases (Paschen, 1906; 1917; 1924).

Paschen himself saw the specificity of his findings not only in the fact that these bodies were regularly detected in clinical samples of smallpox cases, but also in the ability of these bodies to agglutinate under the effect of the specific serum. However, in the Anglo–Saxon literature, the elementary bodies of smallpox agent were frequently referred to as Buist bodies, as he described these bodies in contents of smallpox pustules as early as in 1886 (Buist, 1886).

Despite a long-standing interest to variola virus (VARV), its study until the late 30s of the 20th century (before the practice of cultivating viruses in chick embryos) was hampered due to the absence of any appropriate biological model. That time, monkeys were the only suitable object.

The studies on morphology and structure of VARV virions as well as its morphogenesis are comparatively sparse (Nagler & Rake, 1948; Lepine & Croissant, 1952; Andres K.H. *et al.,* 1958; Avakyan & Bykovsky, 1970). Their authors came to the same conclusion that the size, morphology, and ultrastructure of variola virus virions as well as the stages of development and interactions with the host cell structures were identical to those of vaccinia virus, thoroughly studied by many researchers (see Chapter 3).

Electron micrographs of variola virus are shown in Figure 4.3.

A distinctive (brick-shaped) form and other specific features of variola virus virions (as well as other orthopoxviruses) assisted successful use of electron microscopy for rapid detection of smallpox (in particular, while realizing the program of smallpox eradication).

4.3   Biological Properties

4.3.1   Pathogenicity for Animals

Primates. The first research into susceptibility of monkeys to variola virus was commenced as early as at the end of 19th and beginning of 20th centuries. These studies discovered that inoculation of human material from smallpox cases to certain monkeys caused local lesions and sometimes development of a generalized process resembling the smallpox in the manifestations (Copeman, 1894; Roger & Weil, 1902; Teissier *et al.,* 1911).

Comprehensive studies of Magrath and Brinckerhoff (1904a; 1904b) along with Brinckerhoff and Tyzzer (1906) on experimental smallpox of monkeys, including anthropoid apes has become classical. In particular, the authors imitated respiratory infection route by introducing pustular material into bronchi and succeeded in causing monkey disease similar to smallpox. A smallpox-like disease with rash and fever developed upon intravenous inoculation with human blood of smallpox cases (Kyrle & Morawetz, 1915). Later, these results were confirmed using pure variola virus cultures.

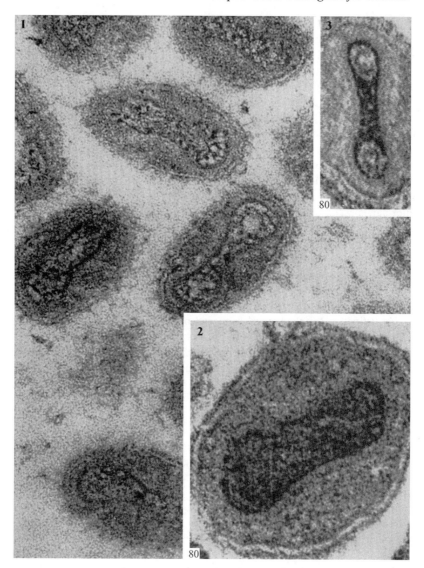

Figure 4-3. Virions of variola virus (electron microscopy; Avakyan & Bykovsky, 1970):
(1) cluster of virions in the cytoplasm, × 230,000; (2) individual virion, × 400,000; and (3)
a virion with an evident S-like structure in the nucleoid, × 300,000.

The information accumulated demonstrated that both lower monkeys—
Cercopithecus (Zuelzer, 1874), *M. nemestrinus* (Teissier *et al.,* 1911),
M. cynomolgus, M. rhesus, baboons *P. cynocephalus* (Heberling *et al.,*

1976)—and anthropoid apes—orangutan *Pongo pygmaeus* (Bras, 1952), chimpanzee *Pan troglodytes* (McConnell *et al.*, 1968; Kalter *et al.*, 1979)— are susceptible to variola virus.

Note that the severity of disease is determined to a considerable degree by the infective dose of the virus. According to Herrlich (1960), a large dose (7.2 log PFU) caused a generalized infection of *M. cynomolgus* (*M. iris*) that followed the course of hemorrhagic smallpox. However, lower doses (4.5–5 log $TCID_{50}$) administered intravenously induced nonlethal generalized infection of *M. rhesus* with a moderate severity (Solov'ev *et al.*, 1962; Gendon & Chernos, 1964). Aerosol infection of *M. iris* at a dose of 4.5 log PFU induced a similar disease course (Hahon & Wilson, 1960). In addition to the virus dose, the age of monkeys influenced the severity and outcome of experimental smallpox: younger animals appeared more sensitive to variola virus (Heberling *et al.*, 1976).

The interest to primate model of smallpox arose at the very end of the 20th century. To a considerable degree, this was directly related to a real menace of bioterrorism attacks involving variola virus and the resulting need in evaluating adequately the efficiency of developed protection agents (new vaccines, antivirals, etc.). Research teams from USAMRIID and CDC performed the main studies. Overall, these studies confirmed the observations of Herrlich (1960) mentioned above on the possibility of causing a fatal infection of *M. cynomolgus* with a hemorrhagic smallpox course using intravenous inoculation of high VARV doses. At a maximal dose (10^9 PFU), the virus was detectable in peripheral blood leukocytes and fauces of the infected animals within 48 h after infection. The authors believe that this allows for diagnosing the disease as early as during the prodromal stage (Jahrling *et al.*, 2002). A combined infection was also studied; this comprised intravenous inoculation supplemented with aerosol inhalation at a dose of 10^8–10^9 PFU. Under these conditions, the disease of the majority of monkeys lasted 3–4 days (rarely, 6–13 days) and was accompanied by hemorrhagic lesions on the skin, mucosae, and in internal organs. In addition, a pronounced coagulopathy along with degeneration and necrosis in internal organs were observed along with bronchiolitis (Huggins *et al.*, 2002).

Note that a variant of variola virus—variola minor alastrim—is also capable of causing generalized infection with characteristic rash on various parts of the body (Torres, 1935–1936; De Jong, 1956), at least when *M. rhesus* were inoculated intravenously and intracardially. However, Noble (1970) reported that several monkey species from South America appeared insensitive to variola minor alastrim virus.

Heberling *et al.* (1976) demonstrated that variola virus might be passaged on skin of baboons. Using pustular material, the authors showed that the virus properties remained unchanged after 10 passages.

As for natural transmission of this infection among monkeys, the available data are rather sparse. Initially, the attempts of experimental transmission of variola virus infection from infected monkeys to healthy animals were unsuccessful (Brinckerhoff & Tyzzer, 1906). A series of experiments with *M. iris* involving both respiratory and contact infections routes was performed in 1969. It was demonstrated that variola virus could be transmitted to healthy animals with a contaminated airflow without any direct contact. Contact infection transmission was also demonstrated. However, the latter transmission route appeared possible only during a limited number of passages: it decreased gradually to cease at passage 6 (Noble & Rich, 1969). Later, Kalter *et al.* (1979) reported a case of successive infection of two other animals of the same species dwelling in the same room from an artificially infected chimpanzee (index case).

Presumably, variola virus could also been transmitted to monkeys from infected humans. Gispen (1949) and Bras (1952) observed such case when orangutans kept in the local zoo got infected during smallpox outbreaks in Jakarta. Several reports coming down to the end of 19th and beginning of 20th centuries describe outbreaks of smallpox-like diseases among wild monkeys, as consolidated by Arita and Henderson (1968); we dwell on this issue in Section 4.4.

White mice. Several authors demonstrated that adult white mice were insusceptible to variola virus independently of the infection route (Nelson, 1939; Andres K.H. *et al.,* 1958; Gutman, 1957; Brown *et al.,* 1960; Marennikova, 1961). Experiments with younger mice, especially, suckling mice, displayed another pattern. For example, VARV intraperitoneal inoculation to suckling mice induced the disease with lethal outcome (Mayr & Herrlich, 1960). Using this model, the authors succeeded in performing serial passages of the virus. Similar results were obtained by Brown *et al.* (1960), who used intracerebral infection. Marennikova and Kaptsova (1965) and Kaptsova (1967) studied the age effect in more detail commencing from the first day of life and inferred that mouse age was the major factor determining susceptibility of mice to VARV. According to their observations, intracerebral infection of mice younger than 14 days old caused the development of a lethal infection. In this process, the virus accumulates in various organs, viremia is observed, and encephalitis with foci of degenerating neurons develops. The infection of adult mice proceeds asymptomatically with an insignificant accumulation of VARV in the brain and development of inflammatory reaction without any signs of neuron degeneration. The animals appeared less sensitive to intranasal infection. In this case, only 1–5-day-old mice developed the disease with a lethal outcome. The disease showed up as viremia, accumulation of the virus in internal organs, and development of smallpox lesions on the skin of the majority of animals.

Murti and Shrivastav (1957b) reported development of focal or diffuse interstitial pneumonia with inclusions in alveolar cells upon intranasal infection with variola virus.

Studies performed at the laboratory of S.S. Marennikova demonstrated that irradiation (Co^{60}) could increase the sensitivity of adult white mice to variola virus. Upon irradiation, both the disease clinical course and virological characteristics appeared similar to those of 10–12-day-old mice, naturally susceptible to variola virus (Shafikova, 1970).

Note that some authors (Murti & Shrivastav, 1957b; Marennikova, 1961; Sarkar & Mitra, 1967; Shafikova, 1970; Shelukhina, 1980) reported different virulences of individual variola virus strains for mice (this issue will be considered in detail below).

White rats. According to Murti and Shrivastav (1957b), upon infection by scarification of the skin, the majority of strains caused no changes, whereas certain strains induced papules. Herrlich (1960) also observed the absence of reaction of rats to this infection route. However, infection with variola virus by scarification of the cornea induced keratitis with intracellular inclusions in corneal lesions. Upon intranasal infection of rats, vague symptoms of the disease became evident on days 2–3 (Marennikova, 1962), whereas Murti and Shrivastav observed specific interstitial pneumonia, as confirmed by the presence of cytoplasmic inclusions in lung tissue cells.

Hispid cotton rats (*Sigmodon hispidus*). Sensitivity of these animals was tested in our laboratory using intranasal and intravenous infections with the virus at high doses (6 and 7 log PFU/dose). In either case, neither any clinical manifestations of the disease were observed nor was the virus detectable in blood and internal organs of rats (isolation was attempted on day 4 after intranasal infection). However, both antihemagglutinin antibodies (AHA) and virus-neutralizing antibodies were found in rat blood serum starting from the second week (L.S. Shenkman, unpublished data; Shelukhina, 1980).

Syrian hamsters (*Mesocricetus auratus*). Similar to hispid cotton rats, these animals developed no clinically pronounced response to intracardial infection with a high dose of variola virus (6.3 log PFU/dose). Nevertheless, they displayed antihemagglutinin and virus-neutralizing antibodies; AHA disappeared after 5–7 weeks (Shelukhina, 1980).

Guinea pigs. Similar to adult white mice, these animals are insensitive to variola virus inoculated by various routes (Andres K.H. *et al.,* 1958). However, when the virus was introduced to the anterior eye chamber, the animals responded by keratitis, although less pronounced compared to the disease caused by the same dose of vaccinia virus (Marennikova, 1961).

Rabbits. The majority of researchers who studied susceptibility of rabbits to cutaneous infection with VARV failed to detect any response connectable with the specific effect of this virus (Nelson, 1939; Darrasse *et al.,* 1958; Marennikova, 1961; etc.). Marennikova (1961, 1962) studied other infection routes (intravenous, intraperitoneal, intracerebral, intratesticular, intradermal, and into the anterior eye chamber) using seven variola virus strains. Only intradermal infection with high doses of the virus induced formation of a small infiltration, which then quickly resolved. Other authors (Gutman, 1957; Bedson & Dumbell, 1964; Shelukhina, 1980; etc.) reported similar response to an intradermal inoculation. Inoculation of other rabbits with suspensions of these infiltrates gave negative results (Defries & McKinnon, 1928; Herrlich, 1960; Marennikova, 1962; Bedson & Dumbell, 1964). Aloof are the results obtained by two research teams (Gispen & Brand-Saathof, 1972; Dumbell & Kapsenberg, 1982): they reported hemorrhagic and even necrotic skin lesions upon intradermal infection with certain variola virus strains. As was mentioned, intratesticular infection fails to cause a pronounced response. However, Nelson discovered one strain among those tested that was capable of reinoculating multiply via the rabbit testis and accumulating in the tissue in question. Previously, this strain underwent 44 passages in chick embryos. However, this factor was hardly crucial, as another strain appeared unable to reinoculation after 183 passages (Nelson, 1939; 1943). Introduction of the virus into the anterior eye chamber is accompanied by development of keratitis (Marennikova, 1961). Infection with variola virus by scarification of the cornea (the method known as Paul's test, proposed for smallpox diagnostics) causes formation of white nodules at the scarification site with formation of intracellular inclusions (Guarnieri bodies).

Herrlich (1960) succeeded in performing four passages of VARV on the scarified rabbit cornea. According to Rao (1952), susceptibility of newborn rabbits to intradermal infection and scarification was as low as the susceptibility of adult animals.

Cocks and chickens. Subcutaneous infection of adult cocks with a suspension of variola virus induced formation of small nodules at the site of infection, whereas a general response was unobservable. The attempts to isolate the virus from blood (days 1 to 9) and feces were unsuccessful (McCarthy & Downie, 1948). The response of cocks to VARV intramuscular inoculation was also absent as well as the response of 5-day-old chickens to intracerebral infection (Marennikova, 1961).

Other animals. Since 19th century, one of the widespread hypotheses concerning the origin of vaccinia virus stated a possible transformation of variola virus into the virus in question. However, improvement of virological research techniques made it more and more apparent that the reported cases when passaging variola virus in various animals, including large animals, resulted in its evolution into vaccinia virus were only the consequence of contamination of the passaged material with vaccinia virus. Herrlich *et al.*

Table 4-3. Susceptibility of laboratory animals to variola alastrim virus*

Animal species	Infection route	Reaction to inoculation
Rabbits	Intravenous	None
	Intracerebral	None
	Intraperitoneal	None
	By skin scarification	None
	Intradermal	Development of a small infiltrate
	Intratesticular	None
	Into anterior eye chamber	Mild keratitis
White mice	Intravenous	None
	Intraperitoneal	None
	Intracerebral	Death of a part of young infected animals
Guinea pigs	Into anterior eye chamber	Mild keratitis

*According to data of Marennikova *et al.*, 1965.

provided convincing evidence for this inference. They performed the corresponding experiments in a specialized facility that excluded any possibility of contamination with vaccinia virus and demonstrated that adult goats, sheep, and pigs were resistant to variola virus. Only in the case of piglets, two passages appeared possible. Only once in the numerous experiments with calves and cows, development of a solitary pock at the site of scarification was observed (Herrlich, 1960; Herrlich *et al.*, 1963). These studies closed the door on a more than one-century-old dispute about the so-called transformation of variola virus into vaccinia virus[1]. Only one addition here is the work of Baxby *et al.* (1975), where the authors report susceptibility of camels to variola virus. Similarly to the animal species listed above, camels appeared resistant to this pathogen.

To conclude the Section on susceptibility of various animal species to variola major virus, find the consolidated data on susceptibility of laboratory animals to variola alastrim virus (Table 4.3).

4.3.2   Behavior in Chick Embryos

Use of developing chick embryos as a biological system for cultivation of viruses (Goodpastur *et al.*, 1935) allowed for a series of studies investigating variola virus in the system in question (Lazarus *et al.*, 1937; Buddingh, 1938; etc.). Consequently, it was found that upon inoculation on CAMs, variola virus induced formation of characteristic small white dome-shaped distinct lesions—pocks (Figure 4.4). Earlier works described pock size as equaling pinprick or pinhead (Buddingh, 1938; Irons *et al.*, 1941). Later, the pock size was determined more precisely by numerous direct measurements. For example, Marennikova and Shafikova (1969) studied this characteristic in 14 variola virus strains (infection of 12-day chick embryos; incubation for 72 h at 35°C) and found that the majority of strains formed pocks with an average diameter of 0.47–0.52 mm. Two strains induced larger pocks

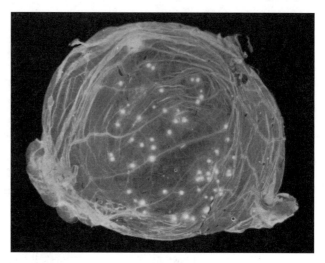

Figure 4-4. The lesions (pocks) on chick embryo CAMs induced by variola virus.

(average diameter, 0.56–0.61 mm) and four, smaller (average diameter, 0.32–0.43 mm). Note nonetheless that the size of pocks varied in both directions within the same strains. As a rule, the lesions appear already 24 h after infection and are localized mainly to the inoculation site. By 48–72 h, the pocks increase in number and size and spread over the CAM surface. The lesions cover mainly CAM ectoderm, where proliferative changes take place; mesoderm and endoderm are involved to a lesser degree and display an inflammatory response (Downie & Dumbell, 1947; Herrlich, 1960).

The time course of virus accumulation depends on various factors: incubation temperature, infective dose, and the age of chick embryos. Comparing various incubation temperatures (35, 37, and 39°C), Hahon *et al.* (1958) found that the temperature of 35°C was optimal for virus accumulation. A lower temperature (33°C) slowed the virus accumulation (Gutman, 1957), which reached its maximum only by 96–120 h of incubation. When the temperature was higher than the optimal, virus accumulation was slower, virtually ceasing at 39°C (Hahon *et al.,* 1958). Later, Downie and Dumbell (1961) demonstrated that 38.5°C was the ceiling temperature for variola virus to induce lesions on CAMs; 37.5°C, for variola alastrim virus. These are the temperatures when pock formation is dramatically inhibited (83–95% for variola virus and 62–80%, for alastrim). The size and density of pocks decreased too (Marennikova & Shafikova, 1970). Hahon *et al.* (1958) also reported the dependence of virus accumulation on the infective dose.

When variola virus is introduced into CAM, the resulting infection develops according to a generalized pattern and is not limited to the membrane. It spreads to the embryo, resulting in virus accumulation in all internal organs. In particular, Hahon *et al.* (1958) discovered an active accumulation of the virus in the embryo's body. Helbert (1957) reported not only a quick and intensive reproduction of variola virus in the chick embryo liver, but also pronounced distinctions between individual strains with reference to this feature. These distinctions were most pronounced in the case of variola major and alastrim viruses. However, the latter virus accumulated in a significantly lower amount, allowing for using this trait for differential diagnostics (Helbert, 1957). A practical value of this test was confirmed in several subsequent studies (Andres K.H. *et al.,* 1958; Marennikova & Shafikova, 1969; Shelukhina *et al.,* 1979a; etc.). Other manifestations of the infection in embryos include pocks on the skin, diaphragm, heart, and in spleen as well as hemorrhagic lesions in the myocardium and other organs (Marennikova & Shafikova, 1969). Earlier, Gutman (1957) paid attention to the presence of pustule-type lesions on the embryo's body and hemorrhages in the regions of head and thorax. The majority of researchers who studied the behavior of variola virus in chick embryos inoculated into CAM noticed that unlike vaccinia virus (see Chapter 3), the former virus was not fatal to the embryos (Buddingh, 1938; Irons *et al.,* 1941; Nelson, 1943; Downie & Dumbell, 1947; Marennikova, 1962; etc.). However, some authors reported death of a part or even all the infected chick embryos. Undoubtedly, this discrepancy to a certain degree may stem from the experimental conditions and criteria used for evaluation of the results (age of embryos, virus dose, incubation time post infection, etc.). However, later studies demonstrated that the pathogenicity of variola virus for chick embryos depended to a considerable degree on specific features of individual strains (Murti & Shrivastav, 1957a; Sarkar & Mitra, 1967; Marennikova & Shafikova, 1969; Shelukhina *et al.,* 1979a). This issue is considered in detail below when describing VARV intraspecies variation (see Section 4.3.5).

The infection proceeds less intensively upon inoculation into the allantoic cavity of chick embryos at a dose of 3 log PFU/ml: the virus accumulates at a considerably lower titers and the embryos remain alive even when infected with highly pathogenic strains. The increase in the dose to 6 log PFU/ml results in development of lesions in embryo's internal organs; however, fail to kill the embryo (Marennikova & Shafikova, 1969). When using this infection route, the highest virus titer appears in the allantoic fluid; somewhat lower, in CAM and yolk membrane.

When embryo is infected in the amnion, the maximal virus amount is detected in the embryo's body; lower, in CAM.

Inoculation into the yolk sac results in maximal accumulation of variola virus in the yolk membrane (Hahon *et al.,* 1958). Intravenous infection of chick embryos induced the pattern similar to that caused by CAM infection.

The effect of dose on the lethality rate in this process was apparent (Marennikova & Shafikova, 1969).

4.3.3   Cultivation in Cell Cultures

Boue and Baltazard (1956) were the first to cultivate VARV in tissue cultures; they demonstrated that the virus was capable of inducing a cytopathic effect (CPE) in the explants of human embryo skin and kidney, newborn rabbit kidney and heart, and adult human kidney. Further studies demonstrated that both primary and continuous cell cultures of various origins were sensitive to variola virus. Only in nine years after the research of Boue and Baltazard over 40 cultures sensitive to variola virus were found, including in addition to human cells, cells of various organs and tissues of the animal species both susceptible and insusceptible to VARV. According

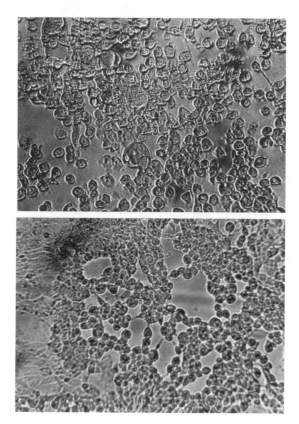

Figure 4-5. CPE caused by variola virus in Vero cell culture on day 5 post infection (magnification, 9 × 12.5).

to the opinion of the majority of researchers studying this problem, primary cultures of human and primate cells displayed the highest sensitivity to variola virus. Marennikova and Stepanova (1958), Solov'ev and Mastyukova (1958) and Vieuchange *et al.* (1958) investigated the behavior of variola virus in monolayer cell cultures. It was demonstrated that the time course and pattern of the cytopathic effects caused by VARV depended to a certain degree on the infective dose and type of culture (see Figures 8.3 and 4.5). However, specific of this virus compared to vaccinia virus are the rate of development and mainly focal pattern (at medium and low doses). The foci develop in a yet unchanged cell monolayer and initially are represented by separate round cells with distinct outlines. Later, the adjacent cells fuse to form large multinuclear polykaryocytes (Figure 4.6), and a sort of proliferative or hyperplastic effect, which is actually an aggregation of the cells in the foci (Kitamura, 1968), becomes evident. With time, the size and

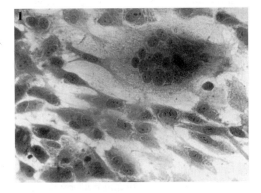

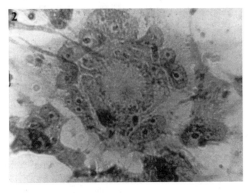

Figure 4-6. Giant cells (polykaryocytes) in human embryo fibroblast culture infected with variola virus (1) 46 h post infection (magnification, 7 × 20) and (2) 96 h post infection (× 300).

number of foci increase, gradually covering the entire cell layer. The development of CPE (from the first foci to total degeneration of the monolayer) caused by variola major and alastrim viruses require twofold and more time compared with vaccinia virus (Al'shtein *et al.,* 1961; Marennikova, 1962; Gurvich, 1964; etc.).

Specific features of the cell culture influence to a certain degree the pattern of developing changes. For example, cell fusion is less pronounced in chick embryo fibroblasts; consequently, the polykaryocyte counts are lower. The proliferative component is predominant in the foci, and polykaryocytes are also rare in continuous cell cultures (Gurvich, 1964). Study of Yumasheva (1968), who observed stimulation of mitotic activity in highly sensitive cultures during the first hours after infection, suggests that this phase is prolonged in less sensitive continuous cell lines, resulting in an apparent proliferative effect.

The CPE pattern remains stable while passaging the virus in cell cultures. The only distinction observed is a shorter period to CPE development; in addition, it is better observable (Marennikova, 1962; Gurvich, 1964).

According to the data obtained by Gurvich, VARV is detectable in the cell fraction 8 h after infection of sensitive cells at a dose of 3 log $TCID_{50}$. The virus titer increases noticeably between 24 and 48 h, reaching the maximum during next two days. In liquid fraction of the culture, small amounts of VARV appear 24 h post infection followed by a slow increase and remaining approximately by 2 log PFU/ml lower than the titer of intracellular virus. Yields of the strains adapted to cell culture are slightly higher.

The CPE pattern of variola alastrim virus is similar to that of variola major virus and also depends on a particular culture; however, accumulation of the former virus is slower (Figure 4.7). The phenomenon of hemadsorption is observed in cell cultures infected with both variola major and alastrim viruses (see also Figure 8.5).

Appearance of cytoplasmic inclusion bodies, known as Guarnieri bodies or type B inclusions, is the earliest display of variola virus infection in cultures. These inclusion bodies play a key role in replication of the virus and due to that are metaphorically called "the virus factories" (Cairns, 1960). Their present name is virosomes. These bodies are detectable in individual cells 3–4 h post infection, and their number increases progressively: 3% cells contain inclusions B after 10 h; 30%, after 24 h; and 60–70%, after 48–72 h of incubation (these data were obtained infecting human embryo fibroblast culture at a dose of 3 log $TCID_{50}$). One cell usually contains one to five type B inclusion bodies; however, some cells may contain a considerable number of virosomes. The inclusions may vary in shape and size (Figure 4.8). A specific feature of the pattern of cell cultures infected with variola virus is the presence of type B cytoplasmic inclusions on the background of relative

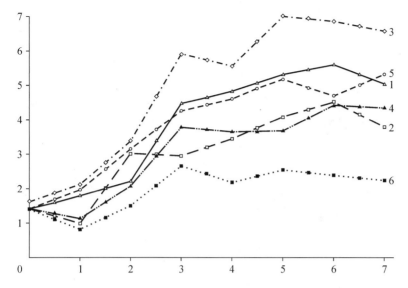

Figure 4-7. The time courses of variola alastrim virus accumulation in various cell cultures (Marennikova *et al.*, 1965): (1) primary culture of human embryo fibroblasts; (2) HeLa cell culture; (3) primary culture of monkey kidney; (4) Hep-2 cell culture; (5) A-1 cell culture; and (6) primary culture of chick embryo fibroblasts (the ordinate, log TCID50/0.1 ml; the abscissa, incubation time post infection, days).

integrity of individual cells and the entire monolayer. Note that the inclusion bodies are detectable in cells even when subthreshold virus doses, incapable of inducing CPE, are used (Marennikova, 1962; Marennikova *et al.*, 1963).

Immunofluorescence technique localizes distinctly the smallpox antigen in cells. Individual cells displaying fluorescent band around the nucleus are detectable 5–6 h post infection (Kirillova *et al.*, 1961; Gurvich, 1964). Later, the fluorescence spreads as "tongues" covering large areas of the cytoplasm either retaining the perinuclear localization or occupying the entire cytoplasm. Brightly fluorescent polykaryocytes supplement the picture (Figure 4.9).

Kirillova *et al.* (1961) discovered fluorescence, indicating the presence of the specific antigen, in nucleoli at later stages of infection. In this relation, of interest are the data of Yumasheva (1968), who observed regular changes in nucleoli in the cells infected with variola virus.

Similar to variola major virus, variola alastrim virus forms small plaques of hyperplastic type in various cell cultures. However, the data related to formation of plaques by these viruses in the culture of chick embryo fibroblasts are ambiguous. Mika and Pirsch (1960) as well as Gendon and Chernos (1964) failed to obtain plaques in this culture. On the other hand, several papers (Solov'ev & Bektemirov, 1962; Gurvich & Marennikova,

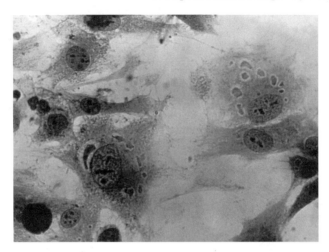

Figure 4-8. Type B cytoplasmic inclusion bodies (Guarnieri bodies) in
the human embryo fibroblast culture 48 h post infection with variola
virus (magnification, 7 × 40).

1964; Mal'tseva, 1965) reported plaque formation when infecting chick
fibroblasts with both variola major or alastrim viruses. The plaques formed
by these viruses appeared considerably later compared with the plaques
induced by vaccinia and cowpox viruses and were smaller (Table 4.4).

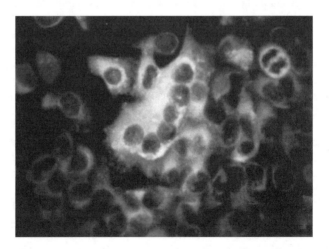

Figure 4-9. A fluorescent polykaryocyte, treated with FITC conjugate
of anti-vaccinia antibodies, in the PEK cell culture 24 h post infection
with variola virus.

Table 4-4. Plaque formation in chick embryo fibroblast culture*

Virus	Time of emergence	Size, mm
Variola major virus	96–120 h	1.0
Variola alastrim virus	96–120 h	1.0
Vaccinia virus	48 h	2.96–3.95
Cowpox virus	48 h	3.16–3.29

*According to the data of Mal'tseva (1965). The results were recorded after 120-h incubation at 34°C under agar overlay.

Similar to the behavior in chick embryos, incubation temperature influences the VARV behavior in cell culture. In particular, the temperature affects the rates of CPE development, virus accumulation, and the number of plaques formed. For example, Solov'ev and Mastyukova (1958) demonstrated that the virus accumulated more intensively at 34.5°C than at 37°C. It was shown that the ceiling temperature for CPE development and plaque formation depended on the type of cell culture and the degree of virus adaptation to the culture. In the case of chick fibroblast culture, the ceiling temperature amounts to 38°C for variola major virus and 37°C for variola alastrim virus. Under the same experimental conditions, the ceiling temperatures in human embryo fibroblast and monkey kidney cell cultures were higher. The ceiling temperature of the VARV strains that underwent several passages in human embryo cells amounted to 40°C (Gurvich & Marennikova, 1964). This observation allowed Dumbell and Huq (1975) to develop a specialized test for characterizing variola virus strains according to the hemadsorption intensity in human embryo fibroblast culture incubated at 40°C.

4.3.4   Stability of Variola Virus

This question attracted the attention first due to its practical value, as the cases of human infection without a direct contact with smallpox patients, i.e., via contaminated things or objects, were reported. Certain interest to this issue remains now, when smallpox as a disease is eradicated. In particular, some researchers consider that the virus may remain in tissues of smallpox victims buried in the permafrost region or in tubes left or lost in laboratory refrigerators.

Research into VARV preservation in various materials from human cases demonstrated its high stability to environmental factors. Presumably, this is the result of a sort of drying of the virus in the protective medium .

The data obtained by various researchers on the period of variola virus preservation rather disagree. For example, the work of Wolff and Croon (1968), which attracted a great attention, reported survival of variola virus in scabs kept for 13 years at a room temperature. However, other papers reported smaller survival terms. According to Downie and Dumbell (1947b), the maximal period when variola virus was detectable in scabs kept at 18–

20°C was 13.5 months. Akatova-Shelukhina (1962) obtained similar results. In addition, she found that storage of scabs at a higher temperature (34°C) accelerated inactivation of the virus to 3–5 months. Note also the data of McCallum and McDonald (1957), who demonstrated that the survival period of variola virus kept at 30°C in cotton fibers depended on the moisture content and amounted to 70–185 and 60–90 days at humidities of 58 and 73–84%, respectively.

Several authors reported a long preservation of viable variola virus at low temperatures (Hahon, 1959; Herrlich, 1960; Marennikova, 1962; etc.). Perhaps, recent results of Belanov *et al.* (1997), who examined the scabs stored at –20°C for 26 years, are most impressive. The viable virus was isolated from all the studied seven scab samples of different human cases. The average amount of infective virus was 5–10 PFU/mg. Calculations of the authors allowed them to infer that the virus kept under these conditions might survive 100 and more years and, therefore, there is a real chance to discover a viable virus in corpses of the smallpox casualties buried in the permafrost.

Experimental studies of variola virus (Akatova, 1958; Akatova-Shelukhina, 1962) demonstrated that the virus in CAM suspension stands a 2-h heating at 50°C without a pronounced drop in its infectivity. However, further increase in temperature appeared critical: the virus was inactivated after 10 min at 60°C and 5 min at 70°C. Boiling destroys the virus virtually immediately. Drying of the virus-containing cultures (frozen cultures in vacuum) increases the stability of virus to high temperatures. The data obtained by Hahon (1959) confirmed these results.

Comparative studies demonstrated that the thermal stability of variola major virus (using single heating at 50, 60, and 70°C) is higher than the stabilities of variola alastrim and vaccinia viruses. The inactivation rate (decrease in infectivity, log $TCID_{50}$/min) of variola major virus was two–fourfold slower compared to both viruses. Comparison of the sensitivities of these three viruses to UV irradiation and disinfectants (phenol or chloramine) demonstrated the same pattern. As for pH, the optimal range appeared close to neutral, whereas deviations to both acid and alkaline regions in certain ways decreased the VARV infectivity. The virus is completely inactivated during 1 h at pH 3.0 (Akatova-Shelukhina, 1962; Marennikova *et al.*, 1965).

4.3.5   Intraspecies Variation

Characteristic of variola virus is a high genetic stability and low rate of changes. As reported by many authors, VARV serial passages on chick embryo CAMs failed to change its properties (Lazarus *et al.*, 1937; Nelson, 1943; Downie & Dumbell, 1947; etc.). Some authors noted an increase in variola virus pathogenicity for chick embryos as well as certain elevation in its infectivity and hemagglutination activity (North *et al.,* 1944; Marennikova, 1962; Herrlich *et al.,* 1963; Shafikova, 1970; etc.).

Tsuchiya and Tagaya (1972) produced a peculiar variant of variola virus by cloning in cell culture the strain that underwent 35 passages in chick embryo CAMs. One of the clones obtained (G) transformed the cell monolayer (monkey kidney cell culture; dose, 1 PFU/cell) into a one giant multinucleated structure. Other cultures displayed a less pronounced effect; although the monolayer nonetheless split up into giant separate cells containing at least 100–200 nuclei. However, a number of cultures failed to display this effect. The clone in question, which lost its hemagglutination activity, retained the properties in question when passaging in both CAMs and cell culture. Nelson (1939) discovered that variola virus after 64 and more passages in chick embryos acquired the ability to induce pneumonia foci in the lungs of adult white mice.

The properties of variola virus did not change considerably during passages in various cell and tissue cultures, except for quicker CPE onset (Baltazard *et al.,* 1958; Higashi & Ichimiya, 1959).

An attempt of Shafikova (1970) to change antigenic properties of variola virus by passaging in chick embryos in the presence of specific antibodies was unsuccessful: 50 passages failed to cause any changes in the VARV properties, except for a certain increase in the pathogenicity for chick embryos. The properties of variola virus strains studied also remained unchanged upon 40 intracerebral passages in partially immune white mice irradiated with Co^{60} (Shafikova, 1970).

In certain cases, researchers succeeded in increasing the ceiling temperature of lesion development during adaptation of variola virus. For example, Gurvich and Marennikova (1964) found that the strains adapted to human embryo cells were capable of inducing lesions in the culture in question at a temperature of 40°C. Dumbell *et al.* (1967) described a strain that commenced causing lesions at 39°C after passaging at an elevated temperature (the initial ceiling temperature was 38°C). Certain data suggest that the observed insignificant changes in the virus properties that arose during adaptation are not profound and stable. For example, observations of Marennikova (1961) on restoration of the initial VARV pathogenicity for chick embryos, which increased upon passaging on CAMs only after a single passage in allantoic cavity, confirm this inference. Similarly, Dumbell and Huq (1986) reported that the strains that acquired the ability to induce hemagglutination and hemadsorption already after the second passage in Hep-2 cell culture lost these properties when returned to chick embryo CAMs.

When discussing the variation of variola virus, we cannot but return to the issue of the possibility of VARV transformation into VACV during passages in various animals. This question was a matter of heated debates and disputes over several dozens of years. Impossibility of such transformation was finally proved in 1963 thanks to irreproachable studies by Herrlich *et al.* (1963).

The data described above, which illustrate the stability of variola virus during cultivation under various conditions, in no way mean the absence of intraspecies variation of this virus, especially taking into account the occurrence of clinical variants of smallpox—*variola major, variola minor,* and *variola alastrim* (see also Section 4.4).

Dinger (1956) and Helbert (1957), who paid attention to differences in time points when the viruses disappeared in the survived chick embryos, values of LD_{50}, and intensities of virus accumulation in the embryo's liver, were the first to demonstrate the distinctions between the agents of these smallpox clinical variants. Then, different sensitivities of variola major and variola alastrim viruses to increased temperatures during incubation in chick embryos were demonstrated and the ceiling temperature of lesion development—37.5°C for variola alastrim virus and 38.5°C for variola major virus—were determined (Nizamuddin & Dumbell, 1961). Later, Dumbell and Huq (1975, 1986) studied isolates of *variola minor* cases and discovered certain distinctions between the biological properties of this pathogen isolated on the African continent and in South America. Profile of *Hin*dIII digests of variola alastrim virus DNA also distinguished this pathogen from the other two (Fenner *et al.,* 1988), allowing the authors to propose the terminological demarcation, i.e., name the viruses isolated in South America as variola alastrim virus and the African viruses, variola minor virus.

Bedson *et al.* (1963) described a group of strains isolated in Tanganyika (now Tanzania) from human smallpox cases with a milder course. Compared to the classical strains, these strains differed in a higher sensitivity to increased temperatures and lesser pathogenicity for chick embryos; however, they were not identical with the alastrim virus strains studied earlier.

Dumbell and Huq (1975) also confirmed interstrain differences of variola virus using a large set of strains and an additional marker (hemadsorption at 40°C).

Different pathogenicities for suckling mice and chick embryos were also found for strains isolated in other regions of the world (Sarkar & Mitra, 1967; Shafikova & Marennikova, 1970; etc.). Sarkar and Mitra believed that pathogenicity of strains was the main reason underlying severity of the disease course. However, it appeared that, on the one hand, such highly pathogenic strains were isolated from varioloid cases and, on the other, several strains isolated from hemorrhagic and confluent smallpox cases displayed low pathogenicities in the laboratory tests (see 4.1.1).

Shelukhina *et al.* performed a wide comparative study of VARV strains isolated during 1969–1975 from smallpox cases with various degrees of severity using nine markers, including Helbert's test, pathogenicities for chick embryos and 10-day-old mice infected intracerebrally, differences in infectivity titers of the strains when chick embryos were incubated in 35 and 38.5°C (Shelukhina *et al.,* 1979a; Shelukhina, 1980). Overall, 60 strains were examined; of them, 20 strains came from Asia, 27 from Africa, 6 from countries of the Middle East, and 5 from South America. The results demonstrated variation of VARV properties, such as pathogenicities to chick embryos and white mice, accumulation in chick embryo's liver, and thermal sensitivity. The variation ranges were rather wide. For example, the pathogenicity for chick embryos changed from 1.2 to 4.5 log LD_{50}; accumulation in chick embryo's liver, from 4.0 to 8.8 log PFU/g; neuropathogenicity for white mice, 2.8 to 6.3 log LD_{50}; and thermosensitivity, from the sensitivity tenfold lower than that of the reference strain Harvey to a complete incapability of replication at 38.3°C.

Typical of the strains isolated in Asia were high pathogenicities and high ceiling temperatures of pock formation on CAMs. The subgroup of African strains appeared pronouncedly nonuniform; however, the strains with lower pathogenicity for laboratory animals and lower ceiling temperatures of pock formation on CAMs compared to Asian strains were predominant on this continent. Among the studied African strains, the strain Ethiopia-4 occupies a special place. It was weakly pathogenic for chick embryos (3.8 log LD_{50}), accumulated at a lower amounts in chick embryo's liver (4.0 log PFU/g), and failed to induce pocks at 38.3°C. These properties allowed the strain in question to be classified as variola minor virus.

The strains from Middle East region displayed a low pathogenicity for chick embryos accompanied by the thermostability virtually equal to that of the reference strain Harvey. In the ability to accumulate in chick embryo's liver, they occupied an intermediate position between Asian and African strains. One among the strains of this subgroup, Kuw-5-67 appeared antigenically distinct (Marennikova *et al.,* 1976a). However, it was similar to Middle Eastern and Asian strains in the rest properties analyzed.

Of the five Brazilian strains, four were typical variola alastrim viruses, as demonstrated by their low pathogenicity for chick embryos, lower indexes of Helbert's test, and most important, inability to replicate at 38.3°C. Only one of the five Brazilian strains displayed the last ability.

Table 4.5 lists the characteristics of the strains with outermost and intermediate pathogenicities and the abilities to replicate on CAMs at elevated temperatures.

Consolidated materials characterizing all the strains studied with reference to the markers revealing most pronounced distinctions are shown in Figure 4.10.

Comparison of the determined characteristics of strains with the geographical regions where particular strains were circulating and the

Table 4-5. Intraspecies variation of variola virus by the examples of individual strains*

Markers Strains	LD$_{50}$ for chick embryos (log @in PFU/0.1 ml)	Accumulation in chick embryo's liver (log, in PFU/g)	Thermal sensitivity**	LD$_{50}$ for white mice***
India-385	1.5	7.3	1.0	3.5
Botswana-6	4.6	7.4	0.6	5.8
Congo-9	3.3	4.8	1.1	5.2
Ethiopia-4	3.8	4.0	Fails to reproduce at 38.3°C	5.1

*According to Shelukhina (1980).
**Thermal sensitivity was determined as a ratio of the difference between logarithms of infectivity titers at 35 and 38.3°C to that of the reference strain Harvey.
***Mice were infected intracerebrally at an age of 10 days.

smallpox mortality rates in the regions in question revealed certain correlation. For example, a high (15–44%) human mortality rate in Asian countries corresponded to a higher frequency of strains displaying increased pathogenicity for chick embryos and white mice and temperature tolerance. This ratio—prevalence of certain strains in particular geographic areas—yet should not be considered as an absolute rule, as strains with various characteristics were isolated both in Asia and Africa. Computer analysis of the data obtained (Shelukhina, 1980) detected correlations between each of the three laboratory markers (lethality for chick embryos, accumulation of the virus in embryo's liver, and thermal sensitivity), on the one hand, and between all the three markers together and the smallpox mortality rate in the countries where particular strain circulated, on the other. However, the degree of this correlation was determined as moderate. In the same work, by the example of two isolated smallpox outbreaks, the author failed to confirm the correlation between properties of the strain and the severity of disease as well as the time of its isolation during the disease course (at the beginning and at the convalescent stage).

Overall, the results on intraspecies variation of variola virus obtained by Shelukhina *et al.* (1979a) were confirmed in later studies, where the strains were estimated with another set of markers (Dumbell & Huq, 1986).

Intraspecies variation of variola virus may be evaluated by restriction assay of the viral DNA. However, this analysis is rather complex due to a large size of VARV DNA (197 kbp). Hydrolysis of VARV DNA with the restriction endonucleases recognizing 6-bp sequences is relatively uninformative (Esposito & Knight, 1985). The hydrolysis with restriction enzymes recognizing 4-bp sites produces a tremendous set of fragments preventing informative analysis.

The approach realized by researchers from CDC (Atlanta, GA, USA) and SRC VB Vector (Koltsovo, Novosibirsk oblast, Russia) is more progressive. Oligonucleotide primers allowing for extended PCR (ePCR) with VARV

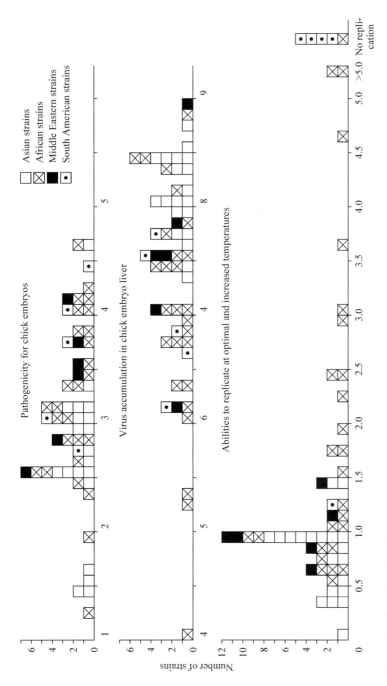

Figure 4-10. Properties of variola virus strains and geographic areas of their circulation (according to Shelukhina *et al.*, 1979a). Each marker is expressed in the same units as in Table 4.5.

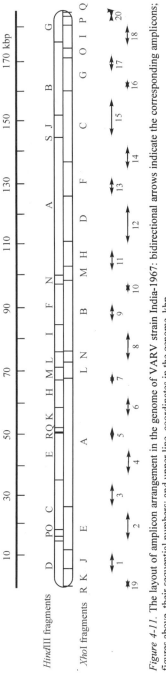

Figure 4-11. The layout of amplicon arrangement in the genome of VARV strain India-1967: bidirectional arrows indicate the corresponding amplicons; figures above, their sequential numbers; and upper line, coordinates in the genome, kbp.

DNA as a template were calculated basing on the determined genomic sequences of VARV strains India-1967, Bangladesh-1975, and Garcia-1966 (see Section 4.5). This ePCR provides amplification of 20 overlapping amplicons covering the entire genome of variola virus (Figure 4.11). These amplicons with lengths of 2.5 to 16 kbp were produced by ePCR using DNA preparations of 64 VARV strains. Each of the 20 amplicons was digested with restriction endonucleases recognizing 4-bp sites (*Hpa*II or *Bst*FNI). Overall, these enzymes have over 300 sites in the complete VARV genome. Hydrolysis products of each amplicon were separated by gradient PAGE (4–20%) concurrently with DNA length standards. Gel images were computer-processed using BioNumerics software. A phylogenetic tree was constructed by neighbor joining method using the totality of obtained ePCR–RFLP (restriction fragments length polymorphism) assay data for *Hpa*II and *Bst*FNI hydrolysates (Figure 4.12).

All the variola virus strains studied fall into three major groups. The group comprising the variola minor alastrim strains and the Western African strains isolated in Benin (V68-59) and Sierra Leone V68-258) displayed the maximal distinctions. These data suggest the inference that Western African strains of variola minor and the strains of variola alastrim belong to a separate VARV subtype and that the alastrim strains evolved from the Western African VARV strains. This complies with the historical facts that the slaves to America were predominantly

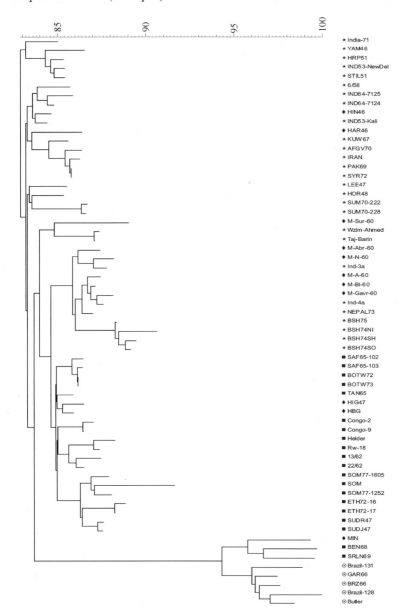

Figure 4-12. The dendrogram constructed by neighbor joining method basing on the results of combined assay of *Hpa*II and *Bst*FNI hydrolysates of all the 20 amplicons of each VARV strain: asterisks indicate Asian VARV strains; rectangles, African; circles, VARV alastrim strains; and diamonds, VARV strains imported from other countries; the top scale shows similarity values for the dendrogram (%).

brought from the western coast of Africa. It is known that the direct massive transfer of slaves to America commenced in the middle of the 16th century. Presumably, smallpox entered the American continent during that particular period. Basing on this assumption, we investigated the evolution of VARV strains using BioNumerics software package. This allowed us to estimate the time of separation of the variola alastrim virus subtype (Western African variola minor virus) as 1100–1300 years ago.

The rest African strains form a separate branch on the phylogenetic tree in question with the subgroups clustering according to their geographic origin: the strains isolated in Central Africa (Congo), in South Africa (Republic of South Africa, Botswana, Tanzania, and Rwanda), and in the Horn of Africa (Somalia, Ethiopia, and Sudan). The strains HIG47 and HBG, isolated during imported smallpox outbreaks in Europe, cluster with the African strains (Figure 4.10).

When studying 35 Asian strains, 5 clusters in the phylogenetic tree were found. Note that the strains HIN46 and HAR46, isolated during imported smallpox outbreaks in the UK (1946), display a high homology to Asian strains.

The VARV strains isolated from human cases during the smallpox outbreak in Moscow (1960), imported from India, also cluster with Asian strains. This small outbreak was initiated by one infected person and, thanks to intensive antiepidemic activities, was eradicated at the third generation of human-to-human transmission chain (Golubchikova & Samvelova, 1961). Noteworthy that all the six studied VARV strains isolated from different human cases display differing restriction patterns in the ePCR–RFLP assay performed. Presumably, the discovered differences of the genomes of Moscow VARV strains may result from heterogeneity of the initial virus material, which manifested itself in the first generation of the disease. If this is the reason, this suggests existence of a set of VARV evolutionary variants in infectious discharges of one smallpox case.

The strains from China, Korea, and Sumatra form a separate cluster. The rest clusters comprise the strains with an Indo–Pakistani origin. Japanese, Middle Eastern, and Bangladesh strains display a high degree of homology within their subgroups.

Thus, the data of ePCR–RFLP assay suggest a trend of clustering according to the geographical origins of VARV strains. Moreover, the analysis performed allowed us to discover close relations between Western African VARV strains and the alastrim strains.

4.4   Ecology of Variola Virus

As is known, one of the main prerequisites for the program of global smallpox eradication was based on the concept that human was the only carrier of variola virus and the reservoir of this pathogen was absent in wild nature. This was suggested by a tremendous epidemiological experience accumulated while studying and analyzing smallpox outbreaks. The fact that no smallpox case was recorded since the moment when the natural transmission of this infection was interrupted (1977) is a convincing confirmation of this opinion. Nevertheless, both the World Health Organization and individual research teams considered it necessary during the entire eradication program to obtain additional confirmations of these facts by examining wild-living animals, mainly monkeys. As a result of these studies, several isolates similar in their properties to variola virus were obtained from monkeys and rodents and named "variola-like" or "white wild" (Marennikova *et al.*, 1971a, c; 1972b; 1976c). The term "white wild" viruses was introduced by the WHO Bureau of Smallpox Eradication. It was meant here that these viruses differed from variola virus, were isolated in wild nature, and, unlike variola virus, formed white pocks on chick embryo CAMs. Two such viruses were isolated in the Netherlands from kidney tissue culture derived from outwardly healthy *M. cynomolgus* monkeys and were first classified by the authors (similar to other strains they isolated that time) as monkeypox virus (Gispen & Kapsenberg, 1966). However, it appeared later that the isolates in question differed essentially from the rest group of strains and were similar to variola virus in their biotype. Another two strains—Chimp-9 and MK-7—were isolated at the WHO Collaborating Center in Moscow from wild-living monkeys, chimpanzees (*Pan troglodytes*) and redtail monkeys (*Cercopithecus ascanius*), trapped in Zaire. Two additional isolates were recovered at the same Center from wild rodents also trapped in Zaire—red-legged sun squirrels (*Heliosciurus rufobrachium*) and multimmamate rat (*Mastomys natalensis*).

Naturally, these findings caused serious anxiety and concern as well as stimulated a series of control studies. One of the studies analyzed the circumstances of isolation of the variola-like viruses from monkey kidney culture in the Netherlands (Dumbell & Kapsenberg, 1982). The authors inferred that the medium while it was changed was accidentally contaminated with a variola virus strain when isolating the white wild strains in question. They made this conclusion basing on the identity of DNA restriction profiles of variola-like virus isolates and an Indian isolate of variola virus, with which the authors was working that time. Occurrence of other variola-like viruses was explained in a similar way (Esposito *et al.*, 1985; Fenner *et al.*, 1989).

4.5   Genetic Organization of Variola Virus

The program of global smallpox eradication was officially completed in 1980. However, the WHO Committee decided in 1986 that this program should be finalized with destroying of the collections of variola virus strains, which are confined only to the two WHO Collaborating Center on Smallpox and Related Infections—in the USA (CDC, Atlanta, GA) and Russia (at the Institute for Viral Preparations until 1994; later transferred to the State Research Center of Virology and Biotechnology Vector, Koltsovo, Novosibirsk oblast, Russia). To preserve the information about this unique virus, WHO considered it necessary to sequence its genome before destroying. Consequently, the international project under the aegis of WHO was performed by the team of Russian and US researchers since 1991.

4.5.1   Organization of Variola Major Virus Genome

At the first stage of research under the international project in question and the *National Program on Conservation of Genetic Material of the Russian Collection of Variola Virus Strains,* approved in Russia in 1990, the team from SRC VB Vector and Institute for Viral Preparations should obtain and characterize libraries of DNA fragments of a variola virus strain and sequence the genomic fragments of the virus.

A strain of the highly virulent variola major virus, isolated from pustular fluid samples by N. Mal'tseva on day 11 of the disease of a 50-year-old man from a village in the Maharashtra state (India) in August 1967, was selected as the object for this study. The fatality rate of the smallpox epidemic in India in 1967 amounted to 31% (Fenner *et al.,* 1988). The virus was isolated in 1967 at the WHO Regional Center on Smallpox with the Institute for Viral Preparations (under the guidance of S.S. Marennikova) by titration on developing chick embryo CAMs. Previously to DNA isolation, the strain, named India-1967, underwent five passages on chick embryo CAMs. The culture of variola virus strain India-1967 (VARV-IND) was produced on CAMs; virions were purified by centrifugation in a stepwise sucrose gradient; and DNA preparation was isolated by phenol deproteination according to the conventional technique (Nakano *et al.,* 1982).

The fragments obtained by hydrolysis of VARV-IND DNA with restriction endonucleases *Hind*III or *Xho*I were cloned within the vector plasmids pUC18 and pUBS19. Large *Hind*III fragments of VAR-IND genome were cloned in the cosmid cosH (Shchelkunov *et al.,* 1992a).

The complete genomic sequence of VARV-IND except for terminal regions with a length of about 600 bp each), cleaved by *Xho*I restriction endonuclease (Figure 4.13), amounted to 185,578 bp (Shchelkunov *et al.,* 1993f). The G + C composition of the sequenced DNA amounted to 32.7%.

The experiments performed allowed the researchers to construct and characterize the required collections of hybrid DNA molecules (Shchelkunov *et al.,* 1991; 1992a), so that the DNA fragments cloned covered the complete VARV-IND genome (Figure 4.13). These collections

HindIII

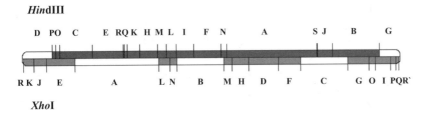

*Figure 4-13. Hind*III and *Xho*I restriction enzyme maps of VARV-IND DNA. The fragments cloned in plasmids and cosmids are crosshatched.

formed the background for a successful sequencing of the complete variola virus genome. Sequencing was performed manually according to Maxam and Gilbert (1980) with certain modifications. By mid-1992, sequencing of VARV-IND genome was completed as well as the first stage of analysis of the data obtained. The sequence data were annotated and submitted to GenBank. Oral presentation of the results of this analysis (Shchelkunov *et al.,* 1992b) opened the IX International Conference on Poxviruses and Iridoviruses (Les Diablerets, Switzerland, September 1992) and the data than were reported at the International Symposium *100 Years of Virology* (St. Petersburg, Russia, September 1992).

In parallel, a research team from CDC and the Institute for Genomic Research (Gaithersburg, MD, USA) performed similar experiments. As the object for sequencing, they chose a variola virus strain isolated by I. Arita from skin lesions of a 3-year-old girl during the last natural *variola major* outbreak in Kuralia, South Dingaldi, Bhola Island, Bangladesh in October 1975 (Fenner *et al.,* 1988). The mortality rate of this smallpox outbreak in Bangladesh was 18.5%. The virus was obtained by titration on chick embryo CAMs to isolate an individual pock from the membrane and infect chick embryo fibroblast (CEF) culture with the resulting virus. This virus isolated was named Bangladesh-1975 (VARV-BSH). The viral DNA was isolated from CEF culture at the second passage of VARV-BSH. The fragments of viral DNA generated with *Hind*III, *Sac*I, and *Bst*EII restriction endonucleases were cloned within plasmids. Sequencing was performed according to Sanger *et al.* (1977). By mid-1993, the complete genome of this strain (186,103 bp) was sequenced (Massung *et al.,* 1994). The G + C composition of this DNA amounted to 33.7%.

Thus, the genomic sequences of two Asian variola major virus strains became available for comparative analysis by 1994. Note that Asian VARV isolates are the most virulent among all the strains studied in the 20th century (Dumbell & Huq, 1975; Fenner *et al.,* 1989).

At the beginning of computer analysis of the sequenced VARV-IND and VARV-BSH DNAs, potential open reading frames (ORFs) encoding polypeptides with a minimal length of 60 amino acid residues (aa) were calculated. The viral ORFs were designated according to the rules accepted that time (Rosel *et al.,* 1986)—the capital letter was assigned to each ORF according to the letter assigned to the *Hin*dIII fragment of the genome (Figure 4.13) where the ORF either began or was located. The ORFs in each fragment were numbered from left to right. The letters R (right) or L (left) designated the ORF orientation—from left to right or *vice versa,* respectively.

Overall, 200 potential ORFs were identified in each VAR-IND and VAR-BSH genomes (Shchelkunov *et al.,* 1995); of them, 108 are leftward-oriented and 92, rightward-oriented (see Appendix 1).

Unlike the other orthopoxviruses pathogenic for humans, the inverted terminal repeats in the genome of variola viruses are very small, amounting to 518–1051 bp (725 bp for VARV-BSH) and do not code any proteins (Massung *et al.,* 1995).

The lengths of coding sequences of VARV-IND and VARV-BSH amount to 185,463 and 184,661 bp, respectively. The genome of VARV-BSH contains two relatively long deletions compared with VARV-IND. One deletion with a length of 210 bp is localized to the noncoding region adjacent to the right end of the genome; the second deletion is 569 bp long and is located in *Hin*dIII-J genomic fragment. Comparison of VARV-IND and VARV-BSH DNA sequences revealed their extremely high identity (99.3%). If the regions of 210- and 569-bp deletions are excluded from the calculations, the identity of the sequences compared increases to 99.7%. The rest distinctions between VARV-IND and VARV-BSH sequences are represented by deletions or insertions not exceeding 50 bp in length and point substitutions in the genome of one virus relative to the other.

Of the 200 proteins encoded by potential VARV-IND and VARV-BSH ORFs, 122 are identical; 42 carry single amino acid substitutions; and 11, double. A large number of amino acid differences were detected in the rest proteins; however, they frequently consist in substitution of amino acids with other isofunctional residues (for example, Leu→Ile). In addition, some ORFs are truncated in the genome of one virus relative to the other.

Distinctions between ORF organization in VARV-IND and VARV-BSH genomes concentrate mainly in the right and left terminal regions, whereas the central region, spanning 148 ORFs (from VARV-IND C8L to B7R), is highly conservative (with some exceptions—VARV-IND E3L, I5R, A27L, A42R, and J6R/J7R; Appendix 2). The conservative region is 135 kb long. Assuming that the short ORFs distinguishing these strains (A42R, J6R, J7R, etc.) are nonfunctional (fragments of genes), the genetic maps of VARV-IND and VARV-BSH are virtually coinciding.

For a more illustrative detection of conservative and variable regions of the viral genome, all the VARV-IND and VARV-BSH ORFs were divided into 10 groups from left to right, each group consisting of 20 potential ORFs, to determine in each group the percent of proteins with the degree of identity exceeding 99% (Shchelkunov *et al.*, 1995). The results obtained (Figure 4.14) also demonstrate that the terminal genomic regions of the strains compared are more variable than the central region. Of interest is the VARV region A33L–A49R (Figure 4.14), displaying a very high conservatism. Presumably, this region is important for manifestation of VARV species-specific properties. This hypothesis is confirmed, in particular, by the virtual identity of the region A30L–J9R of variola major virus strain Harvey (VARV-HAR), sequenced earlier (Aguado *et al.*, 1992), and the region in question (28 point substitutions over 21.8 kbp, which corresponds to 99.8% identity). Interestingly, VARV-HAR was isolated in 1944 from a soldier who returned to England with a convoy from Gibraltar.

Thus, comparison of the DNA sequences of VARV-IND, VARV-BSH, and VARV-HAR allowed for detection of an unexpectedly high

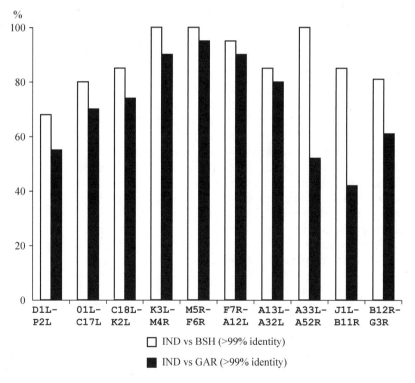

Figure 4-14. Percent of the VARV-IND open reading frames (ORFs) encoding the proteins that display the amino acid sequence identity with the corresponding VARV-BSH or VARV-GAR ORFs exceeding 99%.

conservation of variola major virus at the level of both DNA nucleotide sequence and amino acid sequence of the proteins encoded. Note here that VARV-IND, VARV-BSH, and VARV-HAR were isolated at different time and by different scientists and were studied independently by different research teams. These results suggest that variola major virus evolved relatively slowly. Possibly, this virus evolutionarily had so precisely adapted to the human organism that any significant changes in its genomic organization would only lead to a decrease in the viability and/or level of virulence.

4.5.2   Organization of Variola Minor Alastrim Virus Genome

As was noted in Section 4.1, VARV is conventionally divided into two subspecies—variola major virus, causing the disease with a fatality rate of 5 to 40%, and variola minor virus, with a fatality rate of less than 1% (Fenner *et al.*, 1989). At least two subtypes of variola minor virus exist. One subtype was named *alastrim* (Ribas, 1910), and the outbreak caused by this virus was first recorded in 1896 in Florida, USA. Later, this disease quickly spread over the American continent and was recorded in Europe, and Australia (Chapin, 1913; Chapin & Smith, 1932). Smallpox outbreaks with a low mortality rate were recorded approximately at the same time in the south of Africa (De Korte, 1904). Laboratory studies performed later (Dumbell & Huq, 1986) demonstrated that the variola alastrim virus differs distinctly not only from variola major, but also from the studied African isolates of variola minor viruses according to several laboratory tests (see Section 4.2). Analysis of the nucleotide sequences of extended terminal genomic regions (30 kbp each) of various VARV isolates (Massung *et al.*, 1996) demonstrated that the African variants of variola major (Congo-1970) and variola minor viruses (Somalia-1977) displayed no essential distinctions from Asian variola major virus strains. However, it appeared possible to discover principle differences in the DNA structure of these genomic regions between variola alastrim virus and other VARV subtypes (Figure 4.15). Therefore, it was important to sequence the entire genome of this VARV variant and analyze in detail its structure–function organization.

The variola alastrim virus strain Garcia-1966 (VARV-GAR) was chosen as an object for the study performed in collaboration by researchers from SRC VB Vector and CDC. This virus was isolated by J. Noble in Brazil during the *variola alastrim* outbreak in 1966 from skin lesions of a 14-year-old girl with pronounced eruptions on day 8 after the disease was diagnosed. The virus was titrated on chick embryo CAM. After the second passage on CAM, the virus was isolated from an individual pock to infect FL cell culture. Viral DNA was isolated by cytoplasmic extraction (Esposito *et al.*, 1985) in 1985 upon the third passage of the virus in FL cells. In 1993–1995, the fragments of the viral DNA were cloned within plasmids and the complete VARV-GAR genome was sequenced.

The determined complete VARV-GAR DNA sequence confined between the outermost *Xho*I sites, clipping the terminal hairpins (of approximately 530 bp each; Massung *et al.,* 1995), amounted to 186,986 bp (Shchelkunov *et al.,* 2000). The G + C content of this sequence was 32.7%; identities with VARV-IND and VARV-BSH DNA sequences, 98.24 and 98.02%, respectively. The performed analysis identified 206 potential ORFs with the minimal length of 60 aa (Appendix 2). Taking into account that VARV-IND and VARV-BSH display an extremely high identity in both nucleotide and amino acid sequences, only VARV-GAR and VARV-IND genomes were aligned graphically (Appendix 1). The VARV-GAR and VARV-IND translational maps completely coincided in the region bounded by the VARV-GAR ORFs E2L and A25R. This central region of VARV-GAR genome is approximately 104 kbp long. In the long terminal genomic regions, the segments displaying virtually complete coincidence of VARV-GAR and VARV-IND translational maps alternate with the segments showing pronounced distinctions (Appendix 1). Let us consider in more detail the basic differences between VARV-GAR and VARV-IND in organization of individual ORFs.

The VARV-IND ORF D6L in the VARV-GAR genome is divided into the two shorter ORFs B7L and B8L. This region corresponds to the gene *CHO hr* of cowpox virus (CPXV), determining the host range. The gene in question is deleted in the genome of VACV-COP and fragmented into four potential short ORFs in the case of VACV-WR (Appendix 1). Interestingly, this protein contains the so-called ankyrin repeats, present in several orthopoxvirus proteins (see Section 7.4.1) and presumably playing an important role in regulation of intricate functions of viruses, such as, determination of sensitive host range and tissue tropism (Shchelkunov *et al.,* 1993b; Safronov *et al.,* 1996).

The four VARV-GAR potential ORFs adjacent to the region described above—B9L, B10L, B11R, and B12L—were most likely formed due to mutational changes in the structure of a single gene of the ancestor virus. This hypothesis results from the observation that a full-sized ORF is located in this region in the genomes of VACV-COP, CPXV-GRI, and CPXV-BRT (Appendix 1). However, the VACV and CPXV nucleotide sequences in this region differ considerably (79% homology). Moreover, the identity of DNA nucleotide sequence of VARV-GAR in the region in question with that of CPXV-GRI amounts to 89%; with VACV-COP and VACV-WR, only to 73%. Note that the rest regions of VARV-GAR, except for the insert (a 627-bp segment in the right genomic region; Figure 4.15) display at least a 90% identity with VACV-COP DNA. Thus, the region B9L–B12L of VARV-GAR genome presumably originates from the precursor common with CPXV, whereas this region of VACV genome was possibly recombinationally substituted with a genomic fragment of another poxvirus. Relative to VARV-GAR, variola major viruses have the deletion with a length of 898 bp in the region in question (Figure 4.15). Note that similar to the host range gene of CPXV, VACV-COP C9L also contains ankyrin repeats. However, the functions of C9L gene yet require further investigation.

Orthopoxvirus proteins of the kelch family (see Section 7.4.2) are presumably capable of binding to the cytoskeleton and play an important role in the virus–host interactions (Senkevich *et al.*, 1993; Shchelkunov *et al.*, 2002a; Kolosova *et al.*, 2003; de Miranda *et al.*, 2003). Note that CPXV genome contains six full-sized genes of this family and VACV-COP, three, whereas the genes in question are mutationally fragmented in all the VARV strains studied (Shchelkunov, 1995; Shchelkunov *et al.*, 2000, 2002a). However, note that in all the cases (except for VARV-GAR E3L), VARV-GAR has different patterns of fragments in the impaired genes compared with VARV-IND and VARV-BSH (Appendix 1). This may suggest that the variola major and variola minor alastrim viruses evolved independently from an ancestor virus.

It is known that 3β-hydroxysteroid dehydrogenase (3β-HSD) is involved in many stages of steroid biosynthesis and the steroid hormones have multiple physiological activities in the body. VACV codes for enzymatically active 3β-HSD (Moore & Smith, 1992). This ORF is mutationally fragmented in the genomes of variola viruses: into two potential ORFs in VARV-IND and VAR-BSH; into three, in VARV-GAR (A52L–A54L).

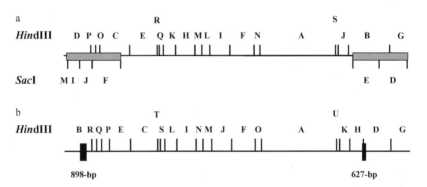

*Figure 4-15. Hin*dIII maps of variola virus genomic DNA: (a) Asian and African variola major and African variola minor viruses have the same *Hin*dIII maps compared with (b) variola alastrim virus, which displays two additional *Hin*dIII sites. The shaded *Sac*I and *Bst*EII fragments on the map in (a) are the genomic regions sequenced and compared for the left ends of strains GAR, SOM, and CNG and right end GAR and SOM. Black bars on the map in (b) represent alastrim-specific DNA segments (Massung *et al.*, 1996).

The VARV-GAR ORF E1L has an unusual structure. The coding part of its DNA sequence contains 31 tandem repeats of the nonamer ATCTATATC, determining the repeat Asp–Ile–Asp/DID (Figure 4.16). The genes VACV-COP F1L and VACV-WR 040 contain only one copy of this nonamer; in the gene in question of VARV-BSH and VARV-IND, the nonamer is repeated 7 and 12 times, respectively. It is yet vague how these repeats influence the properties of the corresponding proteins and what their function is. However, it is clear that this protein has a very high negative charge and the corresponding ORFs of all the variola viruses studied carry these repeats, although their number vary from strain to strain (Massung *et al.*, 1996).

```
GAR-E1L   MYNSMLPMFMCNNIVDYIDDIDDIDDIDDIDDIDDIDDIDDIDDIDDIDDID     52
IND-C5L   ................D.....-------------------------------     22
BSH-C5L   ................D.....-------------------------------     22
COP-F1L   ..S..........V....----------------------------------     18
WR-F1L    ..S..........V....----------------------------------     18

GAR-E1L   DIDDIDDIDDIDDIDDIDDIDDIDDIDDIDDIDDIDDIDDIDDIDDIDDIDDIDD    104
IND-C5L   --------------------------------.....................     44
BSH-C5L   ---------------------------------------------.........     29
COP-F1L   ----------------------------------------------------     18
WR-F1L    ----------------------------------------------------     18

GAR-E1L   IDDIDDIDDKASNNDDHNYVYPLPENMVYRFNKSTNILDYLSTERDHVMMAV    156
IND-C5L   .......E.......-...................................     95
BSH-C5L   ..................................................     81
COP-F1L   NGIVQ..E.E....V..D...........D.....................     70
WR-F1L    NGIVQ..E.E....V..D...........D.....................     70

GAR-E1L   QYYMSKQRLDDLYRQLPTKTRSYIDIINMYCDKVNNDYNRDMNIMYDMASTK    208
IND-C5L   ...........................T.....................E    147
BSH-C5L   ................................................E    133
COP-F1L   R.......................I.....S..................    122
WR-F1L    R.......................I.....S..................    122

GAR-E1L   SFTVYDINNEVNTILMDNKGLGVRLATISFITELGKRCMNPVETIKMFTLLS    260
IND-C5L   ..................................................    199
BSH-C5L   ..................................................    185
COP-F1L   ...............................R......K.........    174
WR-F1L    ...............................R................    174

GAR-E1L   HTICDDCFIDYITDISPPDNTIPNISTREYLKLIGITAIMFATYKTLKYMIG    312
IND-C5L   ..................................................    251
BSH-C5L   ..................................................    237
COP-F1L   ........V.............T.....................    226
WR-F1L    ......Y.V.............T.....................    226
```

Figure 4-16. Alignment of the amino acid sequences of variola virus strains VARV-GAR, VARV-IND, and VARV-BSH and vaccinia virus strains VACV-COP and VACV-WR containing DID repeats. Dots and dashes indicate the coinciding and deleted amino acids, respectively, relative to the VARV-GAR ORF E1L.

Unlike the other variola virus strains studied (Massung *et al.,* 1996), computer analysis allowed for detection of a potential ORF, G4R, in the right terminal region of VARV-GAR genome (Appendix 1). As the main part of its coding sequence is located in the region of terminal short tandem repeats with a length of 69 bp (see Figure 6.15), the putative protein G4R contains seven repeats with a length of 23 amino acid residues. The function of this protein is unknown, and search of the known amino acid sequences for its homologues failed to give any positive result. Additional studies are necessary to clarify whether G4R protein is synthesized by VARV-GAR or not.

The alastrim-specific DNA segment with a length of 627 bp was discovered in the area of H11R and H12R ORFs of VARV-GAR (Appendix 1). This 627-bp segment is absent in the genomes of Asian and African variola major isolates and the studied African variola minor isolates, whereas its presence in DNAs of other variola alastrim strains was confirmed (Massung *et al.,* 1996). This segment was undetectable in various MPXV, CPXV, CMLV, and ECTV strains. Recently, for several isolates of cowpox virus, PCR assay detected the insert of a variola-specific DNA sequence (Meyer *et al.,* 2002a). It was demonstrated that these CPXV strains carry an extended insertion with a length of 6 kbp in this genomic region compared with the other orthopoxviruses studied. We discovered that this insert contains a continuous ORF coding for a protein with a length of 1891 aa (for details, see Section 6.5, Figure 6.20). The ORFs B10R and B11R of VARV-IND and H10R, H11R, and H12R of VARV-GAR are "flinders" of the discovered extended ORF, which most likely originated from the genome of the VARV ancestor. The differences displayed by VARV-GAR and VARV-IND in this region suggest an independent evolution of these viruses.

Thus, the research performed allowed a number of distinctions between variola minor alastrim viruses, on the one hand, and the studied African variola major and variola minor viruses, on the other, to be discovered. Recently performed ePCR–RFLP assay of a large set of VARV strains detected a close relation between the strains of variola alastrim and Western African strains of variola minor (see Section 4.3.5). The most striking difference of VARV-GAR from VARV-IND and VARV-BSH consists in that several genes that are "flinders" of the ancestor virus genes are impaired mutationally in variola alastrim and variola major virus genomes in different ways. These examples also suggest independent evolutionary routes of variola major and variola alastrim from the same ancestor virus. According to our estimations, the evolutionary divergence of these VARV subspecies could occur 1100–1300 years ago. As for the known immunomodulatory factors of orthopoxviruses (see Section 7.5), the computer analysis found no considerable differences between VARV-GAR and VARV-IND/VARV-BSH in their structure. The minor differences detected were, as a rule, point substitutions of some amino acids. The effects of such substitutions on the properties of particular proteins can be clarified only in experiments on expression of individual viral genes in a genetically engineered system. Once individual proteins of various virus strains are obtained, it is possible to study and compare their properties in detail.

4.5.3   Comparison of VARV and VACV Genomes

VARV is the pathogen causing one of the most hazardous human diseases, smallpox, while VACV, belonging to the same genus *Orthopoxvirus* of the family Poxviridae, is an effective live vaccine against this disease (see Chapter 3). Consequently, it is of the utmost importance to find out what are the distinctions in genetic organizations of these viruses and in what they are similar. Such comparison open up the possibilities to detect molecular virulence factors and understand the mechanisms allowing orthopoxviruses to breach numerous protective barriers of the host organism (Shchelkunov, 1995; 2003; Shchelkunov *et al.*, 1992b; 1993c–f; 1994a, b; 1996; Blinov *et al.*, 1995).

Comparison of the translational (genetic) maps of VARV-IND, VARV-GAR, VACV-COP, and VACV-WR genomes (Appendices 1 and 2) demonstrates that the central conservative genomic regions, displaying a high homology between DNA sequences as well as coincidence of VARV and VACV genomic maps, is bounded by the ORFs C12L and A24R of VARV-IND (F8L and A24R for VACV-COP) and has a length of about 100 kbp. This is the genomic region that harbors the majority of the genes determining genus-specific properties of orthopoxviruses. Presumably, the species-specific properties of these viruses are determined mainly by the genes localized to the terminal DNA sequences flanking the central genomic region.

Note the "patchiness" of the VARV genetic map in the left and right terminal regions compared to that of VACV (Appendix 1; Figure 7.1). Relatively short segments virtually coinciding in these maps alternate with regions carrying mutationally changed, deleted, or inserted ORFs. Interestingly, comparison of various strains of within an orthopoxvirus species detected a considerably larger central conservative region. For example, its length is 163 kbp for the pair VACV-COP/VACV-WR and even longer for VARV-IND and VARV-BSH.

The most pronounced distinctions between variola and vaccinia viruses are found in the region of terminal inverted repeats (TIRs), spanning 12 kbp in VACV-COP and 10 kbp in VACV-WR genomes. These viruses contain different sets of genes within TIRs (Appendix 1). Other orthopoxviruses also carry extended TIRs (Esposito & Knight, 1985). The exception is variola virus, whose TIRs have a length of 518 to 1051 bp depending on the strain (Massung *et al.*, 1995). Such short repeats contain no genes and, possibly, represent a minimal telomeric element necessary for replication of the poxvirus DNA. Presumably, such a small TIR size limits essentially the recombinational variation of variola virus compared with the other orthopoxvirus species, which display a relatively high level of genomic rearrangements.

The central conservative region common for VARV and VACV contains all the presently known orthopoxvirus genes controlling biosyntheses of viral DNA and RNA molecules (see Section 3.4) as well as the vast majority of genes encoding virion structural proteins. As a rule, the ORFs localized to these genomic regions of variola and vaccinia viruses display the homology exceeding 95% (Appendix 2). However, the degree of conservation of the central genomic regions of the viruses compared is not absolute. For example, the ORFs C18L of VARV-IND and F14L of VACV-COP, E7R of VARV-IND and VACV-COP, I5R of VARV-IND and H5R of VACV-COP, as well as A13L of VARV-IND and VACV-COP displayed species-specific distinctions.

Of the genes listed, the function of only two last genes is known. VACV-COP gene H5R encodes the transcription factor VLTF-4 (Kovacs & Moss, 1996). Interestingly, only this late transcription factor is essentially different in VARV and VACV. The rest transcription factors VLTF-1, -2, and -3 (genes H8R, A1L, and A2L of VARV-IND) are highly conservative. Earlier, we discovered (Shchelkunov *et al.*, 1992a) that the VARV-IND ORF I5R (VACV-COP H5R isolog) contains three putative Ca^{2+}-binding domains (Figure 4.17). Deletions in VACV-COP DNA relative to VARV-IND destroy two of the three Ca^{2+}-binding domains in the analogous vaccinia virus protein (Figure 4.17). Calcium cations play an important role in regulation of various processes in mammalian organisms. Consequently, we can suppose that I5R protein, encoded by the gene from the conservative genomic region, is able to influence manifestation of variola virus pathogenicity properties (Shchelkunov *et al.*, 1993e).

The VARV-IND and VACV-COP protein A13L is a component of the IMV surface membrane (Takahashi *et al.*, 1994) and, possibly, influences the efficiency of IMV formation, which is important for spreading of the infection in the organism.

The rest principal distinctions between VARV and VACV genetic organizations are clustered in the left and right terminal variable regions of their genomes. Comparison of the genes encoding immunomodulatory proteins of orthopoxviruses is described in Section 7.3.

Inhibitors of apoptosis of infected cells. One of the first nonspecific defense lines of the organism against infectious agents and, likely, the most ancient, is apoptosis or programmed cell death (Cohen *et al.*, 1992; Williams & Smith, 1993). Once a cell is infected with a virus, the function of apoptosis lies in a suicide of this cell, preventing the virus replication in it and, thereby, rescuing the neighboring cells from infection with the virus offspring. Recently, it is discovered that apoptosis is a very widespread biological phenomenon characteristic of multicellular organisms; however, the mechanisms of apoptosis induction and regulation are yet rather vague (Vaux *et al.*, 1994; Hinshaw *et al.*, 1994; Richter *et al.*, 1996; Vito *et al.*, 1996).

The cellular protease cleaving specifically inactive prointerleukin-1β and thereby transforming it into the mature interleukin-1β (IL-1β) plays an important role in induction of the programmed cell death (Gagliardini *et al.*, 1994). This protease got the name ICE (interleukin-1β converting enzyme). However, IL-1β is unrelated to apoptosis induction (Vaux *et al.*, 1994), i.e., another protein (proteins) is the target of ICE in triggering the cell programmed death. It appeared that the gene of SPI-2 protein, discovered initially in cowpox virus and frequently designated CrmA (Pickup *et al.*, 1986), determined ICE protease inhibitor (Ray *et al.*, 1992) and arrested apoptosis (Gagliardini *et al.*, 1994). Comparison of amino acid sequences of SPI-2 proteins demonstrated that these sequences of VARV and VACV displayed

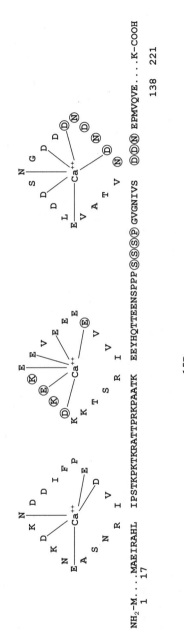

Figure 4-17. Potential Ca²⁺-binding domains of VARV-IND 15R protein. The amino acid residues deleted in the corresponding VACV-COP protein are encircled.

species-specific distinction, whereas the VACV-COP gene in question is damaged.

Another VACV-COP protein, C12L (VARV-IND B25R), assigned to the same family of serine protease inhibitors according to the data of computer analysis of amino acid sequences and designated SPI-1 (Kotwal & Moss, 1989; Smith G.L. *et al.,* 1989c), also inhibits apoptosis (Turner *et al.,* 1995). However, the mechanism underlying inhibition of apoptosis is unclear, as this protein does not inhibit ICE protease. SPI-1 and SPI-2 display 47% identity.

Presumably, double-stranded RNA molecules are also able to induce apoptosis, as was inferred by Kibler *et al.* (1996), who studied the vaccinia virus with mutant E3L gene. It appeared that inactivation of this gene not only increased the virus sensitivity to interferon, but also activated apoptosis of infected cells. Comparison of VARV and VACV demonstrated a relatively high conservation of this gene, suggesting that it is most likely responsible for genus-specific characteristics of orthopoxviruses.

Of interest is the VARV-IND gene D4R. The protein it encodes displays a high degree of homology to ectromelia virus protein p28 (Figure 4.18), which appeared inessential for growth of the virus in cell culture but played an important role in manifestation of ectromelia virus virulent properties in infected mice (Senkevich *et al.,* 1994). The gene of this protein is completely deleted in VACV-COP and truncated by 58 functional aa at the *C*-terminus in VACV-WR. However, vaccinia virus strain IHD-W codes for a full-fledged isolog of VARV-IND D4R. Similar protein NR1 of Shope fibroma virus (genus *Leporipoxvirus*) was discovered (Upton *et al.,* 1994; Figure 4.18). The VARV-IND D4R protein is identical to the analogous protein of VARV-BSH, suggesting that it is important for realizing specific features of this highly virulent virus. The protein analyzed was assigned to the rapidly increasing protein family with the so-called RING zinc finger motif. This large family comprises proteins involved in regulation of gene expression, DNA repair and recombination, and apoptosis inhibition (Freemont *et al.,* 1991; Lovering *et al.,* 1993; Crook *et al.,* 1993; Birnbaum *et al.,* 1994). The *C*-terminal domain, containing the C_3HC_4 motif, is important for specific properties of these proteins. It was recently demonstrated (Ruby *et al.,* 1996) that the ectromelia virus gene of p28 protein inhibited the apoptosis that was induced by either tumor necrosis factor or the ligand for cell surface antigen CD40 (CD40L). Presumably, ECTV p28 protein competes with the signal molecules that bind to cytoplasmic domains of CD40 and TNF receptor (TNFR2) upon their extracellular domains have interacted with the ligands (CD40L and TNF, respectively), thereby transducing the specific signal triggering apoptosis into the cell nucleus. This inference was based on the fact that these endogenous signal molecules for TNFR2 and CD40 are also members of the RING zinc finger family of proteins.

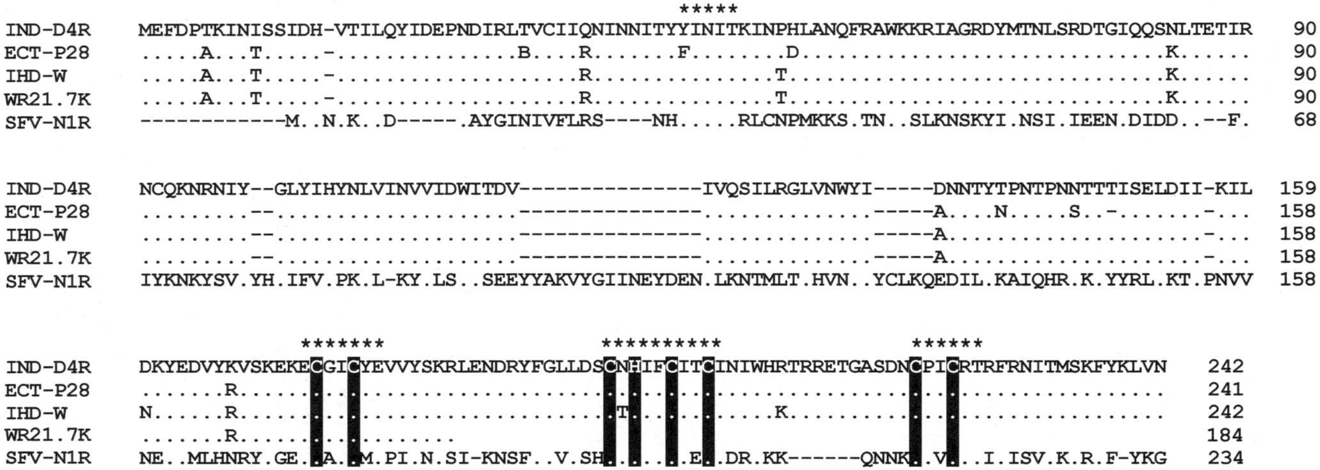

```
                                                     *****
IND-D4R   MEFDPTKINISSIDH-VTILQYIDEPNDIRLTVCIIQNINNITYYINITKINPHLANQFRAWKKRIAGRDYMTNLSRDTGIQQSNLTETIR    90
ECT-P28   ....A..T....-.....................B...R.....F........D..............................K.....     90
IHD-W     ....A..T....-...................R.....................T...........................K.....       90
WR21.7K   ....A..T..-.....................R.....................T...........................K.....         90
SFV-N1R   --------M.N.K..D-----.AYGINIVFLRS----NH.....RLCNPMKKS.TN..SLKNSKYI.NSI.IEEN.DIDD..--F.           68

IND-D4R   NCQKRNRIY--GLYIHYNLVINVVIDWITDV-------------IVQSILRGLVNWYI------DNNTYTPNTPNNTTISELDII-KIL       159
ECT-P28   .......|--|................-----------------------A....N.....S.-...-                            158
IHD-W     .......|--|................-----------------------A.                                           158
WR21.7K   .......|--|................-----------------------A.                                           158
SFV-N1R   IYKNKYSV.YH.IFV.PK.L-KY.LS..SEEYYAKVYGIINEYDEN.LKNTMLT.HVN..YCLKQEDIL.KAIQHR.K.YRRL.KT.PNVV     158

               *******                          **********                        ******
IND-D4R   DKYEDVYKVSKEKE.G.C.YEVVYSKRLENDRYFGLLDS.N.L.FC.TC.INIWHRTRRETGASDN.C.PIC.RTRFRNITMSKFYKLVN     242
ECT-P28   .....R.......................................................................                  241
IHD-W     N.....R..............................T..............K.........................                 242
WR21.7K   .....R...........................................                                             184
SFV-N1R   NE..MLHNRY.GE..A..M.PI.N.SI-KNSF..V.SH.....L.E..DR.KK-----QNNK..VL..I.ISV.K.R.F-YKG            234
```

Figure 4-18. Alignment of the amino acid sequences of the VARV-IND, ECTV, VACV-IND-W, VACV-WR, and Shope fibroma virus ORFs belonging to the protein family with RING zinc finger motif (indicated by vertical black blocks). Asterisks mark the segments displaying high mutual identity; dots, amino acid residues identical to those of VARV-IND; and dashes, deletions.

Thus, already four genes are identified in the orthopoxvirus genome whose protein products inhibit apoptosis via various mechanisms. Presumably, existence of multiple efficient apoptosis inhibitors contributes considerably to manifestation of pathogenic properties of highly virulent viruses, such as variola and ectromelia viruses.

Growth factors. The VACV-COP gene C11R encodes a protein assigned to the family of epidermal growth factor and displaying the highest degree of homology to epidermal growth factor (EGF) and transforming growth factor (TGF; Blomquist *et al.*, 1984; Brown J.P. *et al.*, 1985; Massague & Pandiella, 1993). It was demonstrated that this gene is the virulence factor of VACV (Buller *et al.*, 1988a, b). The protein expressed from C11R is designated as VGF. Initially, it is synthesized as a surface membrane-bound glycoprotein of the infected cell with a molecular weight of 25 kDa (the signal polypeptide is cleaved during secretion). Once the *C*-terminal hydrophobic region (Figure 4.19) is cleaved, VGF with a size of 22 kDa is released into the ambient medium (Chang *et al.*, 1988). It was demonstrated that VGF in certain characteristics was more similar to TGF than to EGF (Schultz *et al.*, 1987; Abdullah *et al.*, 1989). Other poxviruses, such as myxoma, Shope fibroma, and malignant rabbit fibroma viruses also encode an EGF-like protein, which is a virulence factor (Upton *et al.*, 1987, 1988; Opgenorth *et al.*, 1992). The precise function of VGF is thus far unclear (Eppstein *et al.*, 1985; Opgenorth *et al.*, 1993); however, of great interest are the data that substitution of myxoma virus growth factor gene with the gene of rat TGF, VGF, or Shope fibroma virus growth factor restores the pathogenic properties of myxoma virus (Opgenorth *et al.*, 1993). This suggests that the proteins in question, despite noticeable differences in their amino acid sequences, have similar biological effects on the target cells *in vivo*. It is assumed that VGF stimulates the growth and/or metabolic activity of uninfected cells, thereby providing spread of the infection in the body (Buller *et al.*, 1988a).

Comparative analysis of VGF amino acid sequences of two VACV and two VARV strains (Figure 4.19) revealed their high intraspecies similarity. Note that the VARV protein sequence bounded by the second and third conservative cysteine residues is virtually identical to the analogous region of TGF, whereas the VACV VGF in this region is more similar to EGF. Construction of producers of these viral proteins, their isolation, and comprehensive study might assist in clarifying the effects of these distinctions between VARV and VACV on their properties.

3β-Hydroxysteroid dehydrogenase. Computer analysis of the sequencing data (Goebel *et al.*, 1990) demonstrated that VACV-COP ORF A44L displayed 32% identity of amino acid sequence with human 3β-hydroxysteroid dehydrogenase/Δ4-Δ5 isomerase (3β-HSD). Moore and Smith (1992) confirmed experimentally that this viral gene encoded the

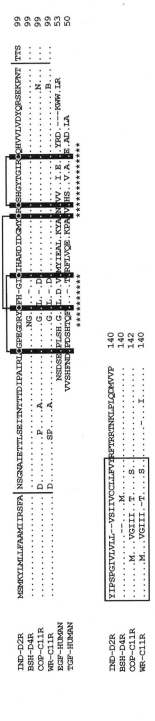

Figure 4-19. Alignment of the amino acid sequences of the VARV-IND, VARV-BSH, VACV-COP, and VACV-WR ORFs encoding proteins from the epidermal growth factor family and cell proteins EGF-HUMAN and TGF-HUMAN (Massague & Pandiella, 1993). Black vertical blocks indicate conservative cysteine residues; square brackets above the alignment, bonds between cysteines in protein molecules; asterisks, segments with the identity between all amino acid sequences exceeding 50%; hydrophobic transmembrane segments of the proteins are framed; mature forms of VACV proteins upon cleavage of N- and C-terminal domains are separated by vertical lines; and the rest designations are as in Figure 4.18.

protein with a putative enzymatic activity. 3β-HSD is a key enzyme in the biosynthesis of steroid hormones and acts at various stages of this pathway. In turn, steroid hormones influence numerous host functions, for example, determine immunosuppression and anti-inflammatory action (Rutherfurd *et al.*, 1991). Using the model of intranasally inoculated white mice, Moore and Smith (1992) demonstrated that a VACV deletional mutant with impaired 3β-HSD gene displayed a decreased virulence. However, the ORF encoding 3β-HSD is mutationally fragmented in VARV-IND, VARV-BSH, and VARV-GAR genomes. Damage of this gene in highly virulent VARV allows us to suggest that the gene in question lost its significance in the case of stringent anthroponosis as VARV due to differences in hormonal regulation in rodents and humans. The gene encoding 3β-HSD is present in not only VACV, but also in cowpox (Appendix 2) and fowlpox virus (genus *Avipoxvirus*; Skinner *et al.*, 1994). The available data suggest that VARV emerged evolutionary later compared with another natural orthopoxvirus, CPXV.

   Proteins of extracellular virion envelope. Extracellular enveloped virus (EEV) is involved in dissemination of daughter virions in the infected host body and cell cultures (Payne, 1980) and is, thus, important for orthopoxviruses to manifest their pathogenic properties.

It is known that the envelope of EEV contains 10 viral proteins absent in IMV (Payne, 1978; 1979); of them, only one protein is not glycosylated. Gene of this acetylated 37K protein (VARV-IND C17L and VACV-COP F13L) was mapped (Hirt *et al.*, 1986; Blasco & Moss, 1992). The 37K protein of VARV-IND displays a high degree of identity with the analogous VACV-COP protein. The rest protein components of EEV envelope are glycoproteins. Payne (1979) discovered that five of these proteins had molecular weights of 20 to 23 kDa; the rest four, 42, 86, 110, and 210 kDa, differing to a certain degree from other data reported (Weintraub & Dales, 1974; Duncan & Smith, 1992). Three ORFs (A34R, A56R, and B5R), encoding all the glycoproteins of EEV lipoprotein envelope except for 110K, were mapped in VACV genome (Smith, 1993).

   It was demonstrated (Roper *et al.*, 1996) that VACV-COP A33R encoded 20K–23K glycoproteins, which are associated with EEV envelope and necessary for efficient dissemination of the virus. Payne (1992) found that these proteins are formed from one precursor protein. They are sulfated, phosphorylated, and capable of forming 55-kDa aggregates via disulfide bonds. Equal sizes and a high degree of identity of VARV-IND ORF A37R and VACV-COP ORF A34R indicate a functional significance of the products of this gene for multiplication of orthopoxviruses.

   The gene B5R of VACV-COP (B7R of VARV-IND) encodes glycosylated and acetylated 42K protein of EEV envelope, which is class I membrane protein (Isaacs *et al.*, 1992b; Engelstad *et al.*, 1992). Presumably, this glycoprotein is important for virus dissemination. It displays a structural homology to members of the complement-binding protein family and thus, may be important for inhibition of complement activation (Engelstad *et al.,* 1992). Deletion of B5R gene results in virus attenuation (Engelstad & Smith, 1993).

The gene A56R of VACV-COP encodes another glycoprotein of EEV envelope, hemagglutinin (HA); its molecular weight estimates are 85 kDa (Shida, 1986) to 89 kDa (Smith, 1993). VACV HA is a multifunctional class I protein (Figure 4.20a; Jin *et al.,* 1989; Shida, 1986; Seki *et al.,* 1990; Oie *et al.,* 1990). This protein contains both O- and N-bound carbohydrate components. VACV HA is sulfated, phosphorylated, and capable of forming 220-kDa complexes due to disulfide bonds (Payne, 1992). HA is synthesized at the late stage of the virus infection, accumulates in the cell membrane, and blocks the virus-induced cell fusion. In addition, HA binds 37K protein on the EEV envelope. Biological activity of the virus-associated HA can modify the virus–cell interactions and influence the virus infectivity and dissemination (Oie *et al.,* 1990; Ortiz & Paez, 1994).

Unlike the envelope proteins described above, the VARV-IND HA (ORF J9R) displays noticeable distinctions from VACV-COP HA in both the primary and secondary structures (Figure 4.20). The proteins in question also differ considerably in their total charges. The isoelectric points of VARV-IND and VACV-COP hemagglutinins amount to 4.56 and 3.84, respectively.

Thus, only hemagglutinin of all the EEV envelope proteins displays evident distinctions in the orthopoxviruses compared. Presumably, of the extracellular virion envelope proteins, HA contributes maximally to the difference between VARV and VACV in their abilities to spread in the human body. However, this question requires further research.

Large transmembrane protein. The region of VARV-IND genome containing ORF B26R (Appendix 1) is very interesting. Organization of 5'- and 3'-terminal DNA regions, flanking ORF B26R, suggests that this is an early gene. Analysis demonstrated that this ORF (1896 aa) coded for a protein with pronounced properties of type III membrane-associated glycoprotein. A.M. Eroshkin discovered a local homology of *C*-terminal VARV-IND B26R region with proteins from the cadherin family (Marennikova & Shchelkunov, 1998), which are integral membrane glycoproteins responsible for Ca^{2+}-dependent cell adhesion (Ringwald *et al.,* 1987; Nagafuchi *et al.,* 1987; Shimoyama *et al.,* 1989; Reid & Hemperly, 1990). It is demonstrated experimentally that cadherin molecules are directly involved in cell–cell binding (Edelman *et al.,* 1987; Nagafuchi *et al.,* 1987). Moreover, they realize their function in cooperation with cytoskeleton components. Thus, the protein in question is also likely to interact with cytoskeleton components in addition to ankyrin-like (see Section 7.4.1) and

a

VAC-WR   MTRLPILLLILSLVIATPPFQ?--SKKIGDDATLSCNRNNTNDYVWSAWYKEFNSIILLAAKSDVLIFDNYTKOKISIDSPYDDLVTTITIKSLPARDAGTVCAFFMTSTNDTDKVDYEKISTELIVNTDSESTIDILLSGSTHSPETSSEKPEDIDNFNCS
VAC-COP   ...................S.Y..QI..K..P........................DY..S......
VAR-IND   ...................S...I..S...........................DY.N.......

VAC-WR   SVFEIATPEPITDNVEDHTDTVTTSDSINTVSASGCESTTDETPEPITDKEEDHTVIDIVSYTTVSTTSSGIVTTKSTTDDADLIYDTTNDNDTVPPTTVGGSTTSISNVIKTKDFVE | IFGITALILILSAVAIFCITYYI | INKRSRKYKTENKV
VAC-COP   ....S.............S.......S........KNI.K..GK.S...Y.K.V..A. | .... | C.............
VAR-IND   ........G......................I...T.R..VK.SG..N.-...........E.......AN.--HND.EPS..S. | .... | C.............

b

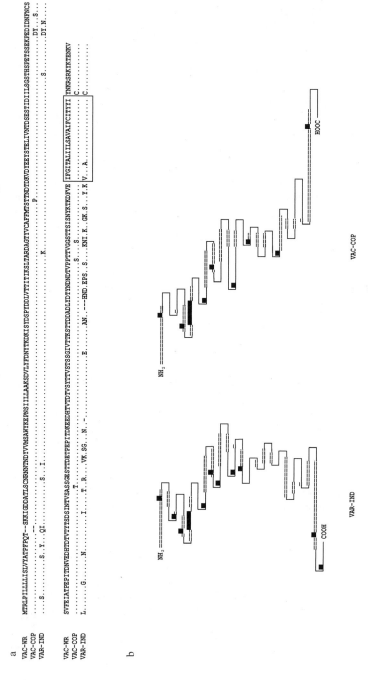

Figure 4-20. (a) Alignment of the amino acid sequences of the VACV-WR, VACV-COP, and VARV-IND ORFs encoding hemagglutinin and (b) comparison of their HA secondary structures: double lines, sheet; single line, helix; black block, turn; and vertical line, coil.

kelch-like proteins (see Section 7.4.2). This protein is highly identical in VARV strains; however, displays multiple differences from the corresponding proteins of other orthopoxviruses (Shchelkunov *et al.*, 1998, 2000, 2002b). The DNA sequence homologous to VARV-IND B26R gene is absent in the genomes of vaccinia and rabbitpox viruses. Neither function nor localization of this protein in the infected cell is known.

"Buffer" genes. The fact that a considerable number of VARV ORFs are smaller (damaged) with reference to the VACV-COP (for example, ORFs A38R–A50L and B2L–B12R of VARV-IND) due to mutational changes arrests the attention when comparing VARV and VACV genomes (Appendices 1 and 2). Presumably, at least some of these genes are "buffer" and their role consists in neutralization of negative effects that develop in the organism during viral infectious disease. A regulation of the virus–host interactions was experimentally confirmed for one of these genes. It was found that damage of the VACV-WR gene encoding IL-1β–binding protein (isolog of VACV-COP B16R) resulted in earlier display of the disease symptoms and death of experimental mice compared with the disease caused by the initial virus (Alcami & Smith, 1992). In VARV genome, this gene is destroyed.

From the evolutionary standpoint, the virus capable of maintaining the balance between pathogenic effect on the host organism and possibility of productive virus development in the body for a relatively long period is most fitted. Such virus can transmit efficiently from animal to animal even if the population density is low. Among orthopoxviruses, cowpox virus, carrying all the orthopoxvirus genes in an undamaged state (see Chapter 6), displays these features most pronouncedly.

4.6   Smallpox Epidemiology

Portal of smallpox. Pharynx, nasopharynx, and respiratory tract are the portal for variola virus infection. The virus can also penetrate into the body through injured skin, for example, during variolation (the practice of protection against smallpox by inoculating the lesion materials of smallpox cases weakened by long-term storage).

Despite the absence of direct evidences, it is likely that infection through the conjunctiva is also possible, as suggested by several successful instances of variola virus isolation from conjunctival swabs and tears of smallpox case contacts.

Alimentary route is denied for smallpox, as no such facts were recorded and the experiments performed by Rao on infecting monkeys orally and intragastrically and were unsuccessful (Fenner *et al.*, 1988). However, this infection route cannot be completely rejected (for example, when material with a high variola virus concentration gets into the mouth), as oral inoculation of susceptible animals allowed clinically pronounced and even lethal forms of other orthopoxvirus diseases (cowpox and monkeypox) to be reproduced.

Infections of fetus through the placenta in the case of pregnant women were reported, although their rate was not high—8.6% according to Rao (1972).

Contagiousness of the smallpox case and his/her excretions. Epidemiological observations and quantitative determination of the virus demonstrate that the contagiousness is highest during the first weak of the disease, when smallpox lesions on oral, faucial, and pharyngeal mucosae open and large amounts of the virus egress into the environment while talking, coughing, and sneezing. The maximal titers of the virus in fauces are detected on days 3–4, also the virus may be isolated from fauces to day 8 in 95–100% cases and through day 14 in 50% (Sarkar *et al.,* 1973a). High concentration of the virus retains until the death in severe smallpox cases. Few observations report isolation of small amounts of VARV from fauces of convalescents on days 24–26 from the disease onset (Shelukhina *et al.,* 1973).

Numerous data unambiguously demonstrate the presence of variola virus at high concentrations in the skin lesions, including scabs. Nevertheless, contagiousness of smallpox cases at later stages of the disease according to epidemiological observations is considerably lower compared with the initial period (Rao, 1972). The most probable reason is the physical state of virus-containing material: suspension of fine drops of highly infective discharge from fauces, pharynx, and oral cavity easily entering the respiratory tract, on the one hand, and the virus incorporated into the compact scab structure, on the other. In the later case, the virus is able to cause the disease only when turning into dust. As for urine, which also may contain the virus, the low virus titers and the type of excrete make this transmission route rather improbable. The epidemiological importance of the smallpox *sine eruptione* is most likely negligible, although transmission of the disease from such cases cannot be excluded, as isolation of small amounts of the virus from fauces of certain smallpox *sine eruptione* cases was reported (Marennikova *et al.,* 1961a; Rao, 1972).

Transmission routes. Infection with smallpox occurred most frequently via a direct contact with smallpox cases (usually, by droplet route or when the virus is brought into the upper respiratory tract with contaminated fingers). Although direct measurements of the virus content in the air of the wards with smallpox patients were mostly negative (presumably, due to imperfection of instrumentation used), the role of indirect infection route—through the air—was convincingly proved by epidemiological observations during several outbreaks (Anders & Posch, 1962; Wehrle *et al.,* 1970; Shooter, 1980). An illustrative example of the role of airborne transmission is provided by the outbreak in a hospital in Meschede, Germany (Wehrle *et al.,* 1970; Gelfand & Posch, 1971), when smallpox cases without any direct or indirect contacts (territorial, via medical or paramedical staff, or objects) with the first case were reported. Further analysis and experiments with smoke demonstrated that the virus discharged during heavy cough (the patient with a confluent rash had a severe bronchitis) due to specific construction of the hospital ventilation system spread with airflows from his isolation ward to other rooms and floors where it infected other individuals. The commission that investigated the outbreak at the University of Birmingham, where the laboratory involved in research into variola virus was the infection source, came to analogous conclusion on the role of infected airflows in smallpox transmission (Shooter, 1980).

One more transmission route is the things and objects used by smallpox patients. Beddings and underwear contaminated with variola virus were most important in smallpox transmission. Rao (1972) report the data of Downie that beddings became infected starting from day 3 of the disease. The virus may be preserved there up to 70 days (when stored in the dark and cool place); however, it is destroyed quickly under sunlight. Analysis of smallpox (first and foremost, imported) demonstrated that persons involved in handling beddings of smallpox patients were frequently met among smallpox cases. Data on the imported smallpox outbreak in Moscow provides such example (Golubchikova & Samvelova, 1961). The entire chain of people involved in processing the beddings (and, in particular, the first case not diagnosed in time)—matron, disinfector, and laundry worker. Contagiousness of beddings results from soiling of the ambient air with fine dust particles contaminated with the dried excreta of the patient. As for other objects (coins, banknotes, books, letters, etc.), we are aware of no reported strictly proven instances of causal relationship between these items and smallpox development. The issue of the so-called "cotton" outbreaks, when cotton contaminated with variola virus was considered the infection source, is also vague.

Carcasses of smallpox victims represent a serious danger. Smallpox infection during autopsies and preparing the bodies for burying were reported (Burgasov & Nikolaevsky, 1972).

Transmissibility. Smallpox is attributed to highly contagious diseases, yielding somewhat only to measles and chickenpox. Tight contacts provided, the secondary attack rate of *variola major* for susceptible population amounted on the average to 58.4% (Fenner *et al.,* 1988a). The data obtained during the global smallpox eradication demonstrate that on the average, one smallpox case infected five people, although in certain instances the rate could be considerably higher. For example, the first smallpox patient infected 19 people during the outbreak in Moscow; 17, in Meschede; and 38, in Yugoslavia (Litvinjenko *et al.,* 1973).

Vaccination against smallpox decreases the susceptibility drastically, reducing the secondary attack rate to 3.8%. Nevertheless, it was demonstrated that smallpox transmission could continue even among the population with a high collective immunity to smallpox. For example, smallpox was transmitted (until the antiepidemic activities were undertaken) via the children that remained nonvaccinated on the background of 94–98% vaccination rate of the population in the Municipality of Jakarta in 1970. Similar situations during the smallpox eradication were recorded in Nigeria, Brazil, and Botswana (World Health Organization, 1980).

Compared to some other infectious diseases, smallpox spread rather slowly—with a 2–3-week interval between each disease generation. Specific of smallpox was also pronounced seasonality: the increase in morbidity rates in the endemic countries was observed during winter–spring period.

4.7   The Fate of Variola Virus Remaining in the Laboratories

Upon the global eradication of smallpox, an unprecedented situation arose when in the absences of the disease itself, collections of the strains of the corresponding pathogen remained and are still reposited in the laboratories previously involved in diagnostics and research of variola virus. Owing to the efforts of WHO the number of such laboratories by 1984 decreased to two (for details of the history of this issue, see Fenner *et al.,* 1988; Marennikova & Shchelkunov, 1998). Both laboratories are the WHO Collaborating Centers on Smallpox and Related Infections (CDC, Atlanta, GA, USA, and State Research Center of Virology and Biotechnology Vector, Koltsovo, Novosibirsk oblast, Russia) have the facilities and used the work regimens meeting completely the international requirements of biosafety and are under constant control of the specialized WHO Committee. All these factors allow the risk of contamination of the environment with the pathogen to be virtually reduced to zero. On the other hand, this system forms the background not only for preservation of variola virus, but also for the necessary research of this pathogen, both basic and applied.

The issue of expediency of preserving variola virus as a biological species after the global eradication of the illness it caused was the topic of numerous specialized debates at WHO at the levels of expert committees, Executive Board, and World Health Assembly. Initially the destroying of laboratory collections of variola virus strains was planned for 1993 upon completion of sequencing of the complete genomes of two variola virus strains and producing of the libraries of their genomic fragments. Announcement about the destroying planned arouse serious disagreements in the scientific community, which spitted into advocates and opponents of this decision. The strongest objection against destroying variola virus collections stemmed from the results obtained while studying the structure–function organizations of the poxvirus genome and, in particular, variola virus genome, commenced in the early 1990s. This research opened a new page in the interactions of the virus and the host macroorganism at the molecular level. Protein products of various genes capable of modifying the immune responses of the organism have been discovered and actively studied. At least a part of these proteins was demonstrated to be species-specific—their activity in the host organism, to which a particular virus had adapted during the evolution, is incomparably higher. From this standpoint, of special interest is variola virus, whose only host is man. A successful study of this new field of virology, molecular biology, and pathology evidently requires the live virus. A number of authoritative experts regard the destroying of variola virus collections as an irreversible action that would forever deprive this and further generations of the possibility to use this natural repository of biologically active proteins and immune modulators. Another essential argument against the destroying is a justified that the virus upon destroying the two official and controlled repositories would not remain forgotten or deliberately kept illegally in some other places.

The main argument in favor of the virus destruction is the risk related to the danger of its accidental release from the laboratories where it is deposited and, as a result, development of a smallpox outbreak. However, we should keep in mind the following. The three documented accidents of laboratory-acquired infection with smallpox occurred before the introduction of the corresponding WHO requirements for handling this pathogen, which implied the system of maximal biosafety in the laboratories (WHO Collaborating Centers) involved in variola virus research. In addition, all the three accidents resulted from serious violations of the work regulations (World Health Organization, 1980).

The lack of consensus in the scientific community on the future fate of variola virus was the reason for repeated postponement of the target data for destruction. Note also that the recent circumstances affected essentially the current situation. The menace of bioterrorist actions with variola virus as a most probable potential agent has become actual (see Chapter 9). Consequently, development the means to defend population from this ganger has become a topical issue. In particular, this includes designing of preparations for emergency smallpox prevention and treatment yet lacking as well as new generation smallpox vaccines with an increased safety with reference to postvaccination complications. Solving of these problems is impossible without the live variola virus required for assessment of efficiency of the new preparations. With this in mind, the WHO experts decided to postpone the planned destruction of the virus until completion of the approved research program. The corresponding studies are being performed under the aegis and control of the World Health Organization.

Chapter 5

MONKEYPOX VIRUS

In recent decades, the nomenclature of nearly every virus taxon was supplemented actively. This refers in full to the poxvirus family, in particular, to the genus *Orthopoxvirus*. Among orthopoxviruses, the most important and interesting was certainly the discovery of monkeypox virus, which is pathogenic for both animals and humans.

5.1   History of the Virus Discovery

In 1958, two outbreaks of a pox-like disease occurred at an interval of 4 months in the State Serum Institute (Copenhagen, Denmark) among the Java macaques imported from Singapore for production of polio vaccine. The outbreaks occurred on days 51 and 62 after the delivery. The main symptom of this disease was a generalized rash (including the soles, palms, face, and tail) that evolved rapidly from petechiae through maculopapular to pustular stage. When the scabs had fallen off, scarring remained at the sites of the lesions.

The disease developed without a pronounced disturbance of the animal general condition and affected 20–30% of the stock. Some monkeys displayed local rash (on the tail). No human cases were observed among the animal maintenance staff in spite of a rather close contact with sick animals. An orthopoxvirus was isolated from the pustule fluid of sick animals on CAMs of chick embryos and in cell culture, and the authors named it monkeypox virus. One of the strains isolated was named Copenhagen and accepted subsequently as the reference strain of this virus species (von Magnus *et al.*, 1959).

In the ensuing years, monkeypox outbreaks among the monkeys in captivity were observed in the USA (Prier *et al.*, 1960; McConnell *et al.*, 1962), the Netherlands (Peters, 1966), and France (Milhaud *et al.*, 1969). Resemblance of the monkeypox clinical pattern to that of human smallpox

155

and insufficient knowledge about its etiological agent came into notice of WHO in the context of possible existence of variola virus natural reservoir outside the human population. In this connection, WHO queried 78 laboratories working with monkeys in various countries and revealed a number of similar outbreaks, which were not reported in the scientific literature (Arita & Henderson, 1968; Arita *et al.*, 1972). It was also found that no human case occurred during these outbreaks.

A new stage of research on this virus and the related pathology started in 1971 after we reported (Marennikova *et al.*, 1971b) its ability to cause a human disease clinically indistinguishable from smallpox (for more detail, see Section 5.6). The comprehensive studies initiated after the discovery of this new disease (Marennikova *et al.*, 1972a; Ladnyi *et al.*, 1972; Foster *et al.*, 1972; Lourie *et al.*, 1972) were extremely important for the smallpox eradication program. The point is that smallpox cases were reported to WHO from the territories where its transmission had already been extinguished. The attempts to find out the source of infection failed. All these issues became a matter of concern because they might indicate either a low efficiency of the current anti-smallpox methodology or the existence of natural reservoir of the smallpox agent. Precluding the doubts, the discovery marked the beginning of research into this previously unknown human disease and stimulated further study of its etiological agent, which is still in progress. Recognition of the importance of human susceptibility to monkeypox virus and its possible consequences were the prerequisites for initiation in 1981 of a WHO project on studying human monkeypox and the ecology of its etiological agent. Particular attention was given to determining the potential of this pathogen to fill the niche vacant after smallpox eradication.

The main properties of the virus in question accumulated so far are detailed below.

5.2   Biological Properties

5.2.1   Behavior in Chick Embryos

After inoculation of monkeypox virus (MPXV) on CAMs of 12-day chick embryos and incubation at 34.4–35°C, small (0.4 to 1 mm in diameter) whitish pocks with hemorrhages in their centers are formed by 72 h. In case there are many discrete and semiconfluent pocks on the membrane, they tinge a slightly pinkish color. Among the bulk of these prevalent pocks, individual (about 1%) white pocks of a larger size without hemorrhages develop (Figure 5.1). A certain similarity between the lesions on CAMs caused by monkeypox and variola viruses may bring to diagnostic errors due to the lack of experience (or when embryos are opened after 48 h,

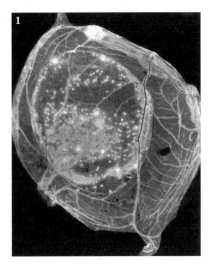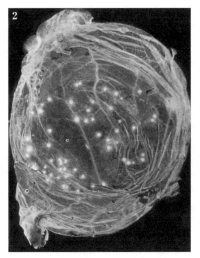

Figure 5-1. Pocks on CAMs caused by (1) monkeypox and (2) variola viruses.

when central hemorrhages are yet undetectable). This was the case with the first human monkeypox cases in West Africa, initially diagnosed as smallpox cases according to the lesions on CAMs (D.A. Henderson, personal communication). A smaller size of pocks, their less distinct raising, and above all, the presence of central hemorrhages are the main characteristics of lesions on CAM typical of monkeypox virus. Note that elevation in incubation temperature by 2–2.5°C made the distinctions less pronounced, as central hemorrhages failed to develop under these conditions. Gispen and Brand-Saathof (1972) found this too while comparing the cultivations at 33 and 38°C. We observed a similar effect of incubation temperature with camelpox virus, which also causes lesions on CAMs resembling those produced by variola virus (Marennikova *et al.,* 1973).

After inoculation on CAM, the virus accumulates in the membrane at very high concentrations (>8 log PFU/ml) both in the cases of primary isolation and further passaging. No significant changes in its properties were observed while passaging.

The LD_{50} values determined in chick embryos for 20 strains of monkeypox virus using inoculation on CAMs and 7-day incubation varied from <1 to 1.6 log PFU/ml. The accumulation rates of these strains in liver of chick embryos (according to Helbert, 1957) were also virtually similar, varying from 9 to 9.4 log PFU/g (Shelukhina, 1980).

The data on behavior of the virus inoculated by other methods in chick embryos are very scanty. Cho *et al.* (1972) reported that inoculation into the yolk sac caused generalized infection leading to death of embryos.

5.2.2   Growth in Cell Cultures

Similar to chick embryos, MPXV accumulates actively in various cell cultures of animal and human origins and attains rather high titers. This was first established by von Magnus *et al.* (1959) and then confirmed in subsequent publications (McConnell *et al.*, 1962; Gurvich, 1964; Mal'tseva, 1980; etc.). Virus accumulation in the culture is accompanied by development of cytopathic effect (CPE); CPE emergence and time course depend on the infective dose. At moderate virus doses, CPE first appears as discrete foci. In the foci, cells become round, the monolayer is disrupted, and cells agglomerate at the periphery of the loci (Figure 5.2). In some cultures, symplasts are formed. Then, CPE spreads over the entire cell monolayer, causing its complete destruction. In infected cultures, the hemadsorption phenomenon, appearing before visual CPE manifestation, is observed (Marennikova *et al.*, 1972a; Mal'tseva, 1980).

Under appropriate conditions, MPXV produces plaques. Unlike variola virus, forming small (less than 1 mm in diameter) plaques, monkeypox virus induces large saw-edged plaques with observable internal structure (Marennikova *et al.*, 1972a; Lourie *et al.*, 1972).

In 1971, it was found that there is a line among the available cell cultures, continuous PEK (porcine embryonic kidney) cell line, which is virtually insusceptible to MPXV and yet permissive with respect to other orthopoxviruses. In this culture, monkeypox virus induced no CPE, plaque formation, or hemadsorption phenomenon (Marennikova *et al.*, 1971c). However, immunofluorescence study of this culture showed the presence of individual luminous cells, which may indicate a slight degree of virus

Figure 5-2. CPE induced by monkeypox virus in Vero cell culture on day 5 after inoculation (strain Copenhagen, magnification 9 × 12.5).

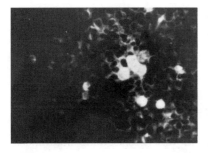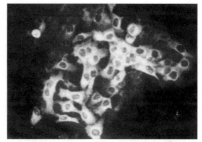

Figure 5-3. Behavior in PEK cell culture of monkeypox (left) and variola (right) viruses, 24 h after infection. The preparations were treated with FITC conjugate of anti-vaccinia antibodies.

replication in these cells (Mal'tseva, 1980). Figure 5.3 illustrates this assumption. For comparison, Figure 5.3 displays features of the focus formed in PEK cells by variola virus under equal conditions. The mechanism providing insusceptibility of PEK cells to MPXV is still unclear. Nevertheless, this feature of PEK culture made it valuable for diagnostic studies aimed at identification of monkeypox virus.

5.2.3 Ceiling Temperatures of Lesion Development on CAM and in Cell Culture

As compared to variola virus, MPXV has a higher ceiling temperature of pock development on CAM and CPE induction in the cell culture. According to Bedson and Dumbell (1961), and Shelukhina (1980), it amounts to 39°C for cultivation on CAMs of chicken embryos. Rondle and Sayeed (1972) reported that the ceiling temperature of pock development on CAM might vary from 39 to 39.5°C for various strains. Marennikova *et al.* (1971c) demonstrated that in cell culture at 39°C the virus is capable of replicating; however, its accumulation is considerably smaller than at 34.5°C. At a multiplicity of infection of 300 TCD_{50}/cell, the viral yield at 39°C was 300–450-fold lower than at 34.5°C.

5.2.4 Pathogenicity for Laboratory Animals

Unlike variola virus, monkeypox virus is highly pathogenic for the majority of laboratory animals.

Rabbits. Intradermal administration of the virus induces large compact infiltrates with necrotic areas in the center. Sometimes, the infiltrates appears slightly hemorrhagic. Their size and density depend on the invective dose. When administered onto scarified skin, the virus induces confluent papular–pustular rash at the inoculation site. In some cases, generalized infection with a rash spreading over the body develops on days 5–6 upon both methods of infection.

Susceptibility of rabbits to MPXV at various infection methods was thoroughly studied in our laboratory (Shelukhina & Marennikova, 1975; Marennikova & Šeluhina, 1976b; Shelukhina, 1980). The main results are summarized in Table 5.1.

It is evident that susceptibility of rabbits to MPXV and the course of disease depend on the infection route and age of the animal. In adult rabbits, clinical reaction to oral infection was absent; while intravenous infection resulted in an acute generalized disease with a rash (Figure 5.4) with the virus isolatable from the blood (at an early stage), lymph nodes, and kidneys. In some recovered rabbits, the virus was isolated from the testis 22

Table 5-1. Susceptibility of rabbits to monkeypox virus

Age and weight	Infection route and dose (PFU)	Results			
		Clinical symptoms	Death rate caused by specific infection	Virus isolation	Anti-body forma-tion
Adult rabbits, 2.5–3 kg	Intravenous, 7 log	Generalized disease with fever, abundant rash, rhinitis, and conjunctivitis	1/12[a]	Within 7 days, from blood; later, from lymph nodes and kidneys; and after 3 weeks, from testis[b]	+
	Oral, 9 log	None	0/20	–	±[c]
	Intradermal, 6 log	Compact large infiltrate with necrosis	0/20	–	+
	On scarified skin, 6 log	Papular–pustular rash with generalized infection	0/20	–	+
10-day old baby rabbits	Intranasal, 6 log	Adynamia, heavy breathing, rhinitis, dyspnea, and acute dystrophy	5/6 (on days 4–5)	From blood, lungs, and other internal organs	+
	Oral, 6 log, 7 log	Acute generalized disease with rhinitis, conjunctivitis, diarrhea, and rash, especially evident on the inner surface of ears, around nostrils, mouth and anus; progressive cachexia	17/20 (on the 4–14th day)	Within 7 days, from blood and internal organs; later, from lungs and kidneys (at titers of 5 and >3 log PFU/g, respectively)	+

Notes: numerator shows the number of dead animals; denominator, the total number of infected animals (time of death indicated in parentheses); (+) the presence of AHA and/or virus-neutralizing antibodies (study with vaccinia virus); (a) the animal died from cachexia 3 weeks post infection; (b) not in all animal; (c) virus-neutralizing antibodies at a titer of 20 were found in one of two rabbits on day 25; and (–) not studied.

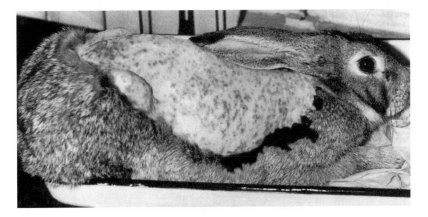

Figure 5-4. Eruptions on the skin of an adult rabbit on day 7 after intravenous infection with monkeypox virus.

days after infection. Antibodies were detected in blood serum starting from day 7.

Young (10-day old) rabbits, which appeared much more susceptible than adult animals, developed severe infection developed upon both intranasal and oral infections. In the former case, symptoms of the respiratory system involvement prevailed, whereas in the latter case, predominant was the involvement of the digestive tract. In addition, in the case of oral infection, the rash developed on the skin. Both inoculation methods gave high mortality rates. In the course of disease, the virus was readily isolated from blood (on day 4), lungs (upon intranasal inoculation, its concentration reached 5 log PFU/g), liver, kidneys, and spleen. Earlier, von Magnus *et al.* (1959) reported a high susceptibility of newborn (2-day-old) rabbits to monkeypox virus at intradermal infection or infection onto scarified skin. The same researchers showed that infection of rabbits onto scarified cornea was accompanied by development of discrete white nontransparent lesions with a craterlike hollow in the center.

Gispen *et al.* (1967) reported that intracerebral inoculation of rabbits with MPXV resulted in development of lethal meningoencephalitis. They also recorded that monkeypox virus was pathogenic for rabbits when inoculated into testis and on cornea.

White mice. Adult mice are highly susceptible to intracerebral infection. In response to virus inoculation, they developed encephalitis with a 100% mortality rate. Two-day-old sucking mice also displayed a 100% mortality 5–14 days after intranasal infection (von Magnus *et al.,* 1959). Oral inoculation of MPXV to 8- and 12-day-old mice caused a generalized infection with death of apart of animals. However, no visible lesions on the skin and mucosae were detectable (Shelukhina & Marennikova, 1975).

Results of the study on mice susceptibility (using various infection routes and age cohorts of animals) are listed in Table 5.2.

Table 5-2. Susceptibility of laboratory rodents to monkeypox virus*

Animal species	Age or weight	Way of infection	Symptoms	Mortality		Antibody formation	Virus isolation from organs
White mice	8–15 days	Intranasal	Flaccidity, adynamia, weight loss, dystrophy	17/17	100		+
	8 days	Oral	Flaccidity, adynamia, weight loss	4/10	40	NS	+
	12 days		Flaccidity, adynamia, weight loss	7/29	24	+	+
	8 days	Intraperitoneal	Flaccidity, adynamia, weight loss, dystrophy	10/10	100		NS
	8 days	Intracutaneous	Local infiltrate, flaccidity, adynamia	5/10	50	+	NS
	8 days	Subcutaneous into the sole	Foot edema, flaccidity, adynamia, dystrophy	18/18	100		NS
	12 days	Subcutaneous into the sole	Foot edema, flaccidity, adynamia	3/5	60	NS	NS
White rats	Adult	Intranasal	None	0/6	0	NS	–
		Intravenous	None	0/6	0	NS	NS
		On scarified skin	None	0/6	0	NS	NS
	1–3 days	Intranasal	Flaccidity, adynamia, dystrophy	24/24	100		+
Guinea pigs	Adult, 250– 300 g	Intracardial	None	0/5	0	+	+
		Intranasal	None	0/5	0	+	–
		Oral	None	0/5	0	–	–
		Subcutaneous into the sole	Foot edema	0/5	0	+	–
Golden hamsters	3 weeks	On scarified skin	None	0/6	0	+	NS
		Intranasal	None	0/6	0	+	NS
		Oral	None	0/3	0	+	NS
		Intracardial	None	0/75	0	+	+
Cotton rat	Adult, 80–90 g	Intranasal	Adynamia, dyspnea, cough, rhinitis, conjunctivitis, acute dystrophy	6/12	50	+	+
		Intravenous	Adynamia, dyspnea, cough, rhinitis, conjunctivitis, acute dystrophy	10/10	100		+

Notes: *According to Marennikova and Šeluhina (1976b) and Shelukhina, 1980; numerator shows the number of dead animals; denominator, the total number of infected animals; (+) the presence of AHA and/or virus-neutralizing antibodies (study with vaccinia virus); (–) absence of antibodies; and (NS) not studied.

As is evident, 8-day-old mice appeared especially susceptible and responded with severe lethal infection to intranasal, intraperitoneal, and subcutaneous (into the pad) infections. Determination of LD_{50} at oral and intranasal inoculation showed that the LD_{50} was lower in the case of intranasal infection as compared to oral route. This difference was most pronounced at inoculation of more adult animals: 12- and 15-day-old mice. Within one week, the virus was isolable from the blood of infected mice; within three weeks, from internal organs. Similar to young rabbits, the greatest amount of virus was detected in lungs after intranasal infection.

Guinea pigs. Prier *et al.* (1960) were the first to report a low susceptibility of guinea pigs to monkeypox virus independently of the route used for its introduction. Only inoculation into the pad caused a local lesion with edema of adjacent tissues. These data were confirmed by Shelukhina (1980); in addition, she found that guinea pigs were also insusceptible to intranasal and intracardial inoculation of the virus (Table 5.2). Nevertheless, the virus was isolated from the lungs of infected animals on day 7 after intracardial infection. Two weeks later, the virus was undetectable in any animal infected by any method used. By that time, AHA were found in the blood serum of all the animals at a titer of 16–64.

Golden hamsters (*Mesocricetus auratus*). Golden hamsters were also unsusceptible to monkeypox virus at various inoculation routes at high doses (7.2–7.8 log PFU/dose). Their external appearance, behavior, appetite, and ability to give healthy offspring remained unchanged. Behavior of the virus was examined in the animal body after intracardial infection. As was shown, the virus was detectable in blood and internal organs of virtually every hamster during 3 days. One week later, MPXV was isolated only from 30% of the animals. Further, the number of virus carriers decreased gradually to 13% after 4 weeks and 9% after 6 weeks. After 7 and 9 weeks (the period of observation), the virus became undetectable. Antibodies to MPXV were found in a half of infected hamsters. AHA appeared during week 2, reached their maximum by the weeks 4–5, and disappeared after 8–9 weeks. Their titers were not high: the maximum titer did not exceed 80 (Shelukhina, 1980).

Morphological examination of internal organs of the hamsters infected intracardially demonstrated a generalized pathological process with predominant injury of the liver, kidneys, and brain. Main symptoms were lymphocyte and histiocyte perivascular infiltration and structural disorder of vascular endothelium. Morphological changes were of a focal nature, localizing mainly around the vessels; destructive changes in the organs were weakly pronounced. The earliest changes (proliferation and desquamation) were observed in the vascular endothelium of parenchyma of the liver, kidneys, and *pia mater encephali*. Subsequent to the changes in endothelium, presumably increasing the vessel permeability, lymphocyte and histiocyte infiltration was observed around the vessels (Shelukhina, 1980).

White rats. Adult white rats, similar to guinea pigs, are insusceptible to monkeypox virus at intravenous, intranasal, and cutaneous inoculation: neither pathological symptoms nor the virus in internal organs of infected animals was detectable (Table 5.2). However, intranasal inoculation of newborn (1–3-day-old) baby rats resulted in a severe generalized infection leading to death on days 5–8. The virus was detected in the lungs and liver of dead animals at a concentration of 2–3 log PFU/g (Marennikova & Šeluhina, 1976b; Shelukhina, 1980).

Cotton rats (*Sigmodon hispidus*). Our experiments showed the cotton rats to be one of the most susceptible to monkeypox virus species among the laboratory animals: intravenous inoculation of the virus at a dose of 5 log PFU/ caused a 100% mortality rate. Acute development of the disease was characterized by heavy breathing, severe dyspnea, cough, sneezing, cyanosis, rhinitis, purulent conjunctivitis, and progressive cachexia ending fatally on days 4–5. At the climax of the disease, the virus was detected at high concentrations in the blood and internal organs (lungs, spleen, liver, and kidneys) of the animals. For example, the virus titer in the lungs of dead animals exceeded 5 log PFU/g. Virtually the same clinical pattern was observed at intranasal inoculation of the virus; however, in this case, only a half of the infected animals died (S.S. Marennikova & L.S. Shenkman, unpublished data; Shelukhina, 1980; Table 5.2).

Multimammate rats (*Mastomys natalensis*). A high susceptibility of a special laboratory line of this African rodent species to intranasal and intraperitoneal inoculations was reported (T. Kitamura, personal communication).

5.2.5   Variability

Analysis of restriction maps of a set of MPXV strains from various geographic regions of Africa isolated from both humans and animals (Esposito & Knight, 1985) demonstrated that the intraspecies variation was rather related to geographical origin, not to the particular mammalian species from which the strain was isolated. Phylogenetic trees constructed basing on the analysis of restriction maps demonstrated that, in general, the MPXV strains fall into two large groups—Central African and West African (Jezek & Fenner, 1988). Sequencing of genes of E5R (Douglass *et al.*, 1994), hemagglutinin (Hutin *et al.*, 2001), and chemokine-binding protein (Mikheev *et al.*, 2004) of various MPXV strains also confirmed this subdivision of the species in question—Central African and West African subtypes. However, note that of the 418 human monkeypox cases reported during 1970–1995, 388 (92.8%) occurred in Democratic Republic of the Congo (DRC). Moreover, the fatal cases were recorded only in this Central African country (Jezek & Fenner, 1988; Breman, 2000; Hutin *et al.*, 2001; Meyer *et al.*,

2002b). In 1996–2002, numerous fatal outcomes of human monkeypox cases with a case-fatality rate varying from 4% (Hutin *et al.*, 2001) to 16% (Meyer *et al.*, 2002b) were identified only in DRC. Recently, experimental infections of monkeys suggested that West African MPXV isolates maybe less virulent for humans than Central African isolates (Chen *et al.*, 2004).

Clinical manifestation of human monkeypox is similar to those of smallpox. The major distinction is lymphadenopathy, i.e., a pronounced inflammatory reaction, in the case of human monkeypox and its absence in the case of smallpox (see 5.5.1).

Complement activation is one of the main inflammation effector mechanisms. All the studied species of orthopoxviruses, such as variola, cowpox, vaccinia, camelpox, and ectromelia viruses, carry the gene of complement-binding protein (VCP; Kotwal & Moss, 1988a; Smith S.A. *et al.*, 2000).

VCP is an important virulence factor of orthopoxviruses. It was demonstrated that recombinant VACV with damaged sequence of this gene became attenuated to a considerable degree—skin lesions of rabbits infected with the mutant VACV were less pronounced compared with the lesions caused by the wild type virus (Isaacs *et al.*, 1992a). By the example of CPXV with mutant VCP gene, it was demonstrated that this viral protein suppressed inflammatory reaction that developed in response to viral infection (Miller *et al.*, 1997; Howard *et al.*, 1998).

We have performed a comparative analysis of VCP gene organization in the genomes of Central African and West African MPXV strains. The following strains isolated from humans in various geographical regions of Africa were used in the work: ZAI-96-I-16, 77-0666, Congo-8, CDC#v79-I-005, and CDC#v97-I-004 from DRC (Central African); CDC#v70-I-187 from Liberia; and CDC#v78-I-3945 from Benin (West African). MPXV DNA samples were kindly provided by J.J. Esposito. While performing PCR, we discovered a deletion of about 1.9 kbp in the region of VCP gene in the genomes of West African strains. The viral genome segments under study were PCR amplified and sequenced by primer walking with an automatic sequencer. Each base was determined one or more times from each strand, with no discrepancies.

We demonstrated that the ORF encoding complement-binding protein in the genome of Central African MPXV strains was truncated (see Figure 7.14; Uvarova & Shchelkunov, 2001). Presumably, this has an essential effect on the property of the viral protein as inhibitor of inflammatory reaction causing lymphadenopathy of humans infected with MPXV and decreasing in the virulence of the virus. In the genomes of West African strains CDC#v70-I-187 and CDC#v78-I-3945, the deletion of 1953 bp was detected (Figure 5.5), that removed not only VCP gene (ORF D14L) from the genome, but also ORFs D15L–D17L, which are fragments of

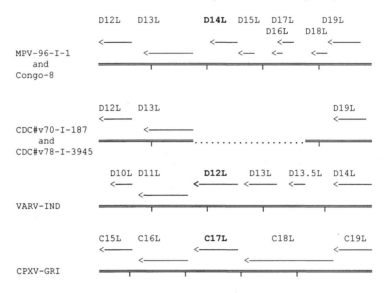

Figure 5-5. Comparison of translation maps of the investigated region for the monkeypox virus strains CDC#v79-I-005 and CDC#v97-I-004 (Central Africa), CDC#v70-I-187 and CDC#v78-I-3945 (Western Africa), variola virus (VARV-IND), and cowpox virus (CPXV-GRI).

kelch-like protein of CPXV and VACV (Shchelkunov *et al.*, 2002b). The deletion of VCP gene in West African MPXV strains should further decrease their pathogenicity for sensitive animals and humans relative to Central African strains (Uvarova & Shchelkunov, 2004).

   In 2003, human monkeypox outbreak was for the first time recorded in the USA (CDC, 2003). The pathogen was imported to the American continent with infected rodents from Ghana. It was proved that West African MPXV subtype caused this outbreak (Reed *et al.*, 2004). The data on deletion of VCP gene in West African MPXV strains could explain the absence of fatal cases among 73 persons who contracted monkeypox during this infectious outbreak in the USA. Evidently, that importation of the Central African MPXV variant to any countries could cause an outbreak of the human infectious disease with more severe consequences. Thus, a better control of all the animals exported from Africa to the other continents for their potential infection with orthopoxviruses is necessary.

5.3   Clinical Pattern and Pathogenesis of the Infection in Monkeys

   Clinical manifestations of the monkeypox in experimental monkey model differed only slightly from those observed during the infection outbreaks in monkey breeding facilities and zoos. Wenner *et al.* (1968) studied most thoroughly the clinical pattern of experimental monkeypox, mainly in *M. cynomolgus* infected intramuscularly. Development of the disease was preceded by a 3–10-day incubation period. The disease began with fever, flaccidity, anorexia, and cough. During the first week, generalized lymphadenopathy developed and remained until the end of the third week; skin eruptions in the form of maculopapular rash appeared at days 7–11 of the illness, spreading all over the body. The rash was most abundant on the palms and soles. The lesions appeared on mucosae of the oral cavity and pharynx. Evolution of the rash (papules—vesicles—pustules—scabs) proceeded rather rapidly. The scabs remained 7–10 days, while enanthema disappeared in 3–4 days. The infection was accompanied by formation of AHA and virus-neutralizing antibodies. In another monkey species, *M. rhesus*, the disease developed in a milder form (the incubation period was somewhat longer, whereas the evolution period of skin lesions was shorter).

Heberling *et al.* (1976) studied the monkeypox clinical pattern in baboons (*Papio cynocephalus*) using inoculation on the skin during 10 passages. The disease developed on days 3–4 manifesting itself as a fever, lymphadenopathy, leukocytosis, and lesions at the inoculation site. The secondary eruptions developed only in the first two passages. The virus was isolatable from the pharynx, blood, and feces. The maximal period of the virus isolation from pharyngeal and rectal lavages was 28 days. Of the 20 infected animals, 5 died. All of them were young (2.5–3.5 months old). Antibodies (AHA) were detected already on day 7 post infection, with a peak on day 14 and a decrease starting from day 30.

Pathogenesis of the disease was studied by Wenner *et al.* (1969a; 1969b) using intramuscular infection. Initially, the virus accumulated at the inoculation site, which was accompanied by a local inflammatory reaction. Through the lymphatic system, the virus entered regional lymph nodes and reached spleen, tonsils, and bone marrow through blood circulation system (primary viremia). This phase manifested itself as elevated body temperature; spreading via the lymphatic system, as lymphadenopathy. The subsequent accumulation of the virus resulted in a secondary viremia and development of specific lesions on the skin and mucosae (Wenner *et al.,* 1969a; 1969b; Cho *et al.,* 1970). The main stages of the monkeypox pathogenesis are shown in Figure 5.6.

The above scheme of monkeypox pathogenesis is similar to that of several other orthopoxviral infections, in particular, mousepox (Fenner, 1949), rabbitpox (Bedson & Duckworth, 1963), and smallpox in monkeys (Hahon & Wilson, 1960).

A number of recent events, including a large human monkeypox outbreak in Democratic Republic of the Congo (1996–1997) and the menace of

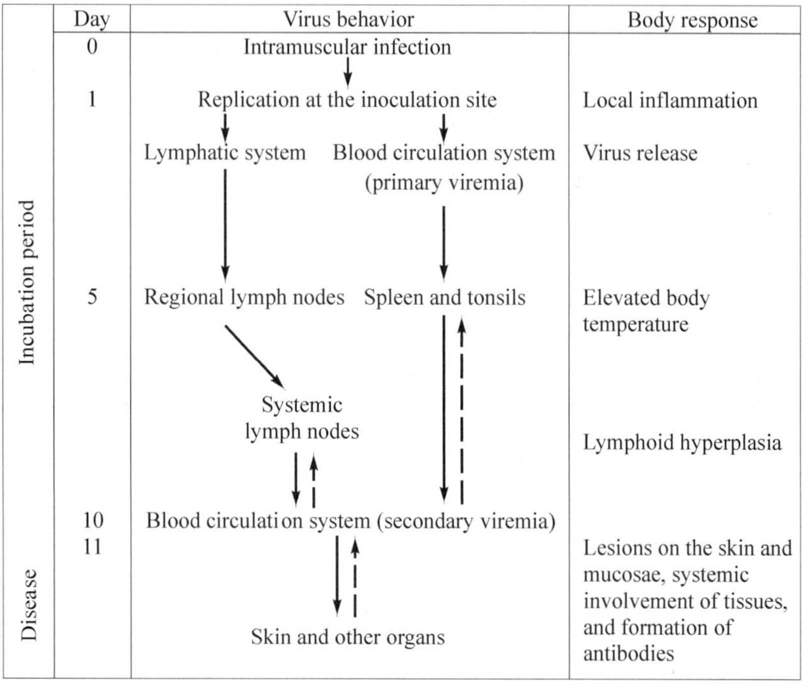

	Day	Virus behavior	Body response
Incubation period	0	Intramuscular infection	
	1	Replication at the inoculation site	Local inflammation
		Lymphatic system Blood circulation system (primary viremia)	Virus release
	5	Regional lymph nodes Spleen and tonsils	Elevated body temperature
		Systemic lymph nodes	Lymphoid hyperplasia
Disease	10 11	Blood circulation system (secondary viremia)	Lesions on the skin and mucosae, systemic involvement of tissues, and formation of antibodies
		Skin and other organs	

Figure 5-6. Main stages of monkeypox pathogenesis in *M. cynomolgus* (Cho & Wenner, 1973).

bioterrorism actions, revived the interest to research of orthopoxviruses pathogenic for humans, the pathologies they caused, and development of new efficient defense means. Huggins *et al.* (2002) studied a model of monkeypox in primates. Infecting *M. cynomolgus* with a fine aerosol of MPXV, they obtained a generalized disease of various degrees of severity accompanied by fever, skin rash, lost of weight, and accumulation of the virus in internal organs. This experimental model allowed them to confirm the effectiveness of cidofovir (and its variants, including the form for oral administration) for prevention of monkeypox. This research team is continuing the study of the pathogenesis of this illness in primates using a wide range of tests as well as upgrading of the cidofovir variants.

5.4   Ecology of Monkeypox Virus

From the standpoint of ecology of monkeypox agent, outbreaks of this disease among the primates in captivity are of doubtless interest.

5.4.1   Monkeys

Analysis of nine monkeypox outbreaks among the primates in captivity allowed the susceptibilities of individual primate species (Figure 5.7) to be determined as well as the clinical courses of the disease and certain issues related to specific features of the disease transmission from ill to healthy animals.

At least 12 primate species appeared susceptible to MPXV—from lower primates to anthropoid apes (gorilla, chimpanzee, orangutan, and gibbon). The severities of disease courses differed, varying from mild to lethal. The rash localized to the trunk, tail, muzzle, and limbs, including palms and soles, was the main clinical symptom for the majority of species (Figure 5.7). Presumably, orangutans are most sensitive to this infection. For example, of the nine orangutans that contracted the illness during the outbreak in the Rotterdam Zoo, six died (Gispen *et al.,* 1967).

Note that this outbreak is distinguished by a high mortality rate—of 23 primates of various species that contracted the disease, 11 animals died (47%). During the other monkeypox outbreaks, the lethality rates were

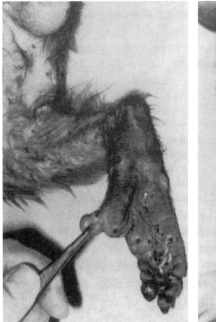

Figure 5-7. Monkeypox of *M. cynomolgus*: skin lesions on limbs (from von Magnus *et al.,* 1959).

lower. According Prier and Sauer (1960), who described the outbreak in the animal breeding facility of Merck, Sharp & Dohme, where over 2000 animals were kept, the mortality rate among *M. philippinensis* and *M. mulatta* did not exceed 0.5%. Note that the morbidity rate of the former species was higher than that of the latter. The researchers paid attention to a specific disease course observed in certain *M. philippinensis* individuals—on the background of rash, they developed edema of muzzle and neck accompanied with heavy breathing and asphyxia, causing the death of animals.

The mortality rate during the outbreak in Copenhagen amounted to 3% (von Magnus *et al.,* 1959). When the number of animals in aperies was large enough, MPXV infection affected 10 to 30% of their population (von Magnus *et al.,* 1959; Prier & Sauer, 1960).

Clarification of the source of infection during these outbreaks appeared more complex that it could be assumed. The disease of monkeys appeared a long time after their delivery, exceeding considerably the incubation period, at least in the case of four outbreaks—51 days, 62 days, 9 months, and 2 years. These facts can be explained only admitting the occurrence of an asymptomatic infection or persistence of the virus in monkeys that recovered from the disease before their delivery. Analysis of the data on an outbreak allowed McConnell *et al.* (1962) to infer that the source of infection was an asymptomatic disease. The outbreak they investigated commenced 10 month after the shipment of monkeys: three animals contracted the disease; of them, two were subjected to total irradiation 1.5 months before. Serological examination of the rest, apparently healthy monkeys detected specific antibodies (AHA) in 11–78% of the animals depending on the species and positions of their cages. Data obtained by von Magnus *et al.* (1959) provided a more heavy proof for the MPXV carriage by apparently healthy monkeys. These researchers succeeded in isolating the virus from kidney cell cultures of these animals. The mechanisms of outbreak development a along time after their shipment may be deduced by the analogy with another orthopoxvirus circulating among wild rodents of Europe and Asia—cowpox virus. Characteristic of the infection caused by CPXV is a postinfection persistence of the virus in its natural carriers, where the virus was detectable in the kidneys and spleen. During experimental infection with CPXV, the virus most frequently was isolatable from the kidneys and urine (see Section 6.3). A similar postinfection carriage of the virus was demonstrated for the white rats infected with CPXV—the animals excreted the virus not only with the urine, but also with feces. The latter fact provides additional possibilities for infecting the susceptible animals. Presumably, a similar mechanism, at least in a part of the cases, formed the background for infection transmission during the mentioned above monkeypox outbreaks with a delayed onset.

Four outbreaks developed soon after the delivery of animals, which presumably had at that moment the disease in an incubation period. Note that the source of one of these outbreaks was not monkeys but another species—giant anteaters.

The question on what geographic zone harbors the natural reservoir of MPXV could not but attract the attention of researchers. Analysis of the published data has demonstrated that pox-like diseases in nature were reported occasionally in the scientific literature starting from the second half of the 19th century (Anderson, 1861; Schmidt, 1870; Councilman, 1907; Bleyer, 1922; etc.). As a rule, these papers reported outbreaks among wild monkeys with a high mortality rate, which were usually regarded as being connected with smallpox outbreaks. However, in certain cases, the people who disposed of numerous monkey corpses without any precautions contracted no disease (Rahman, 1967; cited from Arita & Henderson, 1968). The outbreaks among wild monkeys were recorded in Africa, Central and South Americas, and Southeast Asia. Evidently, it is impossible now to find out the actual sources of those outbreaks. However, it should be kept in mind while analyzing these data that as long ago as at the beginning of the last century Brinckerhoff and Tyzzler (1906) showed that it was extremely difficult to transmit the VARV infection from an ill monkey to a healthy animal when infecting experimentally the anthropoid apes. This was also confirmed by Noble and Rich (1969) using a pure culture of variola virus. They demonstrated that transmission of the infection caused by VARV in *M. cynomolgus* was possible only during a limited number of passages, whereupon the virus disappeared. Thus, this is improbable enough that variola virus caused the outbreaks of a pox-like disease among wild monkeys. At least, it cannot be excluded that monkeypox virus was the agent responsible for a part of these outbreaks.

In the view of the fact that the pathogen underlying these outbreaks was not identified, the above-described data could not but form the reliable start points for localizing the natural reservoir of monkeypox. As the initial data, the information on the sites of trapping the monkeys that were shipped to the vivariums where outbreaks of the disease were later recorded was used. As was found out, the monkeys were delivered from Southeast Asia in all the cases except for one outbreak. This suggested that the natural focus of monkeypox was localized to this region.

In this connection, WHO initiated a serological examination of monkeys for the presence of antibodies to orthopoxviruses. The WHO Collaborating Center on Smallpox and Related Infections in Moscow, Russia, together with four other scientific centers participated in this project (Arita *et al.,* 1972). At the initial stage, 1614 samples of the sera of the monkeys originating from various countries of Asia and Africa stored at various laboratories of Europe and North America were tested. HAIT failed to detect any significant AHA titers in the samples. When neutralization reaction was used, antibodies at a titer of 8–16 were detected only in 2 samples of 1114 tested. Both positive serum samples were obtained from the species *M. iris*, widespread in Southeast Asia. This data was a prerequisite for launching a specialized project on examining this monkey species in Malaysia. To exclude the artifacts (formation of the antibodies due to contacts with animals from other regions), the serum samples were taken at the site of trapping. Of the 481 sera, 22 samples (4.7%) displayed the presence of virus-neutralizing antibodies at a low titer (4–5), which prevented from an unambiguous inference that these sera were positive. A later HAIT examination of the serum samples of 100 *M. mulatta* from another Asian region, India, also gave a negative result (Sehgal & Ray, 1974). Similarly, 147 serum samples of African monkeys (*Erythrocebus patas*), trapped in Republic of Chad, were examined by HAIT with a negative result (Arita *et al.*, 1972).

Once the human monkeypox disease was recorded, we became convinced that examination of the monkeys in the vicinity of the focus of human outbreak in the tome domain close to disease appearance would provide the maximal chances for a successful detection of the natural MPXV reservoir. Overall, we performed four series of examinations during January 1971– October 1973 connected with human monkeypox cases in the Equatorial Province, Democratic Republic of the Congo (Marennikova *et al.*, 1975c). Usually the monkeys were trapped or shot by specialized WHO teams in a radius of 100 km from the village with a recorded human monkeypox case. The corresponding data are listed in Table 5.3.

As is evident, this approach justified the expectations. Even among a rather small number of monkeys examined, carriers of the antibodies to orthopoxviruses were detected at all the foci with a significant level of titers as determined by both HAIT and neutralization reaction. Moreover, precipitins, indicating a recent contact with the virus, were isolated from two monkeys of the nine examined. *Pan troglodytes* and *Cercopithecus mona* Wolfi Meyer were among the animals with a rather high level of antibodies. Note also the effect of the time domain on the results obtained—the number of positive assays was higher when the animals were examined 16–30 days and 4.5 months after the onset of human disease, whereas it decreased in later periods. This study for the first time detected the presence of antibodies to orthopoxviruses in the monkeys inhabiting the regions where human monkeypox cases were recorded. These data were later confirmed by Breman *et al.* (1977), who examined 207 monkey serum samples from Ivory Coast, Mali, and Upper Volta. The animals were trapped in tropical forest, bushes, and savanna. HAIT detected antibodies to orthopoxviruses in 8% of the samples; neutralization reaction, in 21%. The majority of HAIT-positive sera (14 of 15) were obtained from the monkeys belonging to the genus *Cercopithecus*.

Table 5-3. Results of serological examination of the monkeys captured in foci of human monkeypox (Equatorial Province, Democratic Republic of the Congo)

Time of shooting of monkeys		Number of monkeys examined	Of them, with antibodies to orthopoxviruses, as detected by											
			HAIT						Neutralization reaction					Precipitation reaction
			Total	Titer					Total	Titer				
				16–20	32–40	64–80	256	1280		10–20	40–80	160–320		
January 1971	In connection with recent human monkeypox cases	7	2/7	1	0	0	0	1	3/5	1	2	0		2/7
February 1973		5	4/5	0	1	2	1	0	3/5*	0	2**	1		0/5
June 1973	In connection with former human monkeypox cases	13	8/13	1	5	2	0	0	NS					0/13
October 1973	human monkeypox cases	56	0/56	0	0	0	0	0	5/55	1	3	1		NS

Notes: numerator indicates the number of positive results; denominator, the number of sera studied by the corresponding assay; (*) titration of two sera negative for the presence of antibodies was commenced from a dilutions of 1 : 40 and 1 : 80; (**) the titer >40; and NS, not studied.

Referring to unpublished data of J.H. Nakano, Jezek and Fenner (1988) reported results of another study also performed in countries of West Africa. It was found during this project that of the 334 monkey sera examined, 29% contained AHA and 24%, virus-neutralizing antibodies to orthopoxviruses. A part of these serum samples from the two series of serological examination were additionally assayed for the presence of specific antibodies to MPXV. Gispen *et al.* (1976) by immunofluorescent method with a premonitory adsorption of genus-specific antibodies to orthopoxviruses studied three monkey sera from the first screening series and detected the presence of antibodies to MPXV in all the three cases. Fenner and Nakano (1988) used for this purpose the RIA supplemented with serum depletion (adsorption) of genus-specific antibodies (RIA–A). In the second series of 273 sera, they found species-specific antibodies only in 7 samples.

Finally, the last of large-scale serological examinations of African monkeys for the presence of antibodies to orthopoxviruses was performed in Zaire (Democratic Republic of the Congo). It involved 241 sera obtained from 10 monkey species successively studied by HAIT, RIA, and RIA–A.

We summarized in Table 5.4 the data on monkey species inhabiting Central and West Africa that displayed both species-specific antibodies to MPXV and antibodies to orthopoxviruses.

Table 5-4. African primate species that displayed species-specific antibodies to monkeypox virus and/or antibodies to orthopoxviruses*

Primate species	Place of trapping	
	West Africa	Democratic Republic of the Congo
	With antibodies to monkeypox virus	
Allenopithecus nigroviridis		
Cercocebus galeritus		
Cercopithecus ascanius		
Cercopithecus mona		
Cercopithecus pogonias		
Cercopithecus nictitans	+[a, c, d]	
Cercopithecus aethiops	+[a, c, d]	
Cercopithecus petaurista	+[a, b, c, d]	
Colobus badius	+[a, b, c, d]	
	With antibodies to orthopoxviruses	
Cercopithecus neglectus		+[c**]
Cercopithecus pogonias		+[c**]
Colobus pennanti		+[c**]
Perdioticus potto		+[c**]
Cercopithecus mona		+[c, d, e]
Colobus badius		+[c]
Pan troglodytes		+[c, d, e]

Notes: (*) according to Fenner *et al.* (1988), Khodakevich (1990), Marennikova *et al.*, (1972a), and Gispen *et al.,* (1976); (a) species-specific antibodies to MPXV (RIA–A); (b) species-specific antibodies to MPXV (EIA–A); (c) virus-neutralizing antibodies to orthopoxviruses; (d) AHA to orthopoxviruses; (e) precipitins to orthopoxviruses; and (**) RIA- and RIA–A-negative.

5.4.2   Other Animals

Serological examinations for the presence of antibodies to orthopoxviruses performed in the early 1970s in West Africa in several series involved not only monkeys, but also a number of other animal and avian species (Breman *et al.,* 1977). These species included squirrels, several ungulates, and others, which also displayed neutralizing antibodies and AHA to orthopoxviruses. Later (since mid-1979), WHO in collaboration with CDC organized an expedition to Zaire with a special focus on collecting samples from other wild animals in addition to monkeys, in particular, terrestrial and tree rodents. During the first expedition to the north of Zaire, 1331 serum samples were collected from 43 animal species. AHA were detected in 17%, including the red-legged sun squirrels *Heliosciurus rufobrachium* (Jezek & Fenner, 1988). The examination of the samples collected by the expedition repeated later detected the presence of species-specific antibodies to MPXV in several species of squirrels belonging to the genus *Funisciurus.*

The experience accumulated was taken into account during the further expeditions to Zaire. The sites were determined basing on the human monkeypox morbidity rate. In addition, the range of animals with a highest frequency of contacts with the humans was determined basing on specific ethnographic characteristics of the aboriginal population (customs, type of nutrition, methods of hunting, etc.), structure of the surrounding landscape, and the inhabiting fauna (Khodakevich *et al.,* 1987; Khodakevich, 1990). Of these animal species, the species were chosen that due to a high population density, high rate of population renewal, and frequent contacts with humans could be considered as a natural MPXV reservoir in addition to monkeys. These species included terrestrial and tree rodents.

Serological and virological examinations of about 600 terrestrial rodents, including those living in human dwellings, gave negative results by RIA and RIA–A, although 3.1% displayed AHA. However, one of the four examined African brush-tailed porcupines (*Atherurus africanus*) displayed the species-specific antibodies to MPXV. The domestic animals (cats, sheep, goats, etc.) were also negative for the presence of MPXV-specific antibodies. Later, Hutin *et al.* (2001) detected neutralizing antibodies to orthopoxviruses in 15.8 and 33.3% of the terrestrial rodents—Gambian rat (*Cricetomys emini*) and four-toed elephant shrew (*Petrodromys tetradactylus*). However, the most convincing evidence of the role of terrestrial rodents in circulation of MPXV was the importation of this infection into the USA in 2003 with Gambian rats shipped from Ghana (CDC, 2003).

Tree rodents—squirrels—represented in Africa by dozens of species, have a high population density and in some regions, in particular, Bumba, head the list of animals eaten by aboriginal populations (Khodakevich, 1990).

Results of examination of the tropical squirrels most frequently met in Zaire are listed in Table 5.5.

It appeared that of the six squirrel species examined, the MPXV-specific antibodies were detected in all the species except for *Protoxerus strangeri*. As for the last species, it is very likely that the negative results are connected with a very small number of the animals captured. The data listed in Table 5.5 demonstrate also that the MPXV circulation among squirrels occurs in all the four zones, although it is somewhat lower in the Bandundu zone. However, presumably the most interesting fact was revealed when comparing the data of this serological examination with the human monkeypox morbidity rates in these zones. Paradoxical as it may seem at the first glance, the highest percent of seropositive squirrels was recorded in Tshela zone, where no human monkeypox cases were recorded. A special investigation (Khodakevich, 1990) demonstrated that the aboriginal population of this zone did not eat raw meat of wild animals and the children younger than 10 years old were not involved in hunting small rodents. According to the opinion of Khodakevich, hunting animals and eating raw meat of small rodents (including squirrels) underlies an increased morbidity rates in Bumba and Ikela zones along with the fact that the children of the mentioned age cohort formed the main part of human monkeypox cases (80%).

Thus, the overall serological data demonstrate that MPXV is widely circulating among various species of wild-living monkeys and squirrels inhabiting mainly rainforests of Central and West Africa. However, the decisive proof is a direct isolation of the virus from its carriers in wild nature. A number of attempts to isolate MPXV from internal organs of various species caught during the serological project, made for several years,

Table 5-5. Data of serological examination of squirrels trapped in four zones of Democratic Republic of the Congo*

Squirrel species \ Zones	Bumba	Ikela	Tshela	Bandundu	Total
Funisciurus anerythrus	79/320 (24.7)	5/40 (12.5)	–	3/108 (2.9)	87/463 (18.8)
F. congicus	–	11/56 (19.6)	–	0/21 (0.0)	11/77 (14.2)
F. lemniscatus	–	–	58/119 (48.7)	–	58/119 (48.7)
Heliosciurus rufobrachium	6/37 (16.2)	3/23 (13.0)	–	–	9/60 (15.0)
H. gambianus	–	–	6/31 (19.4)	0/10 (0.0)	6/41 (14.6)
Protoxerus strangeri	–	0/2 (0.0)	–	–	0/2 (0.0)

Notes: (*) according to Khodakevich (1990); numerator, the number of animals with antibodies to MPXV (RIA–A); denominator, total number of animals examined; parenthesized, percent of seropositive animals; and (–) not studied.

were unsuccessful. Only in 1985, during a directed search in Bumba zone, an apparently ill squirrel *Funisciurus anerythrus* was found, and the virus was successfully isolated. When the killed animal was examined, skin lesions with a diameter of 2–3 mm were found on the abdomen; they were especially apparent in the front and rear inguinal regions. The virus was isolated in independent and parallel studies at the WHO Collaborating Centers in Moscow, Russia, and Atlanta, GA, USA, from both skin lesions and internal organs (lungs, kidneys, and spleen). The highest MPXV concentration (6 log PFU/ml) was detected in the scrapings of skin lesions; the concentrations in the lungs and kidneys amounted to 5 and 4 log PFU/g, respectively (Marennikova *et al.,* 1986; Khodakevich *et al.,* 1986). The isolate recovered from the squirrel was similar in its biological characteristics and restriction profile to the MPXV isolates recovered earlier from human monkeypox cases. The blood serum of this squirrel contained species-specific antibodies to MPXV.

The attempts to isolate the virus in question from wild-living primates are still unsuccessful.

To understand the mechanisms underlying the MPXV circulation among squirrels and possible routes of its transmission to humans, a series of experimental studies was launched involving five monkey species caught in Zaire. This international WHO project was implemented at the facilities of the National Institute for Biomedical Research in Kinshasa. L. Khodakevich contributed essentially to realization of this project. The virological part of the work was performed both on site (by E.M. Shelukhina) and in the Russian and US WHO Collaborating Centers.

As a preliminary stage, the susceptibility of red squirrels (*Sciurus vulgaris*) infected intranasally, orally, or on scarified skin with the MPXV isolate from wild living African squirrel was studied at the Moscow WHO collaborating Center (Marennikova *et al.,* 1989). Upon inoculation of 6 log PFU/dose, the infected animals developed fever during 1 day; on days 3–5, the animals became flaccid, adynamic, and anorexic. Later, these manifestations were supplemented by rhinitis, cough, and dyspnea. Independently of the way of infection, all the animals died on days 7–8. No skin lesions were detected. The virus was detectable in internal organs and nasal discharge at a high concentration.

Similarly to European squirrels, all the six species of tropical squirrels studied appeared highly sensitive to intranasal infection with high doses of MPXV (6 and 5 log PFU/0.1 ml). The inoculated animals developed an acute generalized infection with a 100% mortality rate (Table 5.6).

Inoculation of smaller MPXV doses detected distinctions in susceptibility of various species. For example, even the minimal doses used (1 and 2 log PFU/0.1 ml) induced the disease of the squirrels *F. pyrrhopus* and *F. congicus* ending in the death of the majority of animals, whereas all the

Table 5-6. Susceptibility of tropical squirrels to intranasal infection with monkeypox virus at doses of 6 and 5 log PFU/0.1 ml (from an unpublished paper of E.M. Shelukhina, L.N. Khodakevich, *et al.*, 1989)

Squirrel species	Clinical manifestations				Presence of MPXV				
	Duration of								
	Incubation period, days	Disease, days	Symptoms	Outcome	Lungs	Liver	Kidneys	Spleen
F. anerythrus	2	4	Adynamia,	100%	+	+	+	+
F. pyrrhopus	2	5	conjuncti-	mortality	+	+	+	+
Protoxerus strangeri	4	7	vitis, rhinitis, cough,	rate	+	+	+	+
F. congicus	6	4	dyspnea,		+	+	+	+
F. lemniscatus	4	6	weight loss,		+	+	+	+
H. gambianus	8	11	and progressing cachexia		+	+	+	+

squirrels *H. gambianus* infected at a dose of 4 log PFU/0.1 ml developed a mild disease and recovered. This drastic difference hardly could be attributed only to a higher body weight of *H. gambianus*.

Microscopic examination of internal organs of the dead squirrels detected a diffuse hemorrhagic pneumonia in the lungs. The majority of animals had liver hyperemia and small necrotic foci in the liver parenchyma as well as frequent cases of steatosis with a predominant interlobular location of the plaques. The kidneys were hyperemic, sometimes with injured glomeruli. An extremely hyperemic spleen had foci of periarteriolar necrosis in follicles and hypertrophy of the white and red pulps with necrosis and destruction of lymph nodes. The intestinal wall was frequently hypertrophied with foci of acute inflammation and cell desquamation. The brain displayed no essential histopathological alterations.

Study of the internal organs for the presence of MPXV confirmed a generalized character of the infection—the virus was isolated at high concentrations (6 log PFU/g and higher) from the liver, spleen, and lungs; at a lower titer (3–4 log PFU/g), from the kidneys, intestines, and brain.

As was found out, the tropical squirrels were also sensitive to oral and dermal infections with MPXV. Upon oral inoculation, the clinical symptoms were similar to the disease induced intranasally. Internal organs of the infected and dead squirrels contained the virus at high concentrations. When infected onto scarified skin, animals developed dense red papules at the inoculation site, and the overall disease pattern was analogous to the courses observed using other infection methods. The virus was detected at a concentration of 6 log PFU/g and higher in the liver, spleen, and lungs; at a titer of 4 log PFU/g, in the brain and intestines.

Note in conclusion that in addition to squirrels and monkeys, antibodies to orthopoxviruses and, in particular, to monkeypox virus were detected in blood of several terrestrial rodent species and other animals inhabiting the regions with recorded human monkeypox cases.

An actual life also provides examples when an accidental contact of a sensitive animal with the virus may not only cause a disease of the animal, but also induce an outbreak among other animals or humans. Such situation took place during the outbreak in the Rotterdam Zoo with an ill South American giant anteater as the source of infection (Gispen *et al.*, 1967). Another example is the outbreak of monkeypox among the Pygmy in Central African Republic with an ill antelope with MPXV as a putative source (Khodakevich *et al.*, 1985).

5.5   Monkeypox in Humans

Before briefing the monkeypox in humans, we considered it appropriate to detail the history of the first human smallpox case.

On September 01, 1970, a 9-month-old boy, Alfonso Ilombe, who on August 22, 1970, displayed an elevated temperature, was placed in the Basankusu Hospital (Equatorial Province of the former Zaire). In 2 days, he developed rash, which later acquired a centrifugal distribution pattern, typical of smallpox, and a hemorrhagic character. A severe disease course was accompanied by otitis, mastoiditis, and, lymphadenitis of cervical glands, which, as was found later (Ladnyi *et al.*, 1972), was characteristic of monkeypox. The initial clinical diagnosis was smallpox. According to the current practice, the material of skin lesions was conveyed to the WHO Collaborating Center on Smallpox and Related Infections in Moscow, Russia, for laboratory examination. Due to emergency, as the suspected smallpox case was recorded on the territory considered free of smallpox transmission, the chick embryos infected with the material from this patient were opened after 48 h. According to examination of lesions on CAMs, which at that time were indistinguishable from the lesions caused by variola virus, and serological identification of the isolate by microprecipitation in agar gel, the pathogen was determined as variola virus. However, by 72 h, the character of CAM lesions changed—hemorrhages appeared in the centers of pocks, which never had been observed in the case of variola virus. In addition, unlike VARV, the isolate displayed a high hemagglutinating activity. These facts suggested another nature of the pathogen in question, and the virus was comprehensively studied and reisolated from the clinical material. Eventually, it was demonstrated that the actual pathogen of the child's disease was monkeypox virus. We reported this discovery at WHO on October 05, 1970.

As was mentioned, this data dissolved the concern of WHO about the situation in question and allowed the four, similarly obscure cases recorded in September–October in Liberia to be retrospectively diagnosed correctly as monkeypox. Initial examination of these four cases in the WHO Collaborating Center in Atlanta, GA, USA, attributed them to smallpox. Upon the report on detection of human monkeypox in Zaire, the isolates of viruses and clinical material of the four cases in Liberia were studied again; it was demonstrated that the disease was caused by MPXV (Foster *et al.*, 1972; Lourie *et al.*, 1972).

5.5.1   Clinical Course

The duration of incubation period of human monkeypox can be estimated only in scarce cases, when the time of a single contact between a human case and an animal is known precisely. Investigation on three such cases by Jezek and Fenner (1988) gave durations of 10 and 14 days. Analysis of the data of human monkeypox outbreak in Kasai Oriental Province demonstrated that the period from contact with a case to beginning of the disease varied in a wider range—from 7 to 21 days (Weekly Epidemiological Record, 1997a).

As a rule, the disease begins with elevation of the body temperature, weakness, malaise, headache, and sometimes throat ache and cough. The specific feature of human monkeypox clinical course, distinguishing it from smallpox, is lymphadenitides, observed in 86.4% of the patients. Lymphadenitides appear as a rule concurrently with fever, 1–2 days before eruption or, rarer, coinciding with the rash onset. In several cases, lymphadenitides were the reason to seek medical assistance. Localization of lymphadenitides vary (submaxillary, cervical, axillary, or inguinal), sometimes bilateral, sometimes unilateral. Lymphadenitides involve lymph nodes of several regions (for example, neck and armpit) in 11% of the cases, whereas the most frequent situation (63.9%) is a generalized lymphadenopathy (Jezek & Fenner, 1988). Rash appears 2–4 days after the onset of the disease, covering the entire body with a predominant involvement of the extremities, including palms and soles. Similarly to smallpox, rash is found on mucosae and the tongue. The lesions are monomorphic and follow the same pattern— from papules through vesicles and pustules to scabs. Once fallen off, the scabs leave depigmented areas on the skin, where distinct but not deep scars or hyperpigmented spots develop further (Foster *et al.*, 1972; Ladnyi *et al.*, 1972; Jezek *et al.*, 1983; etc.). The disease usually continues for 2–4 weeks depending on the clinical course and several other factors.

The overall clinical course of human monkeypox resembles the smallpox pattern that was predominant on the African continent, most frequently close to discrete smallpox type (Figure 5.8). The ratios of various human monkeypox types according to the data on morbidity in Democratic Republic of the Congo during 1981–1986 are listed in Table 5.7.

In addition to these main human monkeypox types, several other were recorded. For example, one case of monkeypox in a child followed the

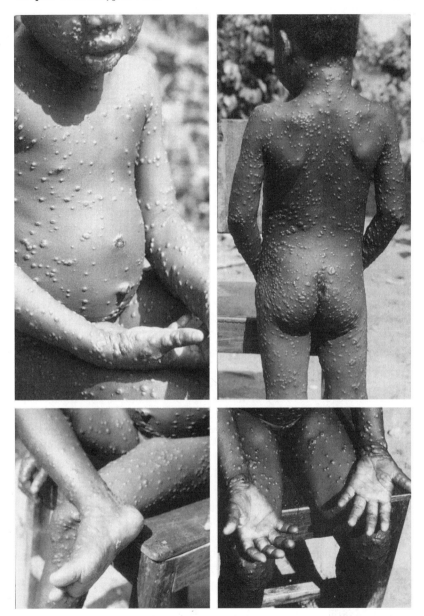

Figure 5-8. Human monkeypox (photos by M. Szczeniowski).

pattern of *purpura variolosa* and had a fatal outcome (Zhukova, 1993). Cases of a very mild monkeypox course resembling varioloid with singular

Table 5-7. Clinical types of human monkeypox

Disease types	In patients	
	Without vaccination scars	With vaccination scars
Discrete	172 (58.3%)	32 (74.4%)
Semiconfluent	93 (31.5%)	10 (23.2%)
Confluent	30 (10.2%)	1 (2.4%)
Total	295 (100%)	43 (100%)

Note: according to Jezek and Fenner (1988).

skin lesions were also observed. A human monkeypox subclinical course, described for smallpox (Heiner *et al.,* 1971), was also reported. This monkeypox type is detected according to the presence of antibodies in the contact cases. According to the serological examination performed by the Moscow and Atlanta WHO Collaborating Centers, of the 734 nonvaccinees who contacted with monkeypox cases, 69 (9.4%) developed a clinically apparent disease, while 27 (3.7%) had a subclinical disease type (Jezek *et al.,* 1986; Jezek & Fenner, 1988). One case of inborn monkeypox of the baby born by a mother who had monkeypox not long before the delivery was reported. Similar situation were repeatedly described for smallpox. Among the monkeypox complications, secondary skin bacterial infections, bronchopneumonia, diarrhea and the related dehydration, and keratites are most frequent. The last complication sometimes ends in blindness or a partial lost of sight.

The case-fatality rate of monkeypox determined under a specialized WHO project (1981–1986), covering over 300 human cases, amounted to 9.8%. However, the data obtained during a large outbreak in Kasai Oriental Province (DRC, February 1996–February 1997) indicated a lower fatality rate, amounting to 3.3–3.7% (Weekly Epidemiological Record, 1997a; Hutin *et al.,* 2001).

Vaccination against smallpox prevents from contracting monkeypox. However, the protection is not absolute, as in the case of smallpox. Analysis by Jezek and Fenner (1988) demonstrated that of the 155 human monkeypox cases, 7% had vaccination scars, while the rest had never been vaccinated. Among the monkeypox cases with old or dubious vaccination scars, 27.3% had a severe disease course, whereas those without vaccination scars developed a severe disease in 61.6% of the cases. The monkeypox in vaccinees never ended fatally. During clinical diagnostics, monkeypox is frequently confused with chickenpox, measles, and syphilis. It is most difficult to differentiate between human monkeypox and chickenpox (Jezek *et al.,* 1988). These difficulties are aggravated by the fact that both infections frequently occur concurrently (Hutin *et al.,* 2001; Meyer *et al.,* 2002b). Before smallpox eradication, monkeypox, a presumably long existing infection, was taken for the former disease.

5.5.2 Epidemiology

Distribution area and abundance. The geographical zone where human monkeypox cases are recorded covers the equatorial part of Central and West Africa, virtually coinciding with the zone of tropical rainforests (Figure 5.9). So far, the reports on human monkeypox cases have come from nine countries of this region—Democratic Republic of the Congo, People's Republic of the Congo, Central African Republic, Gabon, Cameroon, Nigeria, Ivory Coast, Liberia, and Sierra Leone. In addition, an imported outbreak of this infection was recorded in the USA in 2003, which will be detailed below. During the 27 years since the discovery of human monkeypox (1970–October 1997), 917 cases of these disease were reported. Over a half of the human monkeypox cases were those who contracted the disease in 1996–1997 during a large-scale outbreak, unexampled for this infection, in two regions of Kasai Oriental, DRC (Figure 5.10). This outbreak will be described in more detail below. As is evident, the main source of monkeypox is Democratic Republic of the Congo, where 97.8% of the human cases were recorded. Presumably, the fact that the main part of this area on the African continent is covered with tropical rainforests explains this situation. In addition, the WHO project on studying the human monkeypox and provided by specialized epidemiological surveillance teams, laboratory support, and sanitary propaganda, including the bonuses for

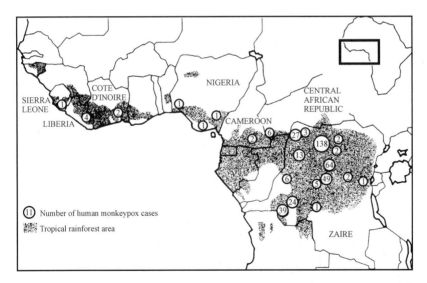

Figure 5-9. West and Central Africa: localizations of human monkeypox cases recorded during 1970–1991 and the zone of tropical rainforests (Courtesy of WHO; Reports of the Moscow WHO Collaborating Center on Smallpox and Related Infections, 1987–1990; 1991).

Figure 5-10. Human smallpox outbreak in Democratic Republic of the Congo in February 1996–October 1997 (Weekly Epidemiological Record, 1997b; 1997c): number of cases is circled. Bokenda is the place where the first human monkeypox case was discovered. In January–February 1997, cases suspicious for monkeypox were recorded in Kinshasa; however, their precise number is unknown.

detecting human monkeypox cases, etc., was performed on the territory of this country.

Note that the recent civil disorders in this country prevent obtaining precise data on the morbidity and other information necessary for understanding the monkeypox epidemiological situation. However, it is known that the human monkeypox outbreak in DRC continued after 1997 (Hutin *et al.*, 2001; Meyer *et al.*, 2002b).

The majority of humans that contracted monkeypox were inhabitants of remote small villages located in dense tropical rainforests. A specific feature of their way of life is frequent contacts with the wild animals of the surrounding forest, as they represent the main animal protein source of their ration. Contacts with animals preceded 72% of the human monkeypox cases. This data in combination with the fact that the pathogen in question was first discovered in animals (monkeys) suggested that monkeypox is a zoonosis.

Season, age, and sex patterns. According to the available data on human monkeypox, the disease takes place round the year with a considerable variation from year to year. According to WHO data obtained over a 6-year observation (1981–1986), 35% of the cases occurred during June–August, coinciding with the activities outside the house (farm works and hunting). In this connection, of interest are the data collected during a large outbreak in Katako-Kombe zone of DRC (Weekly Epidemiological Record, 1997a) and later observations of Hutin *et al.* (2001) that demonstrated seasonal differences in human monkeypox morbidity with a peak in July–August (Figure 5.11).

Monkeypox may affect any age cohorts—the disease was recorded starting from babies of 3 month to elderly persons of 69 years. However, according to WHO data, those younger than 15 years constitute 92.5% of all monkeypox cases, and a half of them are younger than 5 years.

However, this age pattern reflects the situation when the anti-smallpox population immunity was high enough, as it was maintained by repeated vaccinations performed during the campaign on smallpox eradication. The immunization against smallpox in Central and West Africa was officially continued until 1980; however, it was further continued for some period in certain regions. A complete cessation of the vaccination and the resulting decrease in the collective immunity influenced the susceptibility to monkeypox in general and, in particular, the characteristic considered.

No considerable distinctions with reference to sex cohorts were observed.

Sources of primary infection. So far, at least several wild mammalian species may be regarded as the primary sources of MPXV infection.

In the three episodes described below, sufficiently heavy epidemiological evidence was obtained, although the animals could not be examined virologically. Mutombo *et al.* (1983) described a case when a 6-month-old girl was infected. The baby was left by her mother at the edge of field near the forest, where the girl was seized by a chimpanzee and hold by the animal for 40 min. When the animal left the child, a surface wound was found on the girl's

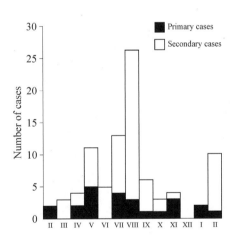

Figure 5-11. Distribution by month of human monkeypox cases during the outbreak in Katako-Kombe, Kasai Oriental Province, February 1996–February 1997 (Weekly Epidemiological Record, 1997b).

body. As for the chimpanzee, it looked uncommonly—was as if powdered with ash, suggesting presumably the stage of scab fall-off. Six days later, the girl developed a fever followed by characteristic eruptions. The diagnosis of monkeypox was confirmed by MPXV isolation.

The second episode was recorded during the monkeypox outbreak among the Pigmy of Central African Republic in the early 1984. Investigation of the origin of this outbreak revealed that shortly before, the Pygmy killed and ate a monkey and a gazelle with pox-like lesions (Khodakevich *et al.,* 1985). According to our observations, the method used for meat processing in this case provided the possibility of infection for those involved in processing and the surrounding people. In particular, they squeezed out the intestine contents by hands and ate a partially raw meat, while children played with the offal of the carcasses (Marennikova *et al.,* 1985).

The third episode occurred in Bumba, the most monkeypox-affected region of DRC. A 6-year-old girl found an apparently ill squirrel (dull and stiff) and carried it to her village for about half an hour. The child developed monkeypox approximately 10 days later. None monkeypox cases were recorded in either her family or the village (Marennikova *et al.,* 1985). As was mentioned above, it was later discovered that tropical squirrels were MPXV carriers and, presumably, important source of human monkeypox infection due to a number of reasons (Khodakevich, 1990).

Gambian rats (*Cricetomys emini*) remained a previously unknown source of human monkeypox until 2003, when this species appeared the cause of an imported outbreak in the USA (see below). The same is true for another rodent species—prairie dogs (*Cynomys* sp.), inhabiting several US states. The prairie dogs, which got infected presumably during joint keeping with imported Gambian rats at pet shops, also became major transmitters of monkeypox to humans (CDC, 2003). Infected humans may also be contagious to other people.

Person-to-person transmission of monkeypox. Before 1981, monkeypox was considered a zoonosis incapable of being transmitted in a person-to-person manner. However, later data demonstrated that the disease contracted by two or more people could not be regarded as a result of infection from the same natural source. All these cases contacted a monkeypox patient and developed the disease in question after a period equal to the incubation period characteristic of monkeypox. As a rule, these were contacts within the family; sometimes, in the hospital or under other circumstances. Such (secondary) cases of infection are designated as the first generation of infection. Nothing was known about the possibility of further person-to-person transmission of monkeypox until 1982. However, second, third, and even fourth generations of human monkeypox cases were later reported. The consolidated data on the cases with respect to the generation number are listed in Table 5.8.

Table 5-8. Morbidity rate of human monkeypox during 1981–1986: sources of infection and infection generations*

Year	Number of cases infected from						Total number of cases
	Animals		Humans with respect to generation				
	Primary	Co-primary**	First	Second	Third	Fourth	
1981	6	–	1	–	–	–	7
1982	22	2	13	3	–	–	40
1983	47	11	19	3	3	1	84
1984	52	10	18	6	–	–	86
1985	40	7	11	4	–	–	62
1986	36	12	7	3	1	–	59

Notes: *According to Jezek and Fenner (1988); **Infected from the same source as the primary cases.

According to WHO data, the number of monkeypox cases resulting from person-to-person transmission constituted 29.6% of the total number of recorded cases by the end of the monkeypox project (1986). Ten years later, this ratio changed drastically. During the outbreak in Kasai Oriental Province, mentioned above, the person-to-person morbidity rate increased 2.5-fold, reaching 73.0% (Weekly Epidemiological Record, 1997a). The number of infection generations in the person-to-person transmission chain increased to eight. Presumably, these changes were provoked by a constantly decreasing level of population immunity to orthopox diseases, connected with the end of massive vaccination in this country at the early 1980s.

Monkeypox beyond its endemic focus. Until 2003, monkeypox was known in Europe and the USA only as a disease sometimes affecting colonized monkeys or the monkeys kept in zoos. The outbreak of human monkeypox in the USA was the first case of this rare zoonosis exported beyond its natural geographic area (the zone of tropical rainforests in Central and West Africa). The most probable source of MPXV was a batch of Gambian rats, which together with 800 other small mammals were imported into the USA from Ghana in April 2003. A part of the rats appeared ill already soon upon arrival and infected the prairie dogs, a rodent species inhabiting several regions of the USA, that were kept together with in pet shops. Later, MPXV was detected in several African rodents—two rope squirrels (*Funiscuirus* spp.) one Gambian rat (*Cricetomys* sp.) and three dormice (*Graphiurus* spp.; Guarner *et al.*, 2004). Since May 15, 2003, human disease cases characterized by fever either preceding or concurrent with skin rash were recorded in various US states; as it was further found out, the patients contacted with prairie dogs or Gambian rats. The rash commenced from papules and evolved following the conventional pattern typical of pox infections (papules through vesicles and pustules to scabs). In the majority of cases, the patients developed lymphadenopathy and displayed throat ache, a set of respiratory syndromes, etc. (CDC, 2003).

Overall, 79 such cases were detected in six US states. Of them, 73 cases were confirmed with monkeypox using various laboratory methods detecting MPXV and/or antibodies to orthopoxviruses. MPXV was also detected in prairie dogs, which were the main infection transmitters, and in Gambian rats. The clinical pattern of human monkeypox cases during this outbreak was virtually the same as described for the African outbreaks. No fatal cases were recorded; although two children's cases developed complications (one, with encephalitis). Six human monkeypox cases were asymptomatic. It was proved that low virulent West African MPXV subtype caused this outbreak (Reed *et al.,* 2004). During this outbreak, no person-to-person transmission was detected. Note that the fact of infection importation and its nature were detected with a certain delay, and this prevents a complete exclusion of the possibility that the virus might enter the population of local rodents.

5.5.3   Prevention

The mean for specific monkeypox prevention among humans is immunization with anti-smallpox vaccine (vaccination or revaccination). However, the vaccination causes a serious concern due to a potential side effects (postvaccination complications), especially in the immunocompromised population cohorts, for example, HIV-infected persons. Nonetheless, the large scale of the human monkeypox in 1996–1997 with a higher (compared with the pervious data) transmission rate makes the issue of vaccination appropriateness the topic of serious consideration. Evidently, while discussing this issue, it is necessary to take into account a particular situation, the possibility of vaccination upon a contact with an infection source, and use of alternative sparing vaccination methods (for example, a two-stage immunization). Recently, promising results of monkeypox prevention using cidofovir (an inhibitor of the viral DNA polymerase) were obtained in an experimental primate model of this infection (Huggins *et al.,* 2002).

As for therapy, the only method for treating complicated monkeypox cases may be, by the analogy with anti-smallpox vaccination, the specific vaccinia immunoglobulin.

The information and education of population in the regions with increased risk of contracting monkeypox also present an important part of the response measures in preventing human monkeypox.

As for the possibility of exportation of this disease beyond its natural focus, the experience gained during the outbreak in the USA demonstrates that it is necessary to provide a more stringent control of the animals imported from the regions of Central and West Africa endemic for monkeypox.

5.6 Genetic Organization of Monkeypox Virus

Previous analyses of the genomic similarity of variola and monkeypox viruses have compared genomic restriction endonuclease maps (Mackett & Archard, 1979; Esposito & Knight, 1985) or nucleotide sequences of individual viral genes (Esposito & Knight, 1984; Douglas & Dumbell, 1992; Mukinda *et al.*, 1997; Hutin *et al.*, 2001). The limited obtained data supported both hypotheses—that MPXV and VARV evolved independently (Douglas & Dumbell, 1992) and that VARV was an ancestor of MPXV (Bugert & Darai, 2000). However, a direct and reliable answer can be obtained only through comparing complete genomes. Therefore, we sequenced the genome of MPXV and analyzed its structure–function organization in comparison with VARV, CPXV, and VACV, the viruses pathogenic for humans.

The MPXV strain ZAI-96-I-16 (MPXV-ZAI), isolated in Democratic Republic of the Congo during an outbreak of human monkeypox in 1996 (Mukinda *et al.*, 1997) was chosen for the study. The 196,858-bp sequence of MPXV-ZAI DNA, comprising the entire genome with the exception of the covalently closed terminal hairpin loops (see below), was determined. The GC content of MPXV-ZAI amounts to 33.1%, similar to the GC content of other orthopoxviruses. Computer-assisted analysis identified 190 potential ORFs containing ≥ 60 amino acid residues (Appendices 1 and 2). The organization of MPXV-ZAI genome was compared with two strains of variola major virus—India-1967 (VAR-IND; Shchelkunov *et al.*, 1993a; 1993b; 1993d; Shchelkunov, 1995) and Bangladesh-1975 (VAR-BSH; Massung *et al.*, 1994)—and variola minor alastrim virus strain Garcia-1966 (VAR-GAR; Shchelkunov *et al.*, 2000). In addition, genomes of CPXV strain GRI-90 (CPXV-GRI; Shchelkunov *et al.*, 1998; see Section 6.5), VACV strain Copenhagen (VACV-COP; Goebel *et al.*, 1990), and genomic regions of other orthopoxviruses were involved in the comparison.

The size of terminal inverted repeats (TIRs) of MPXV-ZAI is 6379 bp, and they include four ORFs (Appendix 1), a set of short tandem repeats, and a terminal hairpin. Using S1 nuclease hydrolysis and DNA polymerase I repair, we succeeded in cloning and determining one sequence of the two incompletely base-paired hairpin loop strands of MPXV-ZAI DNA, whose sequence is conserved within orthopoxviruses (Figure 5.12). The region of tandem repeats adjacent to the terminal hairpin is rather short in the MPXV-ZAI genome and comprises NRI (85 bp) and NRII (322 bp), separated by two 70-bp repeats, one element of 70 bp, and two elements of 54 bp, located between NRII and the TIR coding sequence (Figure 5.13). This region of MPXV-ZAI DNA is most close to VARV-GAR with reference to its organization; however, the TIRs of all the VARV strains studied are very short and lack ORFs, while the sets of repeats at the right and left termini differ (Figure 5.13; Massung *et al.*, 1995).

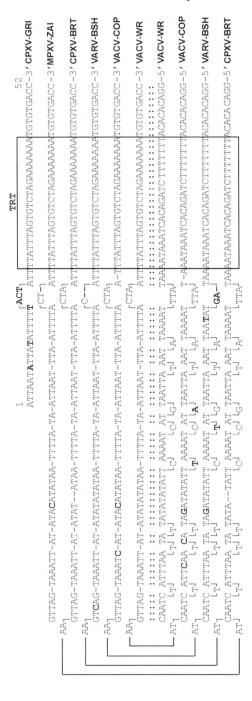

Figure 5-12. Structure of the terminal hairpin (S form) of the orthopoxvirus DNA. The nucleotides differing from the VACV-WR sequence are boldfaced; the two upper lines are the sequenced regions of CPXV-GRI and MPXV-ZAI hairpins; the upper figures indicate the successive number of the sequences; and the sequence of telomere resolution target (TRT) is framed.

The coding sequence of MPXV-ZAI DNA, limited by the leftmost and rightmost ORFs, is 195,118 bp long, whereas the coding sequences of VAR-IND and VAR-GAR are 184,151 and 185,846 bp, respectively. This difference in the genomic DNA sizes between MPXV and VARV results first from the duplication of four ORFs contained in TIRs of MPXV-ZAI (see Appendix 1). Identities of the overlapping DNA sequences of MPXV-ZAI with respect to VARV-IND and VARV-GAR amount to 84.6 and 84.5%, respectively.

The central genomic regions of orthopoxviruses contain mainly the highly conserved essential genes. This region of MPXV-ZAI, which is limited by ORFs C10L and A25R, comprises 101,466 bp with an overall 96.3% identity to the corresponding part of the VARV-IND genome. The virion proteins encoded in this region of MPXV-ZAI are 91.7 to 99.2% identical in amino acid sequences to those of VARV-IND. For comparison, the corresponding ORFs of VARV-GAR are 98.7% to 100% identical to VARV-IND. In contrast to the central region, the two ends of the MPXV-ZAI and VARV-IND genomes exhibit considerable variation resulting from deletions (Figure 5.13) and ORF truncations Appendix 2) in one DNA relative to the other. The presence of DNA in VARV that is absent in MPXV and vice versa indicated that neither virus is a direct ancestor of the other.

The variable terminal regions of orthopoxvirus genomes contain the majority of the virulence and host range genes that were identified earlier for VACV and CPXV. The amino acid sequences of the putative virulence and immunomodulatory factors common to MPXV-ZAI and VARV-IND were 83.5 to 93.6% identical versus 97.3 to 100% identity between the corresponding ORFs of VARV-GAR and VARV-IND.

Notably, MPXV genome carries mutations that affect translation of two interferon resistance genes encoding the intracellular proteins that are intact in VARV and other orthopoxviruses (see Section 7.5). Impairments of these

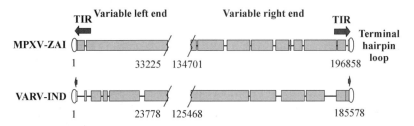

Figure 5-13. Schematic representation of the terminal species-specific variable genomic regions of MPXV-ZAI and VARV-IND. Terminal inverted repeats (TIRs) are designated with arrows; regions of short terminal tandem repeats, with white rectangles; coinciding sequences, with wide gray blocks; deletions in one genome relative to others, as lines. Borders of the variable genomic regions are marked with the numbers of nucleotides corresponding to their positions in the genomes.

genes in MPXV-ZAI genome may result in a decreased virus yield *in vivo* and, consequently, a lower efficiency of virus transmission between humans and sensitive animals via an aerosol route.

Central African MPXV strains encode a form of the complement-binding protein with only three short consensus repeats instead of four found in other orthopoxviruses, while this gene is deleted in West African MPXV strains (see Section 5.2.5). This decreases the MPXV capability for inhibiting the development of immediate inflammatory reactions in the body in response to the virus infection compared with orthopoxviruses studied, thereby reducing the abilities of the virus (especially of the West African variant) to propagate *in vivo*.

On the other hand, MPXV encodes a secreted interleukin-1β binding protein and 3-β-hydroxy-delta-5-steroid dehydrogenase, discovered originally in VACV, whereas the VARV strains lack the intact versions of these ORFs. The species-specific distinctions in the organization of orthopoxvirus immune evasion genes are described in more detail in Section 7.5.

MPXV-ZAI has no unique genes relative to other studied species of orthopoxviruses. On the contrary, MPXV-ZAI is missing 25 ORFs that are found in CPXV-GRI (Shchelkunov *et al.,* 1998) and 19 potential ORFs detected in VARV-IND (Shchelkunov *et al.,* 1993d). While the roles of some of these genes remain entirely unknown, they include the genes that are probably involved in immune evasion, host range, and cell proliferation (see also sections 7.3 and 7.4) Basing on the number of intact and moderately truncated immune evasion genes (indicated in parentheses), we obtain the following order: CPXV-GRI (18) > VARV-IND (11) = VARV–GAR (11) > MPXV-ZAI (10) > VACV-COP (9) > VACV-MVA (6).

Summing up, a comparative analysis of the genetic structures of MPXV-ZAI, VARV, CPXV, and VACV DNAs has demonstrated that MPXV is a discrete species exhibiting multiple distinctions from the other orthopoxviruses pathogenic for humans. This is especially noteworthy as MPXV and VARV cause human diseases that are very similar in their clinical manifestations. A specific set of immunomodulatory proteins encoded by the MPXV genome is predicted that allows the virus to induce a mild generalized infection with a decreased efficiency of person-to-person transmission as compared with smallpox. It seems most likely that MPXV and VARV evolved independently from a cowpox virus-like ancestor that contained a complete set of the orthopoxvirus immune evasion, ankyrin-like, and kelch-like genes (see Section 6.5). While we cannot exclude the possibility that MPXV will undergo certain further adaptation to humans in future bringing forth the necessity of monitoring the infection rates, the virus is unlikely to evolve into VARV. Furthermore, the multiple genetic differences between MPXV and VARV imply that monkeypox may not be a completely valid model of smallpox with respect to understanding the pathogenesis or development of the corresponding antivirals and vaccines.

Chapter 6

COWPOX VIRUS

6.1   Morphology of Virions and General Information on DNA Structure

The virions of cowpox virus (CPXV) possess the brick-like shape, structure, and size typical of orthopoxviruses. No morphological distinctions peculiar to this virus have been found.

The genome of cowpox virus is significantly larger than those of other orthopoxviruses. According to estimates of several authors, its size for various cowpox virus strains ranges from 205 to 229 kbp (Naidoo *et al.*, 1992; Meyer *et al.*, 1999). The restriction maps of DNA molecules of CPXV isolates may differ essentially (Marennikova & Shchelkunov, 1998), which indicates the diversity of this virus. No correlations were found between the genomic distinctions of cowpox virus isolates and the properties of sources wherefrom they were recovered.

6.2   Biological Properties

Cowpox virus (*Orthopoxvirus bovis*) is the first representative of the Vira kingdom that was purposively applied by medical science to the needs of public health. Owing to the discovery of E. Jenner, a remedy for smallpox was found with the aid of this virus and the principle of vaccination became the keystone of the prevention of infectious diseases. The material originally used by Jenner for vaccination was obtained from a skin lesion of a milkmaid infected from a cow with an illness popularly known as cowpox. In his classic work, Jenner (1798) introduced the Latin equivalent of this name, *variolae vaccinae* (from the Latin word *vacca*, cow). Sometimes he used as a synonym the term "vaccine" and called the lesion a "vaccine pustule". In due course, the preparation for inoculation was called smallpox vaccine and later, vaccinia virus. The long, complicated, and rather uncertain history of the derivation of vaccinia virus strains and their use for inoculation led to unexpected results. After development of the method for culturing viruses in chick embryos, which extended the scope for their investigation, Downie (1939) showed that the virus isolated from patients with genuine smallpox or cowpox differed significantly in some important properties from the vaccinia virus used for vaccination. Since that time, cowpox virus and vaccinia virus were considered as separate species of the genus *Orthopoxvirus*. Numerous subsequent studies justified this distinction.

Prior to describing the biology and ecology of cowpox virus, it is necessary to note the following. By now, the strains regarded as cowpox virus have been isolated from many animal species, representing various taxonomic groups (see Chapter 2, Table 2.2). As one could see from the following account, this circumstance manifested itself in a wide intraspecies variation of the properties of this virus.

6.2.1   Behavior in Chick Embryos

Inoculation of large doses (4–6 log PFU/0.1 ml) of cowpox virus on chick embryos CAMs causes the development of semiconfluent or confluent hemorrhagic lesions. At lower doses, discrete lesions develop, which are typical of this virus: superficial grayish pocks 1.2–3 mm in diameter with pronounced hemorrhages (the so-called "red" pocks) (Figures 6.1–6.3). Sometimes, the hemorrhages occupy almost the entire surface of the pock (Figure 6.1). In addition to the red pocks, all strains of cowpox virus form on

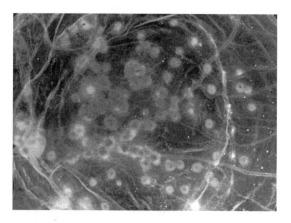

Figure 6-1. Lesions (pocks) on CAM of chick embryos caused by the cowpox virus reference strain Brighton (72 h after inoculation).

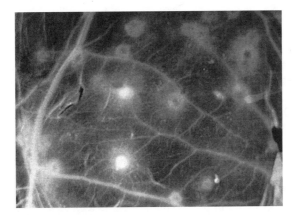

Figure 6-2. White large compact pocks caused by cowpox virus on the background of superficial pocks with hemorrhages (an isolate from an elephant).

CAMs another type of pocks—larger white pocks, slightly rising from the CAM surface (Figure 6.2). Their amount usually does not exceed 1% of the total.

Note that some isolates from animals show only slight hemorrhages during their initial growth in chick embryos. However, further passages made the appearance of pocks typical of cowpox virus (S.S. Marennikova, unpublished data).

Our comparative studies of 30 different cowpox virus strains showed no specific distinctions in the morphology of lesions they produced on CAMs, except for the size of pocks and the rate of their development.

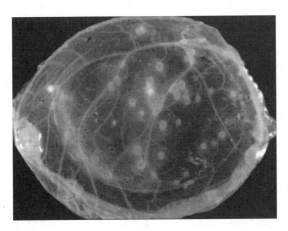

Figure 6-3. Pocks on CAM produced by an isolate from a child.

Among the groups of strains, three were distinguished (including strain Brighton) that produced large mature pocks with bright extensive hemorrhages already at 48 h of incubation. At 72 h, a large number of smaller secondary (daughter) pocks developed. Other strains (the majority of them were recently isolated from humans and animals) produced smaller (1.2–1.5 mm in diameter) subtle pocks, which "matured" only by 72 h (Figure 6.3). In our opinion, these differences should be related to the degree of adaptation of the strains to chick embryos during their passaging in the laboratory. Probably, this is also the reason why the strains of the first group reproduced on CAMs at 40°C, while the rest strains did not produce pocks even at 39.5°C.

After infection on CAMs, all the strains studied caused the death of chick embryos (7 days of observation). The value of LD_{50} for all the strains was less than 1 log PFU/0.1 ml. The Helbert's test (Helbert, 1957)— the amount of virus accumulated in chick embryo liver—showed some differences. For 10 of the 12 strains, the amount of virus in the liver (72 h after inoculation of 5 log PFU/0.1 ml) was 6.9–8.4 log PFU/g. Only for two strains, this parameter was 4.4 and 6.1 log (Shelukhina, 1980). Inoculation into the allantoic cavity (strain Brighton, dose 3 log) caused the death of only a part of the embryos, with the lesions located at the inoculation site.

6.2.2   Pathogenicity for Laboratory Animals

Cowpox virus is pathogenic virtually for all the known laboratory animals.

Rabbits. Endermic inoculation is the most widely used method of infection for these animals, since the character of skin lesions allows for differentiating cowpox virus from vaccinia virus and some other orthopoxviruses. Inoculation into the skin of 5–6 log PFU/0.1 ml of the virus causes the development of a papule, which transforms by day 4 into a compact dark-red infiltrate 26–32 mm in diameter, sharply raising from the surface of the skin. Later, a necrotic zone and hemorrhage appear in the center of the infiltrate, and the adjacent tissue becomes swollen. Involution of the process starts commonly 6–7 days later. Some animals show a generalized infection: after 7–8 days, the temperature elevates, and a pink-red papular rash appears on the body, most noticeable on hairless areas of the skin (the internal surface of ears, etc.). Generally, the rabbits recover from the disease. Inoculation of a similar virus dose on a scarified skin (usually, on an area of 5 cm^2) induces lesions in the form of a massive papular–pustular hemorrhagic rash on a firm and swollen base (Figure 6.4). Similar to endermic inoculation, several rabbits develop a generalized infection.

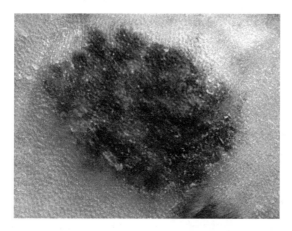

Figure 6-4. Reaction on scarified skin of a rabbit; day 4 post
inoculation with cowpox virus.

Intracerebral infection of rabbits commonly results in fatal outcome from
the specific infection; with large doses of virus, in 48–72 h at a rate of 100%.
Gendon and Chernos (1964) reported that a dose of 4.2 log TCD_{50} caused the
death of a part of infected animals.

Inoculation into the testis of rabbits caused orchitis with generalization of
infection leading to the death of animals.

Shelukhina (1980) tested the susceptibility of 10-day-old rabbits to
cowpox virus. It was shown that intranasal inoculation caused a severe
infection with lesions developing mainly in the upper and lower respiratory
tract (rhinitis, bronchitis, and pneumonia) and the eyes (conjunctivitis and
blepharitis). The disease ended with the death of animals. No differences
were found in the pathogenicity of three cowpox virus strains for baby
rabbits. The value of LD_{50} ranged from 2.6 to 2.9 log PFU/0.03 ml. Using
oral infection of 12-day-old rabbits with the virus isolated from a large
gerbil, Shelukhina induced an acute lethal disease in the animals, with a rash
developed on the nose and lips of one rabbit. These studies demonstrated also
the possibility of a natural transmission of the infection from sick baby
rabbits to healthy ones kept together with the mother in the same cage. The
mother rabbit showed no signs of infection; however, three weeks later, antihemagglutinin (AHA) and virus-neutralizing antibodies were detected
(at titers of 1 : 40 and 1 : 160, respectively).

White mice. Although it is possible to cause the experimental infection of
white mice with cowpox virus independently of inoculation route,
two factors influence significantly the susceptibility of these animals,
namely, their age and specific features of the strain under study. These
factors are least important for inoculation into the paw pad, although even

in this case young animals accumulate the virus more intensely (Subrahmanyan, 1968).

Lethal infection with a high viremia was observed in DBA/2 mice after intranasal inoculation of 6 log PFU/dose of cowpox virus (Reith & Williamson, 1996). Huggins *et al.* (1998), Smee *et al.* (2000 a, b), and Ferrier *et al.* (2002) studied this model more thoroughly using BALB/C mice. Inoculation with high (10^5–10^6 PFU) doses was shown to result in the development of severe bronchopneumonia; animals stopped growing, lost weight sharply, and died on day 8 after infection. The virus was detected in the blood and internal organs, with the highest titer in the lungs (10^6–10^7 PFU/g by day 4 after inoculation). Similar results were obtained with aerosol infection (Bray *et al.*, 2000; 2002). Both intranasal and aerosol infections provided an experimental model of cowpox with predominant involvement of the respiratory tract. Note that a somewhat similar pattern was observed during an outbreak of cowpox in the Moscow Zoo among some feline predators and, in some cases, white rats (see Section 6.3). The above models were used to estimate the prevention and therapeutic efficacies of some promising antivirals, in particular, the DNA polymerase inhibitor cidofovir and ribavirin (Huggins *et al.*, 1998; Bray *et al.*, 2000, 2002). Intravenous inoculation of large doses of the virus caused generalized infection of adult mice and the death of a part of infected animals (Mims, 1968; etc.). Some authors used intracerebral route to infect mice, which allowed them to find distinctions in pathogenicity of individual strains for these animals (Fenner, 1958; Baxby, 1975; Shelukhina, 1980). According to the data obtained by Shelukhina in our laboratory, white mice with a weight of 10 g are most suitable for this purpose. Intracerebral inoculation of such mice showed a much higher pathogenicity for rodents of the isolates recovered from white rats and feline predators. Among other strains studied, most strains exhibited moderate or low pathogenicity for these animals. A similar concurrent study with 10-day-old white mice failed in revealing these distinctions because of an extraordinary sensitivity of mice of this age: the LD_{50} value was less than 1 log PFU/0.03 ml for all strains.

White rats. Sensitivity of white rats to intranasal infection with cowpox virus varies in a wide range, depending on the source wherefrom a strain was isolated. Among the strains of cowpox virus studied in our laboratory, the strains with high pathogenicity for white rats were isolated during an epizootic among white rats in a breeding facility and in an outbreak among feline predators in the zoo (see below, Section 6.2). However, the strains isolated from human cases, cows, and, especially, wild Turkmenian rodents, showed a low degree of pathogenicity—the log LD_{50} (in PFU/dose) for 5–6-week-old white rats varied from 6.3 to 7.1. Even more significant were the distinctions between strains when infecting intranasally 3–5-day-old white rats. The first group of strains, when inoculated at a dose of 5.4 log

PFU/0.02 ml, caused a severe generalized infection with 100% fatal outcome after 11–13 days. The same dose of the virus isolated from Turkmenian rodents caused clinically apparent infection and death of only one of nine infected baby rats. The strain Brighton, even at a larger dose (6.1 log PFU), was virtually apathogenic: none of the 12 animals died during 17 days of observation (Shelukhina, 1980).

Cotton rats (*Sigmodon hispidus*) appeared more susceptible to cowpox virus. Probably this was the reason why the strain distinctions upon intranasal infection were less pronounced in cotton rats as compared to white rats. Clinical picture of the disease was similar to the pattern observed with white rats: adynamia, severe dyspnea, blepharitis, and purulent discharge from the nose and eyes. In sick and dead animals, the virus was detected in the liver, kidneys, spleen, and in most amounts, in the lungs (Shelukhina, 1980).

Monkeys. Gendon and Chernos (1964) studied the susceptibility of rhesus macaques (*M. mulatta*) using intradermal (at a dose of 4 log $TCD_{50}/0.1$ ml) and intracerebral inoculation (at a dose of 5 log $TCD_{50}/1.0$ ml). In the former case, the animals displayed extensive lesions at the site of inoculation; in the latter, one of the four infected monkeys died.

6.2.3 Growth in Cell Culture

Numerous primary and continuous cell cultures are sensitive to cowpox virus. All the conventionally used in laboratories continuous cell lines (Vero, A-1, Hep-2, RK-13, PEK, etc.) and primary cultures (human embryo fibroblasts, monkey kidneys, chick embryo fibroblasts, etc.) are suitable for cultivation and study of the virus. However, some data indicate distinctions in the sensitivity of individual cell cultures. According to observations made by Baxby and Rondle (1967), cells of RK-13 line are more sensitive to this virus compared with chick embryo fibroblasts. Gurvich (1964) estimated the sensitivity of nine different cultures infected with a standard dose of the virus and found the primary culture of monkey kidneys to be the most sensitive; primary culture of human embryo cells and continuous culture of human amnion, highly sensitive cell lines; and HeLa cell culture, the least sensitive. The pattern of CPE caused by cowpox virus depends to a certain degree on the type of cell culture and the infective dose. Moderate doses of the virus at the initial stage cause CPE of a focal type; according to Mal'tseva (1980), proliferative foci are observed along with destruction foci in PEK cell culture. Specific features of the CPE caused by cowpox virus are the presence of a large number of well-defined small rounded cells scattered over the layer (Figure 6.5) and a relatively high rate of CPE development (it takes less than 120 h from the emergence of the first foci to the total degeneration of the layer). At 48–72 h post infection, the CPE of cowpox

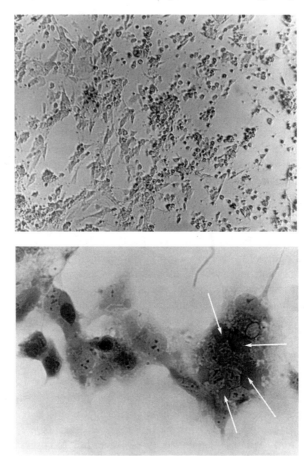

Figure 6-5. The CPE in cell culture of human embryo fibroblasts caused by cowpox virus (upper) and A-type inclusion bodies in syncytium cytoplasm of the same cell culture, ×500 (below).

virus becomes undistinguishable from that caused by vaccinia virus (development of large syncytia, emergence of gaps and holes in the monolayer, and its further degeneration; Marennikova *et al.*, 1964). Some specific features of the CPE foci in the PEK cell culture infected with cowpox virus revealed by immunofluorescent staining are shown in Figure 6.6. Note separate groups of fluorescent cells localized to the center of the focus, comprising 30–50 cells, and several layers of closely arranged rounded fluorescent cells in the periphery and compare it with the CPE pattern of the same culture infected with an equal dose of vaccinia virus.

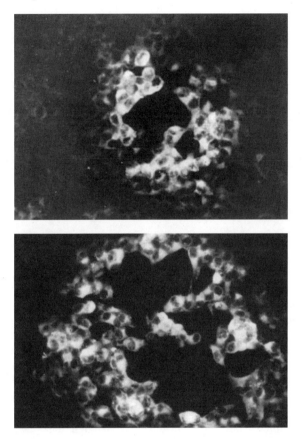

Figure 6-6. Specific features of CPE foci in the PEK cell culture infected with cowpox (upper) and vaccinia (below) viruses. The preparations were treated with FITC conjugate of anti-vaccinia antibodies (24 h post inoculation).

Similar to other orthopoxviruses, the cell cultures infected with cowpox virus clearly display the phenomenon of hemagglutination (Gurvich, 1964; Mal'tseva, 1980).

In monolayer cell cultures, cowpox virus forms species-specific netlike plaques with aggregates of degenerated cells, intensively adsorbing stain, on the background of lysis. According to Mal'tseva (1980), who studied the plaque development in Vero cell culture, the plaques are 1.5–1.8 mm i n diameter. Usually, they emerge already 24 h after infection. The use of agar overlay retards the development of plaques (Gurvich, 1964). A comparative study of 12 cowpox virus strains isolated in Great Britain, the Netherlands, Poland, Germany, Russia, and Turkmenistan did not reveal any significant differences in their behavior in cell cultures (Shelukhina, 1980).

Incubation of cell cultures infected with cowpox virus at 39.5 and 40°C either suppressed the virus replication or decreased its accumulation in cells depending on the strain studied (Mal'tseva, 1980; Gurvich & Marennikova, 1964; Gendon & Chernos, 1964). According to Mal'tseva, only strain Brighton retained its ability to cause CPE, hemadsorption, and plaque formation under these conditions. A decrease in the number and size of plaques already at 39°C were reported by Porterfield and Allison (1960) and later, by Gurvich and Marennikova (1964).

6.2.4   Intracellular Inclusions

Cowpox virus forms two types of inclusions in infected cells—small bodies described by Guarnieri (1892), or B-type inclusions, according to the classification of Kato *et al.* (1959), where the virus replicates, and A-type eosinophilic inclusions. The former type of inclusions is characteristic of all the orthopoxvirus infections, while A-type inclusions were found during cowpox (Downie, 1939, etc.), mousepox (Marchal, 1930), and raccoonpox (Patel *et al.*, 1986). A-type inclusions of cowpox virus are composed of a late viral protein with a molecular weight of 160 kDa.

Meyer *et al.* (1994) studied the structure of the gene responsible for formation of A-type inclusions and discovered deletions in 13 of the 22 strains studied. However, these deletions did not influence the ability of the virus to form inclusions.

Electron microscopy of inclusions demonstrated that various cowpox virus strains cause formation of three A-type inclusions differing in their structure: inclusions containing numerous mature virions, V^+; homogeneous inclusions without virions, V^- (Ichihashi & Matsumoto, 1968; Ichihashi *et al.*, 1971; etc.) and inclusions with the virions localized to their periphery V^i (Baxby, 1975). Among the cowpox virus strains that we studied with reference to this feature, there were strains that formed V^+, V^i, or V^- A-type inclusions (Figure 6.7); and 6 out of the 11 strains examined formed V^- inclusions (see Table 6.1). Note that in similar studies by Baxby (1975), the majority of strains isolated from sick cows and humans infected by them formed V^+ inclusions. This discrepancy may depend in a way on the source of infection, as in our case the majority of strains were isolated from rodents and humans infected directly by them. The mechanisms involved in formation of a certain variant of A-type inclusions yet remain vague. However, it is known that incorporation of virions in V^+ inclusions of type A requires the presence of the virus-encoded factor V0, which is located on the virion surface (Shida *et al.*, 1977; Ichihashi, 1990). According to McKelvey

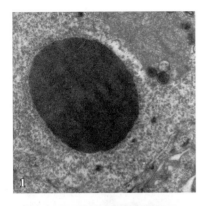

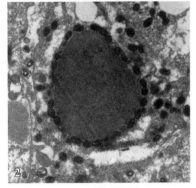

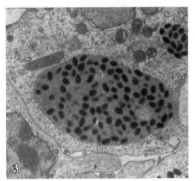

Figure 6-7. Type A cytoplasmic inclusion bodies formed by various cowpox virus strains: (1) V⁻ formed by strain Brighton; (2) V^i formed by strain El-87; and (3) V⁺ formed by GRI-90 strain. Courtesy of E.I. Ryabchikova and V.V. Streltsov (×32.000).

Table 6-1. Biological characterization of cowpox virus strains by markers detecting intraspecies distinctions

No.	Name	Source of infection; place and year of strain isolation	Pathogenicity, log LD$_{50}$ (PFU/dose)[a]		Limiting temperature of pock development on CAM (°C)	Morphological properties of pock development on CAM				
			White mice (intracerebral inoculation)	White rats (intranasal inoculation)		Character of type A inclusions	Production of virions		Infection of meso-dermmal stroma[e]	Hemo-rrhages[f]
							Intracel-lular[c]	Extra-cellular[d]		
1	E-77	Peasant woman, 19 years old (Poland, 1977)	4.0	6.3 (0.5)	39	NS	NS	NS	NS	NS
2	El-87	Vivarium worker bitten by white rat (Moscow, Russia, 1987)	3.2 (0.8)	6.1 (1.0)	39	V^i	++	+++	NS	++

continued

No.	Name	Source of infection; place and year of strain isolation	Pathogenicity, log LD$_{50}$ (PFU/dose)[a]		Limiting temperature of pock development on CAM (°C)	Morphological properties of pock development on CAM					
			White mice (intracerebral inoculation)	White rats (intranasal inoculation)		Character of type A inclusions	Production of virions		Infection of mesodermal stroma[e]	Hemorrhages[f]
							Intracellular[c]	Extracellular[d]		
3	Argo-88	Child, 10 years old, unknown source of infection (Smolensk oblast, Russia, 1988)	2.7 (1.3)	5.8 (1.3)	39	NS	NS	NS	NS	NS
4	GRI-90	Child, 4.5 years old, unspecified source of infection (a Moscow suburb, Russia, 1990)	3.0 (1.0)	5.9 (1.2)	39	V$^+$	++	++	+++	+++
5	Rev-91	Child, 12 years old, after the contact with white rat (Moscow, Russia, 1991)	3.1 (0.9)	5.9 (1.2)	39	V$^-$	+	+	+	+
6	DK-91	Child, 10 years old, after the contact with white rat (Moscow, Russia, 1991)	3.2 (0.8)	6.0 (1.1)	39	NS		NS	NS	NS
7	M-012-73	Sick white rat during an epizootic (a Moscow suburb, Russia, 1973)	3.1 (0.9)	3.3 (3.5)	39	V^i		NS	NS	NS
8	BK-87	Apparently healthy white rat (Moscow, Russia, 1987)	3.0 (1.0)	NS	39	V$^-$	++	++	+	NS

9	WR-R-91	Apparently healthy white rat that infected a child (No. 5; Moscow, Russia, 1991)	2.9 (1.1)	NS	39	V^-	+	+	+	+
10	Turk-74	Wild gerbil (Turkmenistan, 1974)	3.8 (0.2)	7.2 (+0.4)	39[b]	V^-	++	++	+	++
11	Scarboro	Child, 6 years old, unknown source of infection (UK, 1975)	4.9 (+0.9)	NS	39	V^-	++	+++	+++	+++
12	Brighton	Man (UK, 1937)	4.0	6.8–7.1	40	V^-	+	+++	+++	+++
13	Altoff	Sick elephant (Germany, 1971)	5.3 (+1.3)	NS	39	V^i	++	+	+	+++
14	Ham-85	Child, unknown source of infection (Germany, 1985)	NS	6.2 (0.9)	NS	V^+	+	+	+	++

Notes:

[a]The difference between pathogenicity values for cowpox virus reference strain Brighton and a given strain in concurrent studies is given in parentheses the "+" sign in parentheses indicates that pathogenicity of the strain is lower than that of the reference strain by the value specified.

[b]The temperature during isolation at initial passages on CAMs was 38.6°C.

[c]For intracellular virions: + means less than 10 virions in the cell section; ++, 10–50 virions.

[d]For extracellular virions: + indicates solitary extracellular virions between cells; ++, dozens of analogous virions between cells; +++, the intracellular space is filled with the virions.

[e]+ Indicates that infected fibroblasts are observed only near the chorion epithelium; ++, infected fibroblasts and rare extracellular virions are observed to the middle part of the mesoderm; +++, infected fibroblasts and numerous extracellular virions are observed throughout the mesoderm.

[f]+ indicates solitary erythrocytes; ++, 10–30 erythrocytes; +++, pronounced hemorrhages and hemostases.

NS, not studied.

et al. (1992), this factor is a virion protein with a molecular weight of 58 kDa, whose gene was identified by these authors.

6.2.5   Cowpox Virus Antigens

Gispen (1955) was the first to point at specific features of cowpox virus soluble antigens. Using double gel diffusion test, he found that unlike vaccinia virus, cowpox virus failed to form one of the two main precipitation zones. Primarily, this was attributed to the absence of the corresponding soluble antigen in cowpox virus. However, further studies by Rondle and Dumbell (1962) showed this antigen (LS) to exist in cowpox virus, but in a non-diffusive form. This conclusion was supported by two facts—the presence of antibodies to this antigen in anti-cowpox serum and the occurrence of precipitation effect after treatment with trypsin. This protein of cowpox virus was found to be large (160 kDa) and responsible for the formation of A-type inclusions. However, irrespective of the antigen form, the absence of the first main precipitation zone in double diffusion test remains an important criterion for serological identification of cowpox virus isolates (see Chapter 8).

In particular, this was demonstrated for viral isolates from unusual hosts, such as feline predators, wild Turkmenian rodents, white rats, and others (Marennikova *et al.*, 1977b; Shelukhina, 1980; Zhukova *et al.*, 1992). The antigenic identity of these isolates with cowpox virus was most evident in double gel diffusion test (Maltseva & Marennikova, 1976), extending the possibilities for comparative studies of soluble orthopoxvirus antigens.

Antibodies to the 160 kDa protein react with proteins of the A type inclusions formed by other orthopoxviruses (mousepox and raccoonpox) and also with the LS antigen of vaccinia virus, a 94 kDa protein (Patel *et al.*, 1986).

Hemagglutinin is another antigen important for differentiation and study of cowpox virus. Unlike monkeypox, mousepox, and vaccinia viruses, cowpox virus displays a weak hemagglutination activity. Our comparative study of 23 cowpox virus strains isolated during 1937–1991 showed that the hemagglutination activity of CPXV-infected CAM homogenates had titers in the range of 4–8 in 82% of cases. The rest strains had titers of 16–32, with the highest titer observed for only one of the 23 strains. Nevertheless, the anti-hemagglutination activity of the sera of experimental animals obtained to this virus is rather high.

6.2.6   Intraspecies Variability

Comparative study of the properties of cowpox virus isolates recovered from various carriers at different times in many geographic zones demonstrated a wide intraspecies variation of the virus, which we noted when describing its biological properties. Baxby (1975) studied 18 viral strains isolated from humans and cows during 1937–1971 and detected differences between the strains in several phenotypic traits. The difference in pathogenicity for mice upon intracerebral infection appeared most demonstrative—the mortality rate varied from 0 to 100% for different strains inoculated with the same dose of the virus. The strains differed also in the character of A type intracellular inclusions and some other characteristics. Our concurrent studies demonstrated interstrain differences in the limit temperature of pock development on CAMs (Marennikova *et al.*, 1975a). Significant strain differences in pathogenicity for white and cotton rats, especially for 3–5-day-old rats at intranasal infection and for white mice were found later (see Table 6.1 and the appropriate subsections). The intraspecies variation of cowpox virus is not confined to phenotypic traits. In comparative studies of isolates from feline predators and elephants, Baxby *et al.* (1979) discovered that they lacked a 37 kDa polypeptide, which was present in the cowpox virus reference strain Brighton. Cross-neutralization studies revealed certain antigenic distinctions of an isolate recovered of elephant from both strain Brighton and isolates from feline predators. DNA restriction profiles also displayed interstrain distinctions. Differences in these profiles generated by several restriction endonucleases were found for various isolates (Pilaski *et al.*, 1986; Schay *et al.*, 1992; Tryland *et al.*, 1998c; Meyer *et al.*, 1999).

6.2.7   Composition of Cowpox Virus Population

Similar to certain other animal poxviruses, cowpox virus population is genetically heterogeneous. Any strain of this virus contains approximately 1% of the so-called white mutants, emerging due to spontaneous mutation. These mutants produce white pocks, not typical of cowpox virus itself, on chick embryo CAMs. After a prolonged and frequent passaging of strains in chick embryos, the amount of white mutants in the virus population increases, bringing the number of white pocks they produce to 5 and even 13% (E.M. Shelukhina, personal communication). To prevent this, the strain is commonly either cloned or a limited passage scheme is used, returning to the initial preparation of frozen viral seed. Cloning (taking material from one white or red pock and production of subcultures) allows for obtaining both white clones of the virus (derived from white pock) and red clones (derived from the predominant hemorrhagic or red pocks). The red clones are identical in their properties to the initial ("wild-type") cowpox virus (see below). Similar to the wild-type virus, these clones always produce about 1% of white mutants and, correspondingly, white pocks on CAMs. Unlike this, the white clones, derived from white pocks, are genetically stable and form only white pocks while passaging. In addition, the white clones differ from the wild-type virus in the type of skin lesions induced in rabbits, which resembles vaccinia virus (Downie & Haddock, 1952; van Tongeren, 1952).

Reith and Williamson (1996), who studied specific features of the pathogenesis at intranasal infection of mice, confirmed a lower pathogenicity of a white clone for white mice. The authors showed that, unlike the lethal generalized infection caused by a red clone, the white clone did not lead to the death of animals, although producing a moderate viremia. In this respect, the white clone resembled vaccinia virus strain Wyeth (NYCBH).

However, in spite of the phenotypic similarity to vaccinia virus, analysis of the DNA structure of white clones showed that they retained the main characteristics of the parental genome, differing from it by the presence of large deletions in the right end of the genome and sometimes, by transposition of nucleotide sequences (Archard *et al.*, 1984; Pickup *et al.*, 1984). It is known that one of the significant distinctions of white clones is their loss of SPI-2 gene. The absence of this gene (and the corresponding 38 kDa protein it encodes) is responsible for the development of an active inflammatory reaction at the site of inoculation, manifested by the appearance of white pocks on CAMs (Palumbo *et al.*, 1989; Buller & Palumbo, 1991). Involvement of SPI-2 gene suppresses this inflammatory reaction, the absence of which is a characteristic feature of histopathology of hemorrhagic pocks. A possible mechanism underlying suppression of the inflammatory reaction is the action of SPI-2 gene on the metabolism of arachidonic acid (Palumbo *et al.*, 1993) and inhibition of processing of pro-interleukin-1β. On the other hand, as shown in the studies of the same authors, formation of hemorrhagic pocks cannot be explained only by the effect of this gene alone. Presumably, at least four genes are involved in this process (Palumbo *et al.*, 1992).

6.3   Ecology of Cowpox Virus

For 200 years, from the time when Jenner started his study of cowpox virus to the beginning of the 1970s, our conception of this disease remained unchanged. In essence, it was limited to the idea that sporadic outbreaks of the disease among cattle were accompanied by individual human cases of the disease, which developed as a local and, more rarely, generalized infections. The scheme of virus circulation was a classic example of clarity and simplicity: humans were infected from cattle (mainly cows), which were considered to be the natural host of the virus. In turn, a sick human infect cows (usually during milking). However, it was impossible in some cases to find out the source of cowpox outbreaks, as the outbreaks developed suddenly, after a period of many years, with a complete absence of human or cattle cowpox in the country. As a rule, this strange circumstance remained unnoticed, and only a few researchers, for example, Dixon (1962), made it the subject of special examination in his well-known monograph *Smallpox*. Not finding a satisfactory explanation for the mechanism of emergence of such outbreaks, he called them mysterious and suggested that cows were not the reservoir of infection.

In 1960–1970s, the facts that were inconsistent with the traditional conception of cowpox were gradually accumulating. First, reports appeared on spontaneous outbreaks of cowpox among captive animals, in which the disease had never been observed. It was discovered that wild rodents carried the virus, and the transmission of infection by rodents to other animals and humans was reported. These and some other data provided a new look at the ecology of cowpox agents and a reconsideration of the existing ecological scheme.

Materials related to this issue are detailed below.

6.3.1 Cowpox Virus Infection in Cattle

It is deemed that the cowpox virus infection in cattle was first described in 1713 by Sulger (cited by Gubert, 1896). However, a real interest to cowpox arose after the classic works of Jenner on the study of this disease and development of a method for smallpox prevention. Cowpox is an infectious disease caused by the genuine cowpox virus, which infects cattle (predominantly cows) and appears as outbreaks or individual sporadic cases among these animals. The disease is met on a rather limited territory in the countries of Europe—mainly in Great Britain, more rarely in the Netherlands and Germany (Meyer *et al.*, 1999; etc.). However, the etiological agent of this infection is found over a more vast area, both in Europe and Asia (see below). As in the time of Jenner, cowpox in our day is rather uncommon (Dixon, 1962). In England and Wales, 13 outbreaks of cowpox were recorded from 1887 to 1919 and 9 outbreaks, from 1938 to 1958 (Dixon, 1962). Baxby (1977) presents data on 12 cowpox outbreaks that occurred in Great Britain from 1965 to 1976. Numerous reports of "cowpox" cases in various countries are generally not confirmed. Virological examination showed that other viruses caused them. In recent years, the number of cowpox outbreaks among cattle in the countries of Europe decreased considerably (Tryland *et al.*, 1998a).

Clinically, the illness is characterized by fever and the development of skin eruptions, most often on the udder and teats. The cutaneous lesions pass through typical stages of evolution (papules–vesicles–pustules) with subsequent development of scabs and ulcers. The illness usually lasts for 3–4 weeks. The calves suckling infected cows develop lesions on the snout. The infection is transmitted from sick cows to dairy workers, who are able to transmit it in turn to healthy animals. The disease is accompanied by production of specific antibodies. According to Baxby and Osborne (1979), antihemagglutinins persist for 27 weeks after the natural infection and virus-neutralizing antibodies, up to 98 weeks. The authors failed to detect precipitating antibodies, which might relate to a late start of the study (after 6 weeks).

Experimental infection in cows develops after a 2–7-day incubation period. The animals become febrile more often before and sometimes simultaneously with the development of skin lesions. The lesions often have a hemorrhagic character. Experimental infection of cows caused by genuine cowpox virus differs from the infection caused by vaccinia virus, which follows a milder and more prolonged course (Berger & Putingam, 1958).

It is difficult or impossible to distinguish cowpox from several other viral diseases of these animals by the clinical picture. During the period of smallpox vaccination, cowpox was most frequently confused with vaccinia infection transmitted to cows from recent vaccinees. For example, Downie and Dumbell (1956) studied 36 strains isolated during outbreaks of pox infection in cows and detected 8 strains of vaccinia virus among them. Mal'tseva *et al.* (1966), investigated five outbreaks among cows on the territory of the former USSR and found that all these outbreaks were caused by vaccinia virus, carried by recently vaccinated dairymaids. Similar observations were made in Egypt, South America, and other countries (El Dahaby *et al.*, 1966; Lum *et al.*, 1967, etc.). When smallpox vaccination was ceased, this infection has virtually lost its significance as a source of diagnostic errors. Another, much more abundant disease that is easily mistakable for cowpox is paravaccinia or pseudocowpox, whose agent is not an orthopoxvirus, although it belongs to the family Poxviridae (genus *Parapoxvirus*, see Table 2.1).

This virus readily infects humans contacting with sick animals and causes a disease with skin lesions, known as dairymaid's nodes. Outbreaks of pseudocowpox are frequently reported in many countries of the world and are generally accompanied by human diseases. According to our data, only during 1986–1994, such outbreaks occurred in Russia (Smolensk, Vladimir, and Novosibirsk oblasts and Altai krai), Ukraine, Kazakhstan, and Chuvashia (Shelukhina *et al.*, 1986; Zhukova, 1993; E.F. Belanov & S.N. Shchelkunov, unpublished data). See Chapter 8 for the methods of laboratory differentiation of the cowpox agent from vaccinia and paravaccinia viruses.

6.3.2   Cowpox in Exotic Animals in Zoos and Circuses

At least 27 outbreaks or individual cases of cowpox occurred in zoos and circuses of eight European countries during 1960–1986. Except for outbreaks in Russia and Great Britain, the rest outbreaks were recorded in Central

Europe, namely, Germany and adjoining countries (Austria, Czechoslovakia, Poland, the Netherlands, and Denmark). Interestingly, the outbreaks in countries of Central Europe, on the one hand, and in Russia and Great Britain, on the other, differed significantly in the species compositions of infected animals. In Central Europe, predominantly elephants became ill (in 19 out of 22 episodes) as well as rhinoceros and okapis. In Russia and Great Britain, predominantly various species of feline predators contracted the disease (lions, pumas, jaguars, cheetahs, black panthers, ocelots, and Far East cats). Uncommon character of infected objects and the lack of knowledge on such diseases led initially to diagnostic errors. Even when virions morphologically identical to poxviruses were detectable in samples of sick animals or characteristic lesions on CAMs or on rabbit skin were obtained, the disease was considered a poxvirus infection specific of a certain animal species or a vaccinia infection transmitted to animals from vaccinated children (Sprössig *et al.*, 1968; Mayer *et al.*, 1972; Gehring *et al.*, 1972; Mayer, 1973). For example, the isolates recovered from sick elephants during the outbreak in 1971 (Gehring *et al.*, 1972) were classified by the authors as vaccinia virus. Our subsequent study of one of these isolates (Althof-71) using specialized differential diagnostic tests showed explicitly that cowpox virus was actually the agent of the disease among elephants (S.S. Marennikova & E.M. Shelukhina, unpublished data). Baxby and Ghaboosi (1977), who studied another isolate (EP-1) from the same outbreak as well as the isolate EP-2 recovered from an elephant during an outbreak in Nuremberg in 1975, obtained exactly the same results. The data on identifying the causative agents of the majority of the rest outbreaks as cowpox virus (okapis, rhinoceroses, and feline predators) were also confirmed.

Despite the diversity of animal species afflicted, the disease of the majority of animals was severe and had a high mortality rate. Predominantly, the disease pattern was typical of poxviruses, with the development of characteristic lesions (eruptions) on the skin and mucous membranes (Figures 6.8 and 6.9). For elephants, edemas of the head and hind limbs were also observed. Some animals developed generalized process, and the rash became confluent, with ensuing necrosis of the skin and mucosae (Gehring *et al.*, 1972). A cutaneous form of the infection in feline predators as well as in elephants, okapis, and rhinoceroses sometimes had lethal outcome (Marennikova *et al.*, 1975a; 1977b). Both cases of the disease of giant anteaters in the Moscow Zoo were fatal (Marennikova *et al.*, 1976b).

When studying the outbreak in the Moscow Zoo, we were the first to detect a previously unknown form of the disease caused by cowpox virus— the pulmonary form (Marennikova *et al.*, 1975a; 1977b). Some representatives of feline species (lions, cheetahs, and black panther) developed this disease form. The pulmonary form was 100% lethal, and the

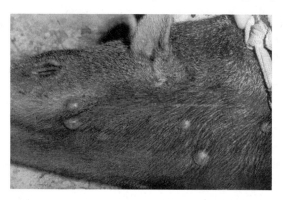

Figure 6-8. Vesicular eruptions on the skin of giant
anteater infected with cowpox virus (day 9 of the disease).

animals died within 8 days after the onset of disease. In some cases, the disease
was fulminant with death on day 3. Its clinical manifestations included
adynamia, refusal of food, difficult breathing, dyspnea, paroxysmal cough, and
cyanosis of mucosae. Autopsy showed serofibrinous pneumonia with massive
exudates (up to 3 liters) in the pleural cavity. As a rule, no skin lesions were
observed during the pulmonary form of cowpox in animals. Later, Baxby
observed the fulminant pulmonary cowpox in cheetah (Baxby *et al.*, 1982).

Therefore, the observations described above demonstrate that the
pathogenicity range of cowpox virus proved to be much wider than it was
thought previously. The list of susceptible animals continues to grow; cases
of severe generalized infection caused by cowpox virus were reported in zoo
animals: alpacas (*Lama glama pacos;* Schüppel *et al.*, 1997), Canadian
beavers (*Castor fiber canadensis*), and small pandas (*Ailurus fulgens;*
Hentschke *et al.*, 1999). Severe illness was reported in many animal species
listed above, but large cats and beavers showed the highest mortality rate.
Horizontal transmission of infection was also observed. However, these
exotic animals cannot be considered as the actual hosts of the virus.

Since the mid-1980s, when preventive vaccination with vaccinia virus
was practiced for captive elephants, feline predators, and some other animals,
the number of cowpox outbreaks in zoos and circuses decreased.
Nevertheless, at least eight similar episodes were reported until 2002. Of
them, noteworthy are two lethal elephant cowpox cases at a traveling circus,
and the observation by Wisser *et al.* (2001), who described a case of
congenital cowpox transmission. A vaccinated and revaccinated female
elephant with no signs of infection delivered a dead calf with a generalized
cowpox.

Note here that cowpox cases were recorded among both African elephants
(*Loxodonta africana*) and Asian elephants (*Elephas maximus*). Greiffendorf

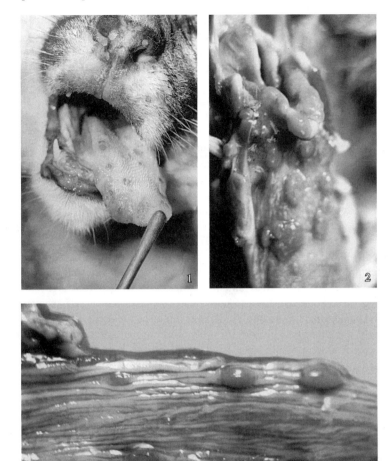

Figure 6-9. Eruptions of young puma (5 months old), which died on day 7 of cowpox infection: (1) lesions on the tongue; (2) enanthema in epiglottis area; and (3) rash elements on esophageal mucosa.

et al. (1998) reported that cowpox cases were more frequent and severe in Asian elephants compared with African elephants.

Interestingly, the source of infection of the animals remained unclear in all the outbreaks of cowpox in circuses and zoos. The only exception was an outbreak in the Moscow Zoo, where white rats infected with cowpox virus became such a source when used for feeding some predatory species (Marennikova & Shelukhina, 1976a). Most probably, some rodent species, being virus carriers, were the source of infection for animals (or humans) in other similar outbreaks, either directly or through their egesta.

6.3.3   Cowpox in White Rats

Natural infections of rodents caused by poxviruses were unknown until the beginning of 1970s, with the only exception of mousepox. Moreover, rodents such as white rats were considered to be among the species of laboratory animals least susceptible to representatives of this virus group. In adult animals, only preliminary suppression of immunity (for example, with ^{60}Co γ-irradiation; Svet-Moldavskaya, 1968) allowed for inducing clinically apparent form of generalized infection by inoculation with, for example, vaccinia virus. However, Krikun (1974) recovered an isolate in 1973 during an epizootic of vague etiology in a colony of white rats and named it the "pneumotropic virus of rats". We identified this isolate and showed its identity with cowpox virus and the isolates recovered from sick animals during the outbreak in the Moscow Zoo. The viral isolates we recovered a year later during the second similar epizootic in the same rat colony were exactly the same type. Since 1974, we observed and studied eight similar outbreaks among white rats in various animal breeding facilities and vivariums in Moscow and Moscow oblast (Marennikova *et al.*, 1994). The outbreaks were virtually similar in their pattern and duration. Minor differences observed were related to the number of animals and conditions of their housing. In several animal breeding facilities, epizootics lasted 4–5 months, with the peak case rate occurring at 1.5–2 months. The average mortality rate during these outbreaks and epizootics amounted to 30% or more.

The disease followed various clinical courses. The pulmonary form was the first to engage attention. Sick animals stopped eating, became inactive, developed difficult breathing, dyspnea, and progressive abdominal swelling to die after 3–4 days. Autopsy showed total or focal pneumonia with serous or hemorrhagic exudates in the pleural cavity; the intestines and stomach were empty and distended, their walls were significantly thinned. This form of the disease had 100% mortality rate. The corpses of these animals were never eaten by other rats (unlike when rats die from other causes). Cutaneous form of illness followed a much milder pattern; red papular rash appeared predominantly in the areas with little or no hair (the tail, paws, and snout; Figure 6.10). The rash dried rapidly (in 1–2 days) and scabs appeared. This form of illness was also characterized by a pronounced conjunctivitis with bloody exudate (red circles shaped as spectacles around the eyes) and rhinitis. This form was not accompanied by a severe disturbance of the

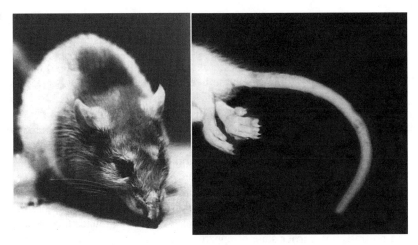

Figure 6-10. Infection of white rats caused by cowpox virus: a sick white rat (left) and eruptions (at scab stage) on the tail and hind limbs of the animal that died from the disease (right).

general condition of animals and was rarely lethal. In some cases, infected areas of the tail and paws necrotized and then fell off. A mixed form of infection also occurred, when skin lesions developed concurrently with pathological changes in the lungs. In some cases, the disease took a suppressed form with recovery in 2–3 days. During the disease, the animals had depressed appetite and were inactive, with ruffed fur. The course and severity of disease depended on the age of animal. For example, suckling rats developed the disease and died without any visible signs of being sick, while the disease in adult females was mild, followed the cutaneous course, and gave no lethal outcomes (Marennikova *et al.*, 1978b; Shelukhina, 1980).

Table 6.2 lists the data of selective virological and serological tests that we performed with sick rats during outbreaks and with apparently healthy

Table 6-2. Results of virological and serological examinations of white rats during and after outbreaks of the disease caused by cowpox virus

Animals	Study for the presence of						
	Cowpox virus in*				Antibodies by*		
	Lungs	Kidneys	Liver	Spleen	HAIT	EIA	Gel precipitation
Sick	24/42	19/42	9/37	8/36	27/33	26/29	20/24
	(57.1)	(45.2)	(24.3)	(22.2)	(81.8)	(89.6)	(83.3)
Apparently	11/109	2/131	0/26	0/31	37/121	15/28	15/96
healthy	(10)	(1.5)	(0)	(0)	(30.6)	(53.5)	(15.6)

*Positive animals/total number of animals; the percent of animals with positive test is given in parentheses.

animals selected after the outbreaks on the background of continuing rare individual cases of disease.

It is evident that more than a half of the rats with evident signs of the disease had virus in the lungs; 45.2% of such animals, in the kidneys; and in a lower percentage of animals, in the liver and spleen. Interestingly, various serological tests detected antibodies to the orthopoxviruses in 81.8–89.6% of rats already at the acute stage of the disease. As noted above, up to 10% of the virus carriers were found among apparently healthy animals examined after the outbreaks were over.

By the example of one outbreak, the viral carrier state was shown limited to a certain postinfection period. For example, 6 weeks after the beginning of the outbreak, virus was detected in the urine of 50% of rats and feces of 2 out of the 12 animals examined. However, the virus could not be recovered from any of the 41 rats tested 4 weeks later. These facts give grounds to infer that the postinfection carrier state of apparently healthy rats and their ability to excrete the virus into the environment are the main causes of prolonged epizootics in animal breeding facilities, even with strict rejection of all sick animals and their cagemates and other anti-epizootic activities.

The role of the postinfection carrier state in the outbreak maintenance manifested itself also when a new shipment of animals was delivered to the breeding facility and became infected after some time, although no animals with visible signs of disease were present in the breeding facility. Most likely, the postinfection carrier state played a key role in transmission of the infection from rats to other animal species. In particular, we succeeded in showing this when analyzing the causes of two outbreaks among healthy animals (mainly feline predators) in the Moscow Zoo, as is described above.

It appeared that white rats were used for feeding some species of feline predators (wild cats, cheetahs, etc.). Serological studies of a shipment of rats delivered from a breeding facility to the zoo as animal feed detected the presence of anti-orthopoxvirus antibodies (AHA and precipitins) in many animals. Further investigation found out that the rats were received from a breeding facility where a large epizootic had occurred not long before. The virus was isolated from four rats (three of them were apparently healthy and one had evident signs of disease) out of the cohort of animals from this breeding facility selected for study. The virus isolates were identical in their biological characteristics to that isolated from sick animals at the Moscow Zoo. All these isolates were similar to the cowpox virus reference strain, except for their higher pathogenicity for white rats and a lower ceiling temperature of pock development on CAMs (Marennikova & Shelukhina, 1976a; Marennikova *et al.*, 1978b).

In an analogous situation one year later, we distinctly traced the connection between the disease of white rats in the breeding facility and the infection of feline predators at the Moscow Zoo (Figure 6.11). In this second outbreak, the infection struck a new generation of pumas and Far East cats (Marennikova *et al.*, 1977b).

Experimental infection of white rats with one of the isolates recovered from them induced the development of disease with the clinical picture depending on the dose. In this case, the courses of disease were similar to those observed for natural infection of rats.

Of special interest was the possibility to demonstrate experimentally the asymptomatic course of infection caused by this virus. The infection was simulated in experiments with intranasal inoculation of small doses of the virus (45 PFU) to highly susceptible animals, such as 5-day-old suckling rats. In

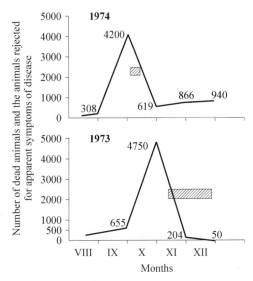

Designations: Duration of smallpox outbreak in the zoo: from the onset of the first case to the onset of the last case (without regard to retrospectively diagnosed cases, which occurred approximately at the same time)

Figure 6-11. Time courses of outbreaks in the white rat breeding facility and timing of the outbreak development among zoo animals (the facility supplied the zoo with white rats for feeding some animal species).

this form of infection, in the absence of any visible clinical signs of the disease, the virus persisted in the kidneys of several animals for a rather long time, up to 4 weeks (the period of observation), with antihemagglutinin present in blood serum (Shelukhina, 1980). An analogous study involving white rats 1.5–2 months old with even lower dose of the virus (30 PFU) also demonstrated the possibility to induce development of asymptomatic infection, as was confirmed by pathohistological alterations in the organs and serologic shifts (Maiboroda & Lobanova, 1980).

6.3.4   Virus Carrier State among Wild Rodents

It is known that white rats are bred in specialized facilities and vivariums, so their population cannot be considered as a part of the wild nature. In this connection, the above-mentioned epizootics among these animals caused by cowpox virus were thought to result from an importation of this virus from the outside. Taking into account some factors, in particular, the uncontrolled penetration of wild-living rodents into the breeding facility, we assumed that the latter might be the source of infection of white rats. This circumstance stimulated initiation of manifold studies in search for the natural reservoir of cowpox virus among wild-living rodents. Results of this work are detailed below.

The studies covered four geographic zones: (1) Nebit-Dag and Krasnovodsk districts of Turkmenistan; (2) four districts in Georgia; (3) two districts in the Baltic states and Poland; and (4) the central region of the European part of Russia (Moscow and Smolensk oblasts). This work involved serological examination (HAIT and neutralization reaction, NR) and isolation of the virus from organs of captured animals. In more detail, we present the materials obtained in the expedition to Turkmenistan, which was specially organized by I.D. Ladnyi according to our suggestion (Table 6.3).

As is evident from Table 6.3, antihemagglutinin and virus-neutralizing antibodies to orthopoxviruses were detected in four species of Turkmenian rodents: large gerbils, yellow ground squirrels, midday gerbils, and Libyan jirds. The first two species were positive for antibodies much more often and displayed higher antibody titers (Ladnyi *et al.*, 1975; Marennikova, unpublished data).

Study for the presence of the virus in organs of animals (Table 6.4) resulted in recovery of three isolates; two of them, Turk-74 and 4981-85, were recovered from the kidneys and spleen of large gerbils and the third, from the kidneys of a yellow ground squirrel (Marennikova *et al.*, 1978a; Shelukhina, 1980). All the three virus-positive animal carriers were captured to the west, north, and south of the city of Nebit-Dag at distances of 36, 130,

Table 6-3. Data of serological examination of wild-living Turkmenian rodents

Animal species	Test	Number of sera tested	With the presence of antibodies	At a titer of		
				2–4 10–20	8–16 40–80	32–64 (HAIT) 160 (NR)
Large gerbil (*Rhombomys opimus*)	HAIT	306	57 (19%)	38	16	3
	NR	258	43 (17%)	35	7	1
Yellow ground squirrel (*Citellus fulvus*)	HAIT	163	25 (15%)	15	7	3
	NR	103	9 (9%)	6	2	1
Midday gerbil (*Meriones meridianus*)	HAIT	35	2 (6%)	0	2	0
	NR	35	2 (6%)	0	2	0
Libyan jird (*Meriones libycus*)	HAIT	32	1 (3%)	0	1	0
	NR	32	1 (3%)	0	1	0
Other species	HAIT	34	0			

Note: HAIT was performed with 2 hemagglutinating units of vaccinia virus; NR, suppression of 50% of pocks on chick embryo CAMs.

Table 6-4. Results of examination of Turkmenian Rodentia and Logomorpha species for the presence of orthopoxviruses

Animal species	Organs			
	Lungs	Liver	Kidneys	Spleen
Large gerbil (*Rhombomys opimus*)	0/603	0/1102	1/1102	1/1102
Yellow ground squirrel (*Citellus fulvus*)	0/173	0/173	1/173	0/173
Midday gerbil (*Meriones meridianus*)	0/133	0/133	0/133	0/133
Libyan jird (*Meriones libycus*)	0/184	0/184	0/184	0/184
Long-clawed ground squirrel (*Spermophilopsis leptodactylus*)	0/2	0/2	0/2	0/2
Three-toed jerboa (*Dipus sagitta*)	0/3	0/3	0/3	0/3
Afghan pika (*Ochotona rufescens*)	0/12	0/12	0/12	0/12
Mouselike hamster (*Calomyscus bailwardi*)	0/2	0/2	0/2	0/2
Gray hamster (*Cricetulus migratorius*)	0/3	0/3	0/3	0/3
House mouse (*Mus musculus*)	0/8	0/8	0/8	0/8

Note: the number of organs where an orthopoxvirus is present to the number of organs tested are shown.

and 70 km, respectively. All of them were apparently healthy. Note that virus isolation was performed in a laboratory where no work with poxviruses had been previously performed. The isolates were identified in Moscow, at the WHO Collaborating Center for Smallpox and Related Infections. These studies provided the first direct evidence of the circulation of cowpox virus in wild-living rodents. Since the isolated viruses were similar in their properties, only one of them—the isolate Turk-74, recovered from the kidneys of a large gerbil—was examined in detail. Along with the cowpox virus reference strain Brighton, we took ectromelia and camelpox viruses for comparative study. Isolate Turk-74 proved to be essentially different in its phenotypic traits from both ectromelia and camelpox viruses (Marennikova *et al.*, 1978a; Shelukhina, 1980). The isolate appeared most close to cowpox virus. The revealed minor distinctions of the isolate were within the range of intraspecies variation of cowpox virus.

Examination of rodents in Georgia showed that Libyan jird was a carrier of antibodies to orthopoxviruses in this region too. Interestingly, the rate of seropositive individuals of this species was higher than in Turkmenistan (6.2% according to HAIT, 11.3% according to neutralization reaction). An orthopoxvirus isolate recovered from the lung tissue of this animal species (370 animals examined) appeared to be similar in the main biological properties (pock type on CAM, skin test in rabbits, hemagglutination activity, and the type of gel precipitation reaction) to the isolates from Turkmenian rodents (Tsanava *et al.*, 1989). Results of examination of rodents from Transcaucasia and other European regions are listed in Table 6.5, except for the data on Libyan jird shown above.

Thus, this study increased the number of the known carriers of antibodies to orthopoxviruses by gray rats and two species of mice (Marennikova *et al.*,

Table 6-5. Examination of rodents from various regions of the European part of the former USSR and Rzeszów province of Poland for the presence of orthopoxviruses and the corresponding antibodies

Family	Species	Number of serum samples		Number of animals	
		Tested	Containing antibodies	Tested	Containing the virus
Muridae	Grey rat (*Rattus norvegicus*)	608	18 (2.9%)	464	1 (0.2%)
	House mouse (*Mus musculus*)	385	1 (0.25%)	Not studied	
	Striped field mouse (*Apodemus agrarius*)	33	1 (3%)	40	0
	Wood mouse (*Apodemus sylvaticus*)	250	0	259	0
	Common vole (*Microtus arvalis*)	411	0	254	0
	Social vole (*Microtus socialis*)	240	0	265	0
	Robert's vole (*Microtus roberti*)	155	0	Not studied	
	Forest dormouse (*Dryomys nitedula*)	180	0	Not studied	

1984a; 1994; Tsanava *et al.*, 1989). Moreover, an orthopoxvirus isolate obtained from gray rats was later identified as cowpox virus. It was recovered from 1 of the 19 animals caught in Moscow oblast near a breeding facility for white rats. Probably, a disguised poxvirus infection smoldered in the breeding facility, since outbreaks of the disease often occurred among the animals shipped to Moscow institutions from this source. Our data on the presence of antibodies to orthopoxviruses in wild-living rodents were confirmed in subsequent studies. As for gray rats, it was confirmed by Pilaski & Jacoby (1993) in Germany and by Tryland *et al.* (1998e) in Norway. Kaplan *et al.* (1980) reported detection of antibodies to orthopoxviruses in some species of voles and mice in Great Britain. Later, these observations were extended and made more exact. Crouch *et al.* (1995) showed that the number of seropositive individuals among various species of mice and voles in Great Britain varied in the range of 2.5–30%. Chantrey *et al.* (1999) and Hazel *et al.* (2000) discovered that red bank vole (*Clethrionomys glareolus*), a species with the highest number of seropositive individuals, along with wood mouse (*Apodemus sylvaticus*) and short-tailed field vole (*Microtus agrestis*) were the main natural reservoir of cowpox virus in this region. The domination of seropositive state of red bank voles was reported also in Germany (Pilaski & Jacoby, 1993) and Finland (Pelkonen *et al.*, 2003). It was shown that the rate of seropositive animals varied with season and attained its maximum during autumn. For example, the fall peak in Great

Britain reached 80% on the background of the annual average amounting to 10%. Although less pronounced, the same pattern was observed with wood mice. These authors suppose that, unlike red bank vole, wood mice cannot maintain circulation of the virus by themselves. The absence of cowpox infection in the adjacent Ireland, where wood mouse is abundant and red bank vole is not found, is another confirmation of this assumption. However, an analogous examination of rodents in Norway demonstrated that the detected rate of seropositive wood mice was nearly twofold higher compared with red bank voles. In addition, antibodies to orthopoxviruses (and/or orthopoxvirus DNA sequences) were found in some other species of small rodents: gray red-backed vole (*Clethrionomys rufocanus*), northern red-backed vole (*Clethrionomys rutilus*), etc. However, the maximal rate (50%) of seropositive animals in Norway was reported for lemmings (*Lemmus lemmus*), a rodent species that has never been known previously as a carrier of antibodies to orthopoxviruses. It was shown for the first time that not only small rodent species can be the carriers of such antibodies, but also certain representatives of insectivores of the shrew family, namely, common shrew (*Sorex araneus;* Tryland *et al.*, 1998c). Unfortunately, these authors (except for those from the former USSR) did not succeed in recovering virus isolates from these wild-living animals. However, DNA sequences specific of the orthopoxviruses and cowpox virus were detected in blood cells and internal organs by various modifications of PCR (Tryland *et al.*, 1998e; Chantrey *et al.*, 1999). According to Sandvik *et al.* (1998), 15% of the small mammals examined in different regions of Norway proved to carry orthopoxvirus DNA in one or more internal organs (most often, in the lungs). Further studies of seropositive samples allowed for identifying the DNA sequences present as specific of cowpox virus.

Thus, the data reported in this Subsection demonstrate that at least 14 species of wild rodents and representatives of insectivores living over a vast territory from Great Britain to Central Asia deserts are natural carriers of cowpox virus. However, it does not often occur that the infection from this natural reservoir is transmitted directly to humans. There is another group of mammals, predators, which use rodents as a part of their ration. While catching rodents, some of these predators (domestic cats and, more rarely, dogs), are infected from virus carriers and may further transmit the infection to humans.

6.3.5 Experimental Reproduction of the Infection in Wild Rodent Species wherefrom Cowpox Virus was Isolated or where Markers of its Presence were Detected

Availability of isolates obtained from wild-living animals provided a unique opportunity to study the course and specific features of the infectious process induced by the virus in its carriers. Seronegative captured animals from each of the four species were used for this purpose. They were infected intranasally, orally, onto scarified skin, or by other methods (Marennikova *et al.*, 1978a; Shelukhina, 1980; Maiboroda, 1982; S.S. Marennikova & E.V. Efremova, unpublished data; Tsanava, 1990).

Data of these studies showed that large gerbils and yellow ground squirrels were highly susceptible to Turk-74 virus. Large gerbils upon infection by any methods except for cutaneous developed an acute disease characterized by depressed activity, emaciation, and, in some cases, cough, dyspnea, and conjunctivitis.

Intranasal or oral infection with the virus at a dose of 6 log PFU caused the disease with a fatal outcome in 66 and 100% cases on days 3–12, respectively. Infection onto scarified skin induced only local lesions, and all the animals recovered, although the virus was detectable in their blood, lungs, liver, kidneys, lymph nodes, and urine on day 5 post infection.

Yellow ground squirrels were even more sensitive to Turk-74 virus. With the same dose of virus, they developed acute disease with a 100% mortality rate irrespectively of the infection route. In individual cases, intranasal infection of yellow ground squirrels caused skin eruptions near the nose and lips. Significant amounts of virus (2–3 log PFU/g) were detected in the lungs, liver, and kidneys upon intranasal or oral infection; in addition, the virus was isolated from the spleen, blood, testis, and urine. The virus was detectable in both the survived rodents upon disappearance of clinical symptoms and the animals infected with low doses of the virus and remained apparently healthy. After 3 weeks, the virus was detected in the urine; 5 weeks post infection (the period of observation), in tissues of the kidney and testis in some of the recovered large gerbils and yellow ground squirrels.

From day 10 of the disease, specific antibodies were found in the blood serum of infected rodents. After 3 weeks, the AHA titer varied in the range of 80–320; titer of virus-neutralizing antibodies, in the range of 320–1280. After 6 weeks, the titers of antihemagglutinin and virus-neutralizing antibodies somewhat decreased (40–80 and 160–640, respectively) and precipitins were detected.

Autopsy of large gerbils died from the infection showed hyperemia and edema of the lungs with the presence of necrotic areas, multiple point foci of necrosis in the liver, hyperemia and enlargement of inguinal lymph nodes, orchitis, pronounced hyperemia of the intestines and peritoneum, and in some cases, distension of the large intestine. The pathologic changes detected in ground squirrels were similar, except for the absence of necrotic foci in the liver.

Analogous experiments with Libyan jirds infected with the virus isolated from the same species gave the results similar to the experiments with Turkmenian rodents (Tsanava, 1990).

Infection caused by the cowpox virus isolated from white rats was reproduced both in wild-living grey rats (Maiboroda, 1982) and in animals of this species raised in captivity (S.S. Marennikova and E.V. Efremova, unpublished data). It was shown that the sensitivity of these animals as well as the display of disease and its mortality rate virtually coincided with the pattern observed in white rats. The disease was especially severe in the case of intranasal infection. Transmission of the infection without direct contact with sick animals (when housing healthy rats in the cages previously occupied by sick animals) was observed. The duration of excretion of the virus with the urine and feces varied from 11 to 35 days post infection.

All these materials unambiguously indicate a high sensitivity of the above listed virus-positive animals to the pathogen in question. They respond to infection by developing acute lethal disease accompanied by massive accumulation of the virus in internal organs and its excretion into the environment, occurring as well during the period of recovery (up to 5 weeks). Under these conditions, the infection can be transmitted from rodents not only through contacting with sick animals, but also indirectly through the etiological agent in egesta (feces and urine). Unfortunately, less informative were the data obtained on reproduction of infection in small rodents, which are recognized as natural carriers of cowpox virus in some European countries. Red bank voles and short-tailed field voles as well as wood mice were infected with cowpox virus in the paw pad or oronasally at doses varying from 2 to 5000 PFU (Bennett *et al.*, 1997). No clinical symptoms of the disease were observed except for the lesion and pad edema (in the case of infection in the paw pad). The virus was isolated from tissues of the inoculated paw only in some cases. Upon oronasal infection, the virus was detectable only in individual cases (nasal discharge, lungs, and spleen). Virtually all the animals infected at a dose exceeding 5 PFU responded with seroconversion starting from days 8–10. Note that although this study explains the presence of antibodies in some rodents upon various kinds of contact with the pathogen, this is yet insufficient for a clear understanding of the mechanism providing transmission of infection within populations of these animals. Probably, the absence of apparent clinical picture is related rather to the use of a low virus dose than to a low sensitivity of the animals.

6.3.6 Cowpox in Domestic Cats

In 1978, the first report of isolation of cowpox virus from a sick domestic cat appeared in Great Britain (Thomsett *et al.*, 1978). Later, such reports appeared not only in Great Britain, but also in other countries—Austria (Schönbauer *et al.*, 1982), the Netherlands (Willemse & Egberink, 1985), Germany (Bomhard *et al.*, 1989; Mahnel *et al.*, 1989; Pfeffer *et al.*, 1999), and Norway (Tryland *et al.*, 1998c).

According to the information available at the beginning of the 1990s, over 150 cases of cat cowpox cases were reported (Naidoo *et al.*, 1992). Analysis of these cases recorded in Great Britain showed that the overwhelming majority of sick cats lived in rural areas, and the rest in the suburbs and cities; nearly all of them, were rodent hunters according to their behavioral features (Bennett *et al.*, 1986; 1990; Schay *et al.*, 1992). The highest occurrence of cat cowpox cases was observed in the fall, which coincides with increased population and activity of rodents. As with other animals, the disease of cats was accompanied by skin eruptions. Therewith, in more than a half of the cases, various primary skin lesions, supposedly, at the sites of bites, preceded rash widely distributed over the body. These lesions complete their development during 4–16 days until emergence of a secondary rash and are located predominantly on the cranial part of the body, neck, and anterior limbs. Most cats recover with no symptoms of systemic involvement. Nevertheless, some animals developed small vesicles and sores on the mouth mucosa. Approximately 20% of sick cats showed signs of upper respiratory tract infections (Bennett *et al.*, 1990). A severe course or lethal outcome occurred rarely, only in combination with a concurrent disease, in particular, FeLV or FIV (Brown A. *et al.*, 1989). Note that antibodies to orthopoxviruses were detected more often in the cohort of cats infected with FIV (30% cases versus 17%). This fact suggests the inference that FIV increases the susceptibility of cats to cowpox virus (Tryland *et al.*, 1998d).

Transmission of infection during joint housing of several cats is also rare (1 case of the 15) and manifests itself in the recipient usually as a subclinical form (Bennett *et al.*, 1986; 1990).

According to our observations, experimental infection of cats induced a much more severe course compared with the natural disease (Zhukova *et al.*, 1992). For example, inoculation of cowpox virus isolated from white rats at a dose of 4 log PFU/0.1 ml to 1.5-2-month-old kittens orally, intranasally, or onto scarified skin developed severe general infection. All the kittens infected orally or onto scarified skin died. Of the three kittens infected intranasally, two survived. Only the kittens infected by scarification developed skin lesions. The changes induced were most pronounced in the lungs, where the highest virus titers were observed (>6 log PFU/g).

In terms of the cowpox virus ecology, of special interest are the data on serological examination of cats as well as other mammals that hunt small rodents. In Germany, serological examination of 2175 cats detected antibodies to orthopoxviruses in 2% of the animals (Czerny *et al.*, 1994). In another report, this rate was much higher, amounting to 10% (Zimmer *et al.*, 1991). According to Nowotny (1994a, b) and Nowotny *et al.* (1994), the seropositive individuals make up 4% of cats in Austria, and 0 to 2.2% in Great Britain (Yamaguchi *et al.*, 1996). This rate attained 10.1% in Norway in the region where two cases of human cowpox were recorded (Tryland *et al.*, 1998a), and 3.9% in Finland (Pelkonen *et al.*, 2003). These data indicate that cats come into contact with the virus more frequently than it could be judged from the number of reported cat cowpox cases. However, a low transmissibility of the natural infection of domestic cats and the fact that, unlike rodents, domestic cats do not live in communities, where virus may circulate, suggest that they are not the natural reservoir of cowpox virus. Thus, cats may be considered only as intermediate hosts of cowpox virus.

6.3.7   Cowpox in Other Animals

Already Jenner believed that that the genuine source of cowpox was sick horses, wherefrom the infection was transmitted to cows (Jenner, 1798). However, there have been virtually no well documented reports of cowpox cases in horses for more than two centuries. Pfeffer *et al.* (1999), who observed a generalized form of cowpox in a circus horse, reported the first such case confirmed by identification of the etiological agent. The source of infection remained vague. Antibodies to orthopoxviruses were detected in 2 of the 27 horses (Pelkonen *et al.,* 2003) examined in Finland in connection with human cowpox cases.

Older authors referred to sporadic cases of pox infection in dogs (Miller, 1887). However, until recently, these observations found no confirmation. Bomhard *et al.* (1991) reported the detection of cowpox virus in a dog with skin lesions, which previously was in close contact with a cat infected with cowpox. Sensitivity of dogs to cowpox virus was confirmed also by Pelkonen *et al.* (2003), who found a high titer of anti-orthopoxvirus antibodies in a dog housed on a farm in Finland. The authors believe that this animal transmitted cowpox infection to a four-year-old girl through a close contact. In spite of a rather broad range of wild animals sensitive to cowpox virus, some species seem to be only slightly susceptible or completely resistant to this virus. For example, during the outbreak in the Moscow Zoo among feline predators and giant anteaters, brown bears, hyena, and genets housed within the same premises did not show any outward signs of disease. Nevertheless, serological examination of the genets (*Genetta genetta*) one year after the outbreak detected high titers of anti-orthopoxvirus antibodies, which confirmed their contact with the virus.

In recent years, red foxes (*Vulpes vulpes*) abundant throughout Europe attracted attention of researchers. According to Boulanger *et al.* (1995), these animals possess a low sensitivity to cowpox virus. However, serological examination of red foxes in Germany showed the presence of antibodies to orthopoxviruses (in some cases, neutralizing antibodies) in 6.3 to 19% of individuals (Czerny *et al.*, 1994; Henning *et al.*, 1995; Müller *et al.*, 1996).

Orthopoxvirus antibodies were detected by ELISA in 11% of red foxes in Norway and in 50%, in Finland (Tryland *et al.*, 1998b).

Recently, attention was directed also to other animal species (mainly predators), whose ration may include rodents. It turned out that anti-orthopoxvirus antibodies were found in lynxes (*Lynx lynx*) captured in Finland and Sweden. However, the number of seropositive animals differed considerably, amounting to 1 and 29%, respectively. Moreover, 2% of brown bears (*Ursus arctos*) in Sweden appeared to be the carriers of antibodies to orthopoxviruses. All the wolverines (*Gulo gulo*) examined were seronegative, which may be related to a small number of animals involved (Tryland *et al.*, 1998b). In Germany, antibodies to orthopoxviruses were detected in 4% of wild boars (*Sus scrofa;* Mayr *et al.*, 1995).

* * *

The data presented in Section 6.3 supported and considerably extended the main postulates of a new conception of the ecology of cowpox virus that we elicited from the studies we were performing for several years (Marennikova *et al.*, 1984b). It is now hardly possible to doubt that the natural hosts of cowpox virus are small rodent and entomophagous mammals living over a vast territory of Eurasia. By now, at least 14 animal species have been relegated to this group. Direct evidence of their role in circulation of cowpox virus has been found not for each species, in some cases, only anti-orthopoxvirus antibodies were detected. Consequently, it should be kept in mind that the antibodies in question might result from the contact of animals with other representatives of orthopoxviruses, either known to us or unknown. The isolation of orthopoxviruses from gerbils in Dahomey (Lourie *et al.*, 1975) and from California voles in the USA (Regnery, 1987) confirms this possibility. Another orthopoxvirus was isolated from tundra voles in Russia (L'vov *et al.*, 1988). A relation of these antibodies to ectromelia virus cannot be excluded completely, although its existence in the wild nature remains unproved.

Nevertheless, these assumptions seem unlikely to us. Repeated cases of cowpox infection in various animals, including cats that hunt and consume small rodents, white rats in breeding facilities, and animals in circuses and zoos, where rodents can intrude, suggests rather that the anti-orthopoxvirus antibodies detected in rodents and the above mentioned animals are directly connected to cowpox virus.

Thus, the natural hosts of this virus include a wide range of species of the rodent and entomophagous orders, which represents one of the mechanisms assisting survival of the pathogen in the natural environment.

An even more important mechanism that provides the circulation of cowpox virus in nature is the postinfection carrier state, with excretion of the virus into the environment. This was demonstrated both under natural conditions (Marennikova & Shelukhina, 1976a; Marennikova *et al.*, 1978a; Tsanava, 1990) and experimentally (Marennikova *et al.*, 1978a; Shelukhina *et al.*, 1979b; Shelukhina, 1980; Maiboroda, 1982). This is observed both during asymptomatic infection (inoculation of small doses of the virus) and in recovered individuals.

Implementation of this mechanism is facilitated not only by a high susceptibility of rodents to cowpox virus, but also by their wide abundance and versatile biological relations with many animal species, including cattle. As for cats, white rats, and other predators hunting for small rodents as well as exotic animals in zoos and circuses, they are actually the casual or intermediate hosts of the virus. Cows and humans should be also relegated to this group. However, this does not exclude a possible transmission of infection within a species and its transfer to representatives of other taxonomic groups. In this case, of great importance are specific susceptibility, nature of the contact, and many other factors. The effect of these factors may be exemplified, on the one hand, by the above described outbreaks among white rats, when almost all the population was involved in the infectious process, and on the other, a low transmission rates of the infection in domestic cats and, especially, in humans, where secondary transmission is extremely rare (see Section 6.4).

6.4   Cowpox in Humans

The first descriptions of this disease appeared in the 18th century and are mainly associated with Jenner (1798). Since then, cowpox was considered an occupational illness of dairymaids, and sick cows were regarded as the only source of infection. Transmission of the infection occurred mainly while milking and taking care of such cows. However, Jenner himself believed that horses in addition to cows could also be a source of infection. In the latter half of the 20th century, the number of cowpox outbreaks in cattle decreased sharply, and increasingly noticeable became the cases of cowpox in humans under the conditions when contact with cattle was completely excluded, while the source of infection remained unknown. By now, such cases have been recorded in Poland, Great Britain, the Netherlands, Germany, Russia, Norway, Sweden, and Finland (Cywicki & Michowicz, 1968; Baxby, 1977; Marennikova *et al.*, 1984a; Wienecke *et al.*, 2000; etc.). Reports of cowpox cases in humans appeared also in some other countries, but they were insufficiently documented (Baran & Zubritsky, 1997; etc.).

Versatile studies, commenced in 1974 and currently in progress, discovered the genuine natural reservoir of cowpox virus among wild-living rodents and insectivores as well as its intermediate carriers (domestic cats, sometimes dogs, white rats, and others, let alone cows, horses, and humans). Despite no less than 14 species of wild-living rodents are carriers of the

Table 6-6. Human cowpox cases in Russia with confirmed or suspected transmission of infection from rodents*

Patient	Male/female	Age (years)	Previous vaccination against smallpox	Occupation	Source of infection	Site and time of infection	Clinical symptoms	Presence of	
								Cowpox virus in skin lesions	Antibodies to virus (EIA-A) in blood serum
E.N.	Female	34	+	Vivarium worker	Bite of sick white rat	Moscow, January 1987	Day 3: malaise, adynamia, headache, temperature elevation to 38°C, and lesion at the bite site	Virus was recovered (isolate EI-87) and identified as cowpox virus	Species-specific antibodies to cowpox virus
S.O.	Female	31	+	Veterinary with vivarium	Bite of sick white rat (through rubber glove)	Moscow, April 1987	Day 3: adynamia, malaise, subfebrile temperature, small lesion at the bite site, and swelling of regional lymph node	Poxvirus virions were detected	Species-specific antibodies to cowpox virus
G.A.	Male	10	+	Student	Unspecified, presumably, small field rodents (or their excrements infected with the virus)	Smolensk oblast, August 1988	Days 3–4: adynamia, malaise, temperature to 39°C, submaxillary lymphadenitis, and pain and lesions in the region of right ear; Days 12–13: eruptions on lower extremities	Virus was recovered (isolate Argo-88) and identified as cowpox virus	Species-specific antibodies to cowpox virus

G.E.	Female	4.5	–		Unspecified (close contact with a mole)	Moscow oblast, August 1990	Day 3: adynamia, malaise, temperature up to 40°C; Day 5: pain and regional lesions in the right axilla, right axillary lymphadenitis; Day 10: secondary eruptions around the primary focus	Virus was recovered (isolate GRI-90) and identified as cowpox virus	Species-specific antibodies to cowpox virus
R.E.	Male	12	–	Student	Close contact with a virus-carrying white rat**	Moscow, April 1991	Day 2–3: adynamia, malaise, temperature to 38°C, local lesions on the right shoulder and neck, right axillary lymphadenitis	Virus was recovered (isolate Rev-91) and identified as cowpox virus	Species-specific antibodies to cowpox virus
D.K.	Male	10	–	Student	Close contact with a virus-carrying white rat**	Moscow, May 1991	Day 3: adynamia, malaise, temperature to 38–40°C, lesions on the body; Day 6–8: lesions on face and arm, and bilateral lymphadenitis	Virus was recovered (isolate DK-91) and identified as cowpox virus	Species-specific antibodies to cowpox virus

*According to the data of Marennikova *et al.* (1988b) and Zhukova (1993).
**The cowpox virus carrier state of rat was established retrospectively (the virus was isolated from its organs).

virus, yet there is only one confirmed report of cowpox infection of a 14-year-old girl who was bitten by a wild rat (*Rattus norvegicus*) in the Netherlands. The fact of infection transmission was proved by isolation of the virus from the sick girl and the rat (Wolfs *et al.*, 2002). The virus carrier state of this animal species was first reported at the beginning of 1990s (Marennikova *et al.*, 1994).

In some cases, investigation of the circumstances of cowpox infection in the persons with unknown infection source gives grounds to assume that the transmission might have occurred through an unaccounted contact with small field rodents in a carrier state or their excrements. An example of such case is presented in Table 6.6, where patient G.A. before the onset of illness was camping for three days on a peninsula with a high abundance of voles (Zhukova, 1993).

As for domestic cats, this animal species is now the main carrier of infection transmitted to humans in Europe (Willemse & Egberink, 1985; Baxby & Bennett, 1990; Czerny *et al.*, 1991, etc.). Similar cases in Great Britain, the Netherlands, and Germany make up no less than 50% of the total number of human cowpox cases. In very rare instances, humans may be infected from dogs (Pelkonen *et al.*, 2003). White rats represent yet another source of infection for humans (as indicated above).

In 1977–1978, during the cowpox outbreaks among felines in the Moscow Zoo and among white rats in a breeding facility, we observed five cases of infection of humans who cared for sick animals. Although the attempts to isolate the virus failed (all of the infected persons were vaccinated and one or more times revaccinated against smallpox), the type and location of eruptions developed by three of the five human cases were characteristic of the generalized cowpox (Marennikova *et al.*, 1977b; 1978b). Later, a direct evidence of possible infection transmission from white rats was found: four persons were infected. Two of these cases (vivarium workers) were bitten by rats; the rest two—10- and 12-year-old children for whom the white rats were bought as pets—had close contacts with the animals before the disease (Table 6.6; Figure 6.12). Cowpox virus was isolated from both sick humans and animals, and the cowpox-specific antibodies were detected by EIA-adsorption in the patients' sera after recovery (Marennikova *et al.*, 1988b; Zhukova, 1993).

Exotic animals in zoos and circuses infected with cowpox virus can be also considered as a potential source of human infection. This is attested by our observation of a case of infection of a worker caring for sick animals during the cowpox outbreak in the Moscow Zoo (Marennikova *et al.*, 1977b).

Along with the animals whose role as a source of infection for humans has been proved unconditionally, there are reports of others whose involvement in infection transmission to humans is suspected. In particular,

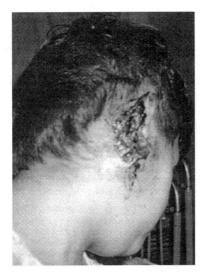

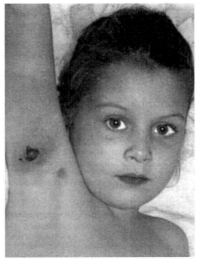

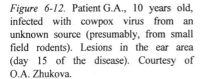

Figure 6-12. Patient G.A., 10 years old, infected with cowpox virus from an unknown source (presumably, from small field rodents). Lesions in the ear area (day 15 of the disease). Courtesy of O.A. Zhukova.

Figure 6-13. Patient G.E., 4.5 years old, infected with cowpox virus from an unknown source (presumably, from a mole upon a close contacts directly before the onset of disease). Lesions in axilla. Courtesy of O.A. Zhukova.

a young girl who contracted cowpox had no contacts with any animals before the disease except for a mole with which she played (Figure 6.13).

In most cases, human cowpox infection follows a benign course, characterized by the development of solitary local lesions. According to Stolz *et al.* (1996), such course of infection is observed in 72% cases. The lesions are most often localized to the hand, forearm, or face, and more rarely to other parts of the body. Commonly, the lesions appear at sites of microtraumas of the skin surface (scratches, cracked skin, etc.). Downie (1965) noted that lesions often occurred on the thumb and index finger, and on the web of skin between them.

No less than 2–3 days usually elapse from the moment of infection to the appearance of the skin eruption—a reddish spot transforming into a nodule (papule). By day 4–7, the papule evolves into a blister with clear contents—a vesicle, which gradually enlarges in size, suppurates, and becomes a pustule. A reddened, edematous, thickened area appears in the surrounding tissue; lymphadenitis and lymphangitis develop, with a feeling of malaise and elevated temperature. The center of the blister "sinks down". Characteristic features of the lesions caused by cowpox are their pronounced hemorrhagic character, a trend to necrotizing and ulceration of the main focus, and development of secondary lesions around it (in some cases). While drying, the pustule or ulcer becomes covered with a black (hemorrhagic) scab located deep in the skin. Then (after 3–4 weeks or even later), the scab falls off, leaving a characteristic scar.

Infection may be mechanically transmitted from the original lesion to other areas of the body. For the first time such case was described as early as by Jenner (1798), who observed transfer of infection from the hands to the face of a woman who had a habit of rubbing her forehead.

Sometimes generalization of the process occurs (Verlinde, 1951; Björnberg & Björnberg, 1956; Zhukova, 1993; etc.).

Typical of generalized human cowpox is appearance of papules with a size of a grain of rice, red or red-brown in color. The rash is localized to the hands, forearms, legs, and neck. The interval between development of the original lesion and appearance of the rash varies from 4 to 20 days. Sites of eruption are itching.

In the cases when the disease develops in patients with eczema, atopic dermatitis, or immunocompromised persons, its course becomes much more severe and may take on a dramatic character (Leroy *et al.*, 1953; Eis-Hübingen *et al.*, 1990; Pfeiff *et al.*, 1991; Czerny *et al.*, 1991; Blackford *et al.*, 1993; Pelkonen *et al.*, 2003). In particular, Czerny *et al.* (1991) observed a generalized cowpox in an 18-year-old patient with severe eczema; the cowpox clinical course resembled the hemorrhagic form of smallpox and was fatal.

Another severe form of human cowpox is connected with the development of encephalitis. Jansen (1949) and Verlinde (1951) described such fatal cases. Sometimes, location of the primary focus determined the severity of disease. Here, an illustration is provided by a case of eye infection caused by cowpox virus (Marennikova *et al.*, 1984a). In the course of this disease of a 19-year-old woman, which was initially considered a phlegmon of the left eyelid, the process spread to the surrounding tissues and affected the other eye. Treatment lasted more than three months, and the patient was discharged with scars on the eyelid and cornea. Klingebiel *et al.* (1988) and Tryland *et al.* (1998a) described similar cases of severe eye involvement. In the first case, the course of disease was complicated also by generalization of the process.

Person-to-person transmission of cowpox is observed very rarely, occurring only in the case of tight contact with the infected individual. In current practice, human cowpox cases are quite a rare pathology. Frequently, paravaccinia (milkmaid's nodes), which occurs more often and has a similar clinical picture, is taken for cowpox. According to Baxby (1981), for example, the reported rate of paravaccinia infection in England and Wales is 30–50 annual cases versus only 1–2 cases of human cowpox. Until the cessation of vaccination against smallpox, cowpox was often confused with vaccinia (Mal'tseva *et al.*, 1966). Anthrax may sometimes be considered as a possible initial diagnosis in the case of uncommon location of the primary lesion (Lewis-Jones *et al.*, 1993; Schupp *et al.*, 2001).

Vaccination is the most effective method for preventing human cowpox infection. However, since immunization against smallpox was stopped all over the world after the global smallpox eradication, now potential vaccinees can receive the vaccine only voluntarily and in the absence of contraindications. The risk cohort for advisable vaccination comprises first of all the persons who are in constant contact with wild-living rodents (in the regions where this infection is met) and the auxiliary staff of the white rat breeding facilities, circuses, and zoos.

Any persons of a close circle with a human cowpox case to say nothing about those with skin diseases (eczema and atopic dermatitis) as well as immunocompromised individuals should avoid close contacts with the patient.

The treatment of cowpox infection with a common course is generally confined to the drugs preventing bacterial complications. In the case of a severe course (eczema, immunodeficiency, etc.), vaccinia immunoglobulin can be used. Effective chemotherapy authorized for clinical use is not available yet. However, rather promising results were obtained on a pronounced prophylactic and therapeutic effects of cidofovir and ribavirin in the model of experimental mouse infection with cowpox virus (Smee *et al.*, 2000a, b; Bray *et al.*, 2000; 2002).

6.5   Genetic Organization of Cowpox Virus

Comparative study of the properties of cowpox virus (CPXV) isolates recovered from various carriers at different time and in different geographic zones allowed for detecting a pronounced intraspecies variation of the virus (see Section 6.2).

Study of the genomes of CPXV isolates recovered from various carriers (Mackett, 1981; Marennikova *et al.*, 1996; Marennikova & Shchelkunov, 1998) demonstrated that preserving the overall common pattern, nevertheless, the DNA restriction maps of the majority of isolates differ from one another.

Presumably, the data accumulated suggest the existence of a natural CPXV reservoir, where the virus is represented by manifold strains with various biological characteristics, including those determining the virus–host relationships.

Finding out the boundaries of CPXV intraspecies variation required sequencing of both individual genes of many strains and complete genomes of several isolates.

Shchelkunov *et al.* (1998) were the first to publish sequencing data of the terminal genomic regions of cowpox virus strain GRI-90 (CPXV-GRI) and results of their computer analysis. Further, the central genomic region of CPXV-GRI was sequenced. This strain was isolated in 1990 from a girl (Figure 6.13) who got infected from an ill mole (Marennikova *et al.*, 1996). The overall length of the sequenced CPXV-GRI genomic DNA amounted to 223,666 bp (EMBL Data Library accession number X94355). This is the complete nucleotide sequence of CPXV-GRI genome except for 30–35 nucleotides in the region of terminal hairpins (Figure 5.14). These regions were lost upon treatment of the terminal fragments with S1 nuclease followed by filling up the single-stranded regions using Klenow fragment of *E. coli* DNA polymerase I (Shchelkunov *et al.*, 1998). Computer analysis of the genomic sequence detected 212 potential ORFs with length of at least 60 aa. The nucleotide sequence of CPXV-GRI genome was compared with genomic sequences of CPXV strain Brighton (CPXV-BRT; EMBL accession number AF482758), VACV-COP, VACV-WR, VARV-IND, VARV-GAR, and MPXV-ZAI (see Appendices 1 and 2).

CPXV-BRT has the longest genome—224,501 bp; however, CPXV-GRI has a longer coding region, amounting to 220,881 bp. The coding region of CPXV-BRT is by 1639 bp shorter, i.e., 219,242 bp. The other orthopoxvirus species yield to CPXV in the lengths of their coding regions, namely, MPXV-ZAI is 195,118 bp; VARV-GAR, 185,846 bp; VARV-IND, 184,151 bp; and VACV-COP, 183,190 bp. The A + T content is a highly conservative characteristic of orthopoxviruses and amounts to 66.3% for CPXV-GRI, 66.6% for CPXV-BRT, 66.6% for VACV-COP, and 67.3% for VARV-IND.

Comparison of translational (genetic) maps of CPXV-GRI, CPXV-BRT, and other orthopoxviruses (see Appendix 1) demonstrates that the central conservative region of the genome, displaying a high homology of DNA sequences, is bounded by the ORFs G9L and A26R for CPXV-GRI and amounts to about 100 bp, i.e., 44 to 53% of the genome depending on the virus species. This region lacks extended deletions, and genetic maps of different viruses virtually coincide. The only exception is VARV-IND ORF E7R, which is mutationally disrupted in all VARV strains. The central conservative region of the orthopoxvirus genome contains all thus far known orthopoxvirus genes controlling biosyntheses of viral DNA and RNA molecules and a vast majority of genes encoding virion proteins. As a rule, ORFs of the orthopoxviruses compared in this genomic region display the mutual homology in amino acid sequences exceeding 95% (see Appendix 2).

However, conservation of the central genomic region is not absolute. For example, the ORFs homologous to CPXV-GRI G14L, F5R, J5R, A5L, A10L, and A14L exhibit species-specific distinctions. The proteins coded for by CPXV-GRI A5L, A10L, and A14L are virion proteins. Presumably, an increased rate of distinctions between these proteins may be explained by evolutionary selection of the variants with altered antigenic determinants.

The VACV-COP homologue of J5R gene encodes VLTF-4 transcription factor (Kovacs & Moss, 1996). Note that only this late transcription factor displays essential differences between various orthopoxvirus species. Other transcription factors VLTF-1, -2, and -3 (CPXV-GRI genes H9R, A1L, and A2L) are highly conservative.

Functions of the CPXV-GRI genes G14L and F5R are unknown. Interestingly, CPXV-GRI G14L exhibits rather high identity to VACV-COP F14L (98.6%), whereas the similarities between this ORF and the homologous ORFs of CPXV-BRT and other orthopoxviruses are the lowest among all ORFs of this conservative region (78.1–87.7%; see Appendix 2). Presumably, this indicates different origins of the gene in question of CPXV-GRI and CPXV-BRT.

The terminal inverted repeat (TIR) may be considered a hypervariable region in the orthopoxvirus genome, since numerous structural distinctions not only between various strains of the same orthopoxvirus species, but also between subclones of the same strain are reported (Fenner *et al.*, 1989). The lengths of long terminal inverted repeats of various orthopoxviruses are the following: CPXV-GRI, 8303 bp; CPXV-BRT, 9710 bp; MPXV-ZAI, 6379 bp; and VACV-COP, 12,068 bp. The TIR lengths of various VARV strains amount to 583–1051 bp (Massung *et al.*, 1995). Along with these drastic differences, the region in question carries a number of typical structures, namely, the terminal hairpin, two conservative nonrepeated regions NRI and NRII, and two sets of short tandem repeats flanking NRII region.

TIR starts from the terminal hairpin, which links the strands of genomic DNA. The length of VACV-COP hairpin is 102 nucleotides (Goebel *et al.*, 1990); of VACV-WR, 104 nucleotides (Baroudy *et al.*, 1983); and VARV-BSH, 103 nucleotides (Massung *et al.*, 1995). Presumably, the terminal hairpin of CPXV-GRI is highly homologous to the hairpins of orthopoxviruses studied (Figure 5.14).

The lengths of NRI and NRII regions in CPXV-GRI genome amount to 86 and 328 bp, respectively. The presence of these regions in all the orthopoxvirus genomes studied, their similar sizes, and a high homology (93–98%) between these regions and terminal hairpins of various orthopoxviruses suggest their important functional role, most likely, in replication of the virus DNA. For example, a sequence necessary for resolution of concatamer junction during replication of orthopoxviruses was discovered in NRI region (Merchlinsky & Moss, 1989; Figure 5.14).

The regions of short tandem repeats are extremely heterogeneous in the length (Figure 6.14); however, they consist of rather conservative subunits with lengths of 70 and 54 bp. In addition, both VACV-COP and VACV-WR each contain two repeats with a length of 125 bp (Goebel *et al.*, 1990). The 54- and 125-bp subunits may be obtained by superposition of individual

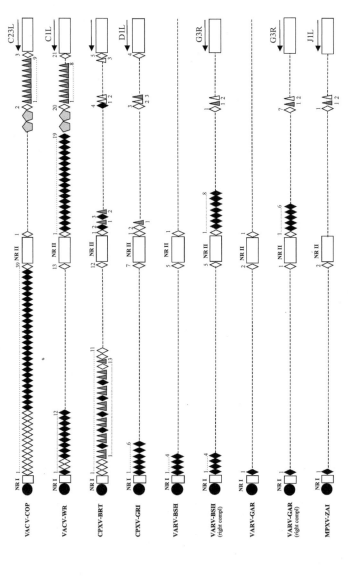

Figure 6-14. Patterns of tandem repeats within the TIR region of orthopoxviruses. White rectangles indicate unique TIR sequences—NRI and NRII—and the coding region; black diamonds, 70-bp tandem repeats with their numbers shown above; gray triangles, 54-bp repeats with their numbers shown below; pentagons in the TIR sequences of VACV-COP and VACV-WR, 125-bp repeats; unfilled diamonds and triangles indicate that these repeats differ from the consensus sequence by deletions, substitutions, or inserts; black circle is the terminal hairpin; and dotted lines, absence of the corresponding DNA.

fragments of the 70-bp subunit, suggesting an intricate alternation of deletions and duplications in the region of tandem repeats. The complex pattern of these processes is especially evident by the example of CPXV-BRT, as the alternation and number of subunits in its first and second blocks of tandem repeats are incomparable even with the structure of these regions in CPXV-GRI genome (Figure 6.14). Interestingly, the strains that underwent a larger number of passages, for example, CPXV-BRT, VACV-COP, VACV-WR, and, especially, VACV-MVA (Mayr *et al.,* 1978) display increased copy number of tandem repeats. Presumably, this provides a selective advantage while replicating the virus DNA *in vitro,* as telomeric regions are known to play an important role in replication initiation (Merchlinsky & Moss, 1989; Baroudy *et al.,* 1982).

TIR variation is not confined to the differences in the region of short tandem repeats. The coding sequence of TIRs is also prone to intricate rearrangements. Five ORFs (D1L–D5L) were found in this CPXV-GRI genomic region, yet nine potential ORFs (C23L–C16L) were discovered in the homologous VACV-COP region, carrying no large insertions or deletions. Except for C16L, the rest VACV-COP ORFs are fragments of larger CPXV-GRI ORFs (Appendices 1 and 2) that appeared due to multiple short deletions causing frameshift.

Note especially that TIRs of various VARV strains lack ORFs. Apparently, the small size of VARV TIRs essentially limits the recombination-based genetic variation of this virus compared with other orthopoxvirus species, which display rather high levels of genomic rearrangements.

The VACV-COP TIR regions contain ORFs absent in the CPXV-GRI TIRs (Appendix 1). Recombinations causing translocation–inversion of genomic regions located in the immediate vicinity of TIRs may underlie such alterations in TIR boundaries. As a result of translocation that occurred while build-up of VACV-COP genome, a region from the right end of the ancestor virus was transferred to the left end. The VACV-COP translocated fragments with a length of over 3 kbp, encoding C15L–C12L genes, correspond to two regions of the CPXV-GRI right end—B20R–B21R (homologues of VACV-COP C14L–C12L) and the 5'-end of K1R (homologue of VACV-COP C15). Interestingly, the homologues of C15L and C14L–C12L in the right end of CPXV-GRI genome are separated by a 5.6-kbp sequence. Presumably, a longer fragment (for example, the fragment bounded by B22R and I1R according to CPXV genome), which increased considerably the TIR region, was initially translocated to the left end of the VACV-COP ancestor genome. Subsequent deletions resulted in loss of various regions of the translocated fragment at the left and right genomic regions; however, the VACV-COP TIR compared with CPXV-GRI appeared enlarged only by C15L gene (Figure 6.15). Note that such rearrangements are a frequent event for

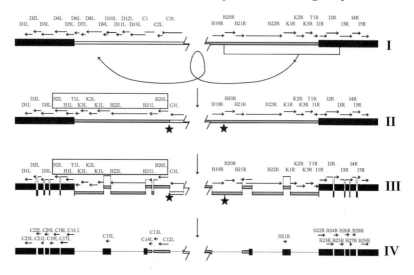

Figure 6-15. Layout of the assumed inversion–translocation of the terminal DNA fragment ir VACV-COP genome. (I) The initial ORF location in a CPXV-like ancestor: ORF name: correspond to the homologous CPXV-GRI ORFs; wide black rectangles indicate TIR regions narrower rectangles, the adjacent DNA sequences; the white rectangle corresponds to the lef genomic end; and gray, to the right. (II) ORFs localized to the left end of the virus genom were translocated to the right end: these ORFs are framed; asterisks indicate the new boundaries of TIR regions. (III) Asymmetric deletions at the right and left ends of the viru: genome: the deleted regions are shown as segments shifted down. (IV) The final layout o: ORF location in terminal genomic regions of VACV-COP and their names. The TIR region: are enlarged only by the fragments containing ORFs C15L (right genomic end) and B21R (lef genomic end).

orthopoxviruses (Moyer *et al.*, 1980a, b; Moyer & Graves, 1982; Pickup *et al.*, 1984).

Unlike TIR regions, the sets of genes of CPXV-GRI and CPXV-BRT are rather conservative. Only two relatively large deletions are detected in the coding region of CPXV-BRT compared with CPXV-GRI. The fragment of 759 bp corresponding to CPXV-GRI ORF D7L, is deleted from the left end of CPXV-BRT; at the right end, deletion of 1030 bp resulted in loss of two genes, K3R and T1R, from CPXV-GRI (Appendix 1). In addition, three CPXV-BRT genes—D9L, C15L, and A27L—disappeared or were disrupted due to mutations. Thus, of the 211 ORFs of CPXV-GRI, CPXV-BRT lacks only 6 ORFs. Apparently, the loss of these genes is a consequence of multiple passages of the strain CPXV-BRT in laboratory (Fenner, 1958).

An interspecies comparison of orthopoxviruses reveals alterations of a larger scale. The sequence with a length of 14,171 bp, coding for 12 potential ORFs (D6L–C3L) and having no analogues in VACV, MPXV, and VARV genomes (Appendix 1), starts immediately after the TIR in the left part of

CPXV-GRI genome. Due to deletion, this region of CPXV-BRT is by 759 bp shorter and encodes 10 ORFs. In addition, the left end of CPXV genome contains all the DNA sequences previously regarded as unique for particular VACV or VARV strains.

Similar pattern is observed in the right genomic region of CPXV-GRI. It contains 1579- and 3585-bp sequences, including ORFs B8R, B9R, K1R–K3R, and T1R, respectively, which are unique for CPXV. As was mentioned above, CPXV-BRT lost ORFs K3R and T1R due to deletions. Compared to CPXV-GRI, the number of deleted or mutationally disrupted genes varies in different orthopoxviruses, amounting to 37, 42, and 51 ORFs in the genomes of MPXV-ZAI, VACV-COP, and VARV-IND, respectively.

The overall data described suggest that CPXV has the most complete genome compared to the other orthopoxviruses pathogenic for humans (VARV, MPXV, and VACV). Therefore, VARV, MPXV, and VACV may be regarded as variants originated from CPXV or a CPXV-like virus as a result of deletions and mutations impairing or altering certain genes of the ancestor virus.

Analysis of VACV and VARV genomes demonstrated the amino acid sequences of the potential ORFs of the strains within the species were highly identical (the homology is below 97% in very rate instances; Shchelkunov *et al.*, 2000; Marennikova & Shchelkunov, 1998). However, comparison of the two CPXV strains—GRI and BRT—gives a different picture. For example, of the 50 genes localized to the left variable genomic region, only 13 ORFs display the homology exceeding 97%. As for the right variable genomic region, the number of such ORFs is 17 of the 65. Moreover, identities of the majority of ORFs of these two CPXV strains in these regions are comparable or even lower than those obtained while comparing CPXV-GRI with VACV, MPXV, or VARV (Appendix 2). This may indicate a high heterogeneity of CPXV strains circulating in various geographical regions as well as explain the variation of their biological properties.

Interestingly, the ORFs of the two compared CPXV strains that encode IL-18–binding protein and hemagglutinin display very low identities in their amino acid sequences (Appendix 2). Presumably, this is a result of adaptation of either CPXV strain to certain animal species.

The most unexpected result when comparing ORFs of these two CPXV strains was the discovery of three ORFs with a very low mutual homology. For example, the amino acid identities of the CPXV-GRI ORFs C2L, C11L, and B8R with the homologous CPXV-BRT ORFs amounted to 58.7%, 67.2%, and 66.2%, respectively; moreover, the corresponding nucleotide homologies appeared only slightly higher—67%, 76%, 74%, respectively. However, interspecies comparison of these three ORFs (Figure 6.16) demonstrated that it was possible to find an orthopoxvirus species whose analogous ORF was highly homologous to either CPXV-GRI or CPXV-BRT for each ORF in question. For example, the identities of CPXV-GRI ORFs C2L and B8R with MPXV-ZAI amount to 97.2% and 97.3%, respectively (Figure 6.16a, b). For CPXV-BRT, a highly homologous ORF (analogue of CPXV-GRI B8R) was discovered in CMLV-CMS (95.6%; Figure 6.16c). In

a
```
CPXV-GRI  M-DSRIAIYVLVSASLLSLVNCHKLVHYFNLKINGSDITNTADILLDNYPIMTFDGKDIYPSIAFMVGNKLFLDLYKNIFEEFFRLFR-V
CPXV-BRT  .HLQYY..-S...--.V..ID....AFN...E.....THS.V.VY..DSQ.I.......R.T.P..I.DEI..PF...V.S...S...R.
MPXV-ZAI  .-........................................................................V........-.

CPXV-GRI  SVSSQYEELEYYYSCDYTNNRPTIKQHYFYNGEEYTEIDRSKKAANKNSWLITSGFRLQKWFDSEDCIIYLRSLVRRMEDSNKNSKKLST*
CPXV-BRT  PT.TP..D.T.F.E....D.KS.FD.F.L........--VKTQE.T...M..T..E...K....G....MH......K....KR.TG----*
MPXV-ZAI  ...........................................................T....................--.*
```

b
```
CPXV-GRI  MVVWDSMLYDSCKTF--DACSAQSLVERNENS---------LKAYVTKKNKNIKTDVVMSLLSSANYKNINDFDIFEYIESDNIDVELLR
CPXV-BRT  .-IN.KI........NI........I.SGA.PLYEYNGETP.........N...N...IL....VD............VC...V.I...K
VACV-COP  ..-N.KI........NI..S.....I.SGA.PLYEYDGETP.........N...N...IL....VD............LC....ID..K
VARV-IND  .....M........--.........---------.NV...........V....T..........................

CPXV-GRI  LLIAKGLEINSCKNGINIVEKYATTSNPNVDVFKLLLDQGIPMCSNVSYGYKIIIEKVTGFSSVYDDDDYYQDYIINI----DDKIGKTA
CPXV-BRT  ...S....I.H.................K...T...IQ.....K..QIRRAGEY.NW..ELD..DYDYTTDY..RMC...V
VACV-COP  ...S..I....I...............K...T...IQ.....K..QIRRAGEY.NW..ELD..DYDYTTDY..RM...V
VARV-IND  ...T.................................................KLLT.Q.GTMITITISI.LLLYR*

CPXV-GRI  LYYYIITRSRD--KLSLDVINCLISYEKEILYYTYRSYTTLHYYVGRCDVKREIFDALYDNNYQNNERMNILHNYLRIRYKNKNP-IDNY
CPXV-BRT  .........Q.GYAT......Y...H...MR......EH...Y..LDK..I.........F.S..SGH.L.H..S...KQFR...YK....
VACV-COP  .........Q.GYAT......Y...HK..MR......EH...Y..LDK..I.........F.S..SGH.L.....S....KQFRK..HK....
VARV-IND

CPXV-GRI  ILDRLLDGCGSKDILVLFNMARYNIIYTSI-KRYKYSIQDLLTRYISHGIVRTNIIKCMIDEGAVLYRYKHVNEYFINIHKVDPKVVEYI
CPXV-BRT  .V.K..SEHDTFY..E.C.SL.N...IST.L...TD.......SE.V.YHT.YI.V.........T...F..I.K..QKFDNR.......
VACV-COP  .V.Q..FDRDTFY..E.C.SL.N..LIST.L...TV......LE.V.YHT.YI.V.........T...F..I.K..QKFGNR.......
VARV-IND

CPXV-GRI  LKNGVDVVEDDN--NTINIMPLFPVYVHNNHDEDVLSILKLCKPYIDDINKIDTRGRSILYYCVYYHNTILVEWLVDNGADINIVTKCGY
CPXV-BRT  ....NV..-N..-D.I.........--TLSIRESE......I.............KH.C....H.IES.SVS.....I.......T.TY.S
ECTV-MOS  ..............M....--TFSIRAL.......................KH........IES.SVA.....I.......T.-Y.S
VACV-COP  ....NL..DN..DD.L.........--TFSMREL.......................KH.C....H.IKS.SVS.....I.......I..Y.F
VARV-IND

BOK-GRI   TCIGICIILAYGCIPEIAELYIQILECILSRLPTIECIKSTVENI-KRRYYVYI-YNKSLIEMCIRYFILVDYKYTCDTYPSYIEYITEC
BOK-BRI   ......V.M.HA........I..K...I...K.........K..DYLSND.HLLIGNKT...LKI..K....................D.
BE-MOS    ...S..VMM.DKY.........K...I...K.........Y..K..CYLENN.L.SANIR..RIL.T..K..............L.....D.
BOB-COP   ...T..V...DKY.........K...I...K.........K..DYLDDH..LFIGGN...LKI..K............SM......F..D.
BHO-IND

CPXV-GRI  EKEIADMRQIKINGTDMLTVMYKLNKPTKKRYVNNPIFTDWANKQYKFYNQIIYNANKLIEQSKKIDDMIEEVSIDDNRLSTLPLEIRHL
CPXV-BRT  .............M.IF.I......H......RH.V..E..KR.........E.N...N...A..A.N.....I........
ECTV-MOS  .........V....M...........H....LQIGLTSNISFIIK*
VACV-COP  ....................M.....................................
VARV-IND  MC.......................IE.TKQ.....................N..N...V.N............L....

CPXV-GRI  IFSYAFL*
CPXV-BRT  .......*
VACV-COP  .......*
VARV-IND  ....V..*
```

c
```
CPXV-GRI  MRSLIIVLLFPSIIYSMSIRRCEKTEEETWGLKIGLCIIAKDFYPERTDCSVHRPTASGGLITEGNGFRVVIHDQCTEPHDFIITDTQQT
CPXV-BRT  ....V............V.....M.....K....M..Q.....SK.........DVG........Y..V...E..NP...AT.K..
VACV-COP  .....................................................L....E.........DIRNTDKL*
CMLV      ......F........V.....M.....K....M..Q.....SK.........SDVG........Y..V...E..NP...AT.K..
MPXV-ZAI  ..................................................................Y...............

CPXV-GRI  RLGSSHVYIKFSNMNTGAPSSIPKCSRTLSISVYCDQEAGDLKFEEYTQ-ESSDISIRVKYDSSCIDYLGINQSFMNECIRRIT---TWD
CPXV-BRT  HF.VT.S..E....S......EN..D..KHIL........SG.D.HTLKYV..NYLH.T....T...NH..V.Y......E.KL.SIYET.
CMLV      HF.VT.S..E....S..V.EN..D..KHIL..I......SG.D.HTLKYL..NYLH.T....T...NH..V.Y......N.KLSSVYES.
MPXV-ZAI  ......T..........V.............I.......-...........................................T---..

CPXV-GRI  RESCVGIDTQTINKYLKSCTNTKFDRSVYKRYILKSKALHAKTEL*
CPXV-BRT  TLT.GAK.I..RD....T............THMQ...I..V....*
CMLV      TLT.GAK.I..RD....T............THMQ...I..V....*
MPXV-ZAI  .....R...................N...................*
```

Figure 6-16. Comparison of amino acid sequences of the CPXV-GRI and CPXV-BRT ORFs displaying the lowest homologies: (a) C2L (GRI/BRT, 58.7%); (b) C11L (GRI/BRT, 67.2%) and (c) B8R (GRI/BRT, 66.2%). Asterisks indicated the C-ends of calculated polypeptides.

the case of CPXV-GRI ORF C11L, fragments of this ORF in VARV-IND genome (D6.5L and D7L) display a high homology (86.8% and 96.2%, respectively), whereas the corresponding ORF of CPXV-BRT is highly homologous (88.9%) to analogous VACV-COP ORFs (Figure 6.16b). Note that the identities of the CPXV ORFs flanking these genes do not differ from the average degree of ORF homology in these regions (Appendix 2). These results suggest that the altered variants of CPXV genes might emerge due to recombination with an essentially diverged orthopoxvirus species or a cellular gene through the process of retrotransposition.

Cowpox virus produces in large amounts the late protein of 160 kDa, which forms cytoplasmic A-type inclusions (ATI; Shida *et al.*, 1977). ATI were also detected during mousepox (Marchal, 1930) and raccoonpox (Patel *et al.*, 1986). Variola, monkeypox, and vaccinia viruses synthesize truncated forms of ATI protein due to mutational frameshift and, thus, fail to form inclusion bodies (Fenner *et al.*, 1989). Sequencing of CPXV ATI gene allowed for detecting ten tandemly repeated sequences in this protein (Figure 6.17; Funahashi *et al.*, 1988), which are believed (DeCarlos & Paez, 1991) responsible for aggregation of the protein and formation of visible ATI. Studying structure of the gene responsible for ATI formation, Meyer *et al.* (1994) found there deletions in 13 of the 22 CPXV strains studied. However, these deletions in neither instance influenced the ability of the virus to form inclusion bodies. When sequencing PCR fragments amplified from genomic DNA preparations of Swedish and Norwegian CPXV isolates, deletion of the third tandem repeat in their ATI was detected (Hansen *et al.*, 1999; Figure 6.17—CPXV-SWE and CPXV-NOR). Deletions were also discovered in the homologous protein of ECTV-MOS; these deletions were localized to the region of repeats. Consequently, the number of ECTV-MOS repeats decreased to five. In the case of VACV-WR, the truncated protein isolog contains four repeats of the ten (DeCarlos & Paez, 1991; Gubser & Smith, 2002; Figure 6.17). Only three repeats are present in the analogous VARV-IND and MPXV-ZAI proteins (Shchelkunov *et al.*, 1994b; 2002b). Note that the C-terminal domain located after the repeats is present only in the ATI protein of the viruses forming A type inclusion bodies (CPXV and ECTV). Presumably, this domain is necessary for forming ATI bodies to a greater degree than the repeats considered above.

The protein P4c with a molecular weight of 58 kDa (McKelvey *et al.*, 2002), encoded by the ORF adjacent to ATI gene (CPXV-GRI A27L), is required to direct the IMV particles into ATI bodies. This protein is the major component of IMV tubules. The gene in question is disrupted in the genomes of CPXV-BRT, ECTV-MOS, and VACV-COP (Figure 6.18); however, all the viruses retained a truncated ORF, corresponding to the P4c *N*-terminal domain. This suggests that the domain in question provides a certain function important for the virus, which is different from the function of

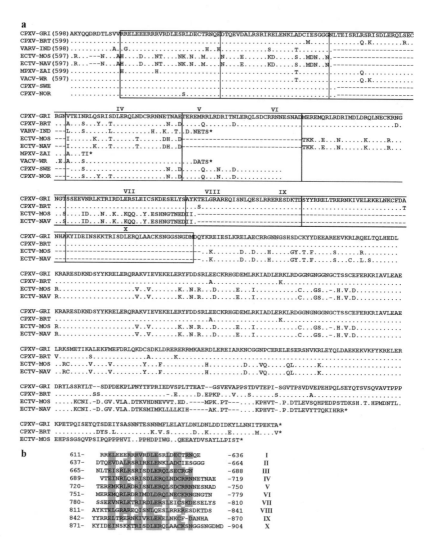

Figure 6-17. (a) Comparison of amino acid sequences of the C-terminal regions of the orthopoxvirus ORFs encoding A-type inclusion bodies: the number of amino acids not included into the alignment is shown in parenthesis to the left of the sequence (first line); repeats are framed; and the Roman numerals above the frame indicate their ordinal numbers. (b) Comparison of ten tandem repeats of CPXV-GRI: numbers of the first and last amino acids of each repeat are given to the left and right of the sequence; Roman numerals indicate their ordinal numbers; coinciding or functionally related amino acids are indicated with dark blocks. The genes of CPXV-SWE and CPXV-NOR are sequenced only partially.

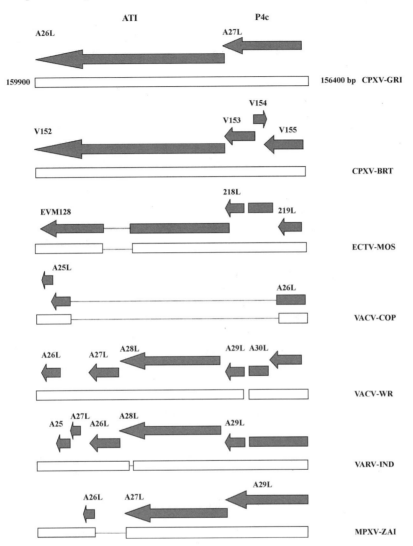

Figure 6-18. Layout of orthopoxvirus ORFs encoding the proteins necessary for production of A-type inclusion bodies (ATI) and direction of IMV particles into these bodies (P4c): arrows indicate the sizes and directions of the corresponding ORFs; ORF names are shown above the arrows; coordinates of the region in question in CPXV-GRI genome, to the left and right; and deletions, by dot lines.

the *C*-terminal domain. The *C*-terminal domain of P4c protein is necessary for the virus for active transport of IMV particles and their direction to ATI (McKelvey *et al.*, 2002), and its absence in CPXV-BRT results in

ATI bodies lacking the virus particles. The P4c protein was not detected in EEV, suggesting Ulaeto *et al.* (1996) to assume that incorporation of this protein into the IMV particle prevents it from further maturation into IEV and EEV.

The phylogenetic analysis of individual genes of a large set of orthopoxvirus strains (see Section 7.5), performed at the Department of S.N. Shchelkunov, demonstrated that unlike the rest species belonging to the genus *Orthopoxvirus,* the strains assigned to cowpox virus display an unusual diversity in the nucleotide sequences of the genes studied. This data allows for hypothesizing that VARV, MPXV, and VACV could have evolved from different CPXV subspecies.

The fact that the sequences homologous to VARV-IND B10R and B11R genes (Meyer *et al.*, 2002a), considered previously unique for VARV, were recently detected in 6 of the 15 CPXV strains studied is an additional illustration to this hypothesis. Shchelkunov *et al.* sequenced this genomic region of two such CPXV strains—91-1 and EP-2—and analyzed its organization. The strain 91-2 was isolated from skin lesions of a domestic cat; strain EP-2, from an elephant during an outbreak of orthopoxvirus infection in a zoo. It appeared that the strain EP-2 encoded in this genomic region an ORF of 5679 bp in length; strain 91-1, of 5676 bp (Figure 6.19). The ORFs differ in 10 nucleotides only, resulting in four amino acids changes (identity of 99.8%). These ORFs were named according to the nomenclature of the CPXV-GRI as B7.5R-EP-2 and B7.5R-91-1. Searching GenBank failed to detect any ORFs or sequences of similar size with a high degree of similarity. However, the ORFs 182R (162 aa), 183R (255 aa), and 184R (223 aa) of CMLV strain CMS showed a high level of identity (97.3, 97.3, and 95.9%, respectively). In a similar way, the ORFs B10R (97 aa) and B11R (65 aa) of VARV-IND and the corresponding ORFs H10R (76 aa), H11R (76 aa), and H12R (138 aa) of VARV-GAR had also an identity of over 90% as compared to EP-2 and 91-1. Previously, these ORFs were considered to contain either CMLV- or VARV-specific sequences.

A considerably less degree of identity, however, spanning virtually the entire protein length was detected when comparing the ORF B7.5R with the largest protein of orthopoxviruses (CPXV-GRI B22R) and its isologs in genomes of poxviruses belonging to other genera (Table 6.7). Properties of the proteins belonging to the B22R family are unknown. However, it is clear that they most likely are transmembrane proteins (Shchelkunov *et al.,* 1993). Isologs of CPXV-GRI B22R are also present in the right terminal region of EP-2 and 91-1 genomes, i.e., each of these CPXV strains contains two genes of B22R family.

Interestingly, fowlpox virus genome harbors a set of six genes belonging to this family (Table 6.7). The rest poxviruses encode one gene homologous to B22R.

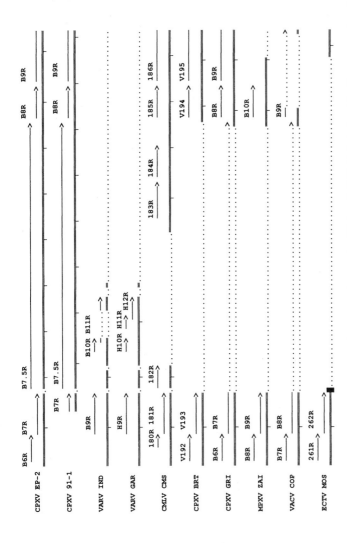

Figure 6-19. Graphic alignment of the genomic region of cowpox virus strains EP-2 and OPV 91-1, isolated in Germany, and analogous regions of cowpox virus GRI-90 (CPXV-GRI) and Brighton (CPXV-BRT); variola major virus strain India-1967 (VARV-IND) and variola minor virus strain Garcia-1966 (VARV-GAR); camelpox virus (CMLV-CMS); monkeypox virus strain Zaire-96-I-16 (MPXV-ZAI); vaccinia virus strain Copenhagen (VACV-COP); and ectromelia virus strain Moscow (ECTV-MOS). A unique species-specific insert in the sequence of ectromelia virus (102 bp) is marked as a black rectangle. ORFs of CPXV EP-2 and 91-1 are designated according to CPXV-GRI.

Table 6-7. Amino acid identity of the ORF B7.5R of cowpox virus strains EP-2 and 91-1 as compared to B22R isologs of other orthopoxvirus species and chordopoxviruses

Genus and species	Strain	ORF	Length (amino acids)	Accession No.	Amino acid identity (%) as compared to EP-2 B7.5R*
OPV CPXV	EP-2	B7.5R	1892	AY519982	100
OPV CPXV	91-1	B7.5R	1891	AY519983	99.8
OPV CPXV	GRI-90	B22R	1933	CAD90748	27.7
OPV CPXV	Brighton Red	CPXV219	1919	NP_619999	28
OPV VARV	India-1967	B26R	1896	NP_042238	28.3
OPV VARV	Bangladesh-1975	B22R	1897	T28621	28.1
OPV VARV	Garcia-1966	D15R	1896	B72175	28.2
OPV MPXV	Zaire-96-I-16	B21R	1879	NP_536609	28.3
OPV CMLV	CMS	CMP202R	1869	AAG37713	27.8
OPV CMLV	M-96	CMLV207	1869	NP_570597	27.8
OPV ECTV	Moscow	EVM169	1924	NP_671688	28
OPV ECTV	Naval	E199	1924	-	28
OPV CPXV	EP-2	B22R	1929	AY519984	27.9
Avipox FWPV	FCV	FPV097	1912	NP_039060	25.2
Avipox FWPV	FCV	FPV098	1802	NP_039061	26.9
Avipox FWPV	FCV	FPV099	1949	NP_039062	27.9
Avipox FWPV	FCV	FPV107	1777	NP_039070	25.2
Avipox FWPV	FCV	FPV122	1870	NP_039085	26.4
Avipox FWPV	FCV	FPV123	1766	NP_039086	25.8
Molluscipox MOCV	SB1	MC035R	2133	NP_043986	30.9
Leporipox SFV	KAS	S134R	1939	NP_052023	38.1
Leporipox MYXV	Lausanne	M134R	2000	NP_051848	37.3
Capripox SPPV	TU-V02127	SPPVgp129	2026	NP_659705	37.5
Capripox LSDV	Neethling 2490	LSDV134	2025	NP_150568	37.5
Suipox SWPV	NEB	SPV131	1959	NP_570291	37.9
Yatapox YLDV		135R	1901	NP_073520	37.6
Yatapox YMTV	VR587	135R	1895	NP_938388	38.9

*Identity was calculated for pairwise alignments of the corresponding sequences using FASTA program (Pearson & Lipman, 1988).

Analysis of the poxvirus proteins in question demonstrated (Figure 6.20) that CPXV B22R proteins cluster with proteins of other orthopoxvirus species and fowlpox virus (genus *Avipoxvirus*), whereas CPXV B7.5R is more related to homologous proteins of other genera from the family

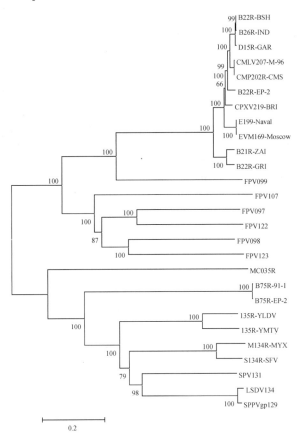

Figure 6-20. Phylogenetic tree based on alignments of amino acid sequences of B7.5R-COW and its homologues in genomes of various chordopoxviruses. Phylogenies were calculated by the neighbor-joining method (Saitou & Nei, 1987) using the software package MEGA (Kumar *et al.*, 1993). Reliability of the phylogenetic relationships was statistically evaluated from 1000 bootstrap replicates. Designations of proteins and viruses: B22R-BSH, variola major virus strain Bangladesh-1975 (GenBank Accession No. T28621); B26R-IND, variola major virus strain India-1967 (NP_042238); D15R-GAR, variola minor virus strain Garcia-1966 (B72175); CMLV207-M-96, camelpox virus strain M-96 (NP_570597); CMP202R, camelpox virus strain CMS (AAG37713); B22R-EP-2, cowpox virus strain EP-2; CPXV219-BRI, cowpox virus strain Brighton Red (NP_619999); E199-Naval, ectromelia virus strain Naval; EVM169, ectromelia virus strain Moscow (NP_671688); B21R-ZAI, monkeypox virus strain Zaire-96-I-16 (NP_536609); B22R-GRI, cowpox virus strain GRI-90 (CAD90748); FPV099, fowlpox virus (NP_039062); MC035R, molluscum contagiosum virus (NP_043986); B7.5R-91-1, cowpox virus strain 91-1; B7.5R-EP-2, cowpox virus strain EP-2; 135R-YLDV, Yaba-like disease virus (NP_073520); M134R-MYX, myxoma virus strain Lausanne (NP_051848); S134R-SFV, rabbit (Shope) fibroma virus (NP_052023); SPPVgp129, sheeppox virus strain TU-V02127 (NP_659705); and SPV131, swinepox virus (NP_570291).

Poxviridae. Possibly, B7.5R gene in genomes of certain CPXV strains was retrotransposed from the host genome or transferred by recombination with a poxvirus. Presumably, further deletion and mutation events could form variola and camelpox viruses from the cowpox virus subspecies containing CPXV-EP-2 and CPXV-91-1. The CPXV subspecies containing the strains GRI-90 and/or BRT could give rise to monkeypox, vaccinia, and ectromelia viruses. If this hypothesis is true, the modern species cowpox virus has subspecies, while their ancestor is more ancient compared to other species of orthopoxviruses.

The performed comparative analysis of the genome organization of various orthopoxviruses (Shchelkunov *et al.*, 1998) has demonstrated that cowpox virus contained the most complete set of the viral genes. Many CPXV genes are unique for this species; however, not a gene is found that would be without its homologue in a CPXV strain. These data suggest that CPXV is the most ancient species among the orthopoxviruses considered, whereas VARV, MPXV, and VACV evolved independently from a common ancestor virus, which very likely may be CPXV or a related orthopoxvirus.

These data also suggest that CPXV comprises several subspecies, whose individual genes may be represented by different variants (an analogy with allelic genes). Occurrence of CPXV subspecies in one geographic area could be of powerful adaptive value for the virus, recombination between various CPXV subspecies upon changes in the environmental conditions could result in emergence of manifold new variants of the virus with altered properties.

Chapter 7

MOLECULAR EVOLUTION
OF ORTHOPOXVIRUSES

7.1   Phylogenetic Interrelations of Orthopoxviruses

With accumulation of sequencing data on complete genomes, clarification of evolutionary interrelations of orthopoxviruses was attempted. At the first stage, it was determined that VARV and VACV evolved from a common ancestor independently from one another (Shchelkunov, 1995). Then, it was discovered that CPXV DNA contained the most complete set of genes found in the viral genomes of this genus (Shchelkunov *et al.,* 1998; Appendices 1 and 2). The rest orthopoxvirus species have lost certain part of the genes with reference to CPXV. Note that VARV contains the least set of actual genes. This observation suggests that CPXV is most close to the ancestor of orthopoxviruses, while the rest species emerged later due to deletions, recombinations, and mutations. Assuming that more virulent virus variants are less effectively preserved in the biosphere, it is logical to infer that variola minor was the first to appear and then evolved in places with a high enough population density into the variola major variant. In addition, the species *Variola virus* is an example of the evolutionary cul-de-sac, as its host range narrowed to one species (humans) and it lost the capability of persisting, which presumably was characteristic of the ancestor virus. Comparison of VARV and MPXV genomes suggested that neither virus is the direct ancestor to the other (Shchelkunov *et al.,* 2001). Of the orthopoxviruses studied, camelpox virus (CMLV; Gubser & Smith, 2002) appeared most closely related to VARV according to phylogenetic analysis of the genomes. However, it was inferred that evolution of one species from the other is unlikely and VARV and CMLV evolved from closely related ancestor, possibly a rodent virus.

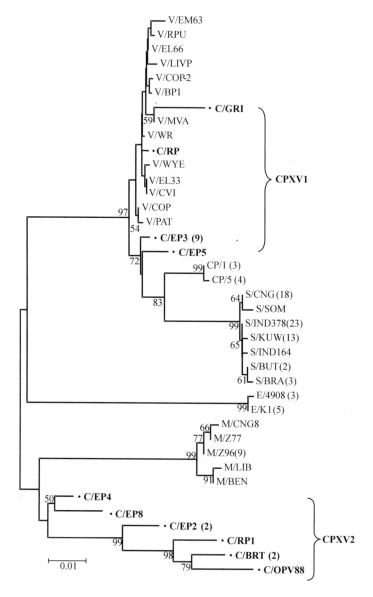

Figure 7-1. Phylogenetic tree based on the coding part of vCCI gene constructed by minimal evolution method. Each branch corresponds to one unique sequence variant. Figures at the branch roots show the results of permutation analysis of statistical significance involving 1000 trees. Only the values exceeding 50% are indicated. The scale shows units of genetic distance. Designations of orthopoxviruses: V, vaccinia virus; C, cowpox virus; S, variola (smallpox) virus; CP, camelpox virus; E, ectromelia virus; and M, monkeypox virus.

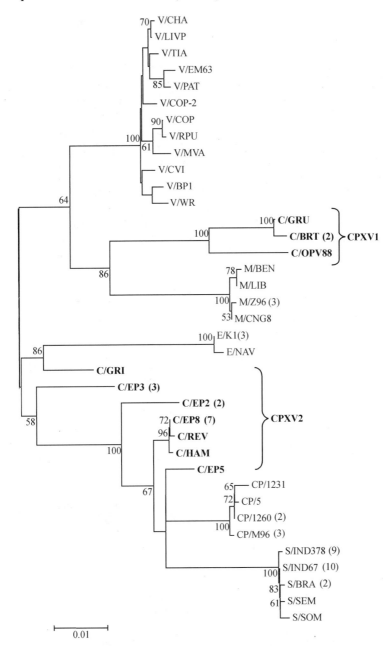

Figure 7-2. Phylogenetic tree based on the coding part of the gene of α/β-IFN-binding protein constructed by minimal evolution method. Designations are as in Figure 7.1.

Comparison of all the sequenced genomes of viruses from the subfamily *Chordopoxvirinae* demonstrated that 90 genes localized to the central genomic region are completely conserved (Upton *et al.,* 2003; Gubser *et al.,* 2004). Within the terminal regions of 10 sequenced orthopoxviruses, only 12 out of about 100 genes are present in every virus (Gubser *et al.,* 2004).

Note that the above conclusions were made basing on comparison of the genomes of a limited set or single strains of each orthopoxvirus species. Consequently, of great interest is expansion of the number of strains of various orthopoxviruses compared. In addition, it is important to find out the phylogenetic patterns of individual viral genes.

The phylogenetic analysis of the genes encoding chemokine-binding protein vCCI (Figure 7.1) and α/β–IFN soluble receptor (Figure 7.2), which we performed, showed that except for CPXV, the rest orthopoxviruses studied form separate compact groups on the phylogenetic trees. In all the cases, CMLV is most closely related to VARV. The phylogenetic relationships for each gene are different. CPXV strains demonstrate a heterogeneity in the sequences of the genes studied that is unusual for the rest species. Arbitrary, all the CPXV genes may be divided according to phylogenetic characteristics into two groups for each gene (Figures 7.1, 7.2). However, it is important that for each gene, the set of strains in any such subgroup is different.

Presumably, these results suggest an independent evolution of virulence genes within individual orthopoxvirus species. It is known that genes of viruses and host cells whose products determine the functionally important ligand–receptor interactions co-evolve relatively quickly (see Section 7.4.3). A considerable heterogeneity of CPXV strains in the structures of such genes may reflect differences in passaging histories of these strains in nature in various animal species. Dissimilar phylogenetic relationships of individual genes of the same CPXV strains may indicate a recombination exchange of these genes between various strains and species.

Note that in the case of A13L and A36L virion proteins, CPXV, similar to other orthopoxvirus species, form separate, comparatively compact groups on the corresponding phylogenetic trees (Pulford *et al.,* 2002).

Thus, the actual evolutionary relationships of orthopoxvirus strains and species may be assessed only by comparing either the complete genomes or the genes from the central highly conservative region of the viral genome.

7.2   Differences in DNA Nucleotide Sequences of Variola and Vaccinia Viruses

Studies performed in various laboratories showed that orthopoxviruses are characterized by high frequencies of both intermolecular and intramolecular homologous recombinations of virus DNA (Fenner &

Comben, 1958; Esposito *et al.*, 1981; Ball, 1987; Kotwal & Moss, 1988b; Fenner *et al.*, 1989; Merchlinsky, 1989; Fathi *et al.*, 1991; Zhang & Evans, 1993; etc.). In addition, cases of nonhomologous DNA recombination of these viruses were described (Ball, 1987; Pickup *et al.*, 1984). Mechanisms of these recombinational events are not always clear. Useful information for elucidating the patterns of recombinational rearrangements of the viral genome can be obtained by comparing DNA sequences of both different strains of the same species and representatives of various species of the *Orthopoxvirus* genus. Thus, we (Shchelkunov & Totmenin, 1995) performed a comparative analysis of DNA nucleotide sequences of two variola virus strains, VARV-IND and VARV-BSH, and one strain of vaccinia virus, VACV-COP, whose genomes have been completely sequenced by 1994.

Our analysis showed various deletions/insertions and regions of multiple mutations in the genome of VARV as compared to VACV. Results of the analysis are shown in Table 7.1. Point mutations and deletions/insertions of 1 and 2 bp are omitted. For the sake of simplicity, hereinafter deletions/insertions will be referred to as deletions, since the available models (see below) imply that the observed rearrangements may occur more often as a result of deletions in pre-existent sequences. In the cases when short tandem repeats are found in one of the genomes compared to the other (asterisked in Table 7.1), it is hard to understand whether these rearrangements resulted from deletion or duplication.

Note that only few short deletions were found in the central conservative region of the orthopoxvirus genome, which generally are a multiple of three, i.e., do not cause a translation frameshift of the corresponding ORFs (Table 7.1). Three deletions not divisible by three are localized to intergenic regions, and only one deletion of 5 bp causes a frameshift of ORF A9L in VARV-IND (A10L in VARV-BSH) and loss of three *C*-terminal amino acids.

In terminal species-specific genomic regions of the viruses in question, eight extended deletions were found (924 to 5894 bp in size) and manifold short deletions (3 to 74 bp). Characteristic of short deletions is that in the virus DNA without these deletions, the corresponding sequences are as a rule flanked by (contain) short direct repeats (Table 7.1), whereas the variant with deletion contains only a single copy of such repeat (Figure 7.3). VACV-COP genome contains 75 short deletions with regard to the two strains of variola major virus compared. VARV-IND and VARV-BSH have 77 identical deletions compared to VACV-COP. More than a third of all short deletions are localized to intergenic regions and, most likely, may influence expression of the correspon ding genes. Each strain of variola virus carries five unique deletions as compared to the other two viruses. In four cases, the deletions VACV-COP and VARV-BSH are identical relative to VARV-IND, whereas in two cases, VACV-COP and VARV-IND have

Table 7-1. Comparison of the genomes of variola major virus strains India-1967 (IND) and Bangladesh-1975 (BSH) and vaccinia virus strain Copenhagen (COP)

Virus(es) with deletion compared to others[a]	Region/ORF with deletion	Deletion size, bp	Repeated sequence (RS)[b]	Number of RS copies in the virus without deletion	Position of the left end of deletion[c]		
					COP	IND	BSH
The left terminal genomic region							
IND/BSH	C23L-C17L compared to COP	5894	–	–	3181	56	723
COP	C17L	7	TATAT	2	9920	251	918
COP	Intergenic region C16L:C17L	6	GAGT	2	9979	317	982
COP/IND	C16L/D1L	23	GTTGAAGACTC-TTCCAGAGAC	2	10110	454	1118
COP	C16L	9	TCC	2	10237	581	1268
COP	C16L	22	GGTTAACC	2	10459	812	1499
COP	Intergenic region C15L:C14L	805	–	–	10532[d]	906	1594
IND/BSH	C14L-C12L compared to COP	3546	–	–	10532[d]	1711	2407
COP	C11R	3	TAT*	2	14505	2137	2833
IND/BSH	D2R/D4R	9	TTA	5	14521	2156	2852
COP	Intergenic region C11R:C10L	12	ATGAGTATTTTT*	2	14636	2261	2957
IND/BSH	D3L/D5L	3	CGT*	3	15240	2875	3572
IND/BSH	Intergenic region D3L:D4L/D5L:D6L	5	TATATA	2	16088	3715	4413
COP	D4R-D6L compared to IND	3188	–	–	16119	3741	4439
IND/BSH	Intergenic region D6L:D6.5L/D8L:D9L	4	CAA	2	16183	6993	7693

IND/BSH	C9L compared to COP	924	—	—	16534	**7340**	**8040**
IND/BSH	Intergenic region D6.5L:D7L/D9L:D10L	14	TGT	2	17555	**7438**	**8133**
BSH	D10L	3	ATA*	5	17633	7504	**8204**
COP	C9L	22	—	—	17671[d]	7540	8237
IND/BSH	D7L/D10L	30	—	—	17671[d]	**7562**	**8259**
IND/BSH	D7L/D10L	27	TAATGGA	2	18026	**7887**	**8584**
IND/BSH	D7L/D10L	6	ATGTT	2	18094	**7922**	**8619**
IND/BSH	Intergenic region D7L:D8L/D10L:D11L	5	TTTC	3	18167	7996	**8693**
IND/BSH	Intergenic region D7L:D8L/D10L:D11L	23	AGACATG	2	18711	**8531**	**9228**
COP	Intergenic region C7L:C6L	4	CGAA*	2	**19275**	9072	9769
COP	C6L	15	CGTCT	2	**19502**	9302	9999
COP	C5L	10	CATAACTA	2	**20068**	9882	10579
COP	C5L	3	ATA*	2	**20462**	10290	10987
COP	C5L	3	TAT	2	**20676**	10507	11204
COP	C4L	9	CTAT	2	**20746**	10579	11276
COP	C2L	14	TTTTTT	3	**23304**	13143	13841
BSH	D16L	14	TTTT	6	—[e]	13149	**13847**
COP	C2L	4	ATCT*	2	23413	**13267**	**13951**

continued

Virus(es) with deletion compared to others[a]	Region/ORF with deletion	Deletion size, bp	Repeated sequence (RS)[b]	Number of RS copies in the virus without deletion	Position of the left end of deletion[c]		
					COP	IND	BSH
IND/BSH	D13.5L/D17L	21	GATATAG	2	23549	13407	14091
IND/BSH	Intergenic region D13.5L:D14L/D17L:D18L	4	AGTT	2	23718	13555	14239
IND/BSH	Intergenic region D13.5L:D14L/D17L:D18L	74	ATAGTTTTT	2	23813	13646	14330
IND/BSH	Intergenic region D13.5L:D14L/D17L:D18L	21	ATAATTTCC	2	24106	13864	14548
COP	C1L	6	TATCTA*	2	24632	14368	15052
COP	C1L	6	TAT	2	24864	14606	15290
COP	C1L	20/18[e]	TTATTTTATT	2	24888	14636	15320
BSH	Intergenic region D18L:P1L	18	TTATTTTTC	2	–[g]	14645	15329
COP	Intergenic region N1L:N2L	17	TTAGT	2	25247	15015	15681
IND/BSH	Intergenic region P1L:P2L	3	AAT	2	25272	15057	15723
IND/BSH	Intergenic region O1L:O2L	8	TATTT	4	27307	17086	17752
IND/BSH	Intergenic region O2L:O3L	4	TGT	2	28091	17863	18529
IND/BSH	Intergenic region C1L:C2L	17	TATA	2	29004	18767	19433
IND/BSH	Intergenic region C1L:C2L	38	GATT	2	29114	18890	19556
COP	K2L	12	CATC	2	30214	19922	20588
IND/BSH	K4L-K5L compared to COP	2009	–	–	30704	20425	21091
IND/BSH	Intergenic region C4R:C5L	11	TTA	3	33379	21092	21758

IND	C5L	3	CAT*	2	34000	21702	22368
COP	F1L	66/45[a]	ATCTATATC	12	34028	21727	22396
BSH	C5L	45	ATCTATATC	12	–[g]	21781	22450
Central conservative genomic region							
COP	F5L	3	ATC*	2	37028	24785	25408
COP	F5L	3	TTA	5	37256	25019	25642
IND/BSH	C9L	3	–	–	37364	25130	25753
IND/BSH	C11L	36	TTCTTA	9	38453	26213	26836
IND/BSH	Intergenic region C11L:C12L	3	AAA*	2	38590	26309	26933
COP	Intergenic region F7L:F8L	9	TAAA	2	38618	26337	26961
COP	Intergenic region F8L:F9L	8	ATTAATTA*	2	38940	26666	27290
IND/BSH	Intergenic region C18L:C19L	3	–	–	45323	33057	33683
IND/BSH	E3L	3	CTA*	2	50887	38611	39237
COP/IND	E3L	6	AGC	2	51255	38976	39602
IND/BSH	Intergenic region E5R:E6R	27	TTATTAG	2	53459	41183	41815
COP	Intergenic region E5R:E6R	7	GTA	3	53511	41208	41840
IND/BSH	Intergenic region E6R:E7R	3	TCA	2	55433	43132	43764
IND/BSH	Intergenic region E6R:E7R	10	TGTAAA	2	55523	43219	43851
IND/BSH	Intergenic region L4R:L5L	24	TATATCT and AATAA[f]	4 2	85960	73639	74272

continued

Virus(es) with deletion compared to others[a]	Region/ORF with deletion	Deletion size, bp	Repeated sequence (RS)[b]	Number of RS copies in the virus without deletion	Position of the left end of deletion[c]		
					COP	IND	BSH
COP	H5R	12	AAAAGGA	2	95224	82882	83515
COP/BSH	H5R/15R	3	AAG*	4	95241	82911	83544
COP	H5R	12	TCTCC	2	95284	82958	83588
COP	H5R	27	TAATGA	4	95356	83041	83671
COP	A4L	3	TGT*	3	116959	104671	105301
IND/BSH	A4L/A5L	27	TAGGAGC	2	116979	104694	105324
IND/BSH	A4L/A5L	6	GTA	2	117014	104702	105332
IND/BSH	A9L/A10L; translation frameshift in IND/BSH	5	GGA	2	121995	109677	110307
IND/BSH	A9L/A10L	3	CTT*	2	122063	109740	110370
COP	A10L	3	AGA*	2	123575	111249	111879
COP	A10L	3	ACC*	2	125225	112902	113532
IND/BSH	A12L/A13L	3	AGT*	2	126209	113889	114519
IND/BSH	A12L/A13L	6	TTT	2	126296	113973	114603
IND/BSH	A13L/A14L	6	GGATT	2	126652	114323	114953
IND/BSH	A16L/A17L	3	ATG*	2	127719	115384	116014
IND/BSH	A19L/A20L	3	ATC*	3	130949	118611	119241

The right genomic terminal region

Strain	ORF	No.	Region of multiple point differences from IND/BSH		137803–137937		
COP	A25R				137803–137937		
COP	A26L compared to IND	2	12	ACGTT	138342	126005	126634
COP	A26L-A29L compared to IND	–	4096	–	138358	126033	126662
BSH	A28L	2	6	TTG	–g	126925	127554
COP	A31R	4	48	TAACAATTA	141382	133155	133777
IND/BSH	A36R	2	3	TTG	142538	134358	134980
IND/BSH	A39R/A38R	2	3	GTA*	144454	136266	136888
IND/BSH	A39R/A38R	2	12	GAACA	144660	136469	137091
COP	Intergenic region A36R:A37R	2	3	TTA	144861	136659	137281
IND/BSH	Intergenic region A40R:A41L	2	23	CAAACGA	145171	136971	137593
BSH	Intergenic region A40R:A41L	2	23	ATTATT	145232	137009	137630
IND/BSH	Intergenic region A40R:A41L	2	3	AAA*	145627	137402	138000
IND/BSH	Intergenic region A40R:A41L	2	6	ATCAAG*	145632	137524	138122
IND/BSH	A42R/intergenic region A41L:A42R; translation frameshift in IND	3	7	GGAT	146908	138673	139271
COP	A39R	2	4	TTAT*	147133	138891	139489
COP	A40R	4	12	GAA	148340	140097	140695
IND/BSH	A46L/A44L	2	3	AAAT	148452	140221	140819
IND/BSH	A46L/A44L	2	3	AGG	148680	140446	141044

continued

Virus(es) with deletion compared to others[a]	Region/ORF with deletion	Deletion size, bp	Repeated sequence (RS)[b]	Number of RS copies in the virus without deletion	Position of the left end of deletion[c]		
					COP	IND	BSH
IND/BSH	A48R/A46R	3	ATG*	4	149778	141541	142439
COP	A43R	3	AAT*	2	149913	141673	142271
COP	A43R	3	ATATTCC	2	149936	141699	142297
IND/BSH	Intergenic region A48R:A49L/A46R:A47L	5	GCTAT*	2	150376	142132	142730
IND/BSH	Intergenic region A48R:A49L/A46R:A47L	3	TAT*	2	150548	142309	142907
COP	Intergenic region A43R:A44L	6	ATGAC	2	150567	142325	142923
IND/BSH	Intergenic region A48R:A49L/A46R:A47L	6	ACATC	2	150740	142503	143101
COP	Intergenic region A44L:A45R	3	–	–	151739	143494	144092
COP	A48R	3	AGA*	2	154282	146037	146635
BSH	A52R-A55R as compared to COP	2498	–	–	157959	149719	150317
IND	J6R; translation frameshift in IND	29	ATATC	2	158129	149887	–[h]
IND	A53R-A55R as compared to COP	1900	–	–	158386	150118	–[h]
IND/BSH	J7R/intergenic region J5R:J6R	9	GCG	2	160498	150329	150358
IND/BSH	J8R:J6R	3	GGA*	3	160935	150754	150782
COP	A56R	6	CAGA	2	161251	151065	151094
IND/BSH	J9R:J7R	3	AGA*	3	161832	151654	151683
COP	A56R	Region of multiple point differences from IND/BSH			162023		

COP	Intergenic region A57R:B1R	6	TAA	162733	152542	152581
COP	B2R	19	CGGC	164519	154332	154361
IND/BSH	Intergenic region B3L:B4L	3	ACC*	164537	154369	154398
IND/BSH	Intergenic region B3L:B4L	3	TGT*	164608	154537	154467
IND/BSH	Intergenic region B5L:B6R/B4L:B5R	5	AACTAAC	165465	155285	155315
IND/BSH	Intergenic region B5L:B6R/B4L:B5R	10	TTAA	165510	155324	155354
IND/BSH	Intergenic region B7R:B8R/B6R:B7R	14	TTAATAA	168370	158170	158200
IND/BSH	Intergenic region B7R:B8R/B6R:B7R	3	—	168574	158358	158388
IND/BSH	Intergenic region B7R:B8R/B6R:B7R	4	GTA	168682	158463	158493
COP	B6R	8	ATCAA	168749	158526	158556
IND/BSH	Intergenic region B8R:B9R/B7R:B8R	3	AAT*	169137	158920	158951
COP	B9R-B11R compared to IND	1434	—	170390^d	160164	160195
IND	Intergenic region B9R:B10R	48	TAATAC	—^{g}	160858	160889
IND/BSH	B9R-B12R compared to COP	1784	—	170390^d	161598	161677
IND/BSH	B12R/B11R	3	GTA	172274	161698	161777
IND/BSH	B12R/B11R	18	GAATTAGT	172366	161787	161866
COP	B12R	5	AGACT*	161838	161838	161917
IND/BSH	B12R/B11R	3	TTA*	172552	161906	161985
COP	Intergenic region B12R:B13R	17	ATCTTC	172560	161960	162039

continued

Virus(es) with deletion compared to others[a]	Region/ORF with deletion	Deletion size, bp	Repeated sequence (RS)[b]	Number of RS copies in the virus without deletion	Position of the left end of deletion[c]		
					COP	IND	BSH
COP	3'-end of B13R/5'-end of B14R; translation frameshift	14	TGTCTCC	2	172902	162319	162398
COP	B14R	27	ATTA	3	173550	162981	163060
IND/BSH	Intergenic region B14R:B15R/B13R:B14R	33	TTTC	4	174081	163539	163618
IND/BSH	B16L/B14L	3	ATG	2	174586	164008	164087
IND/BSH	B16L/B14L	3	AAT	2	174595	164014	164093
COP	B16R	22	ATATATATATA	2	174762	164177	164256
IND	B16L	4	TATA*	8	–[g]	164195	164274
IND/BSH	Intergenic region B17R:B18L/B14L:B15L	15	CGTTTAC	2	175086	164519	164601
COP	Intergenic region B17L:B18R	3	–	–	176261	165680	165762
COP	Intergenic region B17L:B18R	Region of multiple point differences from IND/BSH			176312-176331		
COP	Intergenic region B18R:B19R	7	ATTT	2	178081	167505	167587
IND/BSH	B20R/B17R	6	CATTAT*	2	178187	167615	167697
COP	B19R	6	TAAA	2	178404	167826	167908
IND/BSH	B20R/B17R	6	TCT	2	178455	167883	167965
IND/BSH	Intergenic region B20R:B21R/B17R:B18R	25	CTGAATTATTA-TTA	3	179247	168669	168751
IND/BSH	B21R/B18R	6	AGCGA	2	179578	168975	169057

COP	B21R-B25R compared to IND	4942	–	–	179669	169060	169142
IND	Intergenic region B23R:B24R	3	GAT*	2	–[g]	171930	172015
BSH	Intergenic region B21R:B22R	7	ATAAACT*	2	–[g]	173495	173583
COP	Intergenic region B20R:B21R	9	GAT	2	179747	174081	174220
COP	Intergenic region B20R:B21R	9	TTATTC	2	179792	174133	174214
COP	Intergenic region B20R:B21R	10	TACAAGTA	2	180173	174523	174604
COP	B26R compared to IND	5579	–	–	180346[d]	174706	174787
IND/BSH	B21R-B24R compared to COP	3205	–	–	180346[d]	180283	180370
COP	Intergenic region B24R:B25R		Region of multiple point differences from IND/BSH		183617-183948		
COP	B25R	9	TAAA	2	184188	180899	180983
COP	B25R	11	ACA	3	184300	181020	181104
COP	B25R	12	AAA	2	184454	181186	181270
COP	3'-end of B25R/5'-end of B26R	23	GTTAC	2	184630	181377	181461
COP	B26R	13	CGAG	2	184695	181465	181549
COP	B26R	11	GAA	2	184730	181513	181597
COP	B26R	18	ACAAAA	2	184784	181578	181662
COP	B26R	9	ACG	2	184838	181650	181734
COP	B26R	20	ATAC	2	184900	181723	181807
COP	B27R	9	ATAC	2	184936	181777	181861
COP	B27R	8	AAGAC	2	184970	181820	181904
COP	B27R	13	–	–	185255	182113	182197

continued

Virus(es) with deletion compared to others[a]	Region/ORF with deletion	Deletion size, bp	Repeated sequence (RS)[b]	Number of RS copies in the virus without deletion	Position of the left end of deletion[c]		
					COP	IND	BSH
COP	Intergenic region B27R:B28R	23	GTAT	3	**185306**	182177	182261
COP	Intergenic region B27R:B28R	22	CATC	2	**185497**	182394	182478
COP[i]	Intergenic region B27R:B28R	11	AAAATAT	2	**185565**	182484	182568
IND/BSH[i]	Intergenic region G1R:G2R	3	–	–	185601	**182531**	**182615**
BSH[i]	G2R	3	TAA*	3	185673	182600	**182684**
COP[i]	B28R	15	GTGT	2	**185753**	182695	182776
COP[i]	B28R	22	AGA	2	**185876**	182818	182899
COP[i]	Intergenic region B28R:B29R	15	GGA	5	**186010**	182989	183070
COP[i]	Intergenic region B28R:B29R	22	AAAATA	2	**186299**	183300	183381
COP[i]	Intergenic region B28R:B29R	19	TCTAGCG	2	**186420**	183421	183502
COP[i]	B29R	19	TGCCTG	2	**186742**	183762	183843
IND/BSH	G3R	9	ATCCTC	2	186758	**183797**	**183878**
IND/BSH	G3R	6	AGATCC*	2	186908	**183938**	**184019**

[a] If one or two viruses are marked, it means that deletion is located, respectively, in the genome of one or two viruses indicated.

[b] Sequences of direct repeats in the regions of virus(es) genome without deletions corresponding to the regions of deleted DNA segments of the viruses in question; asterisks indicate tandem repeats.

[c] Numeration of nucleotides accepted for DNA sequences of the viruses compared and deposited with EMBL/GenBank is used; if a corresponding deletion was found in genome of a virus, location of the left deletion end is marked bold.

[d] Location of nonhomologous segments of COP and IND/BSH (See Figure 7.3 for an example for COP, position 17671).

[e] See Figure 7.3.

[f] See Figure 7.3h.

[g] Deletion in the region of large deletion in COP genome as compared to variola viruses.

[h] Region of deletion in BSH genome.

[i] See Figure 7.5A.

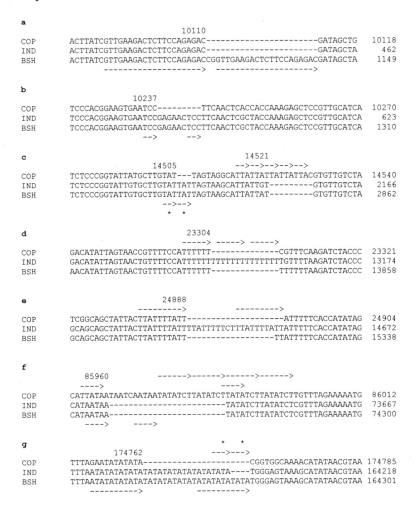

```
a
                        10110
COP    ACTTATCGTTGAAGACTCTTCCAGAGAC---------------------GATAGCTG    10118
IND    ACTTATCGTTGAAGACTCTTCCAGAGAC---------------------GATAGCTA      462
BSH    ACTTATCGTTGAAGACTCTTCCAGAGACCGGTTGAAGACTCTTCCAGAGACGATAGCTA    1149
       ------------------->   ------------------->


b
                        10237
COP    TCCCACGGAAGTGAATCC---------TTCAACTCACCACCAAAGAGCTCCGTTGCATCA    10270
IND    TCCCACGGAAGTGAATCCGAGAACTCCTTCAACTCGCTACCAAAGAGCTCCGTTGCATCA      623
BSH    TCCCACGGAAGTGAATCCGAGAACTCCTTCAACTCGCTACCAAAGAGCTCCGTTGCATCA    1310
                      -->        -->


c                                      14521
                    14505           -->-->-->-->-->
COP    TCTCCCGGTATTATGCTTGTAT---TAGTAGGCATTATTATTATTATTACGTGTTGTCTA    14540
IND    TCTCCCGGTATTGTGCTTGTATTATTAGTAAGCATTATTGT---------GTGTTGTCTA     2166
BSH    TCTCCCGGTATTGTGCTTGTATTATTAGTAAGCATTATTAT---------GTGTTGTCTA     2862
                      -->-->
                        *   *


d                               23304
                            ----->  ----->  ----->
COP    GACATATTAGTAACCGTTTCCATTTTTT--------------CGTTTCAAGATCTACCC    23321
IND    GACATATTAGTAACTGTTTTCCATTTTTTTTTTTTTTTTTTTTTTTGTTTTAAGATCTACCC    13174
BSH    AACATATTAGTAACTGTTTTCCATTTTTT--------------TTTTTTAAGATCTACCC    13858


e                               24888
                           --------->    --------->
COP    TCGGCAGCTATTACTTATTTTATT--------------------ATTTTTCACCATATAG    24904
IND    GCAGCAGCTATTACTTATTTTATTTTATTTTTCTTTATTTTATTATTTTTCACCATATAG    14672
BSH    GCAGCAGCTATTACTTATTTTATT-----------------TTATTTTTCACCATATAG    15338


f
          85960          ------>------>------>------>
          ---->                          ---->
COP    CATTATAATAATCAATAATATATCTTATATCTTATATCTTATATCTTGTTTAGAAAAATG    86012
IND    CATAATAA-----------------------TATATCTTATATCTCGTTTAGAAAAATG    73667
BSH    CATAATAA-----------------------TATATCTTATATCTCGTTTAGAAAAATG    74300
          ---->      ---->


g                                    *   *
           174762               --->--->
COP    TTTAGAATATATATA---------------------CGGTGGCAAAACATATAACGTAA    174785
IND    TTTAATATATATATATATATATATATATATA----TGGGAGTAAAGCATATAACGTAA    164218
BSH    TTTAATATATATATATATATATATATATATATATATGGGAGTAAAGCATATAACGTAA    164301
          ---------->        ---------->
```

Figure 7-3. Examples of short deletions in genomes of viruses VACV-COP, VARV-IND, and VARV-BSH. Arrows indicate the direct repeats. Positions of the left ends of deletions in VACV-COP genome are shown above the sequences (see also Table 7.1). Dash indicates the deletion of a corresponding nucleotide residue with respect to the other sequences compared. Tandem duplications of sequences are asterisked. For each fragment, the position in genome of the last nucleotide residue in each line is shown to the right.

identical deletions compared to VARV-BSH. These data show that virtually all short deletions are species-specific. Moreover, abundance of such deletions, discriminating between VARV and VACV, indicates that these two orthopoxvirus species evolved independently from a common ancestor virus.

In the right terminal region of the viral genomes, four regions of multiple point differences between VARV and VACV were found (Table 7.1). Two of them are localized to intergenic regions. Interestingly, the regulatory sequences of ankyrin-like protein genes are different in both cases. Another relatively short region of multiple nucleotide differences between the viruses in question was found in hemagglutinin gene. Presumably, this indicates that the gene in question is important for the manifestation of species-specific properties of orthopoxviruses. The latter region of multiple mutational alterations corresponds to the potential VACV-COP ORF A25L, which is homologous to the *C*-terminal protein fragment of A type inclusion bodies of cowpox virus and, most likely, not functional in the case of vaccinia and variola viruses.

In terminal genomic regions of the viruses, four regions of nonhomology between variola and vaccinia viruses were also identified (marked with footnote "d" in Table 7.1), which have relatively large sizes and presumably arisen from a series of consecutive deletion/insertion events.

7.3   Mechanisms of Recombinational Rearrangements of Orthopoxvirus DNAs

Taking into account that the mechanisms of recombinational rearrangements of DNA molecules are universal (Ayares *et al.*, 1986), let us consider the data accumulated previously while studying the patterns of deletions in simple laboratory models. In the study of gene *r*II mutants of phage T4, Benzer (1961) found that the layout of mutations in the gene was not random; there are the so-called hot points, where mutations occur with a higher frequency. Sequencing methods allowed for elucidating the molecular basis of such mutational hot points. Using *E. coli lac*I gene, it was shown (Farabaugh & Miller, 1978) that spontaneous deletions and, much more rarely, insertions preexisted between short direct repeats in the initial sequence. Then, deletions (and insertions) between short repeated sequences were found in phages T7 (Studier *et al.*, 1979) and T4 (Owen *et al.*, 1983), in various loci of *E. coli* (Wu *et al.*, 1980; Post *et al.*, 1980; Albertini *et al.*, 1982) and *Bacillus subtilis* (Lopez *et al.*, 1984), in genetic elements of human (Marotta *et al.*, 1977; Efstratiadis *et al.*, 1980), mouse (Brown & Piechaczyk, 1983), etc. It was suggested (Farabaugh & Miller, 1978; Albertini *et al.*, 1982; Owen *et al.*, 1983) that formation of such deletions and inserts may occur due to a slipped mispairing during the DNA synthesis (Figure 7.4). The same works discovered deletions in the regions lacking any repeated sequences (Farabaugh & Miller, 1978; Albertini *et al.*, 1982; Lopez *et al.*, 1984). To explain this type of deletions, the presence of an enzyme was proposed that could bind to certain DNA sequences and produce deletions by a resection and joining mechanism (Lopez *et al.*, 1984).

In recent years, when comparing nucleotide sequences of genome fragments of different orthopoxvirus strains and species, some authors detected short repeated sequences in the region of short deletions and insertions in viral DNAs (Smith *et al.*, 1991; Aguado *et al.*, 1992; Douglass *et al.*, 1994; Shchelkunov *et al.*, 1994a). Analysis of complete coding sequences of two VARV strains and one VACV strain described here demonstrate that deletions (insertions) in the orthopoxvirus genome are relatively frequent (Table 7.1; Figures 7.3 and 7.5A). As a rule, they are flanked by short (3–21 bp) direct repeated sequences (RS). A more detailed analysis allowed us to find (Shchelkunov & Totmenin, 1995) that in many cases short RS in poxvirus DNA do not lead to deletions/insertions (Figure 7.3). We believe that formation of deletions/insertions constantly accompanies the replication of viral DNA and, most likely, is statistically uniform along the overall DNA molecule. However, only those rearrangements are fixed in the genome that do not impair the vital functions of the virus and provide a selective advantage of the emerging mutant over the other variants in the virus population. As a result of such

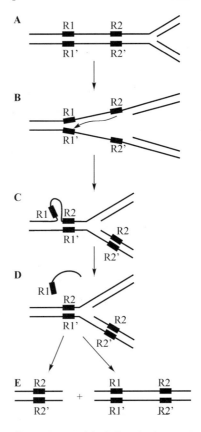

Figure 7-4. Model of slipped mispairing explaining formation of deletions during DNA replication (Efstratiadis *et al.*, 1980). R1 and R2 are short direct repeats; R1' and R2', sequences complementary to R1 and R2; A–E, successive stages of deletion formation.

evolutionary selection, we observe multiple short deletions/insertions in VARV and VACV genomes with respect to each other (Table 7.1).

Of particular importance is the fact that in the regions vital for the virus, RS do not act as hot points of DNA rearrangement (for example, see the highly conservative gene of VACV-COP core protein F17R, Figure 7.5B). On the other hand, in the genes inactivated in VARV or VACV, many deletion events occur in the RS sequences (as an example, see the gene of tumor necrosis factor receptor inactivated in VACV, Figure 7.5B). Presumably, the offspring of any poxvirus always contains subpopulation of

A

```
                 ------>    ------>                                      --
COP   GATAGAAAGAAAATAT-----------CTATATGATTGGAGAAGTAGGAAACAGGAACAC 185598
IND   GATAGGAAGAAAATATTTAAAAAATATCTATATGATTGGAGAAGTAGGAAACAGGAACAC 182528
BSH   GATAGGAAGAAAATATTTAAAAAATATCTATATGATTGGAGAAGTAGGAAACAGGAACAC 182612
                                               ---->    ---->


                 ->  --->                   ***        --->     --->
COP   GACAACGATTACTACATTATTAAATTATGAAGTCCGTATTATACTCGTATATATTGTTTC 185658
IND   GGT---GATTACTACATTATTAAATCATGAAGTCCGTATTATACTTGTATATATTGTTTC 182585
BSH   GGT---GATTACTACATTATTAAATCATGAAGTCCGTATTATACTTGTATATATTGTTTC 182669
                 --->      --->


                              --> --> -->
COP   TCTCATGTATAATAATAAACGGAAGAGATATAGCACCGCATGCACCATCCGATGGAAAGT 185718
IND   TCTCATGTATAATAATAAACGGAAGAGATGCAGCACCGTATACACCACCCAATGGAAAGT 182645
BSH   TCTCATGTATAATAA---ACGGAAGAGATGCAGCACCGTATACACCACCCAATGGAAAGT 182726
      -->-->-->


                                ---->            ---->
COP   GTAAAGACAACGAATACAAACGCCATAATTTGTGT---------------CCGGGAACAT 185763
IND   GTAAAGACACCGAATACAAACGCCATAATCTGTGTTGTTTATCGTGTCCTCCGGGAACAT 182705
BSH   GTAAAGACACCGAATACAAACGCCATAATCTGTGTTGTTTATCGTGTCCTCCGGGAACAT 182786
      -->         -->                         -->-->


                              --> -->                       -->-->
COP   ACGCTTCCAGATTATGCGATAGCAAGACTAACACACAATGTACGCCGTGTGGTTCGGGTA 185823
IND   ACGCTTCCAGATTATGTGATAGCAAGACTAACACACAATGTACACCGTGTGGTTCGGGTA 182765
BSH   ACGCTTCCAGATTATGTGATAGCAAGACTAACACACAATGTACACCGTGTGGTTCGGGTA 182846
      -->        -->                                     -->     -->


                                   --->    --->
COP   CCTTCACATCTCGCAATAATCATTTACCCGCTTGTCTAAGTTGTAACGGAAGA------- 185876
IND   CCTTTACATCTCGCAATAATCATTTACCCGCTTGTCTAAGTTGTAACGGAAGATGCAATA 182825
BSH   CCTTTACATCTCGCAATAATCATTTACCCGCTTGTCTAAGTTGTAACGGAAGATGCAATA 182906
      -->-->                                            -->


                           -->    -->
COP   --------------CGCGATCGTGTAACACTACTCACAATAGAATCTGTGAATGCTCTC 185921
IND   GTAATCAGGTAGAGACGCGATCGTGTAACACGACTCACAATAGAATCTGTGAATGCTCTC 182885
BSH   GTAATCAGGTAGAGACGCGATCGTGTAACACGACTCACAATAGAATCTGTGAATGCTCTC 182966
      -->                 -->->-->                  --->    --->
```

```
COP   CCGGATATTATTGTCTTCTCAAAGGATCATCCGGATGCAAGGCATGTGTTTCCCAAACAA 185981
IND   CCGGATATTATTGTCTTCTTAAAGGATCATCCGGATGCAAGGCATGTGTTTCCCAAACAA 182945
BSH   CCGGATATTATTGTCTTCTTAAAGGATCATCCGGATGCAAGGCATGTGTTTCCCAAACAA 183026
      -->-->


      >        ***                           ---->       ---->
COP   AATGTGGAATAGGATACGGAGTATCCGGA-------------GACGTCATCTGTTCTC 186026
IND   AATGTGGAATAGGATACGGAGTATCCGGACACACGTCTGTTGGAGACGTCATCTGTTCTC 183005
BSH   AATGTGGAATAGGATACGGAGTATCCGGACACACGTCTGTTGGAGACGTCATCTGTTCTC 183086
      -->  -->  -->  -->              -->


          --->    --->                                    --->   --
COP   CGTGTGGTCTCGGAACATATTCTCACACCGTCTCTTCCGCAGATAAATGCGAACCCGTAC 186086
IND   CGTGTGGTTTCGGAACATATTCTCACACCGTCTCTTCCGCAGATAAATGCGAACCCGTAC 183065
BSH   CGTGTGGTTTCGGAACATATTCTCACACCGTCTCTTCCGCAGATAAATGCGAACCCGTAC 183146
                               -->   -->


      ->   -->          -->                              --->    ---
COP   CCAGAAATACCTTTAACTATATCGATGTGGAAATTAATCTGTATCCCGTCAACGACACGT 186146
IND   CCAACAATACATTTAACTATATCGATGTGGAAATTACACTGTATCCAGTTAACGACACAT 183125
BSH   CCAACAATACATTTAACTATATCGATGTGGAAATTACACTGTATCCAGTTAACGACACAT 183206


      >      -->-->   -->                              -->
COP   CGTGTACTCGGACGACCACTACCGGTCTCAGTGAATCCATCTCAACGTCGGAACTAACTA 186206
IND   CGTGTACTCGGACGACCACTACCGGTCTCAGCGAATCCATCTTAACGTCGGAACTAACTA 183185
BSH   CGTGTACTCGGACGACCACTACCGGTCTCAGCGAATCCATCTTAACGTCGGAACTAACTA 183266
                                                 --->          --->


              -->     -->  -->  -->             --->   --->--
COP   TTACTATGAATCATAAAGACTGTAATCCCGTATTTCGTGATGGATACTTCTCCGTTCCTTA 186266
IND   TTACTATGAATCATACAGATTGCAATCCCGTATTTCGTGAGGAATACTTCTCTGTCCTTA 183245
BSH   TTACTATGAATCATACAGATTGCAATCCCGTATTTCGTGAGGAATACTTCTCTGTCCTTA 183326


      >-->                     -->          -->       -->
COP   ATAAGGTAGCGACTTCAGGATTCTTTACAGGAG--------------------AAAGG 186304
IND   ATAAGGTAGCAACTTCAGGATTTTTTACAGGAGAAAATAGATATCAAAATATTTCAAAGG 183305
BSH   ATAAGGTAGCAACTTCAGGATTTTTTACAGGAGAAAATAGATATCAAAATATTTCAAAGG 183386
                                                -----> -----> 
```

```
                                         --->                   --->
COP      TGTGCACTCTGAATTTCGAGATTAAATGCAATAACAAAGATTCTTCCTCCAAACAGTTAA 186364
IND      TGTGTACTTTAAATTTTGAGATTAAATGTAATAACAAAGGTTCTTCCTTCAAACAGCTAA 183365
BSH      TGTGTACTTTAAATTTTGAGATTAAATGTAATAACAAAGGTTCTTCCTTCAAACAGCTAA 183446
           -->            -->     -->          -->-->    -->
           ---> --->
COP      CGAAAGCAAAGAATGATGACGGTATCATGCCGCATTCGGAGACTGTATATCTAGCG---- 186420
IND      CGAAAGCAAAGAATGATGACGGTATGATGTCGCATTCGGAGACGGTAACTCTAGCGGGTG 183425
BSH      CGAAAGCAAAGAATGATGACGGTATGATGTCGCATTCGGAGACGGTAACTCTAGCGGGTG 183506
                   -->--->        -->-->                    ------>
                   ------>------>            -->     -->
COP      ---------------TCGACATCTATATACTATATAGTAATACCAATACTCAAGACTACG 186465
IND      ACTGTCTATCTAGCGTCGACATCTATATACTATATAGTAATACCAATGCTCAAGACTACG 183485
BSH      ACTGTCTATCTAGCGTCGACATCTATATACTATATAGTAATACCAATGCTCAAGACTACG 183566
           ------>
                                        -->           -->
COP      AAACTGATACAATCTCTTATCATGTGGGTAATGTTCTCGATGTCGATAGCCATATGCCCG 186525
IND      AAACTGATACAATCTCTTATCGTGTGGGTAATGTTCTCGATGATGATAGCCATATGCCCG 183545
BSH      AAACTGATACAATCTCTTATCGTGTGGGTAATGTTCTCGATGATGATAGCCATATGCCCG 183626
           -->     -->                                -->     -->
COP      GTAGTTGCGATATACATAAACTGATCACTAATTCCAAACCCACCCGCTTTTTATAGTAAG 186585
IND      GTAGTTGCAATATACATAAACCGATCACTAATTCCAAACCCACCCGCTTTTTA**TAG**TAAG 183605
BSH      GTAGTTGCAATATACATAAACCGATCACTAATTCCAAACCCACCCGCTTTTTA**TAG**TAAG 183686
           -->        -->  -->   -->          ***
```

B

```
                                         ---> --->
                             ------->        ***      ------->
COP      TATAGTAGAATTTCATTTTGTTTTTTTCTATGCTATAA**ATG**AATTCTCATTTTGCATCTG 46854
IND      TATAGTAGAATTTCATTTTGTTTTTTTCTATGCTATAA**ATG**AATTCTCATTTTGCATCTG 34582
BSH      TATAGTAGAATTTCATTTTGTTTTTTTCTATGCTATAA**ATG**AATTCTCATTTTGCATCTG 35208
           ---->                              ---->  -->     -->
           -->   -->                   --->--->          -->
COP      CTCATACTCCGTTTTATATCAATACCAAAGAAGGAAGATATCTGGTTCTAAAAGCCGTTA 46914
IND      CCCATACTCCGTTTTATATCAATACCAAAGAAGGAAGATATCTGGTTCTAAAAGCTGTTA 34642
BSH      CCCATACTCCGTTTTATATCAATACCAAAGAAGGAAGATATCTGGTTCTAAAAGCTGTTA 35268
                             -->   -->          -->        -->
           -->   -->              --->--->          -->
COP      AAGTATGCGATGTTAGAACTGTAGAATGCGAAGGAAGTAAAGCTTCCTGCGTACTCAAAG 46974
IND      AAGTATGCGATGTTAGAACTGTAGAATTCGAAGGAAGTAAAGCTTCCTGCGTACTCAAAG 34702
BSH      AAGTATGCGATGTTAGAACTGTAGAATTCGAAGGAAGTAAAGCTTCCTGCGTACTCAAAG 35328
                 ----> ----> ------>         --->            --->
           --->   --->  ------>                 --->            ----
COP      TAGATAAACCCTCATCACCCGCGTGTGAGGAGAAGACCTTCGTCCCCGTCCAGATGCGAGA 47034
IND      TAGATAAACCCTCATCGCCCGCGAGCGAGAGAAGACCTTCGTCACCGTCCAGATGCGAGA 34762
BSH      TAGATAAACCCTCATCGCCCGCGAGCGAGAGAAGACCTTCGTCACCGTCCAGATGCGAGA 35388
                                                   --->   --->
           --->--->          ---->            --->
           -->         --->                 --->
COP      GAATGAATAACCCTGGAAAACAAGTCCCGTTTATGAGGACGGACATGCTACAAAATATGT 47094
IND      GAATGAATAACCCAGGAAAACAAGTTCCGTTTATGAGGACGGACATGCTACAAAATATGT 34822
BSH      GAATGAATAACCCAGGAAAACAAGTTCCGTTTATGAGGACGGACATGCTACAAAATATGT 35448
                                                   --->--->
           ---->    ---->      -->   -->    *** -->      -->
COP      TCGCGGCTAATCGCGATAATGTAGCTTCTAGACTTTTGAAC**TAA**AATACAATTATATCTT 47154
IND      TCGCGGCTAATCGCGATAATGTAGCTTCTAGACTTTTGTCC**TAA**AATACTATTATATCTT 34882
BSH      TCGCGGCTAATCGCGATAATGTAGCTTCTAGACTTTTGTCC**TAA**AATACTATTATATCTT 35508
           --->         --->  -->    -->
```

Figure 7-5. Comparison of the nucleotide sequences of VACV-COP, VARV-IND, and VARV-BSH in the regions of the genes encoding (A) tumor necrosis factor receptor and (B) 11 kDa core DNA-binding phosphoprotein. Arrows indicate direct repeats; dash, deletion of the corresponding nucleotide residue. Translation initiation and termination codons are bold-faced and marked with three asterisks. In each line, position of the last nucleotide residue of the sequence in the genome of the corresponding virus is indicated to the right.

"new" variants, some of which are nonviable or have neutral alterations in the genome, while others, due to rearrangements in DNA, may acquire useful properties providing their effective reproduction in a certain system (host organism). These new variants may be either point mutants or emerge due to a deletion/insertion. For example, it is known that duplications of genes or parts of genes play an important role in evolution of proteins of both prokaryotes and eukaryotes (Calos *et al.*, 1978; Brown & Piechaczyk, 1983; Clark, 1994). At least one case of gene amplification of VACV under selective pressure was reported (Slabaugh *et al.*, 1988).

Extended deletions in the poxvirus DNA molecules are formed by a different mechanism, as they are not flanked by repeated sequences. Analysis of this type of deletions in VARV and VACV (Table 7.1) as well as in spontaneous white variants of cowpox virus (Pickup *et al.*, 1984) allowed us (Shchelkunov & Totmenin, 1995) to reveal interesting features in the sequences adjacent to the boundaries of deleted region (Table 7.2). We believe that the consensus sequence flanking deletions the type in question looks as follows:

Table 7-2. Junction regions at the sites of nonhomologous recombination of orthopoxvirus DNAs

Virus(es) with rearrangements in DNA	Left end		Right end	Deletion size, bp
Deletions				
IND/BSH	GATGCA*TCAGA*A	//	*TG*TATCGC*ATTTA*TT*GGAG*	5894
IND/BSH	ACCGGC*TGAG*TA	//	CC*TA*CACT*ATTATATA*T*GA*	3188
IND/BSH	TTCGC*ATT*CGGT	//	*TT*TGTCAAGA*TAATAA*TAG	924
IND/BSH	ATAT*TCAGG*AAT	//	*TTTCC*AATATATGT*AAT*CA	2009
COP	TTCA*TGCG*TCTC	//	*TGTG*ATCC*GTTTA*CG*TT*AA	4096
BSH	TATG*TACAG*AGG	//	*TG*GAGGAGT*TGCTGA*T*GAA*	2498
IND	ATA*TTTGT*TATT	//	*TT*CACGTAGA*TATAGGTGT*	1900
COP	AAAG*TACAC*AAC	//	ACAA*TGA*C*TTATAAAAATA*	4942
BR-W10	GTA*TTGGA*TCCT	//	*TC*AT*TACTTCTG*CATC*TAT*	ca. 2000
Deletions/				
duplications				
BR-W5	TAGAGACAAATA	//	*TA*TAC*G*TGG*A*AG*TA*TATGA	
	→ —→		← ←	
BR-W6	AGAAAATTCATA	//	*TA*T*CC*CAA*TT*T*A*CGAGCCC	
	→ —————— —→		← ————— ←	
BR-W8	ACGGATATCTAA	//	*TT*TAT*CCA*T*CCA*GTA*T*GGG	
	→—→ —→		← ← ←	

Notes: In the case of VARV-IND and -BSH and VACV-COP, deletions were found during concurrent comparison of their DNA sequences. BR-W isolates are spontaneous white variants of CPXV Brighton redpox strain characterized earlier (Pickup *et al.*, 1984). Nucleotides at conservative positions for the junction regions compared are marked with bold italics. Arrows below the sequences indicate inverted repeats.

T-x-A-G-A-$(x)_{2-4}$ // T-x-T-$(x)_{1-2}$ -C-$(x)_{2-3}$ -A-T-$(x)_{1-2}$ -T-A-$(x)_{1-2}$ -T,

where "x" is any nucleotide residue.

Thus, the data suggest potential existence of an enzyme of several enzymes that recognizes this consensus sequence with subsequent formation of deletion according to a resection and joining mechanism.

Therefore, our analysis demonstrates that deletions (duplications) play an important role in the evolution of orthopoxviruses. Apparently, two types of deletions occur in the poxvirus DNA. Most frequently, short deletions occur (or preserved), which are generally flanked by RS (Figures 7.3 and 7.5) and result from a slipped mispairing during replication of viral DNA. The second-type deletions are commonly larger, not flanked by RS, and presumably, require a viral enzyme yet not identified.

An important result of this work (Shchelkunov & Totmenin, 1995) is that for the first time the species specificities of orthopoxviruses in the range of short deletions/insertions in their DNAs with respect to each other were discovered.

7.4   Multigenic Families

7.4.1   Ankyrin-like Proteins

Metabolism of animal viruses depending on their specific features suggests the use of structures of cytoplasmic or nuclear protein cell skeleton, as well as cell membranes. In particular, virus-specific cytopathic effect can be caused by a specific reconstruction of the elements of cell skeleton and certain membranes. The aim of this reconstruction is to create the conditions for viral replication (Luftig, 1982). The cytoskeleton proteins are encoded by a large set of genes, which are characterized by tissue-specific expression. This accounts for the difference in the protein contents of "framework" of various cell types, which affects the function of these cells (Steinert & Roop, 1988). Various cell types may also differ in their membrane contents. The above differences can also influence such a parameter of virus propagation inside the organism as tissue tropism.

VACV does not propagate on CHO cell line, although this line is permissive for the other orthopoxvirus, CPXV. It was shown (Spehner *et al.,* 1988) that CPXV gene coding for the protein of 77 kDa was responsible for this property of the virus and when integrated into VACV genome, allowed the latter virus to propagate in CHO cells. Another gene that is necessary for VACV virus propagation in a number of human cell cultures was found in the genome of this virus (Gillard *et al.,* 1986). The results obtained proved that the presence of the gene family that regulates the ability of these viruses to multiply in different cell types is characteristic of large cytoplasmic viruses, such as poxviruses. Computer analysis detected that the above genes of orthopoxviruses code for the proteins containing ankyrin-like repeats (Shchelkunov *et al.,* 1993b). This observation is of considerable interest, since ankyrin and ankyrin-like proteins control the interactions between the integral membrane proteins and elements of cell cytoskeleton (Steinert & Roop, 1988; Lux *et al.,* 1990). Taking into account the importance of this type of proteins in determining the properties of viruses, we performed a computer analysis of nucleotide sequences of VARV, MPXV, CPXV, and VACV. Families of ankyrin-like proteins specific of each virus were discovered.

This is the most numerous family of orthopoxvirus proteins, being encoded exclusively in the left and right variable parts of the genome. For example, only the left genomic regions of CPXV-GRI and CPXV-BRT harbors ten genes of this family each, while the right genomic region—six (two outermost ankyrin like genes in the terminal regions are localized to TIRs; Figure 7.6). It is interesting that in VARV, MPXV, and VACV genomes, most of these genes either are absent at all or are represented by reduced fragments (Table 7.3; Figure 7.6).

Analysis of amino acid sequences of ankyrin-like proteins of orthopoxviruses allowed us to identify in each protein a different number of the so-called ankyrin repeats, typical of proteins from this family (Figure 7.7). Comparison of 54 such repeats, mainly located in tandems, enabled us to establish a consensus sequence of the ankyrin repeat for CPXV proteins (Figure 7.8).

In most conserved regions of the ankyrin repeat (4 to 10 and 20 to 25), amino acid residues are assumed to create α-helical structures (Lux *et al.,* 1990) that may adjoin each other in an anti-parallel way due to hydrophobic interaction of amino acids located at one side of each α-helix (positions 6, 10, and 21, 25, respectively). The regions between these conservative blocks in different repeats vary in the number of amino acid residues and can be characterized as having an increased level of polar and charged amino acids.

Identity rate of different ankyrin-like proteins from the orthopoxvirus family in the regions bearing tandems of ankyrin repeats amounts to approximately 25%. Rather moderate homology level of the *hr* genes and other genes from ankyrin-like protein family, although giving no guarantee of the identity of their biological functions, suggests that the proteins of this family to be involved in determining the virus host range.

Thus, cowpox virus contains in its genome all the known ankyrin-like genes of orthopoxviruses. Probably, just all these genes together underlie an extremely wide CPXV host range in nature and are important for its stable maintenance in the biosphere.

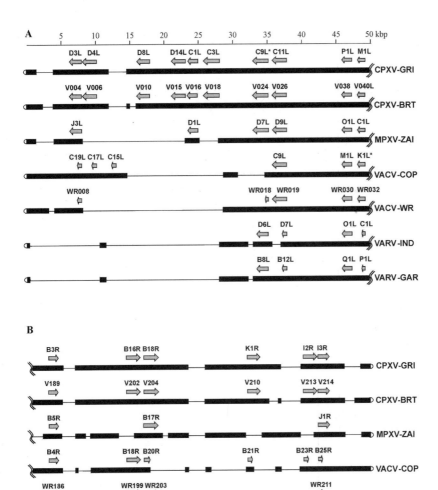

Figure 7-6. Graphical alignment of (A) left and (B) right terminal species-specific genomic regions of CPXV-GRI, CPXV-BRT, MPXV-ZAI, VACV-COP, VACV-WR, VARV-IND, and VARV-GAR. Arrows indicate location, direction, and size of ankyrin-like ORFs. Names of ORFs are given above the arrows. Asterisks mark the genes whose host range function was demonstrated experimentally. Fine lines indicate deletions on one virus relative to the other.

Table 7-3. Orthopoxvirus ankyrin-like proteins

VARV-IND		VARV-GAR		MPXV-ZAI		CPXV-GRI		VACV-COP		VACV-MVA	
ORF	Size, aa	ORF	Size, aa	ORF	Size, aa	ORF	Size, aa	ORF	Size, aa	ORF	Size, aa
–	–	–	–	J3L*	587	D3L*	586	C19L*	259	–	–
–	–	–	–	–	–	D4L*	672	C17L*	386	4L*	233
–	–	–	–	–	–	–	–	C15L*	91	–	–
–	–	–	–	–	–	D8L	661	–	–	–	–
–	–	–	–	–	–	D14L	764	–	–	–	–
–	–	–	–	D1L*	437	C1L	437	–	–	–	–
–	–	–	–	–	–	C3L	833	–	–	–	–
D6L	452	B8L	355	D7L	660	C9L	668	C9L	634	16L	297
D7L	153	B12L	132	D9L	630	C11L	614	M1L	472	–	–
O1L	446	Q1L	449	O1L	442	O1L	474	K1L	284	22L	98
C1L	66	P1L	66	C1L	284	M1L	284	B4R	558	171R	177
B6R	558	H6R	558	B5R	561	B3R	558	B18R	574	186R	574
B19R	574	D8R	574	–	–	B16R	574	B20R	127	–	–
B21R	787	D10R	787	B17R	793	B18R	795	B21R*	91	–	–
–	–	–	–	–	–	K1R	581	–	–	–	–
–	–	–	–	N4R*	437	–	–	–	–	–	–
–	–	–	–	–	–	H2R*	672	B23R*	386	190R*	233
G1R	585	G1R	585	J1R*	587	H3R*	586	B25R*	259	–	–

Note: Asterisks indicate ORFs that are duplicated in the left and right terminal inverted repeat regions (TIRs).

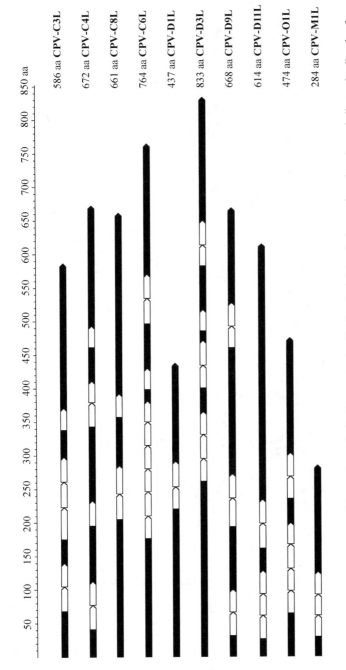

Figure 7-7. Graphical alignment of CPXV-GRI ankyrin-like proteins localized to the left terminal genomic region. Arrow indicates the direction from N- to C-end; light blocks, ankyrin repeats.

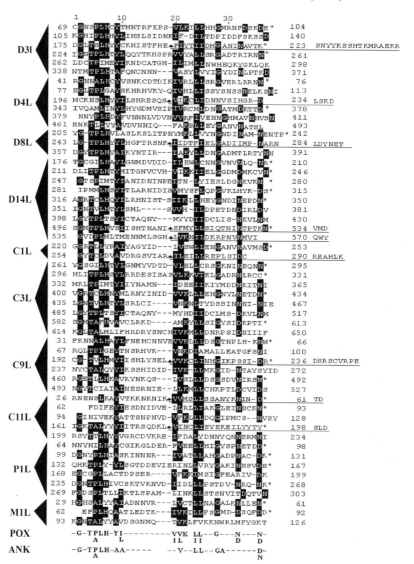

Figure 7-8. Comparison of amino acid sequences of ankyrin repeats of CPXV-GRI ORFs localized to the left terminal genomic region. Black blocks mark conservative amino acid residues in the sequences compared; figures to the left and right, positions of the beginning and end of ankyrin repeats in the corresponding viral proteins. The asterisk to the right of repeat indicates that this repeat is followed by another without interruption. Two lower lines are the conservative sequence of ankyrin repeat of poxvirus protein (POX), deduced in our work, and similar sequence for ankyrin-like proteins published earlier (ANK; Lux *et al.*, 1990).

7.4.2   Kelch-like Proteins

It is believed that poxviruses during their evolution have incorporated into their genome a number of nucleotide sequences encoding various cellular proteins that are most of all necessary for modulating the host immune responses to viral infection, allowing thereby the viruses to overcome various host defense systems (Kotwal & Moss, 1988a; Smith & Chan, 1991; Spriggs, 1996; Alcami & Koszinowski, 2000). In addition, the genes whose products are homologous to various representatives of cellular protein families not directly involved in immunomodulation but regulating cell cycle, cytoskeleton formation, etc., were also discovered in the genome of orthopoxviruses (Shchelkunov *et al.,* 1993b; 2000; Senkevich *et al.,* 1993). One of such families unites the so-called kelch-like proteins (Adams *et al.,* 2000); however, the function of these proteins of orthopoxviruses is still vague. We performed a comparative analysis of the kelch-like proteins and their genes in the orthopoxviruses pathogenic for humans.

Repeated genes displaying a high mutual homology were discovered while sequencing and transcriptional mapping of the terminal genomic regions of Shope fibroma virus (SFV), belonging to the genus *Leporipoxvirus* of the family Poxviridae, and designated T6, T8, and T9 (Upton & McFadden, 1986; Macaulay *et al.,* 1987). Similar genes were then detected in the genome of related myxoma virus (Upton *et al.,* 1990) and vaccinia virus belonging to the *Orthopoxvirus* genus (Goebel *et al.,* 1990). It appeared that specific of the proteins coded for by these genes were five repeats of about 50 aa in their *C*-terminal regions; in turn, typical of these repeats were the presence of glycine–glycine dipeptides (Upton *et al.,* 1990). The function of these viral proteins was not revealed, and the search for cellular homologues was unsuccessful. However, it was soon found that the *kelch* gene of *Drosophila* codes for a protein with a length of 688 aa, whose *C*-terminal region contained elements of about 50 aa repeated tandemly six times and displaying similarity to the repeats of T6, T8, and T9 proteins of poxviruses in their organization (Xue & Cooley, 1993). Kelch protein is a structural component of the so-called ring channels, through which the cytoplasm of nurse cells is transported into the oocyte during *Drosophila melanogaster* oogenesis (Xue & Cooley, 1993). The motif detected was named kelch motif (Bork & Doolittle, 1994), and the block of these repeats, the kelch domain.

It was demonstrated that kelch protein additionally contained the so-called BTB domain with a length of 115 aa, typical of several other *Drosophila* proteins, in its *N*-terminal region (Zollman *et al.,* 1994). Further studies have demonstrated that BTB domains of protein molecules may associate forming homodimers or interact with BTB domains of heterologous protein molecules forming heterodimers (Ahmad *et al.,* 1998).

It was demonstrated experimentally that the kelch domain is necessary for the kelch protein to bind to actin filaments of the cell, while dimerization of BTB domains of two actin-bound kelch proteins resulted in cross-interaction of these filaments and formation of intercellular ring channels of the *Drosophila* oocyte (Robinson & Cooley, 1997).

Search for the kelch repeats and BTB domains in databases on amino acid sequences has demonstrated that both motifs are ancient and had distributed among various types of organisms during the evolution (Ahmad *et al.*, 1998; Adams *et al.*, 2000). Thus, a superfamily of the proteins containing kelch repeats was formed. Proteins of the family are involved in various aspects of the cell function, such as changes in the cytoskeleton and cell plasma membrane, regulation of gene expression, mRNA splicing, etc. The degree of identity between individual kelch repeats appeared to be rather low: six motifs of *Drosophila* kelch protein displayed 25–50% identity, while individual kelch motifs from different proteins showed as little as 11% identity (Bork & Doolittle, 1994; Adams *et al.*, 2000). Duplication of glycine residue and specifically arranged aromatic amino acids are an important pattern characterizing the kelch motif (Xue & Cooley, 1993; Bork & Doolittle, 1994). The number of kelch motifs in different proteins usually varies from four to seven. In addition, localization of the kelch domain may differ, thereby dividing the kelch-like proteins into five groups (Adams *et al.*, 2000). The kelch-like proteins of orthopoxviruses were attributed to the group of the *Drosophila* kelch protein basing on the structural homology (Xue & Cooley, 1993; Bork & Doolittle, 1994; Adams *et al.*, 2000).

The genes of orthopoxvirus kelch-like proteins are localized only to the terminal variable regions of the orthopoxvirus genome (Figure 7.9) and display species-specific differences in the size of potentially encoded proteins (Table 7.4; Shchelkunov *et al.*, 2002b). The genome of CPXV-GRI encodes six proteins of the superfamily in question with a length of about 500 aa and the degree of identity of their amino acid sequences in the range of 22–26%. Interestingly, the identity between T6 protein of SFV and its homologues within the same genome is rather high—59.3% to T9 and 32.8% to T8; the identity between T6 and homologous orthopoxvirus proteins is considerably lower (15.4–20.9%). VACV encodes only three full-size kelch-like proteins (C2L, F3L, and A55R), which are highly homologous to the corresponding CPXV-GRI proteins (99.4, 97.9, and 98.6% identity, respectively). MPXV genome encodes only one full-size kelch-like protein C9L, displaying 97.3% identity to CPXV-GRI protein G3L. In the genome of VARV, all the corresponding potential open reading frames (ORFs) are destroyed due to multiple mutations; therefore, only short potential ORFs that are most likely nonfunctional fragments of the genes of a precursor virus are detected (Figure 7.9; Table 7.4).

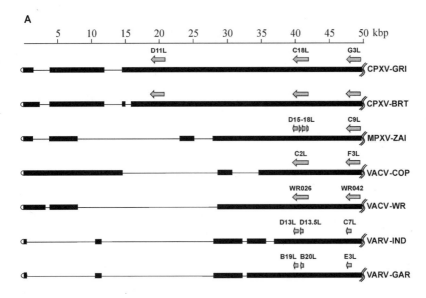

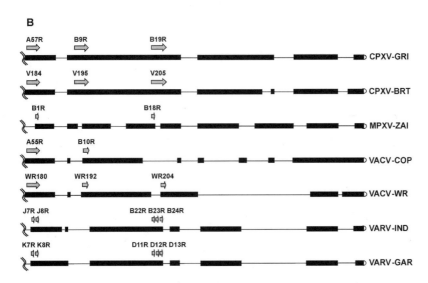

Figure 7-9. Layout of the ORFs of kelch-like proteins of CPXV-GRI, CPXV-BRT, MPXV-ZAI, VACV-COP, VACV-WR, VARV-IND, and VARV-GAR in (A) left and (B) right terminal species-specific regions of the orthopoxvirus genome. Arrows indicate the location and direction of kelch-like ORFs; fine lines, deletions in DNAs of viruses relative to the others.

Table 7-4. Orthopoxviral kelch-like proteins

VARV-IND ORF	Size, aa	VARV-GAR ORF	Size, aa	MPXV-ZAI ORF	Size, aa	CPXV-GRI ORF	Size, aa	VACV-COP ORF	Size, aa	VACV-MVA ORF	Size, aa
—	—	—	—	—	—	D11L	521	—	—	—	—
D13L	201	B19L	154	D15L	105	C18L	512	C2L	512	—	—
—	—	—	—	D16L	77						
D13.5L	79	B20L	65	D17L	98						
—	—	—	—	D18L	107						
C7L	179	E3L	179	C9L	487	G3L	485	F3L	480	31L	476
J7R	71	K7R	71	—	—	A57R	564	A55R	564	—	—
J8R	172	K8R	70	B1R	70						
—	—	—	—	—	—	B9R	501	B10R	166	178R	158
B22R	70	D11R	70	B18R	70	B19R	557	—	—	—	—
B23R	83	D12R	127	—	—						
B24R	88	D13R	88	—	—						

All the six kelch-like proteins of CPXV-GRI contain the *N*-terminal BTB domain (Figure 7.10) and *C*-terminal kelch domain (Figure 7.11). Each kelch repeat comprises four antiparallel β-chains, forming a sheet or the so-called blade of β-propeller, constituted by the entire set of these repeats. It is likely that this β-propeller structure is important for specific protein–protein interactions. Note that each motif in the regions between β-chains 2 and 3 (Figure 7.11) displays differences in both their amino acid sequences and lengths. It is believed that the loops formed by these sequences provide the specificity of protein–protein interactions of each individual β-propeller (Adams *et al.*, 2000).

Note that only ORF A54R of CPXV contains five classic kelch motifs. The fifth *C*-terminal motif of B9R lacks the duplicated glycine; the fourth motifs of G3L and C18L contain single glycine instead of duplicated; and the glycine duplication is absent in the fifth motif of C18L. The kelch motif is damaged in the fourth and fifth repeats of B19R and in the first and third repeats of D11L (Figure 7.11). The kelch motif is destroyed completely only in the *C*-terminal repeats of kelch domains of CPXV-GRI proteins B9R, C18L, and B19R. It is likely that these impairments fail to affect the β-propeller structure formed by the preserved kelch motifs (Adams *et al.*,

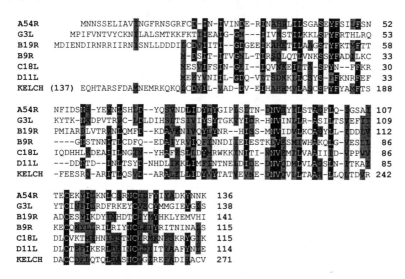

Figure 7-10. Comparison of the amino acid sequences of the *N*-terminal BTB domain of *Drosophila* kelch protein and the kelch-like proteins of CPXV-GRI—proteins A54R, G3L, B19R, B9R, C18L, and D11L. Vertical black and gray blocks mark the conservative amino acid residues and the residues similar in their physicochemical characteristics, respectively, detected in over a half of the sequences analyzed. Left parenthesis contains the number of amino acid residues in the kelch protein before the sequence shown. The number of amino acid residues in the protein sequences is shown to the right of the lines.

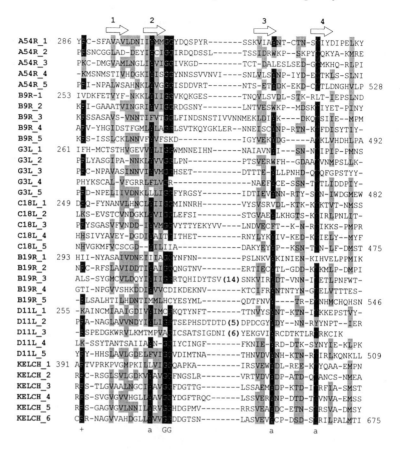

Figure 7-11. Comparison of amino acid sequences of kelch motifs (1-5) from CPXV-GRI proteins A54R, B9R, G3L, C18L, B19R, and D11L with analogous motifs (1-6) of *D. melanogaster* kelch protein. Each kelch motif contains four b-chains, indicated above by arrows and designated as 1-4. Dashes mark deletions of amino acid residues with respect to the other sequences compared. Parenthesis contains the number of extra amino acid residues in a motif. Vertical black blocks show the duplicated glycine (GG) and other conservative residues; vertical gray blocks, residues similar in their physicochemical properties, detected in over a half of the sequences analyzed. Left and right numbers indicate the positions of the first and last amino acid residues in the kelch repeat of each protein.

2000). Thus, the kelch-like proteins of CPXV not only differ in their amino acid sequences, but probably may display considerably different architecture of their β-propeller domain, formed by different number (three to five) of circularly arranged layers/blades.

Despite kelch-like proteins of orthopoxviruses are numerous, their functions are yet vague. It is demonstrated that kelch genes of VACV are inessential for growth in cell culture (Kotwal & Moss, 1988a; Perkus *et al.*, 1991), however, the VACV mutant lacking C2L gene has different plaque morphology on cell culture monolayer due to an altered cytopathic effect on infected cells. Mutant-infected cells showed a reduction in the formation of VACV-induced cellular projections and in the VACV-induced Ca^{2+}-independent cell/extracellular matrix adhesion phenotype (Miranda *et al.*, 2003). In a murine intradermal model, VACV C2L kelch-like gene product reduces the cell infiltrate in VACV-infected ears and reduces pathology associated with VACV infection, promoting healing of lesions.

We constructed mutant variants of a CPXV strain with targeted deletions of one to four genes of this family, namely, D11L, C18L, G3L, and A57R. It was demonstrated that the following effects occurred in a dose-dependent manner with increase in the number of the genes deleted: (1) CPXV virulence for BALB/c mice infected intranasally reduced (according to analysis of LD_{50}); (2) morphology of pocks formed on chorioallantoic membranes of chick embryos changed and the pocks became smaller; and (3) range of sensitive cells altered—deletion mutants lacking three genes displayed a considerably decreased ability to reproduce in MDCK cells; mutants lacking four genes lost this ability at all (Kolosova *et al.*, 2003; Kochneva *et al.*, 2004).

The data of the comparative computer analysis of kelch-like genes of orthopoxviruses allow us to conclude that CPXV is the most ancient and may be considered as an ancestor of the other orthopoxvirus species pathogenic for humans. A weak homology between kelch-like proteins of the same virus and a high homology between isologues of different orthopoxvirus species may have two major explanations. The first explanation involves independent integration of DNA copies of RNAs of different kelch-like proteins belonging to one initial animal host or belonging to various hosts with the expansion of its host range into the ancestral orthopoxvirus genome. Consequently, a variant of cowpox virus appeared to evolve into other orthopoxvirus species. The second explanation implies integration of the coding sequence of a cellular kelch-like protein into the genome of an ancestral orthopoxvirus, its subsequent multiplication in the genome, and long independent evolutions of each copy. If the latter hypothesis is true, than the ancestral orthopoxvirus had a very ancient origin and divided into modern species recently from the evolutionary standpoint.

7.4.3   Tumor Necrosis Factor Receptor Family

Comparison of amino acid sequences of a great number of various types of human and rodent polypeptides revealed most pronounced interspecies differences in the sequences of the proteins forming ligand–receptor pairs of the organismal protective systems of these mammals against infectious agents. In addition, the polypeptide ligands and their receptors proved to be subjected to co-evolution (Murphy, 1993)

Pathogenic microorganisms are assumed to be able to cause an accelerated evolution of the defense system proteins (genes) of infected animal species. Such evolutionary changes in the primary structure of the proteins constituting ligand–receptor pairs were suggested to result in alterations of the quaternary structure of the ligand–receptor contact region (Murphy, 1993; Beutler & van Huffel, 1994). As a result, the species-specific mimicry of mammalian defense system proteins may emerge, providing a narrowed range of hosts sensitive to certain infectious microorganism (Murphy, 1993).

Recently accumulated data demonstrate that cytokines and their specific receptors undergo considerable structural alterations during the evolution. Such alterations may occur relatively rapidly on the evolutionary scale. Thus, mouse γ-IFN interacts weakly with human γ-IFN receptor and vice versa. On the other hand, this cytokine binds to its receptor with a high efficiency within the species (Beutler & van Huffel, 1994). It is evident now that any cytokine can function only in a pair with the receptor located on the plasmatic membrane of the cell. That is why the question of co-evolution of cytokines and their receptors and, on a more global scale, of various receptor–ligand pairs, is of great interest.

Protein families similar to tumor necrosis factor (TNF; Tracey & Cerami, 1994) and its receptor (Smith C.A. *et al.,* 1994), which are recently rapidly expanding, attract considerable interest of researches. It is suggested that a complex network of TNF-like receptor–ligand pairs, which are involved in protective functions in the animal organism, has been formed during the evolution as a result of duplications and subsequent modifications of ancestor genes of a receptor and its ligand (Beutler & van Huffel, 1994). Deletion of the genes for TNF and γ-IFN, the most well-studied cytokines, or the genes for their receptors was demonstrated to increase considerably the sensitivity of animals to pathogenic action of such intracellular parasites as *Listeria,* mycobacteria, and viruses (Havell, 1989; Pfeffer *et al.,* 1993; Dalton *et al.,* 1993; Huang *et al.,* 1993).

TNF-like proteins and/or their receptors may be involved in realizing partially overlapping functions, therefore enabling a relative stability of such multicompetent molecular system and the possibility of its fine regulation (Beutler & van Huffel, 1994).

The interaction network of cytokines and their receptors has been so far studied only to a first approximation, and many discoveries are still awaiting researchers in this direction. Orthopoxviruses can play an important role here. The genes encoding cell-secreted (soluble) receptors for TNF, γ-IFN, α/β-IFN, interleukin 1β (IL-1β), and IL-18 were discovered in these viruses.

Thus, orthopoxviruses are capable of inhibiting a number of endogenous cytokines of the host organism. Note here that different species of orthopoxviruses vary in the set of genes coding for secreted cytokine receptors (see Section 7.5). Thus, variola virus encodes receptors for TNF, γ-IFN, α/β-IFN, and IL-18 while vaccinia virus encodes receptors for γ-IFN, α/β-IFN, and IL-1β. Moreover, some of these viral soluble receptors possess pronounced specific activity only towards the ligands of the animal species in which the virus is propagating, whereas other receptors do not exhibit such properties (Alcami & Smith, 1995a).

Investigation of the genetic organization of CPXV is of special interest from this standpoint, since this species has the widest host range among the poxviruses, i.e. is able to multiply in different molecular environments and overcome protective mechanisms of various mammalian species. We obtained unexpected results while analyzing the CPXV-GRI genes coding for proteins of the TNF receptor family (Shchelkunov *et al.*, 1998). This virus appeared to determine five different proteins of this family, characteristic of which were relatively high homology together with pronounced distinctions (Figure 7.12, Table 7.5).

TNF cell receptors types I and II, belonging to a superfamily of the membrane-associated proteins, consist of four extracellular domains, rich with cysteine, located at the *N*-terminus of the protein, and responsible for binding to TNF hydrophobic transmembrane region and cytoplasmatic domain. Viral receptors of TNF also contain four cysteine-enriched domains homologous to cellular analogues (Shchelkunov *et al.*, 1994a) and, in addition, the *C*-terminal domain with no homology not only to sequences of the TNF cellular receptors, but also any other known proteins (Smith & Goodwin, 1994).

Comparison of the four orthopoxvirus species (Figure 7.12) demonstrated that VACV contained only two mutation-disrupted genes of this family in its genome, and one of them was duplicated (C22L and B28R) as a part of terminal inverted repeats, i.e. VACV was not capable of synthesizing this type of proteins. VARV contains the only gene of the family in question and is likely to synthesize a full-value protein secreted from the infected cells that binds human TNF. MPXV encodes a duplicated copy of the same gene, which displays certain species-specific differences in the sequence (Shchelkunov *et al.*, 2002a).

An unprecedentedly big set of genes of the TNF receptor family within the CPXV genome demonstrates how little we know yet about the network of the mammalian TNF-like receptors and ligands. Considering that functionally useless viral genes are being eliminated in the evolutionary process, we may conclude that each of the TNF receptor homologues containing in the CPXV genome must have its specific protein ligand(s) in the host organism, which are most likely belong to the family of TNF homologues.

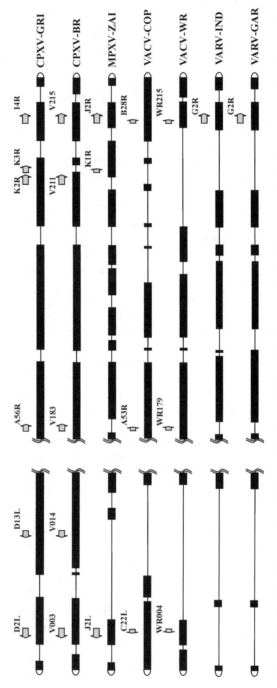

Figure 7-12. Graphical layout of the ORFs encoding viral proteins belonging to the family of tumor necrosis factor receptor. Arrows indicate the directions and sizes of the corresponding ORFs of CPXV-GRI, CPXV-BRT, MPXV-ZAI, VACV-COP, VACV-WR, VARV-IND, and VARV-GAR. Names of the ORFs are shown near the arrows.

Table 7-5. Open reading frames (ORFs) of CPXV-GRI ascribed to the family of tumor necrosis factor receptor

ORF	Size, aa	Identity, %				
		D2L	D13L	A53R	K2K	K3R
D2L/H4R	351		21	31	49	44
D13L	111	21		19	23	26
A56R	186	31	19		32	35
K2R	322	49	23	32		41
K3R	167	44	26	35	41	

For example, we found that CPXV-GRI protein D13L exhibited a maximal homology to the receptor part of cellular protein CD30, which belongs to the superfamily of TNF receptor proteins (Shchelkunov *et al.*, 1998). Then, Panus *et al.* (2002) showed that CPXV produced a soluble secreted form of vCD30, the receptor for CD153. Protein vCD30 is most likely causes the block of mediated CD30L interaction of monocytes and lymphocytes and consequently, hinders the development of both cell-mediated and humoral immune responses to viral infection. Interestingly, the potential ORF D12L encoding *C*-terminal chemokine-binding domain of orthopoxviral TNF receptors (Ruiz-Arguello *et al.*, 2004) is adjacent to ORF D13L of CPXV-GRI (see Appendix 1). It is likely that the ORFs D13L and D12L originated through mutational splitting of the single preexisting ORF.

It is known that TNF receptor functions in a form of a trimer (Smith & Goodwin, 1994). The assumption that viral analogues of TNF receptor are capable of forming receptor molecules consisting of the subunits of different viral proteins implies a considerable expansion of the possibilities of CPXV to produce numerous combinations of such soluble receptors and its adaptation to various hosts.

The question whether all the discovered CPXV genes of the family in question are of cellular origin (for example, they all have been integrated into the viral genome through retrotransposition) or they have originated through duplications of a single ancestor viral gene and subsequent evolutionary divergence is also to be answered.

7.5 Comparison of Immunomodulatory Proteins of Orthopoxviruses

It is known that the animal organism, including humans, possesses a set of defense responses directed against infectious agents. The immediate defense is nonspecific reactions, i.e., they are universally directed against any foreign cells, viruses, and large molecules. The secondary defense response is highly specific and realized by the immune system of the organism. Triggering of this system requires certain time. Viruses during their evolution have mastered various molecular mechanisms to evade the defense reactions of the host organism. The viruses in this respect may be divided into two major groups according to the strategy they use to overcome the protective barriers of the host organism depending on the type of infection the pathogens cause, i.e., acute or persistent. For example, a virus causing persistent or slow infection (retroviruses or herpesviruses) should first possess the mechanism (or several mechanisms) for suppressing the immune system or prevention the recognition of virus-infected cells by the immune system. On the other hand, the viruses causing an acute infection (poxviruses, influenza viruses, or enteroviruses) need to propagate to a high level to provide transmission to another sensitive host. In the case of such viruses, of the utmost importance is the efficient protection against the immediate nonspecific response of the organism to infection.

It is considered that during the co-evolution with the host organism, viruses incorporated in their genomes coding sequences of various cellular genes and modified them to provide their viability and preservation in the biosphere (Alcami & Koszinowski, 2000; Shchelkunov, 2001). When understanding the mechanisms used by viruses to overcome manifold defense systems of the animal organism, represented by molecular factors and cells of the immune system, we would not only comprehend better, but also discover new patterns of organization and function of these most important reactions directed against infectious agents. Here, study of the orthopoxviruses pathogenic for humans may be most important.

The species- and strain-specific distinctions between VARV, MPXV, CPXV, and VACV DNAs are localized to the variable terminal regions. These distinctions comprise not only deletions in DNAs of the viruses compared relative one another (Figure 7.13), but also rearrangements and

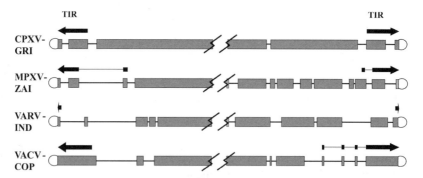

Figure 7-13. Layout of the left and right terminal variable regions of cowpox (CPXV-GRI), monkeypox (MPXV-ZAI), variola (VARV-IND), and vaccinia (VACV-COP) viruses. Terminal inverted repeats (TIRs) are indicated with arrows; terminal variable regions, with gray rectangles; and deletions in the genome of one virus relative to the others, with bold horizontal lines.

nucleotide substitutions. The determined sequences of viral DNAs allow for a comparative analysis of organization of the VARV, MPXV, CPXV, and VACV molecular pathogenicity factors, whose function was verified at various laboratories in experiments mainly with VACV, CPXV, and ectromelia virus (mousepox virus, ECTV). Described in this section is organization of genes of immunomodulatory proteins of variola virus strains VARV-IND (major) and VARV-GAR (minor alastrim), monkeypox virus strain MPXV-ZAI, cowpox virus strain CPXV-GRI, and vaccinia virus strains VACV-COP and VACV-WR.

7.5.1   Inhibitors of Inflammatory Reactions

Inflammatory reactions play an important role in the early nonspecific protection of the organism against the viral infection. They are induced rapidly to limit the virus dissemination during the first hours and days upon infection while the full-fledged immune response is being formed.

It is known that the complement system and the cytokines, such as tumor necrosis factor (TNF), interleukin 1β (IL-1β), gamma-interferon (γ-IFN) and chemokines play the key role in inducing the inflammatory reactions (Tracey & Cerami, 1994). In addition, several other mediators influence either directly or indirectly the development of the inflammatory process (Kotwal, 1996; Moss & Shisler, 2001). Therefore, poxviruses potentially need several genes whose protein products are able to act as inhibitors of various stages of inflammation development to suppress efficiently the inflammatory response.

Orthopoxviruses carry the gene of complement-controlling protein with one of the functions connected with regulation of inflammatory reactions. The complement system comprises over 20 blood plasma proteins. The mechanisms of antiviral action of the complement system include neutralization and opsonization of the virus, lysis of the virus-infected cells, and boosting of the inflammatory and specific immune responses (Kotwal, 1996). Vaccinia virus complement control protein (VCP), secreted from VACV-infected cells and controlling complement activation reactions, was the first identified microbial protein displaying complement-binding properties (Kotwal & Moss, 1988a). This protein is composed of four short degenerate repeats (about 60 aa residues each, designated SCR, short consensus repeats), characteristic of the family of proteins called the regulators of complement activation (RCA; Liszewski & Atkinson, 1998). VCP displays the highest homology to amino acid sequence of the first four of the eight SCRs of the alpha-chain of human complement C4b-binding protein (C4BP; Aso *et al.,* 1991). It is considered that the gene encoding VCP originated initially due to incorporating a part or the complete coding sequence of a protein belonging to RCA family of the host into the viral genome followed by adaptation (alteration) of the gene in question to perform the functions necessary for the virus (Kirkitadze *et al.,* 1999). X-ray structure assay demonstrated that the SCR sequences of VCP formed discrete compact domains successively tightly linked with one another (Smith S.A. *et al.,* 2000).

It is proved that an impairment of VCP gene results in a decrease of VACV virulence (Isaacs *et al.,* 1992a). VCP inhibits development of the *in vivo* inflammatory response, as was demonstrated using CPXV-infected mouse model (Miller *et al.,* 1997; Howard *et al.,* 1998).

VCPs of VARV, CPXV, and VACV contain four SCR each. Interestingly, the VCP sequences of two VACV strains are identical; however, contain differences in 12 amino acid residues compared with VCPs of VARV, which in turn are highly conservative (Figure 7.14). In the recent study of Rosengard *et al.* (2002), individual VARV and VACV VCPs were synthesized in a baculovirus system. The analysis performed demonstrated that VARV VCP inhibits human complement with a considerably higher efficiency compared with the VACV analogue. In particular, this confirms the concept of evolutionary adaptation of viral receptors to the corresponding ligands of the host organism.

We discovered the unique structure of MPXV VCP (Uvarova & Shchelkunov, 2001). A premature ORF termination results in a truncated protein with deleted *C*-terminal SCR4 in MPXV Central African strains (Figure 7.14), distinguishing essentially this species from the other studied orthopoxviruses pathogenic for humans. Moreover, the Western African MPXV subtype does not contain the gene encoding VCP in its genome (see Figure 5.6).

Another viral gene whose product inhibits development of the inflammatory response to the infection was initially discovered in CPXV genome (Palumbo *et al.,* 1989) and named SPI-2 or crmA. The protein SPI-2 inhibits the cellular protease ICE (interleukin-1β converting enzyme), thereby preventing maturation of pro-IL-1β into IL-1β and its secretion from the infected cell; as a result, induction of inflammatory reactions near infected cells is inhibited too. In addition, it appeared that SPI-2 inhibited formation of inflammation mediators (leukotrienes) during metabolism of arachidonic acid (Palumbo *et al.,* 1993). Moreover, SPI-2 is also involved in suppression of apoptosis of the infected cell. Thus, the protein in question is likely to play an important role in determining the pathogenic properties of orthopoxviruses *in vivo,* Note that characteristic of VARV SPI-2 are several species-specific distinctions from the MPXV, CPXV, and VACV homologues, which display minimal differences between each other (Shchelkunov, 2001). In the case of VACV-COP, the gene encoding this protein is damaged (Table 7.6).

Analysis of VACV DNA sequencing data allowed two ORFs homologous to human and murine IL-1 receptors to be detected (Smith &

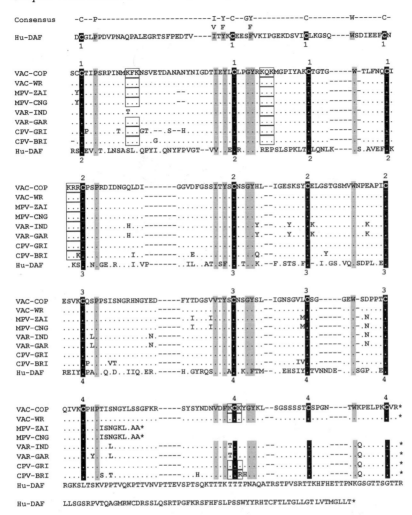

Figure 7-14. Comparison of the amino acid sequences of the orthopoxvirus complement-binding proteins of VACV (strains COP and WR), MPXV (ZAI and CNG), VARV (IND and GAR), and CPXV (GRI and Brighton, BRT). Vertical black blocks mark conservative cysteine residues; vertical gray blocks, the other conservative amino acid residues characteristic of SCRs. The sequences forming potential heparin-binding sites are framed; figures above black blocks indicate the number of SCR; *C*-ends of amino acid sequences of proteins are asterisked.

Chan, 1991). Further, VACV-WR gene B15R was experimentally confirmed to encode secreted glycoprotein with properties of soluble (secreted from the cell) IL-1β receptor (Alcami & Smith, 1992). Unlike the cellular IL-1 receptors, this viral protein binds only IL-1β and fails to bind IL-1α. The

Table 7-6. Immunomodulatory proteins of orthopoxviruses

	VARV-IND		VARV-GAR		MPXV-ZAI		CPXV-GRI		VACV-COP		VACV-WR	
	ORF	Size, aa	ORF	Size, aa	ORF	Size, aa	ORF	Size, aa	ORF	Size, aa	ORF	Size, aa
Growth factor	D2R	140	B3R	140	D3R	142	C5R	138	C11R	142	C6R	140
IL-18-binding protein	D5L	126	B6L	126	D6L	126	C8L	124	–		C9L	68
Complement-binding protein	D12L	263	B18L	263	D14L	216	C17L	259	C3L	263	C21L	263
Homologue of eIF-2α, factor of interferon resistance	C3L	88	P3L	88	–		M3L	88	K3L	88	K2L	88
dsRNA-binding protein, factor of interferon resistance	E3L	190	C3L	192	F3L	153	F3L	190	E3L	190	E3L	190
3-β-Hydroxy-delta5-steroid dehydrogenase	A50L	61	A54L	61	A45L	346	A47L	346	A44L	346	A49L	346
γ-IFN-binding protein	B9R	266	H9R	266	B9R	267	B7R	271	B8R	272	B8R	272
Serine protease inhibitor, SPI-2	B13R	344	D2R	344	B12R	344	B12R	345	B13R	116	B13R	345
IL-1β-binding protein	B15R	63	D4R	63	B14R	326	B14R	326	B16R	290	B15R	326
α/β-IFN-binding protein	B20R	354	D9R	355	B16R	352	B17R	351	B19R	353	B18R	351
TNF-binding protein, CrmB	–		–		J2L*	348	D2L*	351	C22L*	122	C3L*	122
TNF-binding protein, CrmC	G2R	349	G2R	349	J2R*	348	H4R*	351	B28R*	122	B29R*	122
TNF-binding protein, CrmD	–		–				A56R	186	A53R	103	A58R	103
TNF-binding protein, CrmE	–		–		K1R	70	K2R	322				
Chemokine-binding protein							K3R	167				
Chemokine-binding protein	G3R	253	G3R	253	J1L*	246	D1L*	255	C23L*	244	C1L*	244
					J3R*	246	H5R*	255	B29R*	244	B31R*	244
Semaphorin-like protein	A42R	74	A45R	74			A41R	402	A39R	403	A43R	403

Notes: the open reading frames (ORFs) duplicated in the left and right ends of terminal inverted repeats are asterisked; the ORFs differing from the corresponding CPXV-GRI ORFs are marked gray.

second predicted VACV-WR protein B18R binds neither IL-1α nor IL-1β; however, it interacts efficiently with type I interferons (see below). It was demonstrated that production of soluble IL-1β receptor (IL-1βR) by VACV prevents development of systemic reactions (fever and elevated body temperature) in infected mice (Alcami & Smith, 1996). It appeared that damage of VACV-WR gene B15R increased the VACV virulence when administered to mice intranasally (Alcami & Smith, 1992). Further analysis showed that VACV strains causing an increased rate of postvaccination complications in humans did not possess IL-1β-binding activity (Alcami & Smith, 1996). These results comply well with the fact that the corresponding VARV genes are fragmented (disrupted). Earlier, the existence of "buffer" genes in the orthopoxvirus genome was postulated (Shchelkunov, 1995), whose role was to neutralize the negative effects developing in the body during infection. Damage of such genes increases the pathogenic properties of the virus.

Thus, we may hypothesize that VARV suppresses production and secretion of IL-1β by infected cells but does not inhibit the effect of extracellular IL-1β synthesized by other cells of the body. This suggests that VARV is capable of suppressing local inflammatory reactions due to SPI-2 production in the region of virus replication; however, does not inhibit the systemic reactions, as is unable to synthesize IL-1β-binding protein. Decrease in the local inflammatory reactions may assist a more active virus replication, while uncontrolled development of the systemic reactions weakens the overall resistance of the organism to infection. A concurrent development of these reactions is likely to boost the pathogenic effect of the vital infection on the host organism. In the case of MPXV and CPXV, both genes in question are native (Table 7.6).

TNF is also multifunctional (Tracey & Cerami, 1994). In particular, along with IL-1β, TNF is a key cytokine inducing inflammatory reactions in the organism of infected host. VARV-IND carries gene G2R, encoding the protein homologous to type II TNF receptor (TNFrII). This protein was demonstrated to be an important secreted virulence factor of myxoma virus, the genus *Leporipoxvirus* (Upton *et al.*, 1991). Presumably, its analogue, VARV G2R, displays similar properties. MPXV has two copies of this gene within TIRs. An essential difference between VACV and VARV is that the majority of VACV strains lack the genes encoding analogues of TNFr (Table 7.6). In CPXV genome, we found five genes belonging to the family of TNF receptors (Table 7.5; Shchelkunov *et al.*, 1998), and a TNF-binging activity was experimentally confirmed for four of these genes (Smith C.A. *et al.*, 1996; Loparev *et al.*, 1998; Saravia & Alcami, 2001). Presumably, such an unexpectedly large set of TNF-binding proteins underlies the evolutionary fitness of CPXV and provides the ability to replicate in a wide range of animal species (see Section 7.4.3). Analysis of amino acid sequences of

VARV isolog proteins G2R demonstrates their manifold species-specific distinctions (Figure 7.15). In a baculovirus expression system, we produced individual TNFr proteins of VARV, MPXV, and CPXV and demonstrated that they differed essentially in inhibiting the biological activities of human, mouse, and rabbit TNFs (Gileva *et al.,* 2003). Presumably, this is a result of evolutionary adaptation of the viral receptors to the ligands of their hosts.

```
H4R-GRI   MKSVLYSYILFLSCIIINGRDIAPHAPSNGKCKDNEYNRHNLCCLSCPPGTYASRLCDSK
MUN-85    .............................V..Y............................
J2R-ZAI   .R....................L...................RSR...............
J2R-CNG   .R....................L...................RSR...............
G2R-IND   ......L.............A..YT.P......T..K.......................
G2R-GAR   ......L.............A..YT.P......T..K.......................

H4R-GRI   TNTNTQCTPCGSGTFTSRNNHLPACLSCNGRCDSNQVETRSCNTTHNRICECAPGYYCLL
MUN-85    .....................H....................P...........S......
J2R-ZAI   ...--.......D....H....Q....................S......
J2R-CNG   ...--.......D....H....Q....................S......
G2R-IND   ...--.....................N...........S......
G2R-GAR   ...--.....................N...........S......

H4R-GRI   KGSSGCKACVSQTKCGIGYGVSGHTSTGDVVCSPCGLGTYSHTVSSADKCEPVPSNTFNY
MUN-85    .........................I........................
J2R-ZAI   .....RT.I.K..........Y.....I.....P.........T......T......
J2R-CNG   .....RT.I.K..........Y.....I.....P.........T......T......
G2R-IND   .....................V...I....F...........N.....
G2R-GAR   .....................V...I....F....Y...T.......N.....

H4R-GRI   IDVEINLYPVNDTSCTRTTTTGLSESISTSELTITMNHKDCDPVFRDGYFSVLNKVATSG
MUN-85    ............................................................
J2R-ZAI   .........................................AE......N.....
J2R-CNG   .........................................AE......N.....
G2R-IND   .....T................L..........T..N....EE.............
G2R-GAR   .....T................L..........T..N....EE.............

H4R-GRI   FFTGENRYQNISKVCTLNFEIKCNNKDSSSKQLTKTKNDDGIMPHSETVTLVGDCLSSVD
MUN-85    ..........T.N.......................T-..................
J2R-ZAI   .........T..I.......................T-..................
J2R-CNG   .........T..I.......................T-..................
G2R-IND   .....................G..F.....A....M.S......A.......
G2R-GAR   .....................G..F.....A....M.S......A.......

H4R-GRI   IYILYSNTNTQDYETDTISYHVGNVLDVDSHMPGSCDIHKLITNSK-PTRFL*   351
MUN-85    ....................A....Y.............QN..H..*        351
J2R-ZAI   ...................M......N....A.........QN..H-.*      348
J2R-CNG   ...................M......N....A.........QN..H-.*      348
G2R-IND   ........A.........R.....D.........N...P.....-.....*    349
G2R-GAR   ........A.........R.....D.................-.....*      349
```

Figure 7-15. Comparison of amino acid sequences of the orthopoxvirus proteins CrmB, belonging to the family of TNF receptors, involving two strains of each species—CPXV (GRI and 85), MPXV (ZAI and CNG), and VARV (IND and GAR). Amino acid residues identical to CPXV-GRI are indicated with dots; deletions, with dashes. The species-specific distinctions of VARV and MPXV from CPXV are marked with black and gray blocks, respectively.

All the orthopoxviruses considered encode soluble gamma-interferon receptor (γ-IFNr), which is capable of modulating the host inflammatory response to the viral infection (Mossman *et al.*, 1995). VARV major and minor γ-IFNr proteins display a number of structural species-specific differences from the analogous MPXV, CPXV, and VACV proteins, presumably, influencing their functions (Shchelkunov, 2001; Shchelkunov *et al.*, 2002a). By the example of rabbit myxoma virus (genus *Leporipoxvirus*), it was demonstrated that the viral cytokine-binding proteins displayed a pronounced species specificity towards the cytokines of their natural host (wherein they replicate). For example, the TNF receptor of myxoma virus binds specifically the rabbit TNF, however, fails to bind this cytokine of humans and mice (Upton *et al.*, 1992). Similarly, γ-IFN receptor of myxoma virus inhibits the biological activity of rabbit γ-IFN and has no effect on human and murine γ-IFNs (Mossman *et al.*, 1996). On the other hand, the viruses with a wide host range—CPXV and VACV—synthesize γ-IFN receptors that block biological activities of human, bovine, rabbit, and rat γ-IFNs and with a considerably lower efficiency, mouse γ-IFN (Alcami & Smith, 1995b). Thus, poxviruses during the evolution adapt their molecular virulence factors so that the factors would modulate with a maximal efficiency the defense reactions of the host organism. A virus with a wide host range synthesizes immunomodulatory proteins capable of interacting with the defense systems of the overall set of the sensitive animal species. If the host range of a virus narrows to a single species, consequently, its molecular virulence factors underwent evolutionary changes to influence only this particular species (compare rabbit myxoma virus and cowpox virus; see above). Presumably, the ranges of specific activities of viral cytokine receptors may indicate an evolutionary relatedness and origin of various poxviruses within one genus.

Chemokines are chemoattractant cytokines, which control migration and effector functions of leukocytes, thereby playing an important role in development of inflammatory response and protection against pathogens (Rollins, 1997). It was demonstrated that VACV strain Lister at the early stages of infection produced a protein secreted from the cells in large amounts (Patel *et al.*, 1990), which bound a wide range of CC chemokines and inhibited their activities (Smith C.A. *et al.*, 1997). This gene is damaged in many other VACV strains. Presumably, these proteins of various orthopoxvirus species have different functions, as analysis of their amino acid sequences detected considerable species-specific distinctions (Figure 7.16).

It is considered that a secreted VACV protein A39R, which was ascribed to the family of semaphorins (Comeau *et al.*, 1998; Spriggs, 1999), is involved in modulation of inflammatory and/or immune responses.

```
H5R-GRI   MKQYIVLACMCLAAAAMPASLQQSSSS---CTEEENKHHMGIDVIIKVTKQDQTPTNDKI
J3R-ZAI   ...........V....T........---.........................
Zaire-77  ...........V....T........---.........................
Congo-8   ...........V....T........---.........................
G3R-IND   .......................---......Y....................
G3R-GAR   .......................---......Y....................
B29R-COP     MHV.........SSS...........................
B31R-WR      MHV.........SSS...........................

H5R-GRI   CQSVTEITESESDPDPEVESSSDSTSVEDVDPPTTYYSIIGGGLRMNFGFTKCPQIKSIS
J3R-ZAI   .....V..T.D.--------EV.EE.VKG.-.....T.V.A..N.........K.S...
Zaire-77  .....V..T.D.--------EV.EE.VKG.-.....N.V.A..N.........K.S...
Congo-8   ......V..T.D.E--VS.DDEV.EE.VKG.-.....N.V.A..N.........K.S...
G3R-IND   .................--....ED.............................
G3R-GAR   .................--....ED.............................
B29R-COP  ....................ED.............................
B31R-WR   ....................ED.............................

H5R-GRI   ESADGNTVNARLSSVSPGQGKDSPAITHEEALAMIKDCEVSIDIRCSEEEKDSDIKTHPV
J3R-ZAI   ..S......T..................R..........M............
Zaire-77  ..S......T..................R..........M............
Congo-8   ..S......T..................R..........M............
G3R-IND   ...N..A........PL..........RA........L............Q...
G3R-GAR   ...N..A........P...........RA........L............Q...
B29R-COP  ...........................R........................
B31R-WR   ...........................R........................

H5R-GRI   LGSNISHKKVSYEDIIGSTIVDTKCVKNLEFSVRIGDMCKESSELEVKDGFKYVDGSASE
J3R-ZAI   ............K....................E.................
Zaire-77  ............K....................E.................
Congo-8   ............K....................E.................
G3R-IND   .E..............................D...........V..
G3R-GAR   .E..............................D...........V..
B29R-COP  ...............................................
B31R-WR   ...............................................

H5R-GRI   GATDDTSLIDSTKLKACV*    255
J3R-ZAI   ................*     246
Zaire-77  ................*     246
Congo-8   ................*     252
G3R-IND   .V...........S..*     253
G3R-GAR   .V...........S..*     253
B29R-COP  ................*     244
B31R-WR   ................*     244
```

Figure 7-16. Comparison of amino acid sequences of CC chemokine-binding proteins of orthopoxviruses—CPXV (strain GRI), MPXV (strains ZAI and CNG), VARV (strains IND and GAR), and VACV (strains COP and WR). Amino acid residues identical to CPXV-GRI are indicated with dots; deletions, with dashes. The species-specific distinctions of VARV and MPXV are marked with black and gray blocks, respectively.

However, the role of this protein is yet vague. Note that neither VARV nor MPXV synthesizes this semaphorin-like protein (Table 7.6).

Thus, at least five genes capable of controlling development of inflammatory reactions have been detected in the genomes of orthopoxviruses. It is likely that further studies will continue this line. To all appearances, orthopoxviruses display species-specific distinctions not only in the set of these genes, but also in their structures.

7.5.2   Interferon Inhibitors

Interferons (IFNs) are produced and secreted by animal cells in response to double-stranded RNA (dsRNA) molecules synthesized during viral infection. IFNs bind to specific cell receptors and induce an antiviral state of these cells (Samuel, 1991). At least two enzymatic pathways determine an IFN-induced antiviral state of the cell. One pathway involves IFN-inducible dsRNA-dependent protein kinase (the so-called DAI kinase); the second, 2-5A[ppp(A2'p)nA]-synthetase (conventionally called 2-5A-synthetase). The protein kinase is activated due to autophosphorylation upon binding of the enzyme to dsRNA. The activated protein kinase phosphorylates the alpha subunit of eukaryotic translation initiation factor (eIF-2α), thereby inhibiting the protein synthesis. The other enzyme—2-5A-synthetase—upon activation by dsRNA molecules, catalyzes polymerization of ATP molecules into 2'-5'-bound oligoadenylates, which, in turn, activate the latent cellular endo-RNase L. This RNase cleaves mRNA and rRNA molecules, thereby impairing the protein synthesis.

Despite production of large amounts of virus-specific dsRNAs at the late stage of development (Varich *et al.*, 1979), characteristic of orthopoxviruses is a high resistance to the action of IFN (Whitaker-Dowling & Younger, 1984). Watson *et al.* (1991) demonstrated that VACV gene E3L encoded inhibitor of IFN-inducible DAI kinase. This early viral protein is capable of binding to dsRNA molecules, competing with the specific cellular protein kinase, thereby inhibiting activation of the specific enzyme. Another VACV gene, K3L, encodes a homologue of eIF-2α. This homologue competes with the endogenous eIF-2α of the cell for the phosphorylation by activated protein kinase (Davies *et al.*, 1993). A mutant VACV with impaired K3L

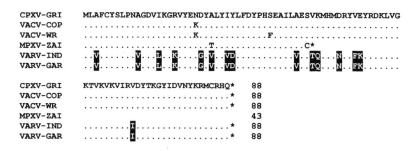

Figure 7-17. Comparison of amino acid sequences of the orthopoxvirus homologues of eukaryotic translation initiation factor alpha subunit (eIF-2a) of CPXV (strain GRI), MPXV (strain ZAI), VACV (strains COP and WR), and VARV (strains IND and GAR). Amino acid residues identical to CPXV-GRI are indicated with dots. The species-specific distinctions of VARV are marked with black blocks. C-end of amino acid sequence of each protein is asterisked.

```
CPXV-GRI  MSKIYIDERSDAEIVCEAIKTIGIEGATAAQLTRQLNMEKREVNKALYDLQRSDMYYS SDDIPPRWFMTTEADKTDADVMADVIIDDVSREKSMRED
MPXV-ZAI  ......................................................................T........ST.MD..TRPT.SD.A........
VARV-IND  ..........................N..L.V.V.............................A.............................P..MT.......
VACV-COP  ..........................A..N................................A.............................P..A........

CPXV-GRI  HKSFDDVIPAKKIIDWKDANPVTIINEYCQITKRDWSFRIESVGPSNSPTFYACVDIDGRVFDKADGKSKRDAKNNAAKLAVDKLLGVVIIRF*  190
MPXV-ZAI  N.....V.....Y..GV....V.........R...............................................S..........*  153
VARV-IND  N...........N.................................................................................*  190
VACV-COP  ..............................................................................................*  190
```

Figure 7-18. Comparison of amino acid sequences of the orthopoxvirus dsRNA-binding proteins. N-terminal Z-DNA-binding domain is marked gray; dsRNA-binding domain is framed.

gene is sensitive to IFN; as a result, the virus yield decreases by two orders of magnitude (Beattie *et al.*, 1991). Thus, VACV synthesizes the proteins that inhibit the effect of IFN-inducible dsRNA-dependent DAI kinase by two ways.

Note that the VARV-IND protein (C3L) differs essentially from VACV-COP protein K3L; however, it is identical in VARV-IND and VARV-GAR. Comparison of ORF K3L in VACV-COP and VACV-WR reveals only a single substitution (Figure 7.17). This protein of CPXV is highly homologous to the VACV isolog. Thus, the viral analogue of eIF-2α is highly conservative within a species, but displays multiple distinctions when comparing VARV and VACV/CPXV. MPXV genome due to manifold mutational alterations in the corresponding gene does not code for intracellular factor of interferon resistance (Table 7.6).

During VACV infection, E3L gene is expressed from the first and second initiation codons and produces concurrently the long and short forms of the protein, respectively (Beattie *et al.*, 1991). The long form contains the *N*-terminal domain (Figure 7.18), which provides the binding of VACV protein to Z-DNA, its nuclear localization, and manifestation of pathogenicity (Romano *et al.*, 1998; Brandt & Jacobs, 2001). Both the long and short forms of this protein contain the *C*-terminal domain, which provides binding to dsRNA and inhibition of IFN-inducible DAI kinase (Chang *et al.*, 1992) and 2-5A-synthetase (Rivas *et al.*, 1998). The first initiation codon of MPXV-ZAI is damaged by mutation; consequently, only a short form of the protein in question is translated (Figure 7.18). Thus, unlike the other orthopoxvirus species considered, MPXV is unique in the organization of the viral intracellular interferon resistance factors, which, presumably, may lead to a decreased *in vivo* MPXV replication and, consequently, a reduced rate of person-to-person respiratory transmission, which is actually observed in the case of human monkeypox compared with smallpox.

Resistance of orthopoxviruses to IFN is also provided by extracellular γ-IFN-binding protein, considered above, and type I IFN-binding protein (α/β-IFN; Symons *et al.*, 1995). We determined pronounced species-specific distinctions in amino acid sequences of α/β-IFN-binding protein of orthopoxviruses (Shchelkunov, 2001). The effects of these distinctions in amino acid sequences on the properties of the corresponding proteins can be clarified only in experiments on expression of individual viral proteins and their comparative study.

Thus, orthopoxviruses have a multigenic system controlling their resistance to interferon. The detected distinctions between the corresponding genes of VARV, MPXV, CPXV, and VACV require further studies of the properties of the corresponding proteins.

7.5.3   Modulators of the Immune Response

The immune response to an infection is an intricate interaction of various cell types controlled by cytokines and resulting in appearance of B lymphocytes, which synthesize specific antiviral antibodies, and virus-specific T lymphocytes (Spriggs, 1996). The specific antibodies can interact with the virus particles or their components either by themselves or in a complex with the complement. Numerous experiments have demonstrated that the specific antibodies to poxviruses are inefficient in suppressing the primary infection, but may be important in prevention of the reinfection (Buller & Palumbo, 1991). The cell-mediated immunity is most important at the stage of specific protection of the organism against poxvirus infection (Demkowicz & Ennis, 1993).

Before the specific antiviral T lymphocytes appear, nonspecific cytolytic natural killer (NK) cells play a certain role in protection of the organism against poxviruses. The proliferation peak of NK cells occurs on days 2–3 post infection; however, these cells alone are unable to prevent completely the dissemination of infection in the body (Buller & Palumbo, 1991). It was discovered that interleukin-18 (IL-18) induced cytotoxicity of NK cells (Okamura *et al.*, 1998). In addition, this cytokine stimulates production of γ-IFN and proliferation of T lymphocytes of Th1 phenotype (Novick *et al.*, 1999). By the example of ECTV, it was demonstrated (Born *et al.*, 2000) that this virus produced secreted protein p13 binding to IL-18 and inhibiting the cytotoxicity activation of NK cells *in vivo*. This protein is synthesized by VARV, MPXV, and CPXV and is absent in VACV (Table 7.6).

The virus-specific T lymphocytes, whose proliferation peak occurs on days 5–6 post infection, are of crucial importance for infection inhibition in the host organism. As was mentioned, TNF, IL-1β, and γ-IFN are most important cytokines, which along with regulation of inflammatory reactions, control the immune response to infection. For example, γ-IFN plays an important role in activation of macrophages, induces expression of antigens of the major histocompatibility complex, and directs development of T cell cytolytic response to infection. TNF and γ-IFN also display a direct antiviral activity towards the infected cells, which is boosted in the case of their joint action.

All the orthopoxviruses considered encode soluble γ-IFN receptor (Table 7.6). VARV, MPXV, and CPXV produce also secreted TNF-binding protein, whereas this gene is damaged in the genome of VACV. It was demonstrated experimentally (Upton *et al.*, 1991) for rabbit myxoma virus (genus *Leporipoxvirus*) that TNF-binding protein was extremely important for inhibiting the antiviral immune response. Presumably, synthesis of this protein is necessary for development of a generalized infection characteristic of VARV. VARV does not produce secreted IL-1β-binding protein, whereas MPXV, CPXV, and VACV synthesize this protein (Table 7.6).

Computer analysis of VACV-COP sequencing data (Goebel *et al.*, 1990) demonstrated that ORF A44L in its amino acid sequence displayed a 31% identity with human 3β-hydroxysteroid dehydrogenase/Δ^5–Δ^4 isomerase (3β-HSD). Moore and Smith (1992) confirmed experimentally that this viral gene encoded a protein with a putative enzymatic activity. 3β-HSD is a key enzyme in biosynthesis of steroid hormones involved in various stages of this pathway. In turn, steroid hormones influence many functions of the host, including the immune system. All the orthopoxviruses considered except for VARV carry the gene of this enzyme; therefore, it may be regarded as a "buffer" gene (see above).

Interestingly, CPXV encodes all these immunomodulatory proteins, whereas VARV, MPXV, and VACV have a reduced set of these genes, characteristic of each individual orthopoxvirus species. In particular, this suggests that CPXV is evolutionary a more ancient orthopoxvirus species compared with VARV, MPXV, and VACV, which evolved independently from an ancestor virus.

From the evolutionary standpoint, the virus that maintains the balance between the pathogenic effect on the host organism and the possibility of productive replication in the host for a relatively long period is most evolutionary fitted. Such virus is able to transmit efficiently from animal to animal under a low population density. Of the orthopoxviruses, this is most typical of CPXV.

Summing up the available data, we may infer that variola virus and other orthopoxviruses possess an unexampled set of genes whose protein products efficiently modulate the manifold defense functions of the host organisms compared with the viruses from other families. It is likely that by the example of orthopoxviruses, it will be possible in the nearest future to trace the patterns of co-evolution of the viral pathogenicity factors and mammalian systems providing defense against infectious agents. The research into application of immunomodulatory proteins of orthopoxviruses, and first and foremost, variola virus, as drugs also deserves attention.

Chapter 8

LABORATORY DIAGNOSTICS
OF HUMAN ORTHOPOXVIRUS INFECTIONS

Despite of the fact that orthopoxvirus infections have specific apparent manifestations, skin lesions, their clinical identification is often inconsistent. Orthopoxvirus diseases with rash throughout the body are most often taken for chickenpox (varicella) or herpes simplex. Large solitary skin lesions caused by cowpox are often diagnosed as anthrax, if unusually located; eye lesions are frequently assigned to phlegmon. On the other hand, diseases caused by parapoxviruses, including milker's nodules, are often taken for cowpox. Monkeypox in man is always suspicious, because it is difficult to distinguish it from smallpox. Note that as is stated by WHO, eradication of smallpox does not eliminate the necessity of laboratory examination of all suspicious cases. This is required not only to assure the world community that smallpox is absent of, but also to recognize the emergencies. It should be kept in mind that, despite the advance in orthopoxvirus study, the further evolution of orthopoxviruses is yet hardly predictable. There is no absolute guarantee that an existing orthopoxvirus or a new variant with the pathogenicity for humans and contagiousness equal to smallpox virus or close to it would not emerge.

In this context, the human monkeypox outbreak that occurred in Democratic Republic of the Congo, former Zaire, in 1996–1997 causes a considerable concern. It demonstrated an increase in the transmissibility of this infection on the background of disappearing immunity against smallpox (Weekly Epidemiological Record, 1997a; 1997b; 1997c; Cohen, 1997).

All these considerations illustrate the importance of laboratory tests as an effective mean for obtaining reliable and objective diagnosis independent of the clinical experience of general practitioners and other circumstances.

The main requirements imposed on the methods of laboratory diagnostics are quick results and high sensitivity, which would ensure the detection of a virus or corresponding antigens even at low concentrations. Moreover, laboratories should have at hand the protocols for reliable differentiation of agents causing clinically similar diseases.

The development and application of methods for laboratory diagnostics of orthopoxvirus infections began in the second decade of the 20th century. At that time, all the three main groups of laboratory methods appeared: serological, morphological, and biological. These methods have been improved since that time. The program of global smallpox eradication was a sort of a test site for the trials of both old and new methods. After completion of the eradication campaign, another, biochemical, group of methods appeared. It involved mainly the data obtained by sequencing orthopoxvirus genomes and polymerase chain reaction (PCR). These methods allow not only the orthopoxvirus species to be differentiated, but also subspecies of variola virus to be identified.

We brief the main methods of each of the groups below.

8.1   Morphological Methods

Viroscopy, one of the earliest methods for laboratory diagnosis of smallpox, based on studies by Paschen (1917; 1924), is detection of the pathogen in clinical samples (smears) by light microscopy upon staining of the preparations. Of a diversity of staining methods, the staining according to Morozov (1924; 1926) became most widespread. This method, modified later by Gispen (1952), was recommended by WHO for smallpox diagnostics (World Health Organization, 1969). The practice of viroscopy showed that, despite the simplicity and rapidity, its efficiency depended on the type of clinical sample. The sensitivity and reliability of the method were significantly lower if pustular liquid or, particularly, scabs were examined. It is also difficult to distinguish Paschen bodies from somewhat smaller Aragao bodies, agent of chickenpox. For these reasons, the method was abandoned by the mid-1960s–early 1970s.

Immunofluorescence allowed for a successful detection of smallpox antigens directly in smears of the contents of skin lesions (Avakyan *et al.,* 1961). Studies performed at our laboratory showed that the efficiency of a FITC-labeled antiserum to vaccinia virus (a direct variant) for this purpose depended on the type of samples examined, as in the case of viroscopy. The virus was detected at a high rate in scrapings of macules and papules and in print smears from the day of vesicle opening (Mal'tseva, 1980). Examination of pustules or scabs was usually unreliable because of the admixture of autofluorescent leukocytes. However, the immunofluorescence technique appeared very useful for a rapid detection of orthopoxviruses when isolated from cell cultures (Kirillova *et al.,* 1961; Gurvich, 1964; etc.).

Electron microscopy. The pioneering studies by Nagler and Rake (1948) introduced electron microscopy into the practical diagnostics, which allowed both the virion shape and exact size to be determined. This allowed poxviruses not only to be detected, but also distinguished from herpes viruses. This method became more powerful after invention of a negative staining by Brenner and Horne (1959), involving a contrasting agent (e.g., sodium phosphotungstate). The dye penetrates into some virions (C form) and visualizes their inner structure. In the virions impermeable for the dye (M form), the negative contrasting elucidates their surface features (Figure 8.1). The significance of electron microscopy for diagnostics was confirmed by other scientists (Cruickshank *et al.*, 1966; etc.). Since 1971, this technique became a mandatory component of diagnostic studies carried out by the WHO Collaborating Centers during the program of global smallpox eradication and post-eradication epidemiological surveillance (Fenner *et al.*, 1988). Its efficiency is illustrated by data accumulated by the Collaborating Centers. According to Nakano (cited from Fenner *et al.*, 1988), electron microscopy of 981 positive samples revealed poxvirus virions in 98.6% of the cases versus 88.7% of samples isolated using chick embryos. Mal'tseva (1980) analyzed the data on smallpox from the Moscow WHO Collaborating Center and showed that the sensitivity of electron microscopy was virtually the same as that of the biological method of isolation on chick embryos. The methods gave 92 and 94% of positive responses, respectively.

The rapid result is an advantage of electron microscopy-based diagnostics. The analysis takes as little as 1.5–2 h.

Another advantage of this method compared with viroscopy is that its effectiveness is independent of the sample kind (vesicular or pustular liquid, scabs, or papule scrapings). In addition, it allows the virus that lost its

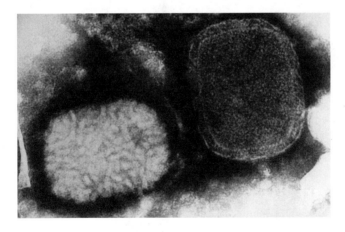

Figure 8-1. M form (left) and C form (right) of orthopoxvirus virions (negative staining, ×240,000). Preparation by N. Yanova.

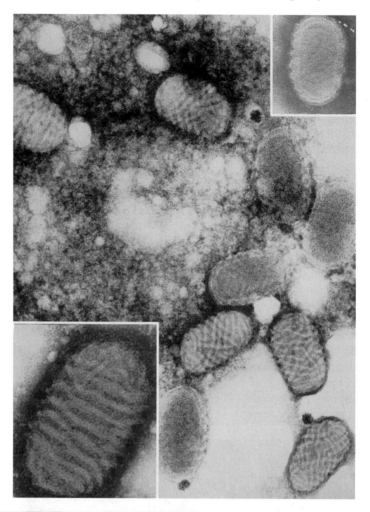

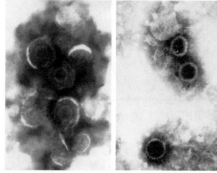

Figure 8-2. Virions of parapoxviruses and herpes viruses in the lesion contents of human cases (negative staining). Upper panel: paravaccinia virus, ×140,000; upper right inset, ×197,000; and lower left inset, ×244,000. Lower left and right panels: varicella zoster virus with the external envelope, ×100,000, and without it, ×168,000. Preparations by N. Yanova.

viability for some reasons to be detected, which is obviously impossible using chick embryos or cell cultures.

In addition to herpes viruses, electron microscopy allows orthopoxviruses and other poxviruses of vertebrates to be distinguished from parapoxviruses, whose virions are ovoid, not brick-shaped and display certain other morphological distinctions (Figure 8.2; Table 8.2).

Such differentiation is required if diseases caused by parapoxviruses occur among the staff engaged in cattle, goat, and sheep farming. These diseases include milker's nodules, or pseudocowpox (caused by paravaccinia virus); bovine papular stomatitis (caused by bovine papular stomatitis virus); and orf, or contagious pustular dermatitis (caused by contagious pustular dermatitis virus or ecthyma contagiosum virus of sheep and goats). All these diseases are accompanied by skin lesions on hands and, sometimes, other parts of the body and are often confused with cowpox (Shelukhina *et al.*, 1986).

A disadvantage of electron microscopy, as of all morphological methods, is that it cannot differentiate the species of orthopoxviruses or other vertebrate poxviruses, except for parapoxviruses (Table 8.1). Vertebrate poxviruses include the causative agents of molluscum contagiosum; tanapox, occurring in Equatorial Africa; and Yaba monkey tumor virus, which can infect humans only by means of occasional virus inoculation.

Immune electron microscopy is a variation of electron microscopy. It is applied for diagnosis of many viral infections. We tested one of the versions of this method involving embedding of specific antibodies into agar according to Anderson and Doane (1973) and Lamontagne *et al.* (1980) with minor modifications. We found that the treatment with antibodies boosted

Table 8-1. Differentiation of orthopoxviruses and some poxviruses from parapoxviruses and herpesviruses by electron microscopy

Virus*	Morphological features of virions
Orthopoxviruses (cowpox, monkeypox, variola, vaccinia, etc.) Yatapoxviruses (tanapox) Molluscum contagiosum virus	Brick-like particles with a size of 250–300 × 200 × × 250 nm**, whose surfaces consist of irregularly arranged tubular elements; virions of M and C forms are present. Virions with outer envelopes were found in clinical samples only in tanapox virus.
Parapoxviruses (paravaccinia, orf, and bovine papular stomatitis viruses)	Ovoid particles with a size of 220–300 × 140– 170 nm; a long helical filamentous structure enwinding the virion is clearly visible on the surface; virions of M and C forms are present.
Herpesviruses (varicella zoster and herpes simplex viruses)	Rounded virions with a size of 120 × 200 nm. Virions with outer envelopes are often found in clinical samples (vesicular and pustular liquids).

*Parenthesized are the viruses that may be required to distinguish between.
**Sizes are indicated according to *Virus Taxonomy* (2000).

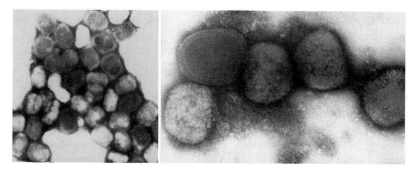

Figure 8-3. Immune electron microscopy (negative staining). Left panel: aggregation of vaccinia virus virions treated with hyperimmune antivaccinia serum, ×66,000. Right panel: cluster of monkeypox virions after treatment with monkeypox-specific monoclonal antibodies, ×192,000. Preparations by N. Yanova.

the efficacy of the method significantly by increasing the number of virions detected in samples examined (Figure 8.3). The method allowed a virus in samples negative by conventional electron microscopy to be detected (Marennikova *et al.*, 1990; Yanova, 1996). In addition, immune electron microscopy reduces the time required for examination and increases the reliability of the result. By the example of monkeypox virus and species-specific monoclonal antibodies to it, the studies showed the potential of this method for species-level identification of orthopoxviruses (Marennikova *et al.*, 1988a).

8.2   Biological Methods

The earliest biological method is the assay on rabbit cornea, proposed by Paul (1915) and widely used for diagnostics previously. It was abandoned after the development of assays on chick embryos and cell cultures, especially because its sensitivity was insufficient, as reported in the late 1950s–1960s.

8.2.1   Chick Embryo Assay

Chick embryos were brought into use for diagnostics in 1937, when it was shown that inoculation of samples from smallpox cases onto CAMs gave rise to virus-specific lesions (Lazarus *et al.*, 1937).

Further studies showed that virus isolation on chick embryos allowed the variola virus to be distinguished not only from chickenpox virus, which did not grow on CAM (Buddingh, 1938), but also from other orthopoxviruses—vaccinia, cowpox, monkeypox, etc., forming lesions different from those formed by vaccinia virus (Downie & Dumbell, 1947; Marennikova *et al.*,

1956; Marennikova, 1960; Figure 8.4, Table 8.2). A high sensitivity of this method independent of the stage of lesion development was recorded by virtually all the researchers who used it. As shown at our laboratory, the virus was isolated at a 100% rate from intact samples of skin lesions of smallpox cases. During the program of smallpox eradication, long transportation of samples from the countries with hot climate reduced the rate to 94% (Mal'tseva, 1980). The method was also applicable to isolation of the virus from matter with its poor content: blood or throat swabs (Downie *et al.,* 1953; Marennikova, 1962). This is of significance in cases with the absence of skin lesions, for example, during primary hemorrhagic purpura or at the pre-eruptive stage. The CAM assay was widely used during the smallpox eradication program. In particular, it allowed for the first detection of the human monkeypox, previously unknown and virtually identical to smallpox. Unlike cell culture, this method is less fastidious to the inoculum quality with regard to its dilution, clarification, and antibiotic treatment for suppressing bacterial flora. So far, it is still the best biological method for diagnostics of orthopoxvirus infections. The protocol of the method was detailed in various guidelines (World Health Organization, 1969). We dwell on two important features, often neglected. Of the three

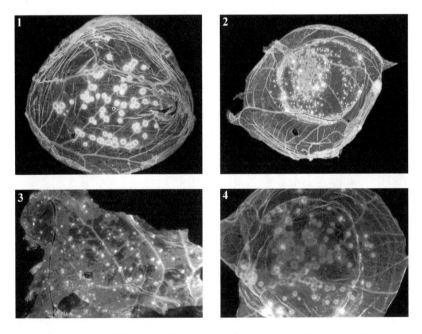

Figure 8-4. Pocks on CAM produced by orthopoxviruses pathogenic for humans: (1) vaccinia virus, (2) monkeypox virus, (3) variola virus, and (4) cowpox virus.

Table 8-2. Species identification of orthopoxviruses and their differentiation from causative agents of clinically similar diseases according to biological and serological markers

Markers	Orthopoxviruses				Yatapoxviruses (Tanapox)	Molluscum contagiosum virus	Parapoxviruses*	Herpesviruses	
	Cowpox	Monkeypox	Variola	Vaccinia				Varicella–zoster	Herpes simplex
CAM lesions	Most pocks are flat, indistinct, hemorrhagic; singular pocks are white	Small pocks with central hemorrhage; singular pocks are larger and white	Monomorphic distinct dome-like white pocks	Large white or flat grayish pocks		None			Small whitish lesions
Hemagglutination activity	Absent or low	High	Absent or low	Pronounced			Absent		
Reaction on scarified rabbit skin	Papular–pustular rash with hemorrhages and edema	Papular–pustular rash, sometimes with generalized process	Absent	Papular–pustular rash			Absent		
PEK test**	Positive	Positive	Positive	Positive			No data		
Intracellular inclusions	Cytoplasmic Types A and B	Type B	Type B	Type B	Similar to inclusions of type B	Hyaline acidophilic granular bodies (molluscum bodies)***	Type B (of virus)	Nuclear	
Gel precipitation assay with anti-vaccinia serum	Positive The major precipitation band is absent	Obligatory presence of the first and third major precipitation bands					Negative		

*Human pathogens are included: paravaccinia, bovine papular stomatitis, and orf viruses;

**CPE and hemadsorption in PEK cells (continuous strain of porcine embryo kidney cells); and

***Moss (1996).

chick embryo ages recommended (World Health Organization, 1969; Nakano, 1978)—11, 12, and 13 days—the age of 12 days is optimal. The reasons are that 11-day embryos are less stable and lesions on CAM are more difficult to differentiate, whereas 13-day embryos often develop hemorrhages on CAM during preparation for inoculation, hampering the development of lesions. The other feature important for differentiation in the appearance of CAM lesions, necessary for distinguishing certain orthopoxviruses, is the temperature of embryo incubation upon inoculation. The required temperature is 34.5–35°C.

8.2.2   Cell Culture Assay

This method was first used for smallpox diagnostics in France (Boue & Baltazard, 1956) and in Russia (Marennikova *et al.*, 1959). The assay utilizes the cytopathic effect of orthopoxviruses.

Orthopoxviruses can be isolated using monolayers of various primary or continuous cell cultures: monkey kidney cells, human embryo fibroblasts, human diploid and continuous cell lines, etc. Inoculation requires a culture with a well-developed cell layer. The culture is incubated at 36°C with daily microscopic examination. Cytopathic effect (CPE) is usually noticed 24–96 h after inoculation (Figure 8.5). The virus can be isolated before the onset of pronounced CPE with the use of hemadsorption and immunofluorescence assay (Figures 8.6, 3.3, 5.7). In addition, intracellular (cytoplasmic) inclusions can be distinguished in infected cells (Marennikova *et al.*, 1959; Gurvich, 1964).

Like the CAM assay, this approach ranks among the most sensitive methods. It is virtually 100% efficient with the vesicular or pustular liquid or scabs of a disease case. The virus can be detected in samples with its low content: blood, throat swabs at the beginning of the disease, etc. However, a disadvantage of this method in comparison with the chick embryo assay is difficulty in differentiating between closely related orthopoxviruses according to the CPE pattern.

To improve differentiation of orthopoxviruses in cell cultures, additional methods were elaborated, including determination of plaque morphology and the time of appearance and ceiling temperature of CPE development (Solov'ev & Bektemirov, 1962; Gurvich & Marennikova, 1964; Gurvich, 1964). However, these methods protract the assay considerably. A simple method is simultaneous use of cell cultures nonpermissive for one or another orthopoxvirus species. For example, we showed that monkeypox virus could be differentiated from variola and other orthopoxviruses in that it demonstrated no CPE, plaques, or hemadsorption in the continuous PEK (porcine embryo kidney) cell line (Marennikova *et al.*, 1971c).

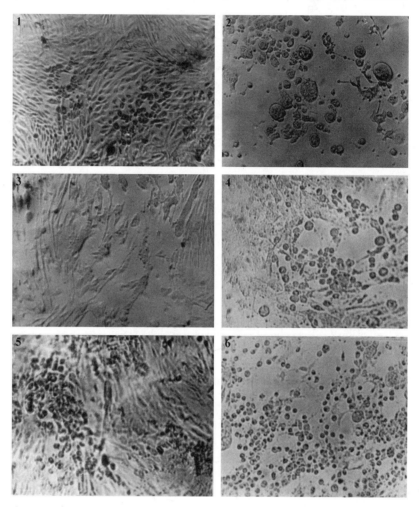

Figure 8-5. Human embryo fibroblast culture. The CPEs of orthopoxviruses and herpesviruses capable of causing human diseases with skin lesions: (1) 24 h after inoculation with variola virus (×35); (2) same culture, after 72 h; (3) 24 h after inoculation with vaccinia virus (×35); (4) 96 h after inoculation with variola alastrim virus (×28); (5) 9 days after inoculation with varicella virus (×28); (6) 72 h after inoculation with herpes simplex virus (×28).

Of herpes viruses, some herpes simplex strains are similar to variola virus in CPE they cause. Isolates are identified by serological assays or histologically according to the type of inclusions in infected cells.

The cytopathic effect of varicella virus in cell cultures of human or monkey origin occurs much later: 5 to 17 days post infection. It appears in

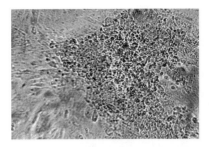 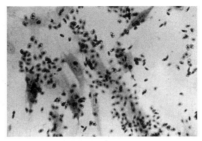

Figure 8-6. Human embryo fibroblast culture inoculated with variola virus. Left panel: hemadsorption in a tube with the culture before the onset of a visible CPE. Right panel: the culture on a stained slide (magnification 7 × 8).

the form of small lesions (Figure 8.5). Cytopathic effect spreads very slowly. Varicella virus, like herpes simplex virus, does not show a hemadsorption phenomenon, typical of orthopoxviruses. Cytological examination by the plate method reveals inclusions in the nuclei of infected cells (Gurvich, 1964; Mal'tseva, 1980).

8.3   Serological Methods

The oldest serological test is the complement binding reaction. Sugai as early as in 1909 introduced this reaction for detection of anti-smallpox complement-binding antibodies, and since the 1930s, it was used for detection of smallpox antigen in clinical samples. However, this reaction lost its significance since the appearance of more sensitive tests (EIA etc.).

8.3.1   Gel Precipitation

The assay utilizes the formation of precipitation zones in agar gel when the variola virus antigen interacts with a specific serum (Nizamuddin & Dumbell, 1961; Marennikova & Mal'tseva, 1961; etc.). In combination with other methods, it was widely used in diagnostics for examination of samples from human cases and serological identification of isolates (validation of their orthopoxvirus nature). However, it ranks below other methods in sensitivity when applied to clinical samples. According to the data from the Moscow WHO Collaborating Center, it showed only 62% of positive results (Mal'tseva, 1980).

Nevertheless, the advantages of the method are simplicity, high specificity, and rapid results obtainable within 4–5 h. In experiments with some forms of smallpox, it allowed the variola virus antigen to be detected in blood. In these cases, the time required for visualization of precipitation bands increased to 24 h or more.

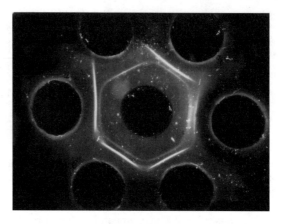

Figure 8-7. Immunoprecipitation reaction in agar gel with antivaccinia serum (in the central well). The major precipitation band is absent in the left upper well with the cowpox virus. The rest wells contain vaccinia, variola, and monkeypox viruses.

The reverse version of gel precipitation can be applied for detection of antibodies to orthopoxviruses. For this purpose, serum of a human case with a known vaccinia virus culture is examined.

Note that neither gel precipitation nor other serological methods allow for identification of orthopoxvirus species without special techniques, except for cowpox virus. In the classic version developed by Ouchterlony (1949), cowpox virus does not show the first major precipitation band, produced by the other orthopoxviruses (Figure 8.7).

In 1976, a two-stage gel precipitation reaction was proposed. It increased the potential of orthopoxvirus identification. The first stage involves adsorption of antivaccinia serum by a virus culture examined. After the appearance of precipitation band(s), usually taking 6 h, the serum is tested with all viruses to be differentiated. The result is estimated according to the appearance or absence of additional precipitation bands. Additional bands indicate that the virus tested is heterologous with respect to the virus that adsorbed the serum (Maltseva & Marennikova, 1976).

8.3.2   Enzyme Immunoassay (EIA)

The assay involves histochemical detection of the antigen–antibody complex. EIA version for antibody and antigen detection were developed (Hutchinson *et al.* 1977; Marennikova *et al.*, 1981; 1988a; 1991; Mal'tseva *et al.*, 1984; 1985; Gates *et al.* 2000).

8.3.3   Radioimmunoassay (RIA)

As mentioned above, RIA was used for detecting antibodies to orthopoxviruses (Ziegler *et al.*, 1975). According to Jezek and Fenner (1988), this assay is the most sensitive. However, studies of human monkeypox cases showed that RIA could not detect antibodies in the first days of the disease. They are detectable only at the beginning of the first or the end of the second week. Their titers vary within 3000–20,000 or more and remain high for 1 or 2 months. Antibodies can be detected for 8–9 years after the disease.

RIA adsorption (RIA-A). This RIA version was developed at the US WHO Collaborating Center (Hutchinson *et al.*, 1977; Walls *et al.*, 1980; 1981) for species-level identification of antibodies. Like EIA-A, it involves pre-adsorption of the sera to be assayed. The practice of using RIA-A showed that it sometimes failed to identify antibodies. In particular, the identification is hampered during the first month of the disease and in human monkeypox cases of vaccinees. In the latter case, the antibodies may behave as vaccinia-specific. Note that RIA-A played a great role in ecological studies of monkeypox. By detection of species-specific antibodies to this virus, it revealed wild animal species involved in monkeypox virus circulation. These data were obtained owing to the modification of the method by Nakano (Fenner & Nakano, 1988).

8.3.4   Hemagglutination Inhibition Test (HAIT)

This method utilizes the phenomenon of suppression by specific antibodies of agglutination of sensitive chick erythrocytes, caused by a number of orthopoxviruses. Collier and Schonfeld (1950) proposed this method for diagnostic purposes. Numerous studies of these authors performed during a smallpox outbreak on the island of Java showed that the human cases displayed high AHA titers, exceeding considerable the titers of recent vaccinees. These differences allowed variola virus to be differentiated from vaccinia in most cases. Later, these data were confirmed by Marennikova (1958) by observing the maximal titers on days 10–15 followed by their gradual decrease. In some cases, however, AHA titers remained low (20–40), and the presence of low AHA titers prevented ruling out smallpox.

An increase in AHA titers is also observed in human monkeypox and cowpox cases. Being simple and readily available, this method, in addition to diagnostic purposes, was widely applied for various serological studies in both humans and animals. In particular, it was used in searching for the natural reservoirs of monkeypox and cowpox viruses (Ladnyi *et al.*, 1975; Jezek & Fenner, 1988; Tsanava, 1990).

8.3.5   Neutralization Reaction

Neutralizing antibodies to orthopoxviruses are usually detected in chick embryos or in cell cultures according to 50% inhibition of the developed pocks on CAMs (Boulter, 1957) or plaques in a cell monolayer (Esposito *et al.,* 1985). The advantage of this method is a possibility to use it for a retrospective diagnostics a long period (months and even years) after the disease. However, the experience of its application demonstrated that the reaction in question is used rather infrequently for diagnostic purposes. Nonetheless, it is very valuable when assessing the reactogenicity of smallpox vaccines, detection of natural carriers, etc.

Comparison of various methods for detection of antibodies to orthopoxviruses performed during serological studies of the populations of several African countries in 1981–1983 confirmed that HAIT ranked much lower in the sensitivity than EIA or neutralization test (Marennikova *et al.,* 1984c). The neutralization test and EIA detected antibodies in 15% and 19% of cases with negative HAIT, respectively. Nevertheless, antibodies were detected by HAIT in 64.2% of cases differing in time elapsed after vaccination. In addition, the presence of AHA at titers of 5 or higher indicates that antibodies are highly probable to be detectable by neutralization or EIA (88.8–93.6%). Thus, HAIT is a conventional method of screening of sera for further studies (identification of antibodies at species level. Recently, detection of IgM and IgG antibodies and assessment of IgG avidity are used for serological diagnostics of the diseases caused by orthopoxviruses pathogenic for humans. The presence of specific IgM (early) antibodies and the IgG antibodies with a low avidity indicate a current or recent orthopoxvirus infection (Tryland *et al.,* 1998a; Hutin *et al.,* 2001; Pelkonen *et al.,* 2003; etc.).

8.4   Biochemical Methods for Differentiation of Orthopoxviruses

The potential increase in the degree of danger of orthopoxvirus infections for humans required development of modern efficient methods for rapid detection and identification of orthopoxviruses pathogenic for humans.

The conventional biological and serological methods used appeared insufficiently effective for rapid diagnostics of orthopoxviruses (see 8.2, 8.3). The biological analysis takes too long time (3–6 days) and involves handling of special viral pathogens. The serological methods, as a rule, allow only for a genus-level identification; moreover, their sensitivities are frequently insufficient for assaying clinical samples.

8.4.1 Restriction Fragment Length Polymorphism Analysis of Viral Genomic DNAs

The methods based on genomic analysis may be regarded as an effective and efficient approach to diagnosing viral infections. It was demonstrated (Esposito *et al.*, 1978; Mackett & Archard, 1979; Esposito & Knight, 1985) that restriction enzyme assay of viral DNAs provided a reliable species-level identification of orthopoxviruses. However, this requires propagation of the virus and its purification, demanding specialized equipment, and is time-consuming.

8.4.2 Virus Identification Using Polymerase Chain Reaction

The advent of the method of DNA fragment amplification by polymerase chain reaction (PCR; Bej *et al.*, 1991; Arens, 1999) formed the background for designing various techniques appropriate for rapid identification of orthopoxviruses. The method of DNA fragment amplification by PCR makes it possible to recover the specific DNA fragments from trace quantities of genetic material under study during a short time. It does not require any cultivation of the virus to be tested, which is especially important when we have to deal with highly pathogenic strains.

Determination of the genomic nucleotide sequences for a number of strains of several orthopoxviral species (Goebel *et al.*, 1990; Shchelkunov *et al.*, 1993f; 1995; 1998; 2000; 2001; Massung *et al.*, 1994; Antoine *et al.*,

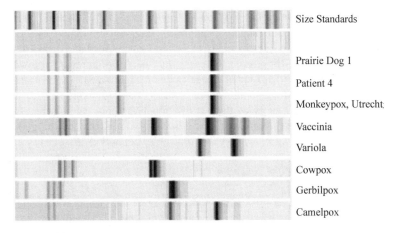

Size Standards
Prairie Dog 1
Patient 4
Monkeypox, Utrecht
Vaccinia
Variola
Cowpox
Gerbilpox
Camelpox

Figure 8-8. PCR amplification of the orthopoxvirus HA gene followed by restriction fragment length polymorphism (RFLP) assay using TaqI restriction endonuclease. Analysis of the samples from patient 4, prairie dog 1, and reference isolates of other orthopoxviruses (Reed *et al.*, 2004; ©2004 Massachusetts Medical Society. All rights reserved).

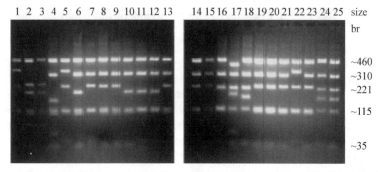

Figure 8-9. NlaIII RFLP assay of PCR-amplified CrmB fragments of DNA of CPXV isolates from human and animals: lane 1, OPV85 (human); lane 2, OPV88/L (cat); lane 3, OPV88/H (cat); lane 4, OPV89/1 (cat)-M5; lane 5, OPV89/2 (cat); lane 6, OPV89/3 (cat); lane 7, OPV89/4 (cat)-M6; lane 8, OPV89/5 (cat)-M7; lane 9, OPV90/1 (cat)-M8; lane 10, OPV90/2 (human); lane 11, OPV90/4 (dog); lane 12, OPV90/5 (cat)-M9; lane 13, OPV91/1 (cat); lane 14, OPV91/2 (human); lane 15, OPV91/3 (cow); lane 16, CATPOX3; lane 17, CATPOX5; lane 18, RAT Moscow (rat); lane 19, EP-1 (elephant); lane 20, EP-2 (elephant)-M1; lane 21, EP-3 (elephant); lane 22, EP-4 (elephant); lane 23, EP-5 (elephant); lane 24, CPV BRT-Atlanta; and lane CPV BRT-Munich (Loparev *et al.*, 2001; reprinted with permission of the American Society for Microbiology).

1998) allowed species-specific differences of some loci to be revealed and new methods for virus identification to be designed. So far, application of PCR for detection of orthopoxviruses using common pair of oligonucleotide primers to the regions of genes encoding hemagglutinin (*HA;* Ropp *et al.*, 1995), A-type inclusion protein (*ATI;* Meyer *et al.*, 1997), and homologue of tumor necrosis factor receptor (*CrmB;* Loparev *et al.*, 2001) has been described. In all these techniques, the DNA fragments obtained by PCR are hydrolyzed with certain restriction endonucleases and separated by electrophoresis; the resulting patterns of subfragments allowed orthopoxviruses to be identified at a species level (Figure 8.8).

However, when a large enough set of isolates of an orthopoxvirus species was analyzed, heterogeneity of their restriction fragments patterns became apparent, making interpretation of the results obtained rather ambiguous (Figure 8.9; Ropp *et al.*, 1995; Meyer *et al.*, 1997; Loparev *et al.*, 2001).

8.4.3   Multiplex PCR Analysis

Specific detection of MPXV by PCR was developed using one species-specific oligonucleotide primer pair for gene *ATI* sequence (Neubauer *et al.*, 1998). However, it would be very important to have at hand a technique that allow for discriminating between orthopoxviral species in a one-step assay.

Therefore, a new method of multiplex PCR assay (MPCR) of orthopoxviruses pathogenic for humans was developed recently (Gavrilova

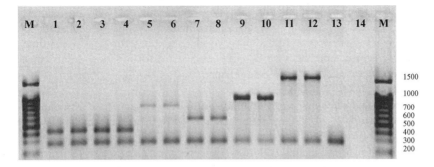

Figure 8-10. Electrophoretic separation in 2% agarose of the amplicons produced by PCR using four pairs of oligonucleotide primers: (1) CPXV strain GRI-90; (2) CPXV strain Puma-73; (3) CPXV strain Turkmenia; (4) CPXV strain EP-2; (5) MPXV strain CDC#v79-I-005; (6) MPXV strain CDC#v97-I-004 (Central African); (7) MPXV strain CDC#v70-I-187; (8) MPXV strain CDC#v78-I-3945 (Western African); (9) VARV strain Congo-9; (10) VARV strain Ind-3a; (11) VARV strain Butler; (12) VARV strain Brazil-128; (13) ectromelia virus strain MP-1; (14) negative control; and M, DNA marker (lengths in bp are shown to the right; Gavrilova *et al.*, 2003).

et al., 2003). This method displays a high specificity and sensitivity of such analysis. The essence of the method designed is that the selected unique oligonucleotide primers allow orthopoxviruses to be identified at a species level in one stage. Four pairs of oligonucleotide primers (three pairs for VARV, MPXV, and CPXV, respectively, and one genus-specific pair) were used in a united polymerase chain reaction producing amplicons of various lengths specific of each orthopoxvirus species in question (Figure 8.10).

Used primers allow for discriminating between the VARV subtypes, such as major and minor as well as Central African and Western African MPXV subtypes. The genus-specific pair was used as an internal PCR control for the presence of orthopoxvirus DNA in the sample and discrimination from other genera of poxviruses. Specificity and sensitivity of the method developed were evaluated using DNAs of 57 orthopoxvirus strains, including the DNAs isolated from human case clinical materials (scabs from skin lesions of smallpox human cases infected in 1970–1975 and deposited with the Russian Collection of Variola Virus). Being simple, quick, and exact, the developed MPCR assay allows separate analysis of each sample using each pair of primers to be avoided and orthopoxviruses pathogenic for humans to be identified at a species level in one stage.

8.4.4   Real-Time PCR Assay

When identifying a particular virus species, a real-time PCR is even more efficient, because it combines amplification and detection of the target DNA in one tube, thereby eliminating any time-consuming post-PCR procedures, and potentially limiting possible contamination events. Recently, three independent research teams developed the procedures of real-time PCR identification of variola virus and its discrimination from other orthopoxviruses pathogenic for humans (Espy *et al.,* 2002; Ibrahim *et al.,* 2003; Olson *et al.,* 2004).

Note that screening large orthopoxvirus strain collections is essential to demonstrate the utility and establish the performance characteristics of assays developed. In a recent paper (Espy *et al.,* 2002), the authors state that mismatches in the FRET (fluorescence resonance energy transfer) probes used in their assay enabled discrimination of VARV from other orthopoxviruses by DNA melting curve analysis of a 204-bp amplicon from the *HA* gene region. Due to new orthopoxvirus sequences in GenBank, the FRET probes display also an identity to camelpox and some cowpox virus strains. Analysis of such strains has to prove whether a reliable identification of smallpox virus is still possible.

A new screening assay for real-time LightCycler (Roche Applied Science, Mannheim, Germany) PCR identification of variola virus DNA was developed and compiled in a kit system under GMP conditions with standardized reagents (Olson *et al.,* 2004). In search of a sequence region unique to variola virus, the nucleotide sequence of the 14 kDa fusion protein gene of 14 variola virus isolates from the Russian WHO Smallpox (Variola) Virus Repository was determined and compared to the published sequences. PCR primers were designed to detect several species of the genus *Orthopoxvirus.* A single nucleotide mismatch resulting in a unique amino acid substitution in variola virus was used to design a hybridization probe pair with a specific sensor probe that allowed for a reliable differentiation of variola virus from other orthopoxviruses via melting curve analysis (Figure 8.11). The applicability of this method was demonstrated by successful amplification of 120 strains belonging to the orthopoxvirus species variola, monkeypox, cowpox, vaccinia, camelpox, and mousepox viruses. The melting temperature (T_m) determined for 46 strains of variola virus (T_m, 55.9–57.8°C) differed significantly ($p = 0.005$) from those obtained for 15 strains of monkeypox virus (T_m, 61.9–62.2°C), 40 strains of cowpox virus (T_m, 61.3–63.7°C), 11 strains of vaccinia virus (T_m, 61.7–62.7°C), 8 strains of mousepox (ectromelia) virus (T_m, 61.9°C), and 8 strains of camelpox virus (T_m, 64.0–65.0°C).

Another highly sensitive and specific assay for a rapid detection of variola virus DNA on both the Smart Cycler and LightCycler platforms was developed (Ibrahim *et al.,* 2003). The assay is based on TaqMan chemistry with the orthopoxvirus HA gene used as the target sequence. The assay was evaluated in a blinded study with 322 coded samples that included genomic DNAs from 48 different isolates of variola virus; 25 different strains of

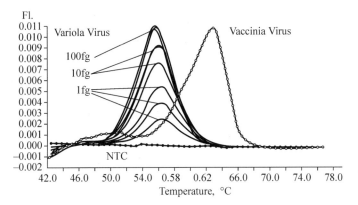

Figure 8-11. Melting curves generated after LightCycler amplification of 100, 10, and 1 fg of DNA prepared from variola virus infected cell culture material (strain Kali Mathu). Each DNA concentration was run in triplicate. The melting curve after amplification of 100 fg of vaccinia virus strain MVA is also shown (Olson *et al.*, 2004); NTC, no-template control.

monkeypox, cowpox, vaccinia, rabbitpox, camelpox, ectromelia, gerbilpox, raccoonpox, skunkpox, myxoma, herpes, and varicella zoster viruses.

Along with the evident advantages of real-time PCR identification, this method yet has certain shortcomings. It allows for analyzing only one relatively short fragment of the virus genome and identifying, as a rule, only one species. Taking into account the natural variation of the orthopoxvirus genetic loci analyzed, it is necessary to assay concurrently several individual loci located in various parts of the virus genome. Therefore, a reliable result on identification of a particular orthopoxvirus species requires a parallel or successive real-time PCR identification involving several different genes.

8.4.5 Oligonucleotide Microarray Analysis

Many of the above-mentioned problems that arise during a species-level detection of the viruses can be solved using hybridization of DNA molecules on oligonucleotide microarrays, frequently called microchips. A method for species-specific detection of orthopoxviruses on an oligonucleotide microchip was described (Lapa *et al.*, 2002; Mikheev *et al.*, 2003). The method is based on hybridization of a fluorescently labeled amplified DNA specimen with the oligonucleotide DNA probes immobilized on a three-dimensional polyacrylamide-gel microchip (MAGIChip™). The probes identify species-specific sites within the viral *CrmB* gene. The microchip contains 14 oligonucleotide probes directed towards 5 species-specific segments of the *CrmB* gene. The probe location relative to the sequence of the gene of the variola virus is shown in Figure 8.12.

```
          Primer TNFR1f
          GCTTCCAGATTATGTGATAGCAAGACTA ->
VARV-IND  GCTTCCAGATTATGTGATAGCAAGACTAACACA------CAATGTACACCGTGTGGTTCGGGTACCTTTACATCT 69
VARV-GAR  ............................-----....................................... 69
MPXV-SL   ............................T..-----..................A................. 69
MPXV-ZAI  ............................T..-----.........G.........A................. 69
CPXV-90   .............C..............AACACA........G...........AC.....C.......... 75
CPXV-85   .............C..............AACACA....................C................. 75
VACV-COP  .............C..............-----....................C................. 69
VACV-T-T  .............C..............-----.........G..C........C................. 69
CPXV-Utr  .............C..............-----.G...........C........C................. 69
CMLV-CP5  .............C..............-----....................C................. 69
CMLV-CP1  .............C..............-----.........G...........C.C............... 69
                              CTAACACAAACACACA            No. 1
                              CTAACACA------CGATGTAC          No. 2
                              CTAACACA------CAATGTAC          No. 3

VARV-IND  CGCAATAATCATTTACCCGCTTGTCTAAGTTGTAACGGAAGATGCAATAGTAATCAGGTAGAGACGCGATCGTGT 144
VARV-GAR  .......................................................................... 144
MPXV-SL   .A.............AG............................TG........................... 144
MPXV-ZAI  .A.............AG............................TG........................... 144
CPXV-90   ...........................................G.............................. 150
CPXV-85   .A.........................................G...................C.......... 150
VACV-COP  ...........................................-------------------------........ 122
VACV-T-T  ...........................................-------------------------........ 122
CPXV-Utr  ...........................................-------------------------........ 122
CMLV-CP5  ...........................................G.............................. 144
CMLV-CP1  ...........................................G.............................. 144
                  TTACAGGCTTGTCT              No. 4
                  TTACCCGCTTGTC               No. 5
                            CGGAAG--------------------CGCGAT  No. 6
                            AAGATGTGATAGTAAT            No. 7
                            AAGATGCGATAGTAAT            No. 8
                            AAGATGCAATAGTAAT            No. 9

VARV-IND  AACACGACTCACAATAGAATCTGTGAATGCTCTCCCGGATATTATTGTCTTCTTAAAGGATCATCCGGATGCAAG 219
VARV-GAR  .......................................................................... 219
MPXV-SL   ....................................A...............C......G...TA..G..T.GA 219
MPXV-ZAI  ....................................A...............C.......A..G..T.GA 219
CPXV-90   ...............T...G................................C..................... 225
CPXV-85   ...................................................C..................... 225
VACV-COP  .....T.............................................C..................... 197
VACV-T-T  .....T.............................................C..................... 197
CPXV-Utr  ...................................................C..................... 197
CMLV-CP5  ...................................A...C................................. 219
CMLV-CP1  ...................................A...C................................. 219
                              GTATTCTCAAAGGA      No. 10
                              GTCTTCTCAAAGGA      No. 11
                              GTCTTCTTAAAGGA      No. 12

VARV-IND  GCATGTGTTTCCCAAACAAAATGTGGAATAGGATACGGAGTATCCGGA    267
VARV-GAR  ................................................    267
MPXV-SL   A.....A....TA.......G............................    267
MPXV-ZAI  A.....A....TA.......G............................    267
CPXV-90   ......A.........................................    273
CPXV-85   ................................................    273
VACV-COP  ................................................    245
VACV-T-T  .................C..............................    245
CPXV-Utr  ................................................    245
CMLV-CP5  ................................................    267
CMLV-CP1  ................................................    267
                  ATTTCTAAAACAAA         No. 13
                  GTTTCCCAAACAAA         No. 14
                       <- CCTTATCCTATGCCTCATAGGCCT
                          Primer TNFR3r
```

Figure 8-12. Aligned sequences of a CrmB gene fragment of various orthopoxviruses. All the sequences are compared with VARV-IND. Dots indicate identical oligonucleotides; dashes, deletions. Numbers 1 to 14 designate sequences of species-specific probes and their location relative to the aligned sequences. Sequences of the primers TNFR1f and TNFR3r for genus-specific DNA fragment amplification are shown (Lapa *et al.*, 2002).

The microchip contains five gel pad columns (Figure 8.13), each column representing a separate interrogated segment of the amplified fragment of viral DNA. Within each column, only one species-specific probe can form a

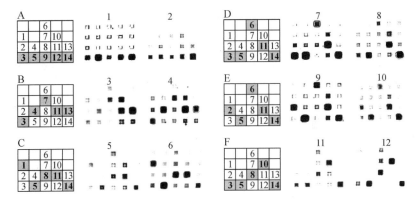

Figure 8-13. Hybridization patterns obtained on the microchip for five orthopoxvirus species: A, variola virus (1, strain Garcia-1966 and 2, strain Semat); B, monkeypox virus (3, strain Congo-8 and 4, strain Zaire-96-I-16); C, cowpox virus (5, strain EP-267 and 6, strain GRI-90); D, vaccinia virus (7, strain WR and 8, strain CVI-78); E, rabbitpox virus (9, strain Utrecht and 10, vaccinia virus strain Elstree/Utrecht); and F, camelpox virus (11, strain CP-1 and 12, strain CP-1260/95). Schemes of the anticipated hybridization patterns are shown in the left column; oligonucleotide numbers in the cells correspond to the numbers of oligonucleotides in Figure 8.12 (Mikheev *et al.*, 2003).

perfect duplex with a viral DNA sample, whereas all other probes form mismatched duplexes. As a result, the hybridization pattern is unique for every tested DNA sample, thus enabling an accurate species assignment.

Overall, 59 samples of orthopoxvirus DNAs representing six different species were analyzed. Different strains of variola, monkeypox, cowpox, vaccinia, and camelpox viruses were successfully identified using hybridization of amplified DNA to the microchip. No discrepancy between hybridization and conventional identification results was observed.

Another kind of oligonucleotide microarray, created on plain glass slides, was developed for discrimination between orthopoxviruses pathogenic for humans involving the virus gene *C23L/B29R*, encoding the CC chemokine binding protein (Laassri *et al.*, 2003).

This microarray-based method detects simultaneously and discriminates between four orthopoxvirus species pathogenic for humans (variola, monkeypox, cowpox, and vaccinia viruses) and distinguishes them from chickenpox virus (varicella zoster virus). The microchip contained several unique 13–21 bases long oligonucleotide probes specific of each virus species to ensure redundancy and robustness of the assay. A region approximately 1100 bases long was amplified from samples of viral DNAs and fluorescently labeled with Cy5-modified dNTPs; single-stranded DNA was prepared by strand separation. Hybridization was performed under plastic coverslips, resulting in a fluorescent pattern that was quantified using a confocal laser scanner (Figure 8.14). Overall, 49 known and blinded samples of orthopoxvirus DNAs, representing different orthopoxvirus species, and two varicella zoster virus strains were tested. The oligonucleotide microarray hybridization technique identified reliably and correctly all the samples. This new procedure takes only 3 h, and it can be used for parallel testing of multiple samples.

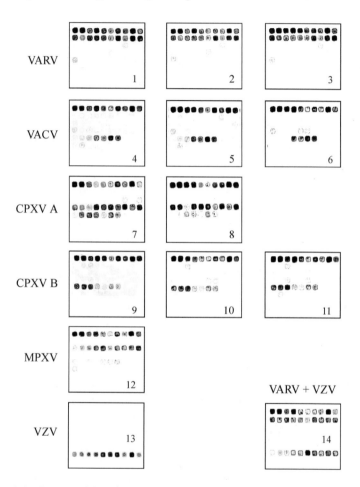

Figure 8-14. Patterns of detection and discrimination between the orthopoxvirus species pathogenic for humans and their differentiation from HHV 3 (varicella zoster virus) by microarray: (1) VARV strain Congo-2; (2) VARV strain Ind-3A; (3) VARV strain Kuw-5; (4) VACV strain Elstree 3399; (5) VACV strain Copenhagen; (6) VACV strain CVI-78; (7) CPXV-A strain Turk-74; (8) CPXV-A strain Brighton; (9) CPXV-B strain EP-1; (10) CPXV-B strain EP-2; (11) CPXV-B strain EP-7; (12) MPXV strain CDC#77-666; (13) varicella zoster virus strain Oka; (14) and pattern of the mixture hybridization of VARV strain Ind-3A and varicella-zoster virus strain Oka (Laassri *et al.*, 2003; reprinted with permission of Elsevier).

Chapter 9

PREPAREDNESS AND RESPONSE
TO POTENTIAL BIOTERRORISM

Bioterrorism is defined as the use of biological agents to inflict disease and/or death on humans, animals, or plants. Thus, crops and livestock as well as human populations are considered possible bioterrorist targets. Bioterrorism acts could have political, religious, or criminal motivation and could conceivably be planned by groups or a single individual or be a part of state-sponsored terrorist activities (Klietmann & Ruoff, 2001).

Many biological agents can cause illnesses in humans, but not all are capable of affecting public health and medical infrastructures on a large scale. According to expert estimations (Rotz *et al.,* 2002), potential bioterrorist agents are divided into three categories (Table 9.1). The agents were categorized according to the overall evaluation of various areas considered (Table 9.2). The category A agents are being given the highest priority for preparedness.

Variola major virus, causing smallpox, is considered the most hazardous potential bioterrorist agent, because it is highly contagious, is transmitted in a person-to-person manner as aerosol, stable in the environment, and can be rather simply produced in large amounts (Henderson, 1999; Berche, 2001).

Smallpox is the first and yet only disease successfully eradicated all over the globe during an international medical campaign under the aegis of WHO (Fenner *et al.,* 1988; Marennikova & Shchelkunov, 1998). The paradox of the current situation is that the triumphant victory over smallpox and the worldwide cancellation of routine anti-smallpox vaccination after 1980 resulted in a gradual increase in the volume of human population cohort susceptible to variola virus, thereby enlarging the menace of variola virus as a potential bioterrorist agent (Bray & Buller, 2004).

Currently, two official repositories of variola virus strains exist at strictly secured facilities inspected by WHO—Centers for Disease Control and

Table 9-1. Critical biological agent categories for public health preparedness (Rotz *et al.*, 2002)

Biological agent(s)	Disease
Category A	
Variola major	Smallpox
Bacillus anthracis Variola major	Anthrax
Yersinia pestis	Plague
Clostridium botulinum (botulinum toxins)	Botulism
Francisella tularensis	Tularemia
Filoviruses and Arenaviruses	Viral hemorrhagic fevers
(e.g., *Ebola virus, Lassa virus,* etc.)	
Category B	
Coxiella burnetii	Q fever
Brucella spp.	Brucellosis
Burkholderia mallei	Glanders
Burkholderia pseudomallei	Melioidosis
Alphaviruses (VEE, EEE, and WEE[a])	Encephalitis
Rickettsia prowazekii	Typhus fever
Toxins (e.g., ricin, staphylococcal enterotoxin B)	Toxic syndromes
Chlamydia psittaci	Psittacosis
Food safety threats (e.g., *Salmonella spp., Escherichia coli* O157:H7, etc.)	
Water safety threats (e.g., *Vibrio cholerae, Cryptosporidium parvum,* etc.)	
Category C	
Emerging threat agents (e.g., *Nipah virus,* hantaviruses, etc.)	

[a]Venezuelan equine (VEE), eastern equine (EEE), and western equine encephalomyelitis (WEE) viruses.

Prevention (Atlanta, GA, USA) and State Research Center of Virology and Biotechnology Vector (Koltsovo, Novosibirsk oblast, Russia). Several experts believe that the main danger of the reemergence of smallpox comes from these official stocks and propose their destruction (Mahy *et al.*, 1993). However, other scientists consider that the proposed destruction of the stocks provides only an illusion of safety, because additional potential sources of smallpox virus might still exist in other laboratories (Joklik *et al.*, 1993; Roizman *et al.*, 1994; Berche, 2001).

Emergence of other orthopoxviruses pathogenic for various animal species that have become adapted to humans (Berche, 2001) might pose a new threat. Here, the most dangerous pathogens are monkeypox (Chapter 5) and cowpox (Chapter 6) viruses. Unlike a strictly anthroponotic variola virus, monkeypox and cowpox viruses are zoonotic infectious agents with wide host ranges including humans. Evolutionary changes of these viruses may lead to an increase in their contagiousness and pathogenicity for humans. This script may cause the situation that is more hazardous to the humankind than the former situation that inspired the campaign of global smallpox eradication. It is possible to control zoonotic infections but virtually impossible to eradicate them.

Table 9-2. Criteria and weighting[a] used to evaluate potential biological threat agents (Rotz *et al.*, 2002)

Disease	Public health impact		Dissemination potential		Public perception	Special preparation	Category
	Disease	Death	P-D[b]	P-P[c]			
Smallpox	+	++	+	+++	+++	+++	A
Anthrax	++	+++	+++	0	+++	+++	A
Plague[d]	++	+++	++	++	++	+++	A
Botulism	++	+++	++	0	++	+++	A
Tularemia	++	++	++	0	+	+++	A
VHF[e]	++	+++	+	+	+++	++	A
VE[f]	++	+	+	0	++	++	B
Q fever	+	+	++	0	+	++	B
Brucellosis	+	+	++	0	+	++	B
Glanders	++	+++	++	0	0	++	B
Melioidosis	+	+	++	0	0	++	B
Psittacosis	+	+	++	0	0	+	B
Ricin toxin	++	++	++	0	0	++	B
Typhus	+	+	++	0	0	+	B
Cholera[g]	+	+	++	+/−	+++	+	B
Shigellosis[g]	+	+	++	+	+	+	B

[a]Agents were ranked from highest threat (+++) to lowest (0);
[b]Potential for production and dissemination in quantities that would affect a large population, based on availability, BSL requirements, most effective route of infection, and environmental stability;
[c]Person-to-person transmissibility;
[d]Pneumonic plague;
[e]Viral hemorrhagic fevers due to Filoviruses (*Ebola* and *Marburg*) or Arenaviruses (e.g., *Lassa, Machupo,* etc.);
[f]Viral encephalitis; and
[g]Examples of food- and water-borne diseases.

As is known, the natural transmission of smallpox was interrupted in 1977; since that time (except for two cases of laboratory-acquired infection in the United Kingdom in 1978), this disease does not actually exist on the globe (World Health Organization, 1980). The preventive routine vaccination against smallpox was stopped and until 2003, was limited to a small cohort of the researchers and assistants handling this agent and other orthopoxviruses pathogenic for humans during their work.

However, an increased danger of potential bioterrorism use of variola virus raised a question on the response strategy in the case of potential smallpox outbreak. The discussions on the optimal scenario for response to this menace considered various approaches (Meltzer *et al.*, 2001; Kretzschmar *et al.*, 2004). The fact that the majority of population (those who were born after cancellation of anti-smallpox vaccination) has no immunity to smallpox at all, while the rest part displays a drastically decreased immunity makes this issue especially critical.

Position of the advocates of restoring the mass immunization against smallpox met serious objections due to the risk of a high rate of postvaccination complications, especially among the elder cohort of nonvaccinees (see Section 3.7). The opponents of this opinion believe that the strategy of surveillance–containment in the case of smallpox outbreak is most reasonable (Foege, 1973; Fenner *et al.,* 1988). This multicomponent strategy is based on a tremendous positive experience of its application during the final phase of the global smallpox eradication campaign. Undoubtedly, this system will be most useful in the case of an attack involving smallpox as a bioweapon. In particular, one of the components of this strategy—early detection of each smallpox case—may appear considerably more important than the overall set of the subsequent complex actions necessary if a smallpox case is detected with a delay. Thanks to this, it is also possible to decrease the cohorts that should be vaccinated, consequently reducing the rate of postvaccination complications. An impressive example of the role of early detection is two smallpox cases imported into Moscow (in 1959 and 1961): the first instance resulted in the outbreak involving 46 cases, whereas the second was stopped at the level of index case. Note that during the first importation, the disease was diagnosed 24 days after its emergence, with appearance of the first generation of infection.

The type and scope of a potential bioterrorist attack remain guesswork to a considerable degree; consequently, a number of questions, first and foremost, the vaccination coverage of the population, have no precise answers in advance. Thus, numerous and not easily confined foci might require a wider vaccination. However, independently of a particular approved response plan, the sanitary services should have at their disposal the necessary amount of smallpox vaccine doses and the corresponding equipment and supplies as well as the diagnostic laboratories, the appropriate reagents and trained staff. The epidemiological service should be prepared to immediate prevention activities according to the arising situation. The issues directly related to vaccination, vaccination immunity, postvaccination complications, etc., are detailed in Chapter 3. However, note here that the protective effect of vaccination (revaccination) is attainable only before a contact with this pathogen or immediately (2–3 days) after infection.

Unfortunately, any reliable therapeutics for treatment of smallpox are yet unavailable. However, there are grounds to expect new generation serological preparations for smallpox prevention and treatment involving humanized monoclonal antibodies to orthopoxviruses (Hooper *et al.,* 2002; Ramirez *et al.,* 2002). Positive results of smallpox serotherapy (especially, at the early stage), when sera of smallpox convalescents, hyperimmune animal sera, or the immunoglobulins isolated from them were used for treatment (Teissier & Marie, 1921; Marennikova *et al.,* 1961b; etc.), corroborate these expectations.

Cidofovir and its derivatives appear rather promising therapeutics, which were recently tested successfully in experimental models of orthopoxvirus infections, in particular, lethal generalized monkeypox of primates (Bray *et al.*, 2000; Baker *et al.*, 2003). With these data in mind, we can agree with the opinion of Breman and Henderson (2002) that "In the event of a smallpox outbreak the drug could be made available under an investigational new drug protocol for smallpox or adverse effects of vaccine".

Appendix 1: Graphic alignment of the genome and ORFs of CPXV-GRI (GRI) with the corresponding genomes of other orthopoxviruses: CPXV-BRT (BRT), MPXV-ZAI (ZAI), VARV-IND (IND), VARV-GAR (GAR), VACV-COP (COP), and VACV-WR (WR). The sizes and directions of potential ORFs are marked with arrows. Deletions in the DNA and proteins exceeding 150 bp and 50 amino acids, respectively, of one virus relative to another are marked with dots. The black blocks mark the sequences within the TIR regions. Nucleotide numbers are shown on the right.

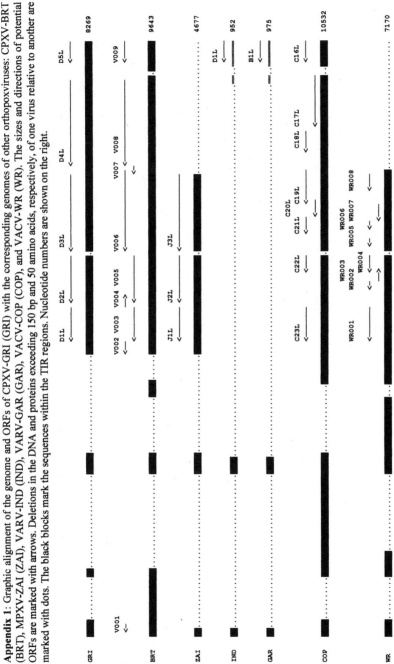

331

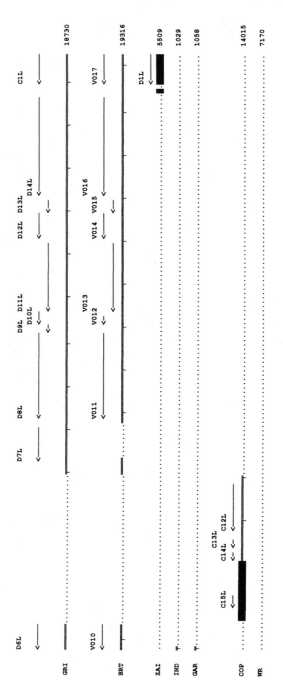

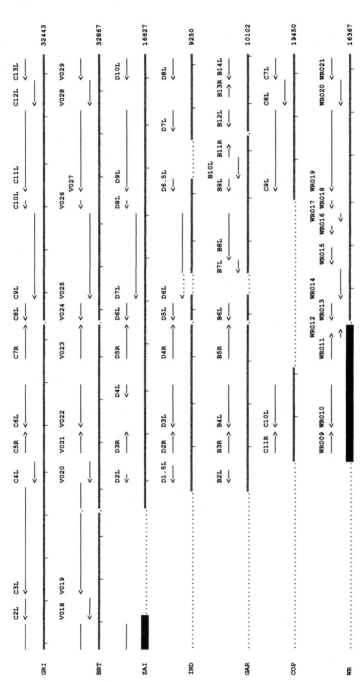

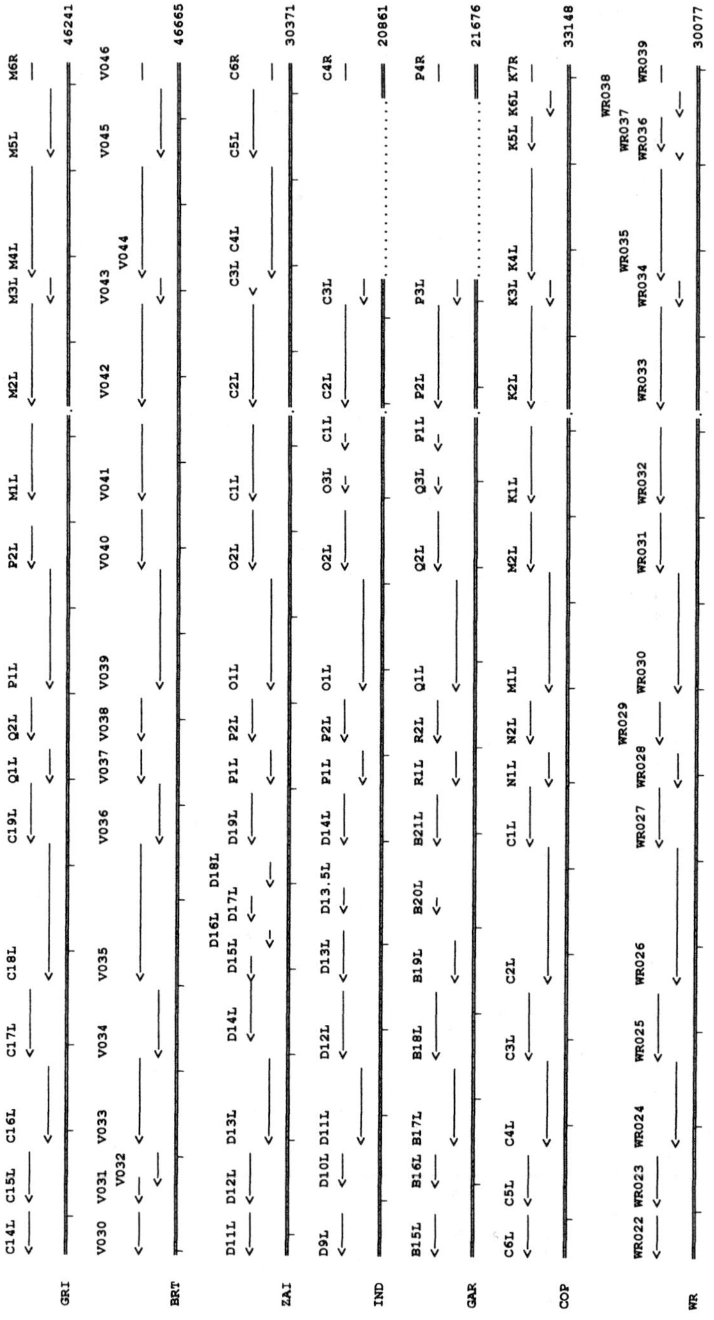
GRI
C14L C15L C16L C17L C18L C19L Q1L Q2L P1L P2L M1L M2L M3L M4L M5L M6R
46241
BRT
V030 V031 V032 V033 V034 V035 V036 V037 V038 V039 V040 V041 V042 V043 V044 V045 V046
46665
ZAI
D11L D12L D13L D14L D15L D16L D17L D18L D19L P1L P2L O1L O2L C1L C2L C3L C4L C5L C6R
30371
IND
D9L D10L D11L D12L D13L D13.5L D14L P1L P2L O1L O2L O3L C1L C2L C3L C4R
20861
GAR
B15L B16L B17L B18L B19L B20L B21L R1L R2L Q1L Q2L Q3L P1L P2L P3L P4R
21676
COP
C6L C5L C4L C3L C2L C1L N1L N2L M1L M2L K1L K2L K3L K4L K5L K6L K7R
33148
WR
WR022 WR023 WR024 WR025 WR026 WR027 WR028 WR029 WR030 WR031 WR032 WR033 WR034 WR035 WR036 WR037 WR038 WR039
30077

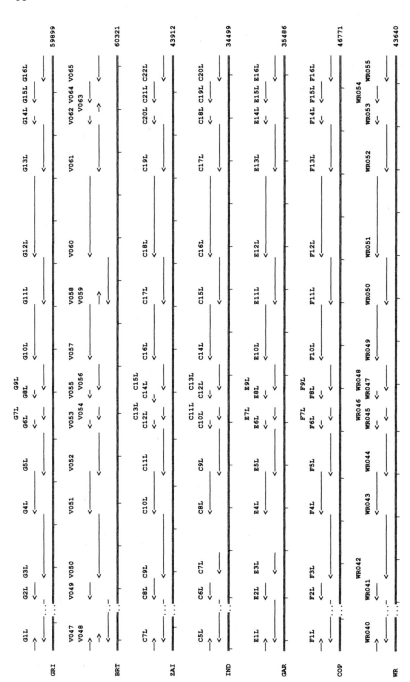

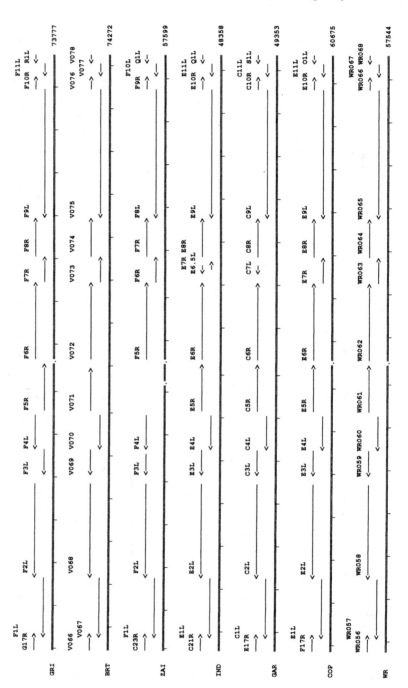

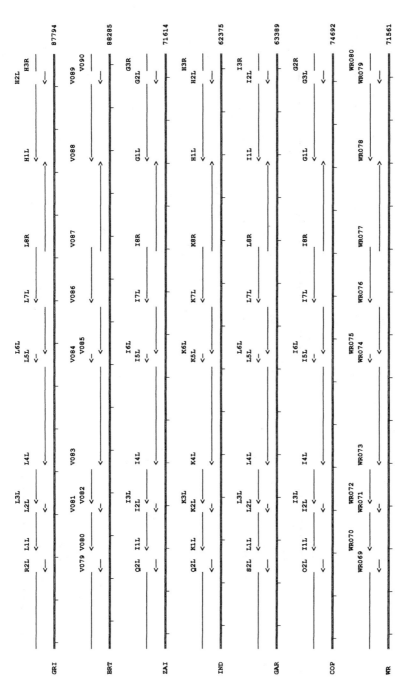

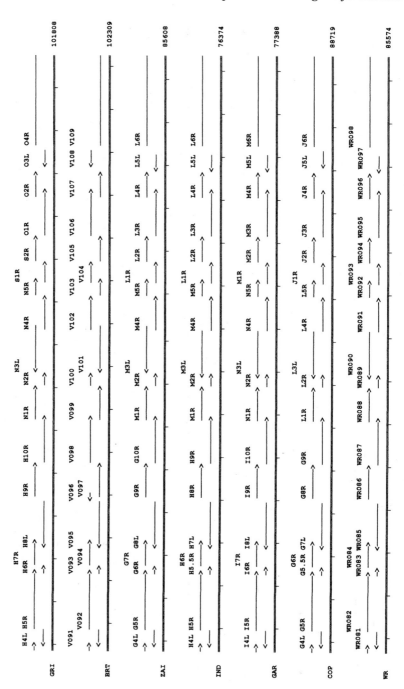

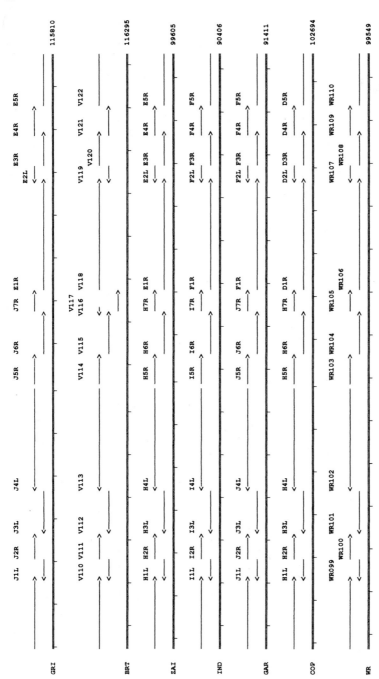

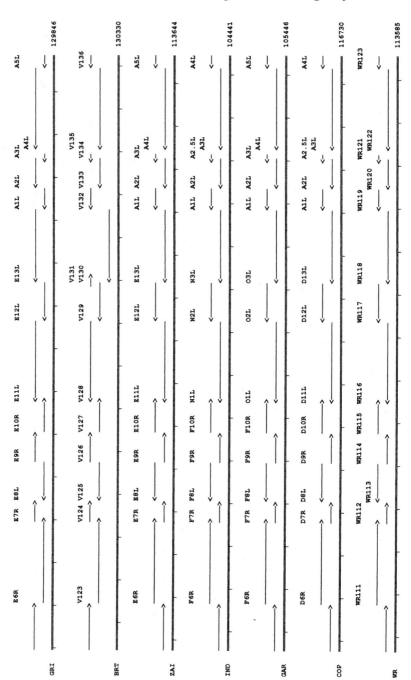

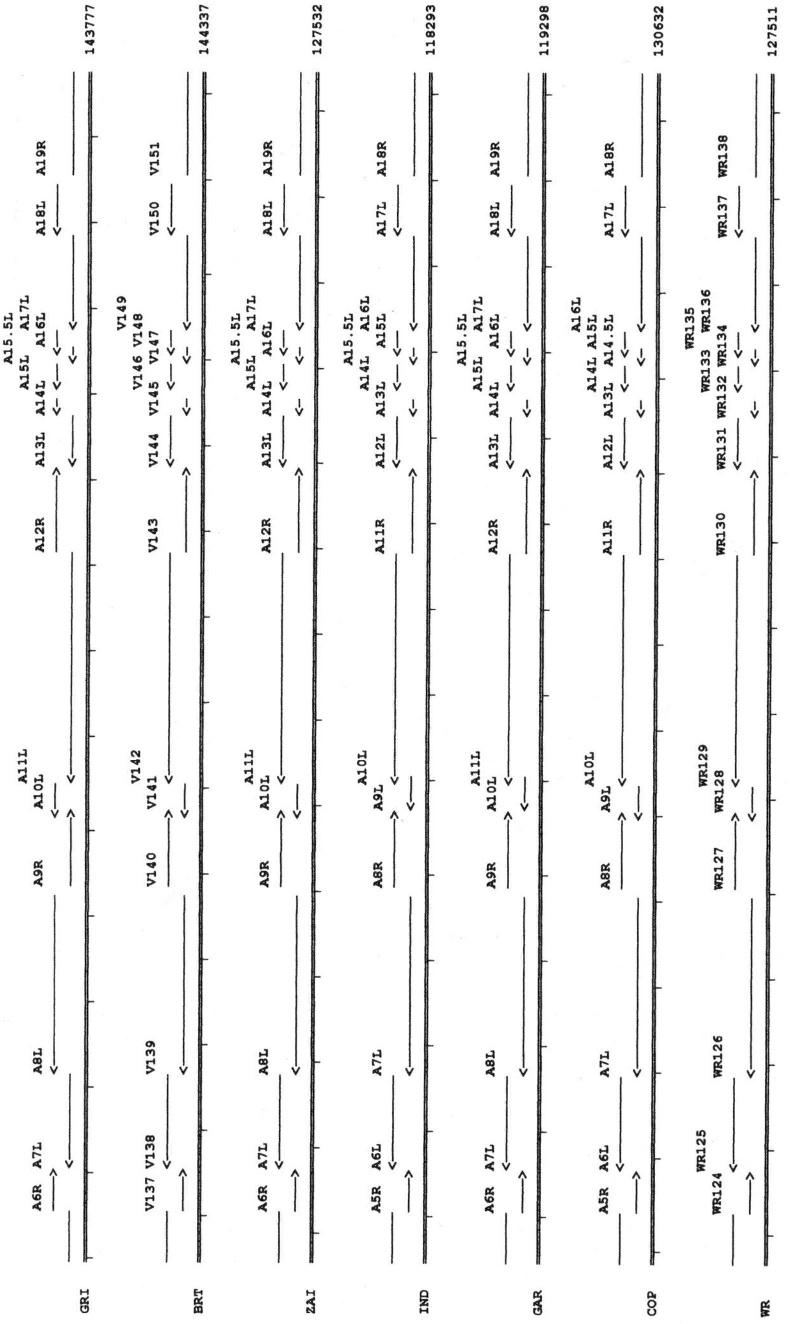

GRI
A6R A7L A8L A9R A10L A11L A12R A13L A14L A15L A15.5L A16L A17L A18L A19R
143777
BRT
V137 V138 V139 V140 V141 V142 V143 V144 V145 V146 V147 V148 V149 V150 V151
144337
ZAI
A6R A7L A8L A9R A10L A11L A12R A13L A14L A15L A15.5L A16L A17L A18L A19R
127532
IND
A5R A6L A7L A8R A9L A10L A11R A12L A13L A14L A15L A15.5L A16L A17L A18R
118293
GAR
A6R A7L A8L A9R A10L A11L A12R A13L A14L A15L A15.5L A16L A17L A18L A19R
119298
COP
A5R A6L A7L A8R A9L A10L A11R A12L A13L A14L A14.5L A15L A16L A17L A18R
130632
WR
WR124 WR125 WR126 WR127 WR128 WR129 WR130 WR131 WR132 WR133 WR134 WR135 WR136 WR137 WR138
127511

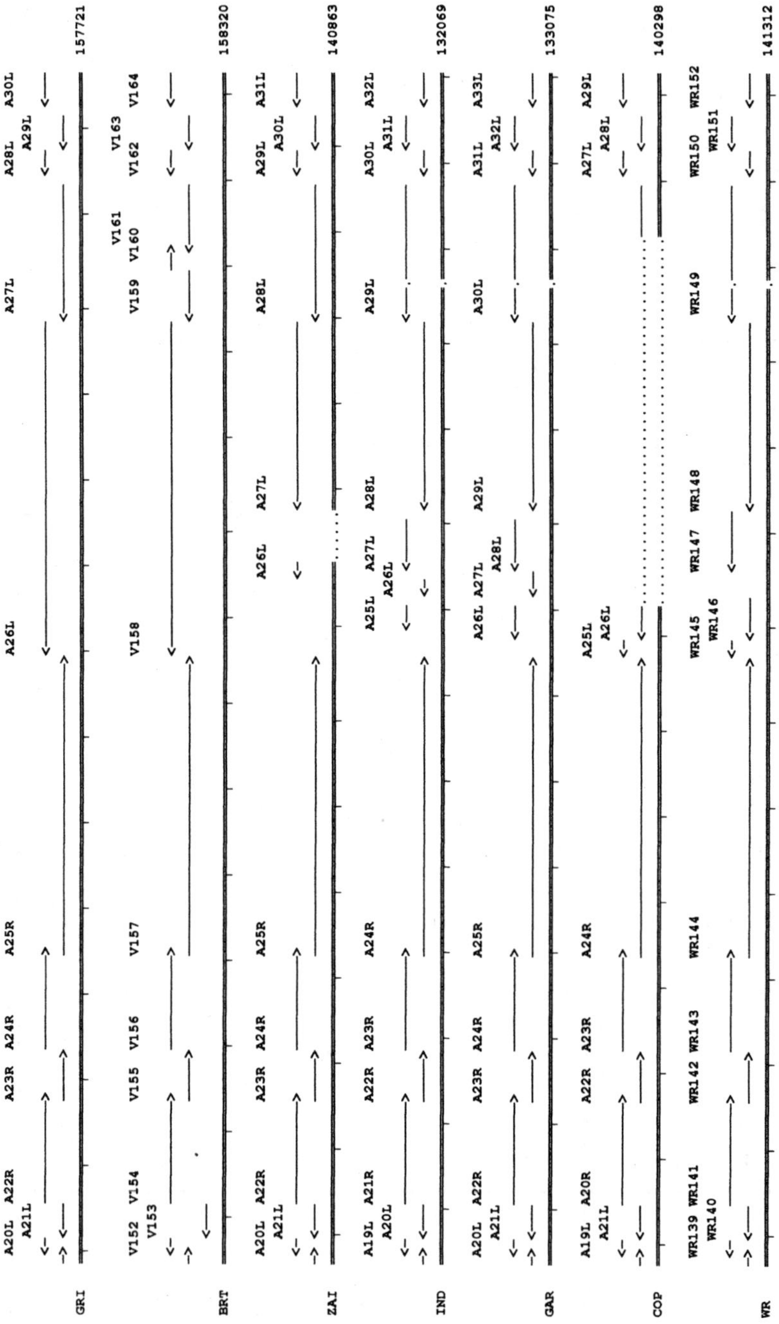
GRI
A20L A21L A22R A23R A24R A25R A26L A27L A28L A29L A30L
157721
BRT
V152 V153 V154 V155 V156 V157 V158 V159 V160 V161 V162 V163 V164
158320
ZAI
A20L A21L A22R A23R A24R A25R A26L A27L A28L A29L A30L A31L
140863
IND
A19L A20L A21R A22R A23R A24R A25L A26L A27L A28L A29L A30L A31L A32L
132069
GAR
A20L A21L A22R A23R A24R A25R A26L A27L A28L A29L A30L A31L A32L A33L
133075
COP
A19L A20R A21L A22R A23R A24R A25L A26L A27L A28L A29L
140298
WR
WR139 WR140 WR141 WR142 WR143 WR144 WR145 WR146 WR147 WR148 WR149 WR150 WR151 WR152
141312

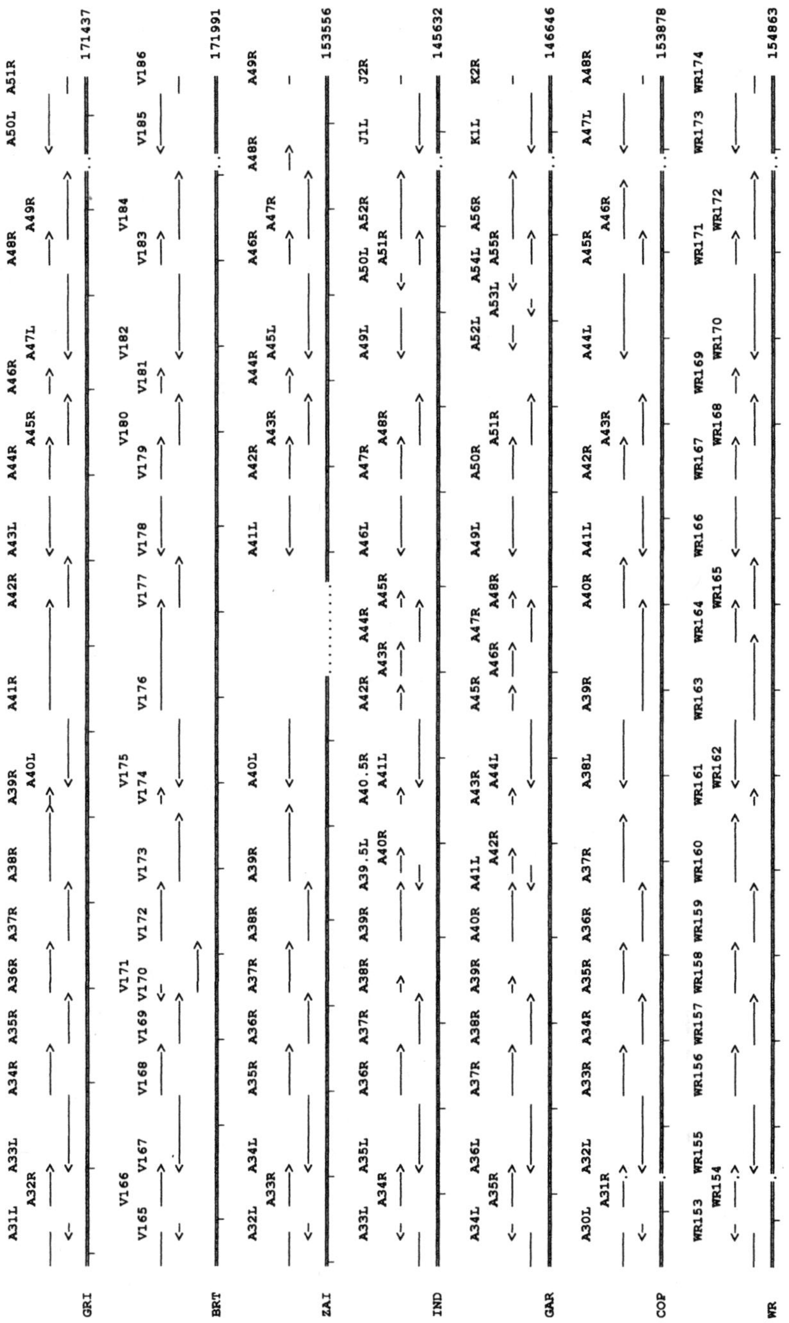
GRI
A31L A32R A33L A34R A35R A36R A37R A38R A39R A40L A41R A42R A43L A44R A45R A46R A47L A48R A49R A50L A51R
171437
BRT
V165 V166 V167 V168 V169 V170 V171 V172 V173 V174 V175 V176 V177 V178 V179 V180 V181 V182 V183 V184 V185 V186
171991
ZAI
A32L A33R A34L A35R A36R A37R A38R A39R A40L A41L A42R A43R A44R A45L A46R A47R A48R A49R
153556
IND
A33L A34R A35L A36R A37R A38R A39R A39.5L A40R A40.5R A41L A42R A43R A44R A45R A46L A47R A48R A49L A50L A51R A52R J1L J2R
145632
GAR
A34L A35R A36L A37R A38R A39R A40R A41L A42R A43R A44L A45R A46R A47R A48R A49L A50R A51R A52L A53L A54L A55R A56R K1L K2R
146646
COP
A30L A31R A32L A33R A34R A35R A36R A37R A38L A39R A40R A41L A42R A43R A44L A45R A46R A47L A48R
153878
WR
WR153 WR154 WR155 WR156 WR157 WR158 WR159 WR160 WR161 WR162 WR163 WR164 WR165 WR166 WR167 WR168 WR169 WR170 WR171 WR172 WR173 WR174
154863

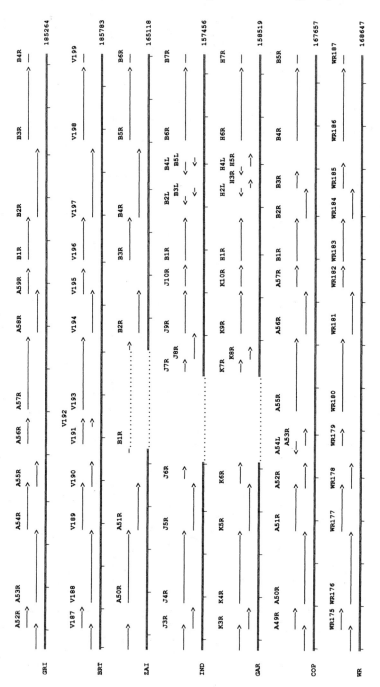

197004

197571

174044

167142

168860

177719

178675

GRI

BRT

ZAI

IND

GAR

COP

WR

B5R B6R B7R B8R B9R B10R B11R B12R B13R B14R B15L B16R

V200 V201 V202 V203 V204 V205 V206 V207 V208 V209 V210 V211

B7R B8R B9R B10R B11R B12R B13R B14R B15L

B8R B9R B10R B11R B12R B13R B14R B15R B17R B18L B19R
B16L

H8R H9R H10R H11R H12R D1R D2R D3R D4R D6R D8R
D5L D7L

B6R B7R B8R B9R B10R B11R B12R B13R B14R B15R B16R B17L B18R

WR188 WR189 WR190 WR191 WR192 WR193 WR194 WR195 WR196 WR197 WR198 WR199

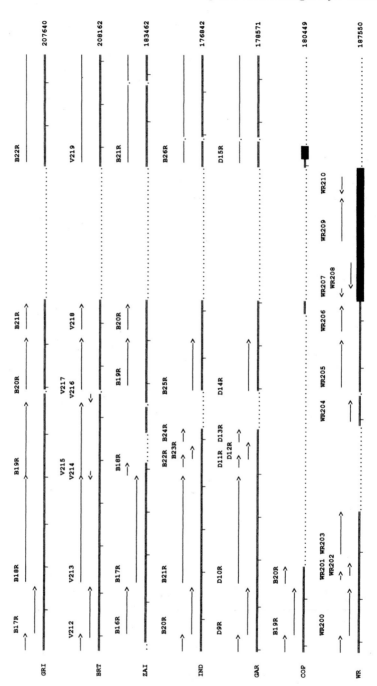

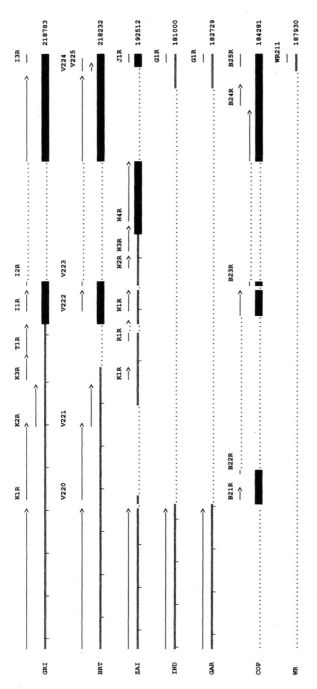

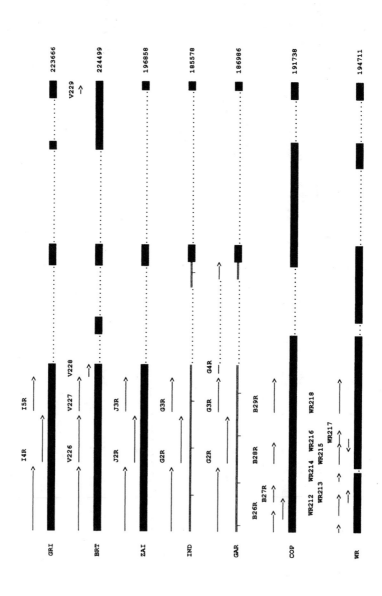

Appendix 2: The potential open reading frames (ORF) of cowpox virus strain GRI-90 and other orthopoxviruses pathogenic for humans

ORF	ORF size (aa)[a]	Type[b]	Function/ feature(s)[c]	Orthopoxvirus isologs[d]		ORF size (aa)[a]	Identity (%)/ overlapping region (aa)[e]	References
1	2	3	4	5		6	7	8
D1L*	255	E–L	Secreted CC-chemokine binding protein	CPXV-BRT	V003	246	80 6 / 258	Graham
				MPXV-ZAI	J1L	246	86.7 / 255	et al., 1997,
				VACV-COP	C23L	247	97 1 / 242	Smith C A
				VACV-WR	WR001	244	97 5 / 242	et al., 1997
				VARV-IND	–	–	–	
				VARV-GAR	–	–	–	
D2L*	351	E	Secreted TNF- and chemokine- binding protein, CrmB	CPXV-BRT	V005	355	90 5 / 357	Upton et al.,
				MPXV-ZAI	J2L	348	90 3 / 351	1991,
				VACV-COP	C22L	122	89 2 / 93	Shchelkunov
				⌐VACV-WR	WR002	61	88 5 / 52	et al., 1993a,
				⌐VACV-WR	WR004	122	89 2 / 93	Hu et al.,
				VARV-IND	–	–	–	1994, Ruiz-
				VARV-GAR	–	–	–	Arguello et al., 2004
D3L*	586		Ankyrin-like	CPXV-BRT	V006	619	89 8 / 588	Shchelkunov
				MPXV-ZAI	J3L	587	96 6 / 586	et al. 1993b;
				⌐VACV-COP	C21L	113	99 0 / 97	Safronov et al.,
				VACV-COP	C20L	103	70 4 / 115	1996
				⌐VACV-COP	C19L	259	84.5 / 233	
				⌐VACV-WR	WR005	49	97.9 / 48	
				VACV-WR	WR006	64	100 / 43	
				VACV-WR	WR007	109	81 8 / 99	
				⌐VACV-WR	WR008	112	89.4 / 104	
				VARV-IND	–	–	–	
				VARV-GAR	–	–	–	
D4L*	672		Ankyrin-like	CPXV-BRT	V008	672	95.4 / 673	Shchelkunov
				MPXV-ZAI	–	–	–	et al., 1993b;
				⌐VACV-COP	C18L	150	94.3 / 141	Safronov
				⌐VACV-COP	C17L	386	91.4 / 374	et al., 1996
				VACV-WR	–	–	–	
				VARV-IND	–	–	–	
				VARV-GAR	–	–	–	
D5L*	153		VACV-COP B15R-like 46.2% in 104 aa	CPXV-BRT	V009	153	92 2 / 153	Goebel et al.,
				MPXV-ZAI	–	–	–	1990
				VACV-COP	C16L	181	96 1 /153	
				VACV-WR	–	–	–	
				VARV-IND	D1L	153	94 8 / 153	
				VARV-GAR	B1L	153	94 8 / 153	
D6L	219			CPXV-BRT	V010	215	92.7 / 219	
				MPXV-ZAI	–	–	–	
				VACV-COP	–	–	–	
				VACV-WR	–	–	–	
				VARV-IND	–	–	–	
				VARV-GAR	–	–	–	
D7L	273		BTB domain of kelch-like protein	CPXV-BRT	–	–	–	Shchelkunov
				MPXV-ZAI	–	–	–	et al., 1998
				VACV-COP	–	–	–	
				VARV-IND	–	–	–	
				VARV-GAR	–	–	–	
D8L	661		Ankyrin-like	CPXV-BRT	V011	658	94 4 / 658	Shchelkunov
				MPXV-ZAI	–	–	–	et al., 1993b;
				VACV-COP	–	–	–	Safronov
				ACV-WR	–	–	–	et al., 1996
				VARV-IND	–	–	–	
				VARV-GAR	–	–	–	
D9L	75		C-type lectin-like (C-end domain)	CPXV-BRT	–	–	–	Shchelkunov
				MPXV-ZAI	–	–	–	et al., 1998
				VACV-COP	–	–	–	
				VACV-WR	–	–	–	
				VARV-IND	–	–	–	
				VARV-GAR	–	–	–	
D10L	96		C-type lectin-like (N-end domain)	CPXV-BRT	V012	69	73.9 / 46	Shchelkunov
				MPXV-ZAI	–	–	–	et al., 1998
				VACV-COP	–	–	–	
				VACV-WR	–	–	–	
				VARV-IND	–	–	–	
				VARV-GAR	–	–	–	
D11L	521		Kelch-like	CPXV-BRT	V013	523	95 2 / 525	Shchelkunov
				MPXV-ZAI	–	–	–	et al., 1998
				VACV-COP	–	–	–	

continued

349

1	2	3	4	5 (virus)	(ORF)	6	7	8
				VACV-WR		–	–	–
				VARV-IND		–	–	–
				VARV-GAR		–	–	–
D12L	202		C-end domain of poxviral TNF-binding protein CrmB/CrmD, chemokine-binding	CPXV-BRT	V014	202	95.1/202	Shchelkunov et al., 1998, Ruiz-Arguello et al., 2004
				MPXV-ZAI		–	–	–
				VACV-COP		–	–	–
				VACV-WR		–	–	–
				VARV-IND		–	–	–
				VARV-GAR		–	–	–
D13L	111		Secreted human CD30 homologue, vCD30	CPXV-BRT	V015	110	90.1/111	Shchelkunov et al., 1998, Panus et al., 2002
				MPXV-ZAI		–	–	–
				VACV-COP		–	–	–
				VACV-WR		–	–	–
				VARV-IND		–	–	–
				VARV-GAR		–	–	–
D14L	764		Ankyrin-like	CPXV-BRT	V016	764	93.6/763	Shchelkunov et al., 1993b, Safronov et al., 1996
				MPXV-ZAI		–	–	–
				VACV-COP		–	–	–
				VACV-WR		–	–	–
				VARV-IND		–	–	–
				VARV-GAR		–	–	–
C1L	437		Ankyrin-like	CPXV-BRT	V017	435	92.7/437	Shchelkunov et al., 1993b, Safronov et al., 1996
				MPXV-ZAI	D1L	437	97.5/437	
				VACV-COP		–	–	–
				VACV-WR		–	–	–
				VARV-IND		–	–	–
				VARV-GAR		–	–	–
C2L	178			CPXV-BRT	V018	171	58.7/167	
				MPXV-ZAI	N3R	176	98.3/175	
				VACV-COP		–	–	–
				VACV-WR		–	–	–
				VARV-IND		–	–	–
				VARV-GAR		–	–	–
C3L	833		Ankyrin-like	CPXV-BRT	V019	796	87.9/826	Shchelkunov et al., 1993b; Safronov et al., 1996
				MPXV-ZAI		–	–	–
				VACV-COP		–	–	–
				VACV-WR		–	–	–
				VARV-IND		–	–	–
				VARV-GAR		–	–	–
C4L	170		VACV-COP C7L-like	CPXV-BRT	V020	170	97.6/170	Shchelkunov et al., 1998
				MPXV-ZAI	D2L	64	95.4/65	
				VACV-COP		–	–	–
				ACV-WR		–	–	–
				VARV-IND	D1 5L	128	91.1/123	
				VARV-GAR	B2L	113	92.0/113	
C5R	138	E	Secreted EGF-like growth factor	CPXV-BRT	V021	139	92.8/139	Buller et al., 1988a
				MPXV-ZAI	D3R	142	91.3/138	
				VACV-COP	C11R	142	90.1/142	
				VACV-WR	WR009	140	91.4/140	
				VARV-IND	D2R	140	85.7/140	
				VARV-GAR	B3R	140	85.7/140	
C6L	331	E	Potential immunosuppressive activity	CPXV-BRT	V022	331	95.8/331	Venkatesan et al., 1982; Kluczyk et al., 2002
				MPXV-ZAI	D4L	83	96.4/83	
				VACV-COP	C10L	331	98.5/331	
				VACV-WR	WR010	331	98.8/331	
				VARV-IND	D3L	330	97.3/331	
				VARV-GAR	B4L	330	97.3/331	
C7R	242	L	Zinc-binding, virulence factor, inhibits UV-induced apoptosis	CPXV-BRT	V023	242	97.9/242	Upton et al., 1994; Senkevich et al., 1994, Brick et al., 2000
				MPXV-ZAI	D5R	242	97.5/242	
				VACV-COP		–	–	–
				VACV-WR	WR011	181	97.3/184	
				VACV-WR	WR012	62	94.7/57	
				VARV-IND	D4R	242	97.9/242	
				VARV-GAR	B5R	242	97.9/242	
C8L	124	E	Secreted IL-18 binding protein	CPXV-BRT	V024	126	79.7/123	Born et al., 2000, Smith S.A. et al., 2000; Calderara et al., 2001
				MPXV-ZAI	D6L	126	79.7/123	
				VACV-COP		–	–	–
				VACV-WR	WR013	126	95.1/123	
				VARV-IND	D5L	126	78.9/123	
				VARV-GAR	B6L	126	78.9/123	
C9L	668	E	Host range, ankyrin-like	CPXV-BRT	V025	668	93.7/668	Spehner et al., 1988, Shchelkunov et al., 1991; 1993b
				MPXV-ZAI	D7L	660	92.4/669	
				VACV-COP		–	–	–
				VACV-WR	WR014	237	97.4/235	
				VACV-WR	WR015	137	92.6/136	
				VACV-WR	WR016	77	91.9/62	
				VACV-WR	WR017	71	98.6/69	

continued

1	2	3	4	5	6	7	8
				VARV-IND	D6L	452	88 5 / 462	
				⌐VARV-GAR	B7L	95	82.8 / 87	
				└VARV-GAR	B8L	355	89 0 / 365	
C10L	62			CPXV-BRT	V026	63	92 1 / 63	
				MPXV-ZAI	D8L	64	93 8 / 64	
				VACV-COP	–	–	–	
				VACV-WR	WR018	60	87 3 / 63	
				VARV-IND	–	–	–	
				VARV-GAR	–	–	–	
C11L	614	E	Ankyrin-like	CPXV-BRT	V027	632	67.2 / 634	Shchelkunov
				MPXV-ZAI	D9L	630	67 8 / 634	*et al.*, 1993b
				VACV-COP	C9L	634	69 8 / 636	
				VACV-WR	WR019	634	69 8 / 636	
				⌐VARV-IND	D6.5L	91	86.8 / 91	
				└VARV-IND	D7L	153	96 2 / 130	
				⌐VARV-GAR	B9L	98	86 6 / 97	
				│VARV-GAR	B10L	172	92 7 / 165	
				└VARV-GAR	B12L	132	96 9 / 130	
C12L	182			CPXV-BRT	V028	185	89.6 / 182	
				MPXV-ZAI	–	–	–	
				VACV-COP	C8L	184	93.4 / 181	
				VACV-WR	WR020	177	87.8 / 181	
				VARV-IND	–	–	–	
				VARV-GAR	–	–	–	
C13L	150	E	Host range	CPXV-BRT	V029	150	97.3 / 150	Chen *et al.*,
				MPXV-ZAI	D10L	150	96.0 / 150	1992
				VACV-COP	C7L	150	100 / 150	
				VACV-WR	WR021	150	100 / 150	
				VARV-IND	D8L	150	99.3 / 150	
				VARV-GAR	B14L	150	99 3 / 150	
C14L	156	E		CPXV-BRT	V030	155	94.1 / 152	Cooper *et al.*,
				MPXV-ZAI	D11L	153	94.7 / 152	1981a
				VACV-COP	C6L	151	94.0 / 150	
				VACV-WR	WR022	151	95.0 / 150	
				VARV-IND	D9L	156	94.9 / 156	
				VARV-GAR	B15L	156	94 9 / 156	
C15L	205	E	BTB domain	⌐CPXV-BRT	V031	69	55.6 / 72	Shchelkunov
			of kelch-like	└CPXV-BRT	V032	125	93 5 / 124	*et al.*, 1998,
			protein	MPXV-ZAI	D12L	206	97 0 / 197	Cooper *et al.*,
				VACV-COP	C5L	204	98 0 / 202	1981b
				VACV-WR	WR023	204	97 5 / 202	
				VARV-IND	D10L	134	90 3 / 134	
				VARV-GAR	B16L	134	90 3 / 134	
C16L	315	E		CPXV-BRT	V033	316	94 6 / 316	Cooper *et al.*,
				MPXV-ZAI	D13L	315	94 3 / 316	1981b
				VACV-COP	C4L	316	99 4 / 316	
				VACV-WR	WR024	316	99.1 / 316	
				VARV-IND	D11L	316	95.9 / 316	
				VARV-GAR	B17L	316	96 2 / 316	
C17L	259	E	Secreted	CPXV-BRT	V034	263	90.1 / 263	Kotwal &
			complement	MPXV-ZAI	D14L	216	93.0 / 214	Moss, 1988a
			binding protein	VACV-COP	C3L	263	95.4 / 263	
				VACV-WR	WR025	263	95.4 / 263	
				VARV-IND	D12L	263	91.3 / 263	
				VARV-GAR	B18L	263	91.3 / 263	
C18L	512		Kelch-like	CPXV-BRT	V035	512	98.0 / 512	Shchelkunov
			protein,	⌐MPXV-ZAI	D15L	105	98.1 / 105	*et al.*, 1998,
			affects calcium-	│MPXV-ZAI	D16L	77	97.3 / 73	Pires de
			independent	│MPXV-ZAI	D17L	98	87 6 / 89	Miranda
			adhesion to the	└MPXV-ZAI	D18L	107	99 1 / 106	*et al.*, 2003
			extracellular	VACV-COP	C2L	512	99 4 / 512	
			matrix and	VACV-WR	WR026	512	99 2 / 512	
			inflammation	⌐VARV-IND	D13L	201	92 6 / 176	
			in a murine	└VARV-IND	D13 5L	79	89 5 / 86	
			intradermal	⌐VARV-GAR	B19L	154	96.1 / 154	
			model	└VARV-GAR	B20L	65	80 8 / 73	
C19L	231	E		CPXV-BRT	V036	231	98.3 / 231	Belle Isle
				MPXV-ZAI	D19L	214	95.8 /214	*et al.*, 1981
				VACV-COP	C1L	224	96.4 / 225	
				VACV-WR	WR027	229	96.5 / 231	
				VARV-IND	D14L	214	95 8 / 214	
				VARV-GAR	B21L	212	95.8 / 212	
Q1L	117	E–L	Secreted	CPXV-BRT	V037	117	94 0 / 117	Kotwal *et al.*,
			virulence	MPXV-ZAI	P1L	117	94.9 / 117	1989
			factor	VACV-COP	N1L	117	94 9 / 117	
				VACV-WR	WR028	117	94 9 / 117	
				VARV-IND	P1L	117	94.9 / 117	
				VARV-GAR	R1L	117	94.9 / 117	

continued

1	2	3	4	5		6	7	8
Q2L	175	E	Putative nucleus interaction control factor during productive infection	CPXV-BRT	V038	177	90 9 / 176	Tamın et al., 1988
				MPXV-ZAI	P2L	177	93.7 / 175	
				VACV-COP	N2L	175	95 4 / 175	
				VACV-WR	WR029	175	98 3 / 175	
				VARV-IND	P2L	177	95 4 / 175	
				VARV-GAR	R2L	177	94 9 / 175	
P1L	474	E–L	Ankyrın-like	CPXV-BRT	V039	473	97 0 / 474	Tamın et al., 1988, Shchelkunov et al., 1993b
				MPXV-ZAI	O1L	442	95.7 / 443	
				VACV-COP	M1L	472	97.9 / 474	
				VACV-WR	WR030	472	97 5 / 474	
				VARV-IND	O1L	446	96 9 / 447	
				VARV-GAR	Q1L	449	96.6 / 447	
P2L	163	E		CPXV-BRT	V040	220	96 3 / 163	Morgan & Roberts, 1984
				MPXV-ZAI	O2L	220	97 5 / 163	
				VACV-COP	M2L	220	100 / 163	
				VACV-WR	WR031	220	99.4 / 163	
				VARV-IND	O2L	220	96.3 / 163	
				VARV-GAR	Q2L	220	95.7 / 163	
M1L	284	E	Host range; ankyrın-like; ınhibits host NF-kappaB activation	CPXV-BRT	V041	284	98.6 / 284	Gillard et al., 1986, Shchelkunov et al., 1993b, Shısler & Jın, 2004
				MPXV-ZAI	C1L	284	93.7 / 284	
				VACV-COP	K1L	284	96 5 / 284	
				VACV-WR	WR032	284	97.2 / 284	
				⌐VARV-IND	O3L	70	90 5 / 63	
				└VARV-IND	C1L	66	95.5 / 66	
				⌐VARV-GAR	Q3L	70	90.5 / 63	
				└VARV-GAR	P1L	66	95.5 / 66	
M2L	373	E	Serıne protease ınhibitor-like, SPI-3, prevents cell fusıon	CPXV-BRT	V042	373	97.3 / 373	Law & Smith, 1992; Turner et al., 2000
				MPXV-ZAI	C2L	375	94 4 / 374	
				VACV-COP	K2L	369	97 1 / 373	
				VACV-WR	WR033	369	97 1 / 373	
				VARV-IND	C2L	373	95 7 / 373	
				VARV-GAR	P2L	373	95.7 / 373	
M3L	88	E	IFN resistance, homologue of eIF-2α ınhibits eIF-2α phosphorylatıon	CPXV-BRT	V043	88	93.2 / 88	Beattıe et al., 1991; Davies et al., 1992
				MPXV-ZAI	C3L	43	97.6 / 42	
				VACV-COP	K3L	88	98.9 / 88	
				VACV-WR	WR034	88	97.7 / 88	
				VARV-IND	C3L	88	83 0 / 88	
				VARV-GAR	P3L	88	83.0 / 88	
M4L	424		Phospholipase D-like	CPXV-BRT	V044	424	97 4 / 424	Cao et al., 1997, Sung et al., 1997
				MPXV-ZAI	C4L	424	97 6 / 424	
				VACV-COP	K4L	424	98 3 / 424	
				VACV-WR	WR035	424	98 1 / 424	
				VARV-IND	–	–	–	
				VARV-GAR	–	–	–	
M5L	276		Lysophospho-lipase-like	CPXV-BRT	V045	276	97 5 / 276	Antoıne et al., 1998
				MPXV-ZAI	C5L	276	97 5 / 276	
				⌐VACV-COP	K5L	136	91.7 / 108	
				└VACV-COP	K6L	81	94 9 / 79	
				⌐VACV-WR	WR036	45	100 / 44	
				│VACV-WR	WR037	134	77.9 / 131	
				└VACV-WR	WR038	81	94 9 / 79	
				VARV-IND	–	–	–	
				VARV-GAR	–	–	–	
M6R	161		VACV-COP B15R-like	CPXV-BRT	V046	149	98 0/ 149	Shchelkunov et al., 2002a
				MPXV-ZAI	C6R	149	97 3 / 149	
				VACV-COP	K7R	149	96 6 / 149	
				VACV-WR	WR039	149	96.6 / 149	
				VARV-IND	C4R	149	100 / 149	
				VARV-GAR	P4R	149	100 / 149	
G1L	238		Mıtochondrial-associated ınhibitor of apoptosıs	CPXV-BRT	V048	251	86 1 / 252	Wasılenko et al., 2003
				MPXV-ZAI	C7L	219	87 2 / 234	
				VACV-COP	F1L	226	89 7 / 234	
				VACV-WR	WR040	226	90 6 / 234	
				VARV-IND	C5L	251	80 6 / 253	
				VARV-GAR	E1L	312	84 8 / 224	
G2L	147	E	Deoxyurıdıne trıphosphatase	CPXV-BRT	V049	147	98.6 / 147	McGeoch, 1990
				MPXV-ZAI	C8L	151	99.3 / 147	
				VACV-COP	F2L	147	98.0 / 147	
				VACV-WR	WR041	147	98 0 / 147	
				VARV-IND	C6L	147	97 3 / 147	
				VARV-GAR	E2L	147	97 3 / 147	
G3L	485	E	Kelch-like	CPXV-BRT	V050	480	96.5 / 479	Xue & Cooley, 1993, Shchelkunov et al., 1998
				MPXV-ZAI	C9L	487	97.3 / 484	
				VACV-COP	F3L	480	97.9 / 479	
				VACV-WR	WR042	480	97 9 / 479	
				VARV-IND	C7L	179	93 3 / 179	
				VARV-GAR	E3L	179	93 9 / 179	

continued

1	2	3	4	5		6	7	8
G4L	319	E	Ribonucleotide reductase, small subunit, R2	CPXV-BRT	V051	333	98 1 / 319	Slabaugh *et al.,* 1988
				MPXV-ZAI	C10L	319	98 4 / 319	
				VACV-COP	F4L	319	98 7 / 319	
				VACV-WR	WR043	319	99 1 / 319	
				VARV-IND	C8L	333	97 2 / 319	
				VARV-GAR	E4L	319	98.1 / 319	
G5L	323			CPXV-BRT	V052	323	90 2 / 326	
				MPXV-ZAI	C11L	343	92 9 / 323	
				VACV-COP	F5L	321	98 5 / 323	
				VACV-WR	WR044	322	97 2 / 324	
				VARV-IND	C9L	348	88 6 / 324	
				VARV-GAR	E5L	348	88 3 / 324	
G6L	74			CPXV-BRT	V053	71	89 2 / 74	
				MPXV-ZAI	C12L	73	94 4 / 72	
				VACV-COP	F6L	74	100 / 74	
				VACV-WR	WR045	74	100 / 74	
				VARV-IND	C10L	72	87 3 / 71	
				VARV-GAR	E6L	72	88 7 / 71	
G7L	80	E		CPXV-BRT	V054	81	90 1 / 81	Panıcalı & Paoletti, 1982
				MPXV-ZAI	C13L	74	93 4 / 76	
				VACV-COP	F7L	92	85.9 / 92	
				VACV-WR	WR046	80	100 / 80	
				VARV-IND	C11L	79	82.5 / 80	
				VARV-GAR	E7L	78	80.0 / 80	
G8L	65			CPXV-BRT	V055	65	98.5 / 65	
				MPXV-ZAI	C14L	64	96.9 / 65	
				VACV-COP	F8L	65	100 / 65	
				VACV-WR	WR047	65	96.9 / 65	
				VARV-IND	C12L	65	96.9 / 65	
				VARV-GAR	E8L	65	96.9 / 65	
G9L	212			CPXV-BRT	V056	212	98 6 / 212	
				MPXV-ZAI	C15L	212	99.1 / 212	
				VACV-COP	F9L	212	100 / 212	
				VACV-WR	WR048	212	100 / 212	
				VARV-IND	C13L	212	97.6 / 212	
				VARV-GAR	E9L	212	97 6 / 212	
G10L	439	L	Serine/threonine protein kinase 2, VPK2, important for early steps in virion morphogenesis	CPXV-BRT	V057	439	99 1 / 439	Lın & Broyles, 1994; Traktman *et al.,* 1995, Betakova *et al.,* 1999; Szajner *et al.,* 2004
				MPXV-ZAI	C16L	439	99 1 / 439	
				VACV-COP	F10L	439	99 3 / 439	
				VACV-WR	WR049	439	99 8 / 439	
				VARV-IND	C14L	439	98.6 / 439	
				VARV-GAR	E10L	439	98 2 / 439	
G11L	354	E		CPXV-BRT	V059	354	98 6 / 354	Golını & Kates, 1984
				MPXV-ZAI	C17L	354	97 2 / 354	
				VACV-COP	F11L	354	100 / 354	
				VACV-WR	WR050	348	96 9 / 354	
				VARV-IND	C15L	354	95 8 / 354	
				VARV-GAR	E11L	354	96 0 / 354	
G12L	634	L	Associated with IEV and required for its microtubule-mediated egress to the cell surface	CPXV-BRT	V060	634	98 1 / 634	van Eıjl *et al.,* 2000
				MPXV-ZAI	C18L	635	98 6 / 635	
				VACV-COP	F12L	635	97.5 / 635	
				VACV-WR	WR051	635	97.6 / 635	
				VARV-IND	C16L	635	95.6 / 635	
				VARV-GAR	E12L	635	95.3 / 635	
G13L	372	L	Major envelope antigen of EEV, wrapping of IMV to form IEV, phospholipase D-like	CPXV-BRT	V061	372	98 7 / 372	Hırt *et al.,* 1986; Baek *et al.,* 1997; Sung *et al.,* 1997; Roper & Moss, 1999
				MPXV-ZAI	C19L	372	98 7 / 372	
				VACV-COP	F13L	372	99.5 / 372	
				VACV-WR	WR052	372	99.7 / 372	
				VARV-IND	C17L	372	97.9 / 372	
				VARV-GAR	E13L	372	98.1 / 372	
G14L	73	E–L		CPXV-BRT	V062	73	87.7 / 73	Golını & Kates, 1984
				MPXV-ZAI	C20L	73	98.6 / 73	
				VACV-COP	F14L	73	98 6 / 73	
				VACV-WR	WR053	73	100 / 73	
				VARV-IND	C18L	73	78.1 / 73	
				VARV-GAR	E14L	73	78.1 / 73	
G15L	158	E		CPXV-BRT	V064	158	96.8 / 158	Golını & Kates, 1984
				MPXV-ZAI	C21L	158	98.7 / 158	
				VACV-COP	F15L	158	99.4 / 158	
				VACV-WR	WR054	147	99.3 / 147	
				VARV-IND	C19L	161	99 3 / 153	
				VARV-GAR	E15L	161	99.3 / 153	
G16L	231	E		CPXV-BRT	V065	231	95.7 / 231	Golını & Kates, 1984
				MPXV-ZAI	C22L	231	97.0 / 231	
				VACV-COP	F16L	231	97.8 / 231	
				VACV-WR	WR055	231	98 3 / 231	
				VARV-IND	C20L	231	98.3 / 231	
				VARV-GAR	E16L	231	98 3 / 231	

continued

1	2	3	4	5		6	7	8
G17R	101	L	Virion core DNA-binding phosphoprotein	CPXV-BRT	V066	101	98 0 / 101	Kao & Bauer, 1987
				MPXV-ZAI	C23R	101	96.0 / 101	
				VACV-COP	F17R	101	99 0 / 101	
				VACV-WR	WR056	101	99 0 / 101	
				VARV-IND	C21R	101	98 0 / 101	
				VARV-GAR	E17R	101	98 0 / 101	
F1L	479	E	Poly-A polymerase, catalytic subunit	CPXV-BRT	V067	479	99 6 / 479	Gershon et al., 1991
				MPXV-ZAI	F1L	479	99 4 / 479	
				VACV-COP	E1L	479	99 6 / 479	
				VACV-WR	WR057	479	99.8 / 479	
				VARV-IND	E1L	479	98 3 / 479	
				VARV-GAR	C1L	479	98 7 / 479	
F2L	737			CPXV-BRT	V068	737	99 1 / 737	
				MPXV-ZAI	F2L	737	97 8 / 737	
				VACV-COP	E2L	737	98 8 / 737	
				VACV-WR	WR058	737	98 9 / 737	
				VARV-IND	E2L	737	98 2 / 737	
				VARV-GAR	C2L	737	98 1 / 737	
F3L	190	E	IFN resistance, dsRNA-binding, inhibits dsRNA-dependent protein kinase and 2-5A-synthetase	CPXV-BRT	V069	190	94 2 / 190	Chang et al., 1992, Rivas et al., 1998
				MPXV-ZAI	F3L	153	86 3 / 153	
				VACV-COP	E3L	190	97 4 /190	
				VACV-WR	WR059	190	96 3 / 190	
				VARV-IND	E3L	190	95 3 / 190	
				VARV-GAR	C3L	192	93 2 / 192	
F4L	259	E–L	RNA polymerase, 30 kDa subunit, intermediate stage transcription factor, VITF-1	CPXV-BRT	V070	261	97 3 / 261	Ahn et al., 1990a; Rosales et al., 1994b
				MPXV-ZAI	F4L	259	98 1 / 259	
				VACV-COP	E4L	259	99 6 / 259	
				VACV-WR	WR060	259	99 2 / 259	
				VARV-IND	E4L	259	97 7 /259	
				VARV-GAR	C4L	259	98.1 / 259	
F5R	331			CPXV-BRT	V071	319	90 9 / 319	
				MPXV-ZAI	–	–	–	
				VACV-COP	E5R	331	95.8 / 331	
				VACV-WR	WR061	341	94 6 / 331	
				VARV-IND	E5R	341	93.7 / 331	
				VARV-GAR	C5R	341	94.0 / 331	
F6R	567			CPXV-BRT	V072	567	99.1 / 567	
				MPXV-ZAI	F5R	567	98 9 / 567	
				VACV-COP	E6R	567	99.5 / 567	
				VACV-WR	WR062	567	99.3 / 567	
				VARV-IND	E6R	567	98.1 / 567	
				VARV-GAR	C6R	567	97 7 / 567	
F7R	166	L	Soluble myristylated protein	CPXV-BRT	V073	166	97.0 / 166	Martin et al., 1997
				MPXV-ZAI	F6R	166	94.6 / 166	
				VACV-COP	E7R	166	98 2 / 166	
				VACV-WR	WR063	166	98 2 / 166	
				VARV-IND	E7R	76	88 3 / 60	
				VARV-GAR	–	–	–	
F8R	273	E	Membrane associated protein packaged into the virion core	CPXV-BRT	V074	273	98 2 / 273	Doglio et al., 2002
				MPXV-ZAI	F7R	273	98 5 / 273	
				VACV-COP	E8R	273	99 3 / 273	
				VACV-WR	WR064	273	99.3 / 273	
				VARV-IND	E8R	273	98 5 / 273	
				VARV-GAR	C8R	273	98.2 / 273	
F9L	1006	E	DNA polymerase, catalytic subunit	CPXV-BRT	V075	1005	99 0 / 1006	Earl et al., 1986; McDonald et al., 1997
				MPXV-ZAI	F8L	1006	98.6 / 1006	
				VACV-COP	E9L	1006	99 1 / 1006	
				VACV-WR	WR065	1006	99 1 / 1006	
				VARV-IND	E9L	1005	98 3 / 1006	
				VARV-GAR	C9L	1005	98 2 / 1006	
F10R	95	L	Protein disulfide bond-forming enzyme	CPXV-BRT	V076	95	97.9 / 95	Senkevich et al., 2000
				MPXV-ZAI	F9R	95	95 8 / 95	
				VACV-COP	E10R	95	98 9 / 95	
				VACV-WR	WR066	95	96 8 / 95	
				VARV-IND	E10R	95	95 8 / 95	
				VARV-GAR	C10R	95	95 8 / 95	
F11L	129	L	Virion core protein	CPXV-BRT	V077	129	98.5 / 129	Wang & Shuman, 1996
				MPXV-ZAI	F10L	129	97.7 / 129	
				VACV-COP	E11L	129	100 / 129	
				VACV-WR	WR067	129	100 / 129	
				VARV-IND	E11L	129	96 9 / 129	
				VARV-GAR	C11L	129	96 9 / 129	
R1L	666	E		CPXV-BRT	V078	666	98 6 / 666	
				MPXV-ZAI	Q1L	665	98.0 / 666	
				VACV-COP	O1L	666	99 4 / 666	
				VACV-WR	WR068	666	99 2 / 666	
				VARV-IND	Q1LS1L	666	94 1 / 666	
				VARV-GAR		666	94 0 / 666	

continued

1	2	3	4	5			6	7	8
R2L	108	L	Virion-associated glutaredoxin	CPXV-BRT	V079		108	99 0 / 108	Ahn & Moss,
				MPXV-ZAI	Q2L		108	99.1 / 108	1992a;
				VACV-COP	O2L		108	99 0 / 108	Rajagopal
				VACV-WR		WR069	108	99.1 / 108	*et al.*, 1995
				VARV-IND	Q2L		108	97.2 / 108	
				VARV-GAR	S2L		108	97 2 / 108	
L1L	312	L	Virosomal telomere binding protein essential for virus multiplication	CPXV-BRT	V080 .		312	99.4 / 312	Ryazankına
				MPXV-ZAI	I1L		312	99.4 / 312	*et al.*, 1993,
				VACV-COP	I1L		312	99.4 / 312	Schmitt &
				VACV-WR		WR070	312	99 4 / 312	Stunnenberg,
				VARV-IND	K1L		312	96 8 / 312	1988, DeMasi
				VARV-GAR	L1L		312	98 1 / 312	*et al.*, 2001
L2L	73	L		CPXV-BRT	V081		73	98 6/ 73	Schmitt &
				MPXV-ZAI	I2L		73	98.6 / 73	Stunnenberg,
				VACV-COP	I2L		73	100 / 73	1988
				VACV-WR		WR071	73	100 / 73	
				VARV-IND	K2L		73	100 / 73	
				VARV-GAR	L2L		73	100 / 73	
L3L	269	E-I	ssDNA-binding P-protein, interacts with R2 subunit of ribonucleotide reductase	CPXV-BRT	V082		268	97.8 / 269	Davis &
				MPXV-ZAI	I3L		269	98.9 / 269	Mathews,
				VACV-COP	I3L		269	99 3 / 269	1993,
				VACV-WR		WR072	269	100 / 269	Rochester &
				VARV-IND	K3L		269	99.3 / 269	Traktman,
				VARV-GAR	L3L		269	99 3 / 269	1998
L4L	771	E	Ribonucleotide reductase, large subunit, R1	CPXV-BRT	V083		771	98.8 / 771	Tengelsen
				MPXV-ZAI	I4L		771	97 9 / 771	*et al.*, 1988
				VACV-COP	I4L		771	99.0 / 771	
				VACV-WR		WR073	771	98.8 / 771	
				VARV-IND	K4L		771	98.1 / 771	
				VARV-GAR	L4L		771	97 8 / 771	
L5L	79	L	IMV surface membrane protein	CPXV-BRT	V084		79	98.7 / 79	Takahashi
				MPXV-ZAI	I5L		79	94 9 / 79	*et al.*, 1994
				VACV-COP	I5L		79	100 / 79	
				VACV-WR		WR074	79	98.7 / 79	
				VARV-IND	K5L		79	96.2 / 79	
				VARV-GAR	L5L		79	94.9 / 79	
L6L	382	E–L	Telomere-binding protein, DNA encapsidation into the virus particle	CPXV-BRT	V085		382	99.2 / 382	Grubisha &
				MPXV-ZAI	I6L		382	98.4 / 382	Traktman,
				VACV-COP	I6L		382	99.7 / 382	2003
				VACV-WR		WR075	382	99 7 / 382	
				VARV-IND	K6L		382	99.2 / 382	
				VARV-GAR	L6L		382	99 0 / 382	
L7L	423	L	Virion core protein, cysteine protein, teinase responsible for cleavage of the major P4a, P4b, and P25K core proteins	CPXV-BRT	V086		423	98.3 / 423	Kane &
				MPXV-ZAI	I7L		423	99.1 / 423	Shuman, 1993,
				VACV-COP	I7L		423	99.5 / 423	Byrd *et al.*,
				VACV-WR		WR076	423	99.1 / 423	2003
				VARV-IND	K7L		423	99.3 / 423	
				VARV-GAR	L7L		423	99.1 / 423	
L8R	676	E–L	Nucleoside triphosphate phosphohy-drolase II, NPH-II, DNA and RNA helicase	CPXV-BRT	V087		676	97.0 / 676	Shuman, 1992;
				MPXV-ZAI	I8R		676	98 2 / 676	Bayliss &
				VACV-COP	I8R		676	99 4 / 676	Smith, 1996;
				VACV-WR		WR077	676	99.3 / 676	Gross &
				VARV-IND	K8R		676	98.4 / 676	Shuman, 1998
				VARV-GAR	L8R		682	97.5 / 682	
H1L	591	L	Virion core protein, putative proteinase	CPXV-BRT	V088		591	99.5 / 591	Whitehead &
				MPXV-ZAI	G1L		591	98 5 / 591	Hruby, 1994b;
				VACV-COP	G1L		591	99 7 / 591	Honeychurch
				VACV-WR		WR078	591	99.3 / 591	*et al.*, 2004
				VARV-IND	H1L		591	98.0 / 591	
				VARV-GAR	I1L		591	98.0 / 591	
H2L	111	E		CPXV-BRT	V089		111	97 3 / 111	Meis &
				MPXV-ZAI	G2L		111	98 2 / 111	Condit, 1991
				VACV-COP	G3L		111	100 / 111	
				VACV-WR		WR079	111	100 / 111	
				VARV-IND	H2L		111	97 3 / 111	
				VARV-GAR	I2L		111	97 3 / 111	
H3R	220	L	Putative transcription elongation factor of intermediate and late genes	CPXV-BRT	V090		220	98 6 / 220	Meis &
				MPXV-ZAI	G3R		220	98.6 / 220	Condit, 1991
				VACV-COP	G2R		220	99 1 / 220	
				VACV-WR		WR080	220	99.1 / 220	
				VARV-IND	H3R		220	98 2 / 220	
				VARV-GAR	I3R		220	98 2 / 220	
H4L	124	L	Virion-associated glutaredoxin, required for disulfide bonds	CPXV-BRT	V091		124	98 4 / 124	Gvakharia
				MPXV-ZAI	G4L		124	99 2 / 124	*et al.*, 1996,
				VACV-COP	G4L		124	98 4 / 124	White *et al.*,
				VACV-WR		WR081	124	98.4 / 124	2000

continued

1	2	3	4	5		6	7	8
			and assembly	VARV-IND	H4L	124	100 / 124	
				VARV-GAR	I4L	124	100 / 124	
H5R	434	E–L		CPXV-BRT	V092	434	98 2 / 434	Meis &
				MPXV-ZAI	G5R	434	98 4 / 434	Condit, 1991
				VACV-COP	G5R	434	98 8 / 434	
				VACV-WR	WR082	434	98 8 / 434	
				VARV-IND	H5R	434	96.3 / 434	
				VARV-GAR	I5R	434	96 8 / 434	
H6R	63	E–L	RNA polymerase, 7 kDa subunit 7 K	CPXV-BRT	V093	63	100 / 63	Amegadzie
				MPXV-ZAI	G6R	63	100 / 63	*et al.*, 1992a
				VACV-COP	G5 5R	63	100/ 63	
				VACV-WR	WR083	63	100 / 63	
				VARV-IND	H5.5R	63	96.8 / 63	
				VARV-GAR	I6R	63	96 8 / 63	
H7R	165			CPXV-BRT	V094	167	94.5 / 165	
				MPXV-ZAI	G7R	165	93 9 / 165	
				VACV-COP	G6R	165	97 6 / 165	
				VACV-WR	WR084	165	97 0 / 165	
				VARV-IND	H6R	165	97 0 / 165	
				VARV-GAR	I7R	165	97 6 / 165	
H8L	371	L	Virion core protein, forms complex with A30L and F10L (VACV-COP), important for early steps in virion morphogenesis	CPXV-BRT	V095	371	99.5 / 371	Takahashi
				MPXV-ZAI	G8L	371	98.7 / 371	*et al.*, 1994
				VACV-COP	G7L	371	99.5 / 371	Szajner *et al.*,
				VACV-WR	WR085	371	99.7 / 371	2004
				VARV-IND	H7L	371	99 2 / 371	
				VARV-GAR	I8L	371	99.2 / 371	
H9R	260	I	Late gene transcription factor, VLTF-1	CPXV-BRT	V097	260	99.6 / 260	Keck *et al.*,
				MPXV-ZAI	G9R	260	100 / 260	1990
				VACV-COP	G8R	260	100 / 260	
				VACV-WR	WR086	260	100 / 260	
				VARV-IND	H8R	260	99.6 / 260	
				VARV-GAR	I9R	260	99 6 / 260	
H10R	340	L	Myristylated protein	CPXV-BRT	V098	340	97.9 / 340	Martin *et al.*,
				MPXV-ZAI	G10R	340	98.8 / 340	1997
				VACV-COP	G9R	340	99.7 / 340	
				VACV-WR	WR087	340	99.7 / 340	
				VARV-IND	H9R	340	99 1 / 340	
				VARV-GAR	I10R	340	99.1 / 340	
N1R	250	L	Myristylated IMV surface membrane protein	CPXV-BRT	V099	250	98.4 / 250	Ravanello &
				MPXV-ZAI	M1R	250	98 4 / 250	Hruby, 1994
				VACV-COP	L1R	250	99 6 / 250	
				VACV-WR	WR088	250	100 / 250	
				VARV-IND	M1R	250	99 2 / 250	
				VARV-GAR	N1R	250	98.8 / 250	
N2R	92			CPXV-BRT	V100	88	91.6 / 83	
				MPXV-ZAI	M2R	92	97 8 / 92	
				VACV-COP	L2R	87	98 8 / 82	
				VACV-WR	WR089	87	97 6 / 82	
				VARV-IND	M2R	87	97 6 / 82	
				VARV-GAR	N2R	87	97 6 / 82	
N3L	350			CPXV-BRT	V101	350	98 9 / 350	
				MPXV-ZAI	M3L	344	97.4 / 350	
				VACV-COP	L3L	350	98.3 / 350	
				VACV-WR	WR090	350	97.7 / 350	
				VARV-IND	M3L	349	95 7 / 350	
				VARV-GAR	N3L	349	95 7 / 350	
N4R	251	L	Virion core protein, ssDNA binding, stimulation of I8R (VACV-COP) helicase activity	CPXV-BRT	V102	251	100 / 251	Yang *et al.*,
				MPXV-ZAI	M4R	251	99 2 / 251	1988,
				VACV-COP	L4R	251	99.6 / 251	Bayliss *et al.*,
				VACV-WR	WR091	251	99 6 / 251	1996
				VARV-IND	M4R	251	99.2 / 251	
				VARV-GAR	N4R	251	99.2 / 251	
N5R	128			CPXV-BRT	V103	128	99 2 / 128	
				MPXV-ZAI	M5R	128	99.2 / 128	
				VACV-COP	L5R	128	100 / 128	
				VACV-WR	WR092	128	100 / 128	
				VARV-IND	M5R	128	98 4 / 128	
				VARV-GAR	N5R	128	99.2 / 128	
S1R	153	L	IMV membrane protein, important for DNA packaging into IMV, interacts with A45R	CPXV-BRT	V104	152	100 / 151	Chiu & Chang,
				MPXV-ZAI	L1R	152	99 3 / 150	2002
				VACV-COP	J1R	153	98 7 / 153	
				VACV-WR	WR093	153	98 0 / 153	
				VARV-IND	L1R	159	98 7 / 153	
				VARV-GAR	M1R	159	99 3 / 153	

continued

1	2	3	4	5		6	7	8
S2R	177	E	Thymidine kinase	CPXV-BRT	V105	177	98 3 / 177	Weir & Moss, 1983
				MPXV-ZAI	L2R	177	98 3 / 177	
				VACV-COP	J2R	177	98 3 / 177	
				VACV-WR	WR094	177	98 3 / 177	
				VARV-IND	L2R	177	97.2 / 177	
				VARV-GAR	M2R	177	97 7 / 177	
O1R	333	E	Poly-A pol stimulatory subunit, cap-specific mRNA (nucleoside-$O^{2'}$-)-methyltransferase	CPXV-BRT	V106	333	99 4 / 333	Gershon et al., 1991, Schnierle et al., 1992
				MPXV-ZAI	L3R	333	99 1 / 333	
				VACV-COP	J3R	333	98 5 / 333	
				VACV-WR	WR095	333	98 8 / 333	
				VARV-IND	L3R	333	98.8 / 333	
				VARV-GAR	M3R	333	98 8 / 333	
O2R	185	E	RNA pol 22 kDa subunit	CPXV-BRT	V107	185	99 5 / 185	Broyles & Moss, 1986
				MPXV-ZAI	L4R	185	99 5 / 185	
				VACV-COP	J4R	185	99 5 / 185	
				VACV-WR	WR096	185	99 5 / 185	
				VARV-IND	L4R	185	98 4 / 185	
				VARV-GAR	M4R	185	98 9 / 185	
O3L	133	L	Essential for virus multiplication	CPXV-BRT	V108	133	99 2 / 133	Zajac et al., 1995
				MPXV-ZAI	L5L	133	100 / 133	
				VACV-COP	J5L	133	98.5 / 133	
				VACV-WR	WR097	133	98.5 / 133	
				VARV-IND	L5L	133	100 / 133	
				VARV-GAR	M5L	133	100 / 133	
O4R	1286	E	RNA pol 147 kDa subunit	CPXV-BRT	V109	1286	99 5 / 1286	Broyles & Moss, 1986
				MPXV-ZAI	L6R	1286	99.1 / 1286	
				VACV-COP	J6R	1286	99.5 / 1286	
				VACV-WR	WR098	1286	99.7 / 1286	
				VARV-IND	L6R	1286	99.5 / 1286	
				VARV-GAR	M6R	1286	99 5 / 1286	
J1L	171	L	Tyrosine/serine protein phosphatase, blocks IFN-γ-signal transduction	CPXV-BRT	V110	171	99.4 / 171	Guan et al., 1991; Najarro et al., 2001
				MPXV-ZAI	H1L	171	99 4 / 171	
				VACV-COP	H1L	171	99.4 / 171	
				VACV-WR	WR099	171	99.4 / 171	
				VARV-IND	I1L	171	98 2 / 171	
				VARV-GAR	J1L	171	98 2 / 171	
J2R	189	L	Required for virion entry into cells	CPXV-BRT	V111	189	98.9 / 189	Senkevich & Moss, 2004
				MPXV-ZAI	H2R	189	100 / 189	
				VACV-COP	H2R	189	98 9 / 189	
				VACV-WR	WR100	189	99 5 / 189	
				VARV-IND	I2R	189	100 / 189	
				VARV-GAR	J2R	189	100 / 189	
J3L	324	L	Immunodominant IMV heparan binding surface membrane protein	CPXV-BRT	V112	325	97 5 / 325	Chertov et al., 1991; Lin et al., 2000, da Fonseca et al., 2000
				MPXV-ZAI	H3L	324	95.4 / 324	
				VACV-COP	H3L	324	98 5 / 324	
				VACV-WR	WR101	324	98 5 / 324	
				VARV-IND	I3L	325	96.9 / 325	
				VARV-GAR	J3L	325	96 6 / 325	
J4L	795	L	RNA pol-associated protein, RAP 94, provides specificity for early promoters	CPXV-BRT	V113	795	99.4 / 795	Ahn & Moss, 1992b, Kane & Shuman, 1992; Zhang et al., 1994
				MPXV-ZAI	H4L	795	98 5 / 795	
				VACV-COP	H4L	795	99 6 / 795	
				VACV-WR	WR102	795	99 4 / 795	
				VARV-IND	I4L	795	98.5 / 795	
				VARV-GAR	J4L	795	98.6 / 795	
J5R	212	E–L	Virosome-associated, late gene transcription factor, VLTF-4, Ca^{2+}-binding motif	CPXV-BRT	V114	206	93 4 / 212	Kovacs & Moss, 1996; Shchelkunov et al., 1993e
				MPXV-ZAI	H5R	213	98 1 / 214	
				VACV-COP	H5R	203	94 8 / 212	
				VACV-WR	WR103	203	94.3 / 212	
				VARV-IND	I5R	221	93.7 / 221	
				VARV-GAR	J5R	218	95.4 / 218	
J6R	314	E	DNA topoisomerase	CPXV-BRT	V115	314	98 7 / 314	Shuman & Moss, 1987
				MPXV-ZAI	H6R	314	99 0 / 314	
				VACV-COP	H6R	314	99 4 / 314	
				VACV-WR	WR104	314	99 4 / 314	
				VARV-IND	I6R	314	98 4 / 314	
				VARV-GAR	J6R	314	98 4 / 314	
J7R	146	L		CPXV-BRT	V117	146	96.6 / 146	Rosel et al., 1986
				MPXV-ZAI	H7R	146	97.3 / 146	
				VACV-COP	H7R	146	97.3 / 146	
				VACV-WR	WR105	146	99 3 / 146	
				VARV-IND	I7R	146	97 3 / 146	
				VARV-GAR	J7R	146	97.3 / 146	
E1R	844	E	mRNA capping enzyme large subunit, RNA 5' triphosphatase	CPXV-BRT	V118	844	99.3 / 844	Morgan et al., 1984, Shuman & Morham, 1990
				MPXV-ZAI	E1R	845	98 9 / 845	
				VACV-COP	D1R	844	99 4 / 844	
				VACV-WR	WR106	844	99 4 / 844	

continued

1	2	3	4	5			6	7	8
			and RNA guanylyl trans-ferase activities	VARV-IND	F1R		844	98 5 / 844	
				VARV-GAR	F1R		844	98 3 / 844	
E2L	146	L	Virion core protein	CPXV-BRT	V119		146	100 / 146	Dyster & Niles, 1991
				MPXV-ZAI	E2L		146	97.9 / 146	
				VACV-COP	D2L		146	99.3 / 146	
				VACV-WR		WR107	146	99 3 / 146	
				VARV-IND	F2L		146	98 6 / 146	
				VARV-GAR	F2L		146	98 6 / 146	
E3R	237	L	Virion core protein	CPXV-BRT	V120		237	99.6 / 237	Dyster & Niles, 1991
				MPXV-ZAI	E3R		233	97 0 / 237	
				VACV-COP	D3R		237	97.5 / 237	
				VACV-WR		WR108	237	96.6 / 237	
				VARV-IND	F3R		237	94 9 / 237	
				VARV-GAR	F3R		237	94 9 / 237	
E4R	218	E	Uracil DNA glycosylase, required for DNA replication	CPXV-BRT	V121		218	98 2 / 218	Upton et al., 1993; Stuart et al., 1993
				MPXV-ZAI	E4R		218	98 6 / 218	
				VACV-COP	D4R		218	99 5 / 218	
				VACV-WR		WR109	218	100 / 218	
				VARV-IND	F4R		218	98 2 / 218	
				VARV-GAR	F4R		218	99 1 / 218	
E5R	785	E–L	Nucleic acid-independent nucleoside triphosphatase, required for DNA replication	CPXV-BRT	V122		785	98 6 / 785	Evans et al., 1995
				MPXV-ZAI	E5R		785	99.2 / 785	
				VACV-COP	D5R		785	99 2 / 785	
				VACV-WR	WR110		785	99.6 / 785	
				VARV-IND	F5R		785	98 6 / 785	
				VARV-GAR	F5R		785	98 7 / 785	
E6R	637	L	Early transcription factor, VETF, small subunit	CPXV-BRT	V123		637	99.7 / 637	Gershon & Moss, 1990; Broyles & Fesler, 1990
				MPXV-ZAI	E6R		637	99 8 / 637	
				VACV-COP	D6R		637	99.8 / 637	
				VACV-WR		WR111	637	100 / 637	
				VARV-IND	F6R		637	99 4 / 637	
				VARV-GAR	F6R		637	99.5 / 637	
E7R	161	E	RNA pol 18 kDa subunit	CPXV-BRT	V124		161	99 4 / 161	Ahn et al., 1990b
				MPXV-ZAI	E7R		161	97.5 / 161	
				VACV-COP	D7R		161	99 4 / 161	
				VACV-WR		WR112	161	100 / 161	
				VARV-IND	F7R		161	96 9 / 161	
				VARV-GAR	F7R		161	97.5 / 161	
E8L	304	L	IMV surface membrane 32 kDa protein, binds cell surface chondroitin sulfate, IMV adsorption to cell surface	CPXV-BRT	V125		304	96 7 / 304	Niles & Seto, 1988; Maa et al., 1990; Hsiao et al., 1999
				MPXV-ZAI	E8L		304	95.4 / 304	
				VACV-COP	D8L		304	95.7 / 304	
				VACV-WR		WR113	304	95 7 / 304	
				VARV-IND	F8L		304	95 1 / 304	
				VARV-GAR	F8L		304	95 1 / 304	
E9R	213	E	MutT-like	CPXV-BRT	V126		213	100 / 213	Koonin, 1993, Lee-Chen et al., 1988
				MPXV-ZAI	E9R		213	97.7 / 213	
				VACV-COP	D9R		213	99.5 / 213	
				VACV-WR		WR114	213	100 / 213	
				VARV-IND	F9R		213	99 1 / 213	
				VARV-GAR	F9R		213	99.1 / 213	
E10R	248	L	Down regulation of gene expression, Mut-like	CPXV-BRT	V127		248	99 2 / 248	Shors et al., 1999, Koonin, 1993; Lee-Chen et al., 1988
				MPXV-ZAI	E10R		248	98 8 / 248	
				VACV-COP	D10R		248	98 8 / 248	
				VACV-WR		WR115	248	99 6 / 248	
				VARV-IND	F10R		248	99 2 / 248	
				VARV-GAR	F10R		248	100 / 248	
E11L	631	L	Nucleoside triphosphate pho-sphohydrolase I, NPH I, DNA-dependent ATPase, early gene trans-cription termination	CPXV-BRT	V128		631	99.8 / 631	Rodriguez et al., 1986, Broyles & Moss, 1987; Christen et al., 1998
				MPXV-ZAI	E11L		631	98.9 / 631	
				VACV-COP	D11L		631	99 7 / 631	
				VACV-WR		WR116	631	99 7 / 631	
				VARV-IND	N1L		631	99 2 / 631	
				VARV-GAR	O1L		631	98 9 / 631	
E12L	287	E–L	mRNA capping enzyme small subunit, mRNA (guanine-N^7-)-methyl-transferase	CPXV-BRT	V129		287	100 / 287	Niles et al., 1989; Shuman & Morham, 1990
				MPXV-ZAI	E12L		287	99.0 / 287	
				VACV-COP	D12L		287	100 / 287	
				VACV-WR		WR117	287	99 7 / 287	
				VARV-IND	N2L		287	99.7 / 287	
				VARV-GAR	O2L		287	99.3 / 287	
E13L	551	L	Needed for immature IMV surface membrane formation	CPXV-BRT	V131		551	99 1 / 551	Zhang & Moss, 1992
				MPXV-ZAI	E13L		551	99 6 / 551	
				VACV-COP	D13L		551	99 3 / 551	
				VACV-WR		WR118	551	99.3 / 551	
				VARV-IND	N3L		551	99.3 / 551	
				VARV-GAR	O3L		551	99.1 / 551	

continued

1	2	3	4	5		6	7	8
A1L	150	I	Late gene transcription factor, VLTF-2	CPXV-BRT	V132	150	98.0 / 150	Keck et al., 1990, 1993b
				MPXV-ZAI	A1L	150	98 7 / 150	
				VACV-COP	A1L	150	98.7 / 150	
				VACV-WR	WR119	150	98 7 / 150	
				VARV-IND	A1L	150	98 7 / 150	
				VARV-GAR	A1L	150	98 7 / 150	
A2L	224	I	Late gene transcription factor, VLTF-3, zinc binding	CPXV-BRT	V133	224	99.6 / 224	Keck et al., 1990, 1993a
				MPXV-ZAI	A2L	224	99.1 / 224	
				VACV-COP	A2L	224	100 / 224	
				VACV-WR	WR120	224	100 / 224	
				VARV-IND	A2L	224	100 / 224	
				VARV-GAR	A2L	224	100 / 224	
A3L	76	L	Redox protein, required for virion morphogenesis and associated with the E10R VACV-COP as a heterodimer	CPXV-BRT	V134	76	97.4 / 76	Senkevich et al., 2002a
				MPXV-ZAI	A3L	77	90 9 / 77	
				VACV-COP	A2 5L	76	97.4/ 76	
				VACV-WR	WR121	76	97 4 / 76	
				VARV-IND	A2.5L	76	96 1 / 76	
				VARV-GAR	A3L	76	96 1 / 76	
A4L	644	L	Major virion core protein p4b	CPXV-BRT	V135	644	98 6 / 644	Rosel & Moss, 1985
				MPXV-ZAI	A4L	644	99 4 / 644	
				VACV-COP	A3L	644	99 1 / 644	
				VACV-WR	WR122	644	99.1 / 644	
				VARV-IND	A3L	644	99 2 / 644	
				VARV-GAR	A4L	644	99 2 / 644	
A5L	283	L	39 kDa immunodominant virion core protein	CPXV-BRT	V136	295	89 5 / 295	Maa & Esteban, 1987, Williams et al., 1999
				MPXV-ZAI	A5L	281	92.9 / 283	
				VACV-COP	A4L	281	95.8 / 283	
				VACV-WR	WR123	281	94 0 / 283	
				VARV-IND	A4L	271	89 8 / 284	
				VARV-GAR	A5L	271	90 1 / 284	
A6R	164	E–L	RNA pol 22 kDa and 21 kDa subunits	CPXV-BRT	V137	164	100 / 164	Ahn et al., 1992
				MPXV-ZAI	A6R	161	97 6 / 164	
				VACV-COP	A5R	164	100 / 164	
				VACV-WR	WR124	164	100 / 164	
				VARV-IND	A5R	164	98.8 / 164	
				VARV-GAR	A6R	164	99.4 / 164	
A7L	372	L		CPXV-BRT	V138	372	98.7 / 372	Weinrich & Hruby, 1986
				MPXV-ZAI	A7L	372	98.4 / 372	
				VACV-COP	A6L	372	99 7 / 372	
				VACV-WR	WR125	372	98.9 / 372	
				VARV-IND	A6L	372	98.1 / 372	
				VARV-GAR	A7L	372	98 1 / 372	
A8L	710	L	Early transcription factor, VETF, large subunit, needed for morphogenesis	CPXV-BRT	V139	710	98.5 / 710	Gershon & Moss, 1990, Hu et al., 1998
				MPXV-ZAI	A8L	710	98 3 / 710	
				VACV-COP	A7L	710	99 7 / 710	
				VACV-WR	WR126	710	99 7 / 710	
				VARV-IND	A7L	710	98 6 / 710	
				VARV-GAR	A8L	710	98.5 / 710	
A9R	288	E	Intermediate transcription factor, VITF-3, 34 kDa subunit	CPXV-BRT	V140	288	100 / 288	Sanz & Moss, 1999
				MPXV-ZAI	A9R	292	99 0 / 288	
				VACV-COP	A8R	288	99.7 / 288	
				VACV-WR	WR127	288	99 3 / 288	
				VARV-IND	A8R	288	99 3 / 288	
				VARV-GAR	A9R	288	99.7 / 288	
A10L	107	L	IMV virion membrane protein, required for an early step in virion morphogenesis	CPXV-BRT	V141	121	84.3 / 121	Yeh et al., 2000
				MPXV-ZAI	A10L	100	91.6 / 107	
				VACV-COP	A9L	99	90.7 / 107	
				VACV-WR	WR128	108	99 1 / 108	
				VARV-IND	A9L	95	87.5 / 104	
				VARV-GAR	A10L	95	88 5 / 104	
A11L	891	L	Major virion core protein p4a	CPXV-BRT	V142	894	97.3 / 894	Van Meir & Wittek, 1988
				MPXV-ZAI	A11L	891	97.4 / 891	
				VACV-COP	A10L	891	98 8 / 891	
				VACV-WR	WR129	891	98.1 / 891	
				VARV-IND	A10L	892	98 0 / 892	
				VARV-GAR	A11L	892	98 2 / 892	
A12R	318	L	Required for the formation of viral membrane structures, not incorporated into virus particles	CPXV-BRT	V143	318	100 / 318	Resch et al , 2004
				MPXV-ZAI	A12R	318	99.7 / 318	
				VACV-COP	A11R	318	99.7 / 318	
				VACV-WR	WR130	318	99.4 / 318	
				VARV-IND	A11R	319	98 7 / 319	
				VARV-GAR	A12R	319	99 1 / 319	
A13L	194	L	Virion core protein	CPXV-BRT	V144	190	95 4 / 194	Whitehead & Hruby, 1994a
				MPXV-ZAI	A13L	190	96.4 / 194	
				VACV-COP	A12L	192	97.4 / 194	

continued

1	2	3	4	5		6	7	8
				VACV-WR	WR131	192	96 9 / 194	
				VARV-IND	A12L	189	94 3 / 194	
				VARV-GAR	A13L	189	94 8 / 194	
A14L	68	L	IMV inner and outer membrane protein	CPXV-BRT	V145	70	91 4 / 70	Takahashi et al., 1994; Salmons et al., 1997
				MPXV-ZAI	A14L	70	88.6 / 70	
				VACV-COP	A13L	70	92.8 / 69	
				VACV-WR	WR132	70	92.8 / 69	
				VARV-IND	A13L	68	91.0 / 67	
				VARV-GAR	A14L	68	89 6 / 67	
A15L	90	L	IMV inner and outer membrane protein	CPXV-BRT	V146	90	98 9 / 90	Takahashi et al., 1994, Salmons et al. 1997
				MPXV-ZAI	A15L	90	98 9 / 90	
				VACV-COP	A14L	90	98 9 / 90	
				VACV-WR	WR133	90	98 9 / 90	
				VARV-IND	A14L	90	98 9 / 90	
				VARV-GAR	A15L	90	98 9 / 90	
A15 5L	53	L	Hydrophobic virion membrane protein that enhances virulence in mice	CPXV-BRT	V147	53	100/ 53	Betakova et al., 2000
				MPXV-ZAI	A15 5L	53	98 1 / 53	
				VACV-COP	A14 5L	53	100/ 53	
				VACV-WR	WR134	53	100/ 53	
				VARV-IND	A14 5L	53	100/ 53	
				VARV-GAR	A15 5L	53	100/ 53	
A16L	94		IMV inner membrane protein	CPXV-BRT	V148	94	100 / 94	Takahashi et al., 1994; Salmons et al., 1997
				MPXV-ZAI	A16L	94	98.9 / 94	
				VACV-COP	A15L	94	100 / 94	
				VACV-WR	WR135	94	100 / 94	
				VARV-IND	A15L	94	97.9 / 94	
				VARV-GAR	A16L	94	97.9 / 94	
A17L	377	L	Soluble myristylated protein	CPXV-BRT	V149	378	98 4 / 377	Martin et al., 1997
				MPXV-ZAI	A17L	377	96 0 / 377	
				VACV-COP	A16L	378	99.2 / 378	
				VACV-WR	WR136	377	99 5 / 377	
				VARV-IND	A16L	377	97 1 / 377	
				VARV-GAR	A17L	377	97 3 / 377	
A18L	203	L	IMV surface membrane protein, early function in virion morphogenesis	CPXV-BRT	V150	202	96 6 / 203	Ichihashi et al., 1994, Rodriguez et al., 1995, Wolffe et al., 1996
				MPXV-ZAI	A18L	204	97 5 / 204	
				VACV-COP	A17L	203	99.5 / 203	
				VACV-WR	WR137	203	99 5 / 203	
				VARV-IND	A17L	203	99 5 / 203	
				VARV-GAR	A18L	203	99.5 / 203	
A19R	493	E–L	DNA helicase, postreplicative negative transcription elongation factor	CPXV-BRT	V151	492	96 6 / 493	Simpson & Condit, 1995, Xiang et al., 1998
				MPXV-ZAI	A19R	492	97.0 / 493	
				VACV-COP	A18R	493	97.6 / 493	
				VACV-WR	WR138	493	98 2 / 493	
				VARV-IND	A18R	493	97 8 / 493	
				VARV-GAR	A19R	493	97 8 / 493	
A20L	76	L	Fowlpox virus homologue is localized around the virosomes	CPXV-BRT	V152	77	97.4 / 77	Laidlaw & Skinner, 2004
				MPXV-ZAI	A20L	77	97 4 / 77	
				VACV-COP	A19L	77	97 4 / 77	
				VACV-WR	WR139	77	97 4 / 77	
				VARV-IND	A19L	76	94.7 / 76	
				VARV-GAR	A20L	76	94 7 / 76	
A21L	117			CPXV-BRT	V153	117	98 3 / 117	
				MPXV-ZAI	A21L	115	97 4 / 117	
				VACV-COP	A21L	117	100 / 117	
				VACV-WR	WR140	117	100 / 117	
				VARV-IND	A20L	117	98 3 / 117	
				VARV-GAR	A21L	117	98 3 / 117	
A22R	426	E	Processivity factor for viral DNA pol	CPXV-BRT	V154	426	99 1 / 426	Ishii & Moss, 2001; Klemperer et al., 2001
				MPXV-ZAI	A22R	426	98.1 / 426	
				VACV-COP	A20R	426	99.1 / 426	
				VACV-WR	WR141	426	99.1 / 426	
				VARV-IND	A21R	426	97 9 / 426	
				VARV-GAR	A22R	426	97 2 / 426	
A23R	187			CPXV-BRT	V155	187	99 5 / 187	
				MPXV-ZAI	A23R	187	97 3 / 187	
				VACV-COP	WR142	176	98 9 / 176	
				VACV-WR	A22R	187	97 9 / 187	
				VARV-IND	A22R	187	95.7 / 187	
				VARV-GAR	A23R	187	95 7 / 187	
A24R	382	E	Intermediate transcription factor, VITF-3, 45 kDa subunit	CPXV-BRT	V156	382	99.7 / 382	Sanz & Moss, 1999
				MPXV-ZAI	A24R	382	99 0 / 382	
				VACV-COP	A23R	382	99.7 / 382	
				VACV-WR	WR143	382	100 / 382	
				VARV-IND	A23R	382	99 0 / 382	
				VARV-GAR	A24R	382	98.7 / 382	
A25R	1164	E–L	RNA pol 132 kDa subunit	CPXV-BRT	V157	1164	99.7 / 1163	Hooda-Dhingra et al., 1990;
				MPXV-ZAI	A25R	1164	99 1 / 1164	

continued

1	2	3	4	5		6	7	8
				VACV-COP	A24R	1164	99.6 / 1164	Amegadzie
				VACV-WR	WR144	1164	99 6 / 1164	et al., 1991b
				VARV-IND	A24R	1164	99 0 / 1163	
				VARV-GAR	A25R	1164	99 2 / 1163	
A26L	1279	L	A-type inclusion body protein	CPXV-BRT	V158	1284	93.6 / 1287	Funahashi
				MPXV-ZAI	A26L	75	79 4 / 63	et al., 1988,
				MPXV-ZAI	A27L	696	96 3 / 695	Shchelkunov
				VACV-COP	A25L	65	65 7 / 70	et al., 1994a
				VACV-COP	A26L	322	90 6 / 127	
				VACV-WR	WR145	65	64 3 / 70	
				VACV-WR	WR146	154	92 9 / 155	
				VACV-WR	WR147	227	94 2 / 223	
				VACV-WR	WR148	725	97.4 / 723	
				VARV-IND	A25L	96	89 0 / 73	
				VARV-IND	A26L	65	92 3 / 65	
				VARV-IND	A27L	194	89.6 / 193	
				VARV-IND	A28L	702	93.1 / 722	
				VARV-GAR	A26L	134	90.1 / 121	
				VARV-GAR	A27L	101	79 4 / 97	
				VARV-GAR	A28L	194	89 6 / 193	
				VARV-GAR	A29L	702	93 1 / 722	
A27L	518	L	Major component of IMV surface tubules, p4c	CPXV-BRT	V159	192	94 3 / 192	Sarov &
				CPXV-BRT	V161	260	96 1 / 255	Joklik, 1972,
				MPXV-ZAI	A28L	520	95 8 / 520	McKelvey
				VACV-COP	A26L	322	93 4 / 212	et al., 2002
				VACV-WR	WR149	500	94.2 / 518	
				VARV-IND	A29L	498	92.7 / 518	
				VARV-GAR	A30L	498	93 2 / 518	
A28	110	L	IMV surface membrane 14 kDa fusion protein, binds cell surface heparan	CPXV-BRT	V162	110	99.1 / 110	Rodriguez &
				MPXV-ZAI	A29L	110	95 5 / 110	Esteban, 1987,
				VACV-COP	A27L	110	98.2 / 110	Chung et al.,
				VACV-WR	WR150	110	99.1 / 110	1998
				VARV-IND	A30L	110	98 2 / 110	
				VARV-GAR	A31L	110	99.1 / 110	
A29L	146		IMV surface protein, needed for core penetration into cells	CPXV-BRT	V163	146	98 6 / 146	Senkevich
				MPXV-ZAI	A30L	146	97.3 / 146	et al., 2004
				VACV-COP	A28L	146	98.6 / 146	
				VACV-WR	WR151	146	99.3 / 146	
				VARV-IND	A31L	146	97 9 / 146	
				VARV-GAR	A32L	146	98 0 / 146	
A30L	305	E	RNA pol 35 kDa subunit	CPXV-BRT	V164	305	98 7 / 305	Amegadzie
				MPXV-ZAI	A31L	305	98.0 / 305	et al., 1991a
				VACV-COP	A29L	305	99 3 / 305	
				VACV-WR	WR152	305	99.0 / 305	
				VARV-IND	A32L	305	97.0 / 305	
				VARV-GAR	A33L	305	97 7 / 305	
A31L	77	L	Forms complex with G7L and F10L (VACV-COP), important for early steps in virion morphogenesis	CPXV-BRT	V165	76	94 8 / 77	Szajner et al.,
				MPXV-ZAI	A32L	77	96 1 / 77	2004
				VACV-COP	A30L	77	100 / 77	
				VACV-WR	WR153	77	100 / 77	
				VARV-IND	A33L	77	100 / 77	
				VARV-GAR	A34L	77	100 / 77	
A32R	145			CPXV-BRT	V166	140	96.6 / 145	
				MPXV-ZAI	A33R	142	96 6 / 145	
				VACV-COP	A31R	124	84.8 / 145	
				VACV-WR	WR154	124	84.8 / 145	
				VARV-IND	A34R	140	93.8 / 145	
				VARV-GAR	A35R	146	96 6 / 146	
A33L	300	E–L	DNA packaging into virion, NTP-binding motif A	CPXV-BRT	V167	311	99 0 / 299	Koonin et al.,
				MPXV-ZAI	A35L	300	99 0 / 300	1993,
				VACV-COP	A32L	300	99.0 / 300	Cassetti et al.,
				VACV-WR	WR155	270	98.9 / 270	1998
				VARV-IND	A35L	270	98.5 / 270	
				VARV-GAR	A36L	270	98 5 / 270	
A34R	185	L	EEV envelope glycoprotein, needed for formation of actin-containing microvilli and cell-to-cell spread	CPXV-BRT	V168	187	96.8 / 187	Roper et al.,
				MPXV-ZAI	A35R	181	96 1 / 180	1996; 1998;
				VACV-COP	A33R	185	98 9 / 185	Katz et al.,
				VACV-WR	WR156	185	98 9 / 185	2003
				VARV-IND	A36R	184	95 1 / 185	
				VARV-GAR	A37R	184	95 1 / 185	
A35R	168	L	EEV envelope glycoprotein, lectin-like, required for infectivity of EEV,	CPXV-BRT	V169	168	99 4 / 168	Blasco et al.,
				MPXV-ZAI	A36R	168	98.2 / 168	1993,
				VACV-COP	A34R	168	97 0 / 168	McIntosh &
				VACV-WR	WR157	168	98 2 / 168	Smith, 1996,
				VARV-IND	A37R	168	97 6 / 168	Wolffe et al.,

continued

1	2	3	4	5		6	7	8
			formation of actin-containing microvilli, and cell-to-cell spread	VARV-GAR	A38R	168	97 6 / 168	1997
A36R	176	E	Acidic hydrophobic protein; associated with intracellular virus particles	CPXV-BRT	V171	176	99.4 / 176	Roper & Moss, 2002
				MPXV-ZAI	A37R	176	98.3 / 176	
				VACV-COP	A35R	176	99 4 / 176	
				VACV-WR	WR158	176	98 9 / 176	
				VARV-IND	A38R	60	95 0 / 60	
				VARV-GAR	A39R	60	94.9 / 59	
A37R	223	E–L	IEV but not CEV envelope protein, provides a direct link between IEV and the micro-tubule motor kinesin	CPXV-BRT	V172	224	96.0 / 225	Parkinson & Smith, 1994; Wolffe *et al.*, 1998, van Eijl *et al.*, 2000, Ward & Moss, 2004
				MPXV-ZAI	A38R	212	95 1 / 204	
				VACV-COP	A36R	221	98.7 / 223	
				VACV-WR	WR159	221	98 7 / 223	
				VARV-IND	A39R	216	94.2 / 223	
				VARV-GAR	A40R	216	94.2 / 223	
A38R	268			CPXV-BRT	V173	263	95.8 / 263	
				MPXV-ZAI	A39R	268	95 9 / 268	
				VACV-COP	A37R	263	96 2 / 262	
				VACV-WR	WR160	263	96.6 / 263	
				VARV-IND	A40R	68	89 6 / 67	
				VARV-GAR	A42R	68	89 6 / 67	
A39R	64			CPXV-BRT	V174	63	90 8 / 65	
				MPXV-ZAI	–	–	–	
				VACV-COP	–	–	–	
				VACV-WR	WR161	62	95 3 / 64	
				VARV-IND	A40 5R	62	96 9 / 64	
				VARV-GAR	A43R	62	96 9 / 64	
A40L	277		Integral membrane glycoprotein, Ig-like, regulates influx of extracellular Ca^{2+}	CPXV-BRT	V175	277	97 5 / 277	Parkinson *et al.*, 1995; Sanderson *et al.*, 1996
				MPXV-ZAI	A40L	277	99 3 / 277	
				VACV-COP	A38L	277	96 8 / 277	
				VACV-WR	WR162	277	96 8 / 277	
				VARV-IND	A41L	277	96 4 / 277	
				VARV-GAR	A44L	277	96 4 / 277	
A41R	402		Semaphorin-like	CPXV-BRT	V176	409	95 5 / 401	
				MPXV-ZAI	–	–	–	
				VACV-COP	A39R	403	95.5 / 404	
				⎡VACV-WR	WR163	295	94 9 / 253	
				⎣VACV-WR	WR164	142	95 1 / 142	
				⎡VARV-IND	A42R	74	91 7 / 72	
				⎢VARV-IND	A43R	122	92 4 / 92	
				⎣VARV-IND	A44R	139	76 6 / 137	
				⎡VARV-GAR	A45R	74	90.3 / 72	
				⎢VARV-GAR	A46R	122	92 4 / 92	
				⎣VARV-GAR	A47R	139	76 6 / 137	
A42R	165		Lectin-like homologue of NK-cell receptor	CPXV-BRT	V177	160	89 8 / 166	Smith G.L *et al.*, 1991
				MPXV-ZAI	–	–	–	
				VACV-COP	A40R	168	97 8 / 136	
				VACV-WR	WR165	159	93 9 / 165	
				VARV-IND	A45R	61	91.5 / 59	
				VARV-GAR	A48R	61	91.5 / 59	
A43L	219	E–L	Secreted protein, reduces influx of inflammatory cells	CPXV-BRT	V178	218	93 2 / 221	Ng *et al.*, 2001
				MPXV-ZAI	A41L	221	94 1 / 222	
				VACV-COP	A41L	219	99.1 / 219	
				VACV-WR	WR166	219	97.7 / 219	
				VARV-IND	A46L	218	95.4 / 219	
				VARV-GAR	A49L	218	95.4 / 219	
A44R	133	L	Profilin-like	CPXV-BRT	V179	133	97 7 / 133	Goebel *et al.*, 1990; Blasco *et al.*, 1991
				MPXV-ZAI	A42R	133	100 / 133	
				VACV-COP	A42R	133	97.7 / 133	
				VACV-WR	WR167	133	97.7 / 133	
				VARV-IND	A47R	133	99 2 / 133	
				VARV-GAR	A50R	133	99 2 / 133	
A45R	196	E–L	Membrane glycoprotein	CPXV-BRT	V180	194	93 4 / 196	Duncan & Smith, 1992
				MPXV-ZAI	A43R	197	94 4 / 197	
				VACV-COP	A43R	194	95 9 / 196	
				VACV-WR	WR168	194	95.4 / 196	
				VARV-IND	A48R	195	96.9 / 195	
				VARV-GAR	A51R	195	96 9 / 195	
A46R	78			CPXV-BRT	V181	81	84 6 / 78	
				MPXV-ZAI	A44R	74	95.9 / 73	
				VACV-COP	–	–	–	
				VACV-WR	WR169	78	100 / 78	
				VARV-IND	–	–	–	
				VARV-GAR	–	–	–	

continued

1	2	3	4	5		6	7	8
A47L	346	E	3-β-Hydroxy-delta5-steroid dehydrogenase	CPXV-BRT	V182	345	97 7 / 343	Moore & Smith, 1992
				MPXV-ZAI	A45L	346	98 8 / 346	
				VACV-COP	A44L	346	99.1 / 346	
				VACV-WR	WR170	346	99 4 / 346	
				VARV-IND	A49L	210	94 9 / 197	
				VARV-IND	A50L	61	88 1 / 59	
				VARV-GAR	A52L	101	92 1 / 101	
				VARV-GAR	A53L	70	95 1 / 61	
				VARV-GAR	A54L	61	86 4 / 59	
A48R	125	L	Superoxide dismutase-like, virion core protein	CPXV-BRT	V183	125	97.6 / 125	Goebel et al., 1990; Almazan et al., 2001
				MPXV-ZAI	A46R	125	98.4 / 125	
				VACV-COP	A45R	125	97 6 / 125	
				VACV-WR	WR171	125	99 2 / 125	
				VARV-IND	A51R	125	98 4 / 125	
				VARV-GAR	A55R	125	98 4 / 125	
A49R	240			CPXV-BRT	V184	242	97.1 / 240	
				MPXV-ZAI	A47R	240	96 3 / 240	
				VACV-COP	A46R	214	92 6 / 204	
				VACV-WR	WR172	240	97.9 / 240	
				VARV-IND	A52R	240	96.7 / 240	
				VARV-GAR	A56R	240	96 7 / 240	
A50L	244			CPXV-BRT	V185	244	97 1 / 244	
				MPXV-ZAI	–	–	–	
				VACV-COP	A47L	244	97.5 / 244	
				VACV-WR	WR173	252	97 9 / 243	
				VARV-IND	J1L	244	95.1 / 244	
				VARV-GAR	K1L	244	95.5 / 244	
A51R	227	E	Thymidylate kinase	CPXV-BRT	V186	227	99 6 / 227	Smith G.L et al., 1989a; Hughes et al., 1991
				MPXV-ZAI	A49R	204	99 0 / 204	
				VACV-COP	A48R	204	99 0 / 204	
				VACV-WR	WR174	227	99.1 / 227	
				VARV-IND	J2R	205	98 0 / 205	
				VARV-GAR	K2R	205	98 0 / 205	
A52R	162			CPXV-BRT	V187	162	97.5 / 162	
				MPXV-ZAI	–	–	–	
				VACV-COP	A49R	162	98 8 / 162	
				VACV-WR	WR175	162	98.1 / 162	
				VARV-IND	J3R	162	94 4 / 162	
				VARV-GAR	K3R	162	94 4 / 162	
A53R	552	E	DNA ligase	CPXV-BRT	V188	554	97.1 / 554	Kerr & Smith, 1989, Smith G.L. et al., 1989b
				MPXV-ZAI	A50R	554	97 8 / 554	
				VACV-COP	A50R	552	97.8 / 552	
				VACV-WR	WR176	552	98 2 / 552	
				VARV-IND	J4R	552	97 1 / 552	
				VARV-GAR	K4R	552	96 9 / 552	
A54R	334			CPXV-BRT	V189	334	96 7 / 334	
				MPXV-ZAI	A51R	334	96 1 / 334	
				VACV-COP	A51R	334	96.1 / 334	
				VACV-WR	WR177	334	95 5 / 334	
				VARV-IND	J5R	334	95 5 / 334	
				VARV-GAR	K5R	334	95 5 / 334	
A55R	190		Blocks the activation of NF-kappa B	CPXV-BRT	V190	190	98.4 / 190	Harte et al., 2003
				MPXV-ZAI	–	–	–	
				VACV-COP	A52R	190	98 4 / 190	
				VACV-WR	WR178	190	97.9 / 190	
				VARV-IND	J6R	71	98.5 / 67	
				VARV-GAR	K6R	168	89.2 / 167	
A56R	186		Secreted TNF binding protein, CrmC	CPXV-BRT	V191	186	87 1 / 186	Smith C A et al., 1996
				MPXV-ZAI	–	–	–	
				VACV-COP	A53R	103	83 6 / 55	
				VACV-WR	WR179	103	83 6 / 55	
				VARV-IND	–	–	–	
				VARV-GAR	–	–	–	
A57R	564		Kelch-like	CPXV-BRT	V193	563	96 1 / 564	Xue & Cooley, 1993; Shchelkunov et al., 1998
				MPXV-ZAI	B1R	70	86 7 / 60	
				VACV-COP	A55R	564	98 6 / 564	
				VACV-WR	WR180	564	98.89 / 564	
				VARV-IND	J7R	71	90.5 / 74	
				VARV-IND	J8R	172	80.2 / 167	
				VARV-GAR	K7R	71	89 2 / 74	
				VARV-GAR	K8R	70	89 2 / 65	
A58R	314	E–L	EEV membrane glycoprotein hemagglutinin, prevents cell fusion	CPXV-BRT	V194	297	84.5 / 316	Shida, 1986; Seki et al., 1990; Brown C K. et al., 1991
				MPXV-ZAI	B2R	313	91.7 / 315	
				VACV-COP	A56R	315	95 9 / 315	
				VACV-WR	WR181	314	95 6 / 315	
				VARV-IND	J9R	313	84 3 / 319	
				VARV-GAR	K9R	318	85 6 / 319	

continued

1	2	3	4	5		6	7	8
A59R	197		Guanylate kinase-like	CPXV-BRT	V195	197	99 0 / 197	Smith G L et al., 1991
				MPXV-ZAI	–	–	–	
				VACV-COP	A57R	151	97.4 / 151	
				VACV-WR	WR182	151	98 0 / 151	
				VARV-IND	J10R	151	95 4 / 151	
				VARV-GAR	K10R	151	94 0 / 151	
B1R	300	E	Serine/threonine protein kinase, essential for DNA replication, intermediate transcription factor	CPXV-BRT	V196	299	96 6 / 298	Banham & Smith, 1992; Lin et al., 1992; Rempel & Traktman, 1992; Kovacs et al., 2001
				MPXV-ZAI	B3R	299	97 3 / 298	
				VACV-COP	B1R	300	99 7 / 300	
				VACV-WR	WR183	300	100 / 300	
				VARV-IND	B1R	300	96.0 / 300	
				VARV-GAR	H1R	300	97 0 / 300	
B2R	503		Schlafen-like	CPXV-BRT	V197	505	90 9 / 505	Schwarz et al., 1998, Shchelkunov et al., 2000
				MPXV-ZAI	B4R	503	89 3 / 503	
				⌐VACV-COP	B2R	219	97 7 / 215	
				⌐VACV-COP	B3R	124	95.1 / 123	
				⌐VACV-WR	WR184	219	96.7 / 215	
				⌐VACV-WR	WR185	167	95 8 / 143	
				VARV-IND	–	–	–	
				VARV-GAR	–	–	–	
B3R	558		Ankyrin-like	CPXV-BRT	V198	558	94 3 / 558	Shchelkunov et al., 1993b; 1998
				MPXV-ZAI	B5R	561	94.5 / 562	
				VACV-COP	B4R	558	96.6 / 558	
				VACV-WR	WR186	558	96 4 / 558	
				VARV-IND	B6R	558	95.0 / 558	
				VARV-GAR	H6R	558	94 8 / 558	
B4R	317	E–L	Palmitylated 42 kDa EEV glycoprotein required for efficient cell spread, complement control protein-like	CPXV-BRT	V199	317	93.4 / 317	Engelstad et al., 1992, 1993, Isaacs et al., 1992b; Wolffe et al., 1993
				MPXV-ZAI	B6R	317	98.4 / 317	
				VACV-COP	B5R	317	96 2 / 317	
				VACV-WR	WR187	317	96.8 / 317	
				VARV-IND	B7R	317	92 7 / 316	
				VARV-GAR	H7R	317	92 7 / 316	
B5R	183			CPXV-BRT	V200	179	89.6 / 183	
				MPXV-ZAI	B7R	176	85 8 / 183	
				VACV-COP	B6R	173	89 1 / 183	
				VACV-WR	WR188	173	89 1 / 183	
				VARV-IND	B8R	56	87.8 / 45	
				VARV-GAR	H8R	65	95 4 / 65	
B6R	182			CPXV-BRT	V201	181	90 1 / 182	
				MPXV-ZAI	B8R	182	96.7 / 182	
				VACV-COP	B7R	182	97.8 / 182	
				VACV-WR	WR189	182	97 8 / 182	
				VARV-IND	–	–	–	
				VARV-GAR	–	–	–	
B7R	271	E	Secreted IFN-γ-binding protein	CPXV-BRT	V202	266	95 1 / 266	Alcami & Smith, 1995b, Seregin et al., 1996
				MPXV-ZAI	B9R	267	97 3 / 263	
				VACV-COP	B8R	272	98.5 / 268	
				VACV-WR	WR190	272	96.6 / 268	
				VARV-IND	B9R	266	92.1 / 266	
				VARV-GAR	H9R	266	91 7 / 266	
B8R	221		Shope fibroma virus T4 protein-like	CPXV-BRT	V203	225	66 2 / 225	Shchelkunov et al., 1998
				MPXV-ZAI	B10R	221	97 3 / 221	
				VACV-COP	B9R	77	97 1 / 69	
				VACV-WR	WR191	77	95.7 / 69	
				VARV-IND	–	–	–	
				VARV-GAR	–	–	–	
B9R	501		Kelch-like	CPXV-BRT	V204	501	96.4 / 501	Shchelkunov et al., 1998
				MPXV-ZAI	–	–	–	
				VACV-COP	B10R	166	92 8 / 167	
				VACV-WR	WR192	166	92 8 / 167	
				VARV-IND	–	–	–	
				VARV-GAR	–	–	–	
B10R	105			CPXV-BRT	V205	90	87 8 / 90	
				MPXV-ZAI	–	–	–	
				VACV-COP	B11R	88	97 7 / 87	
				VACV-WR	WR193	72	95 8 / 71	
				VARV-IND	–	–	–	
				VARV-GAR	–	–	–	
B11R	283	E	Protein kinase-like	CPXV-BRT	V206	285	95 8 / 285	Banham & Smith, 1993
				MPXV-ZAI	B11R	282	96 1 / 283	
				VACV-COP	B12R	283	98 9 / 283	
				VACV-WR	WR194	283	98 2 / 283	
				VARV-IND	–	–	–	
				VARV-GAR	–	–	–	

continued

1	2	3	4	5		6	7	8
B12R	345	E	Serine protease inhibitor-like, SPI-2, inhibits IL-1β converting enzyme, CrmA	CPXV-BRT	V207	341	91.9 / 344	Kotwal & Moss, 1989; Ray et al. 1992, Smith G L. et al., 1991, Palumbo et al., 1993, Kettle et al., 1997
				MPXV-ZAI	B12R	344	97.1 / 344	
				⌐VACV-COP	B13R	116	95.6 / 114	
				└VACV-COP	B14R	222	94 1 / 219	
				VACV-WR	WR195	345	95 1 / 345	
				VARV-IND	B13R	344	95.6 / 344	
				VARV-GAR	D2R	344	94 8 / 344	
B13R	149		Homologue of VACV-COP C16L	CPXV-BRT	V208	149	96.6 / 149	Goebel et al., 1990
				MPXV-ZAI	B13R	149	94 0 / 149	
				VACV-COP	B15R	149	97.3 / 149	
				VACV-WR	WR196	149	98 0 / 149	
				VARV-IND	B14R	149	98.0 / 149	
				VARV-GAR	D3R	149	97.3 / 149	
B14R	326	E	Secreted, IL-1β-binding, inhibits virus-induced fever	CPXV-BRT	V209	326	94.8 / 326	Spriggs et al., 1992, Alcami & Smith, 1992; 1996
				MPXV-ZAI	B14R	326	95 4 / 326	
				VACV-COP	B16R	290	98.3 / 290	
				VACV-WR	WR197	326	96 9 / 326	
				⌐VARV-IND	B15R	63	90 2 / 61	
				└VARV-IND	B17R	69	88 2 / 68	
				⌐VARV-GAR	D4R	63	88 5 / 61	
				└VARV-GAR	D6R	69	89.7 / 68	
B15L	340			CPXV-BRT	V210	340	95 6 / 340	
				MPXV-ZAI	B15L	78	95 9 / 74	
				VACV-COP	B17L	340	98 5 / 340	
				VACV-WR	WR198	340	98 2 / 340	
				VARV-IND	B18L	340	95 0 / 340	
				VARV-GAR	D7L	340	95 0 / 340	
B16R	574		Ankyrin-like	CPXV-BRT	V211	574	92.5 / 574	Shchelkunov et al., 1993b; Safronov et al., 1996
				MPXV-ZAI	–	–	–	
				VACV-COP	B18R	574	97 6 / 574	
				VACV-WR	WR199	574	97 6 / 574	
				VARV-IND	B19R	574	94 6 / 574	
				VARV-GAR	D8R	574	94 4 / 574	
B17R	351	E	Cell surface antigen and secreted IFN-α/β binding protein	CPXV-BRT	V212	366	90.0 / 350	Ueda et al., 1990; Symons et al., 1995
				MPXV-ZAI	B16R	352	92.0 / 351	
				VACV-COP	B19R	353	93.0 / 353	
				VACV-WR	WR200	351	93 4 / 351	
				VARV-IND	B20R	354	90.7 / 353	
				VARV-GAR	D9R	355	90 1 / 354	
B18R	795		Ankyrin-like	CPXV-BRT	V213	800	90 3 / 802	Shchelkunov et al., 1993b, Safronov et al., 1996
				MPXV-ZAI	B17R	793	95.9 / 800	
				VACV-COP	B20R	127	97.6 / 125	
				⌐VACV-WR	WR202	54	88 6 / 44	
				└VACV-WR	WR203	309	94.4 / 305	
				VARV-IND	B21R	787	88 6 / 95	
				VARV-GAR	D10R	787	88 3 / 795	
B19R	557		Kelch-like	CPXV-BRT	V215	557	96 4 / 57	Shchelkunov et al., 1998
				MPXV-ZAI	B18R	70	90 9 / 55	
				VACV-COP	–	–	–	
				VACV-WR	WR204	134	97.0 / 134	
				⌐VARV-IND	B22R	70	82 5 / 63	
				│VARV-IND	B23R	83	92 7 / 82	
				└VARV-IND	B24R	88	96 2 / 79	
				⌐VARV-GAR	D11R	70	84.1 / 63	
				│VARV-GAR	D12R	127	92 5 / 120	
				└VARV-GAR	D13R	88	96 2 / 79	
B20R	375	E	Serine protease inhibitor-like, SPI-1, apoptosis inhibition	CPXV-BRT	V217	372	91 7 / 374	Kotwal & Moss, 1989; Smith G L. et al., 1989c
				MPXV-ZAI	B19R	357	93 9 / 358	
				VACV-COP	C12L	353	94.4 / 358	
				VACV-WR	WR205	353	95.3 / 358	
				VARV-IND	B25R	372	91 5 / 375	
				VARV-GAR	D14R	372	91 5 / 375	
B21R	190			CPXV-BRT	V218	193	89 1 / 193	
				MPXV-ZAI	B20R	190	96 3 / 190	
				VACV-COP	–	–	–	
				VACV-WR	WR206	190	95.3 / 190	
				VARV-IND	–	–	–	
				VARV-GAR	–	–	–	
B22R	1933	E	Putative membrane-associated glycoprotein, cadherin-like domain	CPXV-BRT	V219	1919	87 7 / 1941	Shchelkunov et al., 1994a, Marennikova & Shchelkunov, 1998
				MPXV-ZAI	B21R	1879	92.5 / 1935	
				VACV-COP	–	–	–	
				VACV-WR	–	–	–	
				VARV-IND	B26R	1896	85.8 / 1937	
				VARV-GAR	D15R	1896	85 9 / 1937	
K1R	581		Ankyrin-like	CPXV-BRT	V220	579	94.5 / 581	Shchelkunov et al., 1993b, Safronov et al., 1996
				MPXV-ZAI	–	–	–	
				VACV-COP	B21R	91	93 2 / 88	
				VACV-WR	–	–	–	

continued

1	2	3	4	5	6	7	8
				VARV-IND	–	–	–	
				VARV-GAR	–	–	–	
K2R	322		Secreted TNF- and chemokine-binding protein, CrmD	CPXV-BRT	V221	320	96 6 / 322	Shchelkunov et al., 1998; Loparev et al., 1998; Ruiz-Arguello et al., 2004
				MPXV-ZAI	–	–	–	
				VACV-COP	–	–	–	
				VACV-WR	–	–	–	
				VARV-IND	–	–	–	
				VARV-GAR	–	–	–	
K3R	167		Secreted TNF-binding protein, CrmE	CPXV-BRT	–	–	–	Shchelkunov et al., 1998, Saraiva & Alcami, 2001
				MPXV-ZAI	K1R	70	82 9 / 70	
				VACV-COP	–	–	–	
				VACV-WR	–	–	–	
				VARV-IND	–	–	–	
				VARV-GAR	–	–	–	
T1R	210		Bax-inhibitor, inhibition of apoptosis	CPXV-BRT	–	–	–	Gubser & Smith, 2002
				MPXV-ZAI	R1R	105	82 5 / 103	
				VACV-COP	–	–	–	
				VACV-WR	–	–	–	
				VARV-IND	–	–	–	
				VARV-GAR	–	–	–	
I1R*	153		Homologue of VACV-COP B15R	CPXV-BRT	V222	153	92 2 / 153	
				MPXV-ZAI	N1R	153	92 8 / 153	
				VACV-COP	B22R	181	96.1 / 153	
				VACV-WR	–	–	–	
				VARV-IND	–	–	–	
				VARV-GAR	–	–	–	
I2R*	672		Ankyrin-like	CPXV-BRT	V223	672	95 4 / 673	Shchelkunov et al., 1993b, Safronov et al., 1996
				MPXV-ZAI	–	–	–	
				VACV-COP	B23R	386	91 4 / 374	
				VACV-COP	B24R	150	94 3 / 141	
				VACV-WR	–	–	–	
				VARV-IND	–	–	–	
				VARV-GAR	–	–	–	
I3R*	586		Ankyrin-like	CPXV-BRT	V225	619	89 8 / 588	Shchelkunov et al., 1993b, Safronov et al., 1996
				MPXV-ZAI	J1R	587	96.6 / 586	
				VACV-COP	B25R	259	84.5 / 233	
				VACV-COP	B26R	103	70.4 / 115	
				VACV-COP	B27R	113	99.0 / 97	
				VACV-WR	WR211	112	89 4 / 104	
				VACV-WR	WR212	109	81.8 / 99	
				VACV-WR	WR213	64	100 / 43	
				VACV-WR	WR214	49	97.9 / 48	
				VARV-IND	G1R	585	90.9 / 584	
				VARV-GAR	G1R	585	91.1 / 584	
I4R*	351	E	Secreted TNF- and chemokine-binding protein, CrmB	CPXV-BRT	V226	355	90 5 / 357	Upton et al., 1991, Shchelkunov et al., 1993a, Hu et al., 1994, Ruiz-Arguello et al., 2004
				MPXV-ZAI	J2R	348	90 3 / 351	
				VACV-COP	B28R	122	89 2 / 93	
				VACV-WR	WR215	122	89 2 / 93	
				VACV-WR	WR217	61	88.5 / 52	
				VARV-IND	G2R	349	90 9 / 351	
				VARV-GAR	G2R	349	90.9 / 351	
I5R*	255	E–L	Secreted CC-chemokine binding protein	CPXV-BRT	V227	246	80 6 / 258	Graham et al., 1997, Smith C.A et al., 1997
				MPXV-ZAI	J3R	246	86.7 / 255	
				VACV-COP	B29R	244	97 1 / 242	
				VACV-WR	WR218	244	97.5 / 242	
				VARV-IND	G3R	253	92 9 / 255	
				VARV-GAR	G3R	253	93.3 / 255	

Note: A dash indicates a deletion in the coding sequence of one virus relative to the other; asterisk, ORFs from the inverted terminal repeats of the viral genome. IV, IMV, IEV, CEV, and EEV are immature virion, intracellular mature virion, intracellular enveloped virion, cell-associated extracellular enveloped virion, and extracellular enveloped virion, respectively.

[a] Number of deduced amino acids (aa) encoded within an ORF.

[b] Expression time of the corresponding VACV genes determined experimentally is indicated: E, early; E–L, early–late; I, intermediate; and L, late.

[c] Experimentally determined functions of viral proteins or homologies based on searching of PIR and SWISS-PROT databases.

[d] Comparison was made for the corresponding ORFs of orthopoxviral strains: cowpox Brighton (CPXV-BRT), monkeypox Zaire-96-I-16 (MPXV-ZAI), vaccinia Copenhagen (VACV-COP) and Western Reserve (VACV-WR), variola major India-1967 (VARV-IND), and variola minor alastrim Garcia-1966 (VARV-GAR).

[e] Values of amino acid sequence identity (in percent) are presented and calculated by FASTA analysis (Pearson & Lipman, 1988) for overlapping regions of homologous ORFs.

List of Abbreviations

3β-HSD, 3β-hydroxysteroid dehydrogenase/Δ4-Δ5 isomerase;
AHA, anti-hemagglutinin;
ATI, cytoplasmic A-type inclusions;
CAM, chorioallantoic membrane;
CDC, Centers for Disease Control and Prevention, USA;
CEV, cell-associated enveloped virus;
CMI, cell-mediated immunity;
CMLV, camelpox virus;
CNS, central nervous system;
CPE, cytopathic effect;
CPXV, cowpox virus;
CTL, cytotoxic T lymphocytes;
dNTP, deoxynucleotide triphosphate;
DRC, Democratic Republic of the Congo;
ECTV, ectromelia virus;
EEV, extracellular enveloped virus;
EIA, enzyme immunoassay–adsorption;
EIA–A; enzyme immunoassay–adsorption
ePCR, extended polymerase chain reaction;
FITC, fluorescein-5-isothiocyanate;
HA, hemagglutinin;
HAIT, hemagglutination inhibition test;
IEV, intracellular enveloped virus;
IFN, interferon;
IL, interleukin;
IMV, intracellular mature virus;
IV, immature virion;

IVN, nucleoid-containing immature virion;

LD, lethal dose;

MHC, major histocompatibility complex;

MOCV, Molluscum contagiosum virus;

MPCR, multiplex polymerase chain reaction;

MPXV, monkeypox virus;

NTPase, nucleoside triphosphatase;

ORF, open reading frame;

PCR, polymerase chain reaction;

PFU, pock-forming unit in the case of chick embryo CAMs and plaque-forming unit in the case of cell cultures;

RFLP, restriction fragment length polymorphism;

RIA, radioimmunoassay;

RIA-A, radioimmunoassay–adsorption;

SFV, Shope fibroma virus;

SRC VB Vector, State Research Center of Virology and Biotechnology Vector, Russia;

TCID, tissue cytopathic infectious dose;

TIR, terminal inverted repeat;

TNF, tumor necrosis factor;

TNFR, tumor necrosis factor receptor;

TRT, telomere resolution target;

VACV, vaccinia virus;

VARV, variola (smallpox) virus;

VCP, viral complement-binding protein;

VETF, viral early transcription factor;

VITF, viral intermediate transcription factor;

VLTF, viral late transcription factor;

WHO, World Health Organization

References

Abdullah N.A., Torres B.A., Basu M., Johnson H.M. Differential effects of epidermal growth factor, transforming growth factor-α, and vaccinia virus growth factor in the positive regulation of IFN-γ production. J Immunol 1989; 143:113-7.

Ada G.L., Blanden R.V. CTL immunity and cytokine regulation in viral infection. Res Immunol 1994; 145:625-8.

Adams J., Kelso R., Cooley L. The kelch repeat superfamily of proteins: propellers of cell function. Trends Cell Biol 2000; 10:17-24.

Aguado B., Selmes I.P., Smith G.L. Nucleotide sequence of 21.8 kbp of variola major virus strain Harvey and comparison with vaccinia virus. J Gen Virol 1992; 73:2887-2902.

Ahmad K.F., Engel C.K., Prive G.G. Crystal structure of the BTB domain from PZLF. Proc Natl Acad Sci USA 1998; 95:12123-8.

Ahn B.Y., Moss B. Capped poly(A) leaders of variable lengths at the 5' ends of vaccinia virus late mRNAs. J Virol 1989; 63:226-32.

Ahn B.Y., Moss B. Glutaredoxin homolog encoded by vaccinia virus is a virion-associated enzyme with thioltransferase and dehydroascorbate reductase activities. Proc Natl Acad Sci USA 1992a; 89:7060-4.

Ahn B.Y., Moss B. RNA polymerase-associated transcription specificity factor encoded by vaccinia virus. Proc Natl Acad Sci USA 1992b; 89:3536-40.

Ahn B.Y., Gershon P.D., Jones E.V., Moss B. Identification of *rpo*30, a vaccinia virus RNA polymerase gene with structural similarity to a eucaryotic transcription elongation factor. Mol Cell Biol 1990a; 10:5433-41.

Ahn B.Y., Jones E.V., Moss B. Identification of the vaccinia virus gene encoding an 18-kilodalton subunit of RNA polymerase and demonstration of a 5' poly(A) leader on its early transcript. J Virol 1990b; 64:3019-24.

Ahn B.Y., Rosel J., Cole N.B., Moss B. Identification and expression of *rpo*19, a vaccinia virus gene encoding a 19-kilodalton DNA-dependent RNA polymerase subunit. J Virol 1992; 66:971-82.

Ahn B.Y., Gershon P.D., Moss B. RNA polymerase-associated protein Rap94 confers promoter specificity for initiating transcription of vaccinia virus early stage genes. J Biol Chem 1994; 269:7552-7.

Akatova, Emma M. "Effects of Certain Physical and Chemical Factors and Antibiotics on Variola and Vaccinia Viruses." In *Trudy NIIVS im. I.I. Mechnikova*, Moscow, 1958.

Akatova-Shelukhina, Emma M., *Resistance of Variola, Alastrim, and Vaccinia Viruses to Several Physical and Chemical Factors,* Moscow: Candidate of Science Dissertation, 1962.

Albertini A.M., Hofer M., Calos M.P., Miller J.H. On the formation of spontaneous deletions: the importance of short sequence homologies in the generation of large deletions. Cell 1982; 29:319-28.

Alcami A. Viral mimicry of cytokines, chemokines and their receptors. Nat Rev Immunol 2003; 3:36-50.

Alcami A., Koszinowski U.H. Viral mechanisms of immune evasion. Immunol Today 2000; 21:447-55.

Alcami A., Smith G.L. A soluble receptor for interleukin-1β encoded by vaccinia virus: a novel mechanism of viral modulation of the host response to infection. Cell 1992; 71:153-67.

Alcami A., Smith G.L. Cytokine receptors encoded by poxviruses: a lesson in cytokine biology. Immunol Today 1995a; 16:474-8.

Alcami A., Smith G.L. Vaccinia, cowpox, and camelpox viruses encode soluble gamma interferon receptors with novel broad species specificity. J Virol 1995b; 69:4633-9.

Alcami A., Smith G.L. A mechanism for the inhibition of fever by a virus. Proc Natl Acad Sci USA 1996; 93:11029-34.

Alibek, Ken, *Biohazard.* New York: Random House, 1999.

Almazan F., Tscharke D.C., Smith G.L. The vaccinia virus superoxide dismutase-like protein (A45R) is a virion component that is nonessential for virus replication. J Virol 2001; 75:7018-29.

Al'tshtein A.D., Kirillova F.M., Bykovsky A.F. "Study of Properties of Variola and Vaccinia Virus Strains Isolated in Cell Culture." In *Smallpox,* S.S. Marennikova, ed. Moscow, 1961.

Amegadzie B.Y., Ahn B.-Y., Moss B. Identification, sequence, and expression of the gene encoding a Mr 35,000 subunit of the vaccinia virus DNA-dependent RNA polymerase. J Biol Chem 1991a; 266:13712-8.

Amegadzie, B.Y., Holmes, M.H., Cole, N.B., Jones, E.V., Earl, P.L., Moss, B. Identification, sequence, and expression of the gene encoding the second-largest subunit of the vaccinia virus DNA-dependent RNA polymerase. Virology 1991b; 180:88-98.

Amegadzie B.Y., Ahn B.-Y., Moss B. Characterization of a 7-kilodalton subunit of vaccinia virus DNA-dependent RNA polymerase with structural similarities to the smallest subunit of eukaryotic RNA polymerase II. J Virol 1992; 66:3003-10.

Anders W., Posch J. Die Pockenausbrücke 1961/1962 in Nordrhein-Westfalen. Bundesgesundheitblatt 1962; 17:265-9.

Anderson, I., *Study of Fever.* London: J. Churchill, 1861.

Anderson N., Doane F.W. Specific identification of enteroviruses by immunoelectron microscopy using a serum-in-agar diffusion method. Can J Microbiol 1973; 19:585-8.

Andres K.H., Lieske H., Lippelt H., Mannweiler E., Nielsen G., Peters D., Seelemann K. Variola. Dtsch med Wchschr 1958; 83:12-7.

Andres G., Garcia-Escudero R., Simon-Mateo C., Vinuela E. African swine fever virus is enveloped by a two-membraned collapsed cisterna derived from the endoplasmic reticulum. J Virol 1998; 72:8988-9001.

Andzhaparidze O.G., Unanov S.S., Chernos V.I., Strelkov V.V., Antonova T.P. Moscow Institute of Viral Preparations tissue culture smallpox vaccine in coded control experiments on adult revaccination by scarification. Vopr Virusol 1980; (4):443-6.

Antoine G., Scheiflinger F., Dorner F., Falkner F.G. The complete genomic sequence of the modified vaccinia Ankara strain: comparison with other orthopoxviruses. Virology 1998; 244:365-96.

Appleyard G., Hapel A.J., Boulter E.A. An antigenic difference between intracellular and extracellular rabbitpox virus. J Gen Virol 1971; 13:9-17.

Archard L.C., Mackett M. Restriction endonuclease analysis of red cowpox virus and its white pock variant. J Gen Virol 1979; 45:51-63.

Archard L.C., Mackett M., Barnes D.E., Dumbell K.R. The genome structure of cowpox virus white pock variants. J Gen Virol 1984; 65:775-886.

Arens M. Methods for subtyping and molecular comparison of human viral genomes. Clin Microbiol Rev 1999; 12:612-26.

Arita, Isao, "The Control of Vaccine Quality in the Smallpox Eradication Programme." In *Proceedings of the 37th International Symposium on Smallpox Vaccine Organized by the International Association of Biological Standardization; 1972 October 11–13.* H. Cohen, H. Regamey, eds. Basel, New York: S. Karger, 1973.

Arita I., Henderson D.A. Smallpox and monkeypox in non-human primates. Bull WHO 1968; 39:277-83.

Arita I., Gispen K., Kalter S.S., Lim Teong Wan, Marennikova S.S., Netter R., Tagaya I. Outbreaks of monkeypox and serological surveys in non-human primates. Bull WHO 1972; 46:625-31.

Aso T., Okamura S., Matsuguchi T., Sakamoto N., Sata T., Niho Y. Genomic organization of the alpha chain of the human C4b-binding protein gene. Biochem Biophys Res Commun 1991; 174:222-7.

Avakyan, Arshaluis A., Bykovsky, Al'bert F. *Atlas of Anatomy and Ontogenesis of Human and Animal Viruses.* Moscow: Meditsina, 1970.

Avakyan A.A., Al'tshtein A.D., Kirillova F.M., Bykovsky A.F. The routes for improvement of smallpox laboratory diagnostics. Vopr Virusol. 1961; (2):196-203.

Ayares D., Chekuri L., Song K.-Y., Kucherlapati R. Sequence homology requirements for intermolecular recombination in mammalian cells. Proc Natl Acad Sci USA 1986; 83:5199-203.

Baek S.H., Kwak J.Y., Lee S.H., Lee T., Ryu S.H., Uhlinger D.J. Lipase activities of p37, the major envelope protein of vaccinia virus. J Biol Chem 1997; 272:32042-9.

Baker R.O., Bray M., Huggins J.W. Potential antiviral therapeutics for smallpox, monkeypox and other orthopoxvirus infections. Antiviral Res 2003; 57:13-23.

Baldick C.J., Moss B. Characterization and temporal regulation of mRNAs encoded by vaccinia virus intermediate stage genes. J Virol 1993; 67:3515-27.

Baldick C.L., Keck J.G., Moss B. Mutational analysis of the core, spacer, and initiator region of vaccinia virus intermediate-class promoters. J Virol 1992; 66:4710-9.

Ball L.A. High-frequency homologous recombination in vaccinia virus DNA. J Virol 1987; 61:1788-95.

Baltazard M., Boue A., Siadat H. Etude de comportement du virus de la variole en cultures de tissus. Ann Inst Pasteur 1958; 94:560-70.

Banham A.H., Smith G.L. Vaccinia virus gene B1R encodes a 34-kDa serine/threonine protein kinase that localizes in cytoplasmic factories and is packaged into virions. Virology 1992; 191:803-12.

Banham A.H., Smith G.L. Characterization of vaccinia virus gene B12R. J Gen Virol 1993; 74:2807-12.

Baran V.M., Zubritsky P.K. Human infection with cowpox. Zdravookhranenie (Minsk) 1997; (3):40.

Barbosa E., Moss B. mRNA(nucleoside-2'-)-methyltransferase from vaccinia virus. Characteristics and substrate specificity. J Biol Chem 1978a; 253:7698-702.

Barbosa E., Moss B. mRNA(nucleoside-2'-)-methyltransferase from vaccinia virus. Purification and physical properties. J Biol Chem 1978b; 253:7692-7.

Baroudy B.M., Moss B. Purification and characterization of a DNA-dependent RNA polymerase from vaccinia virions. J Biol Chem 1980; 255:4372-80.

Baroudy B.M., Venkatesan S., Moss B. Incompletely base-paired flip-flop terminal loops link the two DNA strands of the vaccinia virus genome into one uninterrupted polynucleotide chain. Cell 1982; 28:315-24.

Baroudy B.M., Venkatesan S., Moss B. Structure and replication of vaccinia virus telomeres. Cold Spring Harbor Symp Quant Biol 1983; 47:723-9.

Barry M., Bleackley R.C. Cytotoxic T lymphocytes: All roads lead to death. Nat Rev Immunol 2002; 2:401-9.

Baxby D. Laboratory characteristics of British and Dutch strains of cowpox virus. Zbl Vet Med 1975; 22:480-7.

Baxby D. Is cowpox misnamed? A review of 10 human cases. Brit Med J 1977; 1:1379-81.

Baxby, Derrick, *Jenner's Smallpox Vaccine.* London: Heinemann Educational Books, 1981.

Baxby D., Bennett M. Low risk from feline cowpox. Lancet 1990; 336:1070-1.

Baxby D., Ghaboosi B. Laboratory characteristics of poxviruses isolated from captive elephants in Germany. J Gen Virol 1977; 37:407-14.

Baxby D., Hill B.J. Buffalopox virus. Vet Rec 1969; 85:315-9.

Baxby D., Hill B.J. Characteristics of a new poxvirus isolated from Indian buffaloes. Arch ges Virusforsch 1971; 35:70-9.

Baxby D., Osborne A.D. Antibody studies in natural bovine cowpox. J Hyg Camb 1979; 83:425-8.

Baxby D., Rondle C.J.M. The relative sensitivity of chick and rabbits tissues for the titration of vaccinia and cowpox viruses. Arch ges Virusforsch 1967; 20:263-7.

Baxby D., Hessami M., Ghaboosi B., Ramyar H. Response of camels to intradermal inoculation with smallpox and camelpox viruses. Infect Immun 1975; 11:617-21.

Baxby D., Shackleton W.B., Wheller J., Turner A. Comparison of cowpox-like viruses isolated from European zoos. Arch Virology 1979; 61:337-40.

Baxby D., Ashton D.G., Jones D.M., Thomsett L.K. An outbreak of cowpox in captive cheetahs: virological and epidemiological studies. J Hyg 1982; 89:365-72.

Bayliss C.D., Smith G.L. Vaccinia virion protein I8R has both DNA and RNA helicase activities: implications for vaccinia virus transcription. J Virol 1996; 70:794-800.

Bayliss C.D., Wilcock D., Smith G.L. Stimulation of vaccinia virion DNA helicase I8R, but not A18R, by a vaccinia core protein L4R, an ssDNA binding protein. J Gen Virol 1996; 77:2827-31.

Beattie E., Tartaglia J., Paoletti E. Vaccinia virus encoded eIF-2α homolog abrogates the antiviral effect of interferon. Virology 1991; 183:419-22.

Bedson H.S., Duckworth M.J. Rabbitpox: an experimental study of the pathways of infection in rabbits. J Path Bacteriol 1963; 85:1-20.

Bedson H.S., Dumbell K.R. The effect of temperature on the growth of poxviruses in the chick embryo. J Hyg 1961; 59:457-69.

Bedson H.S., Dumbell K.R. Hybrids derived from the viruses of variola major and cowpox. J Hyg 1964; 62:147-58.

Bedson H.S., Dumbell K.R., Thomas W.R.S. Variola in Tanganyika. Lancet 1963; II:1085-8.

Bej A.K., Mahbubani M.H., Atlas R.M. Amplification of nucleic acids by polymerase chain reaction (PCR) and other methods and their applications. Crit Rev Bioch Mol Biol 1991; 26:301-34.

Bektemirov T.A., Shenkman L.S., Marennikova S.S. Interferon-inducing ability of vaccinia virus strains differing in pathogenicity. Vopr Virusol 1971; (5):555-60.

Belanov E.V., Gus'kov A.A., Sokunova E.B., Marennikova S.S., Repin V.E., Sandakhchiev L.S. Survivability of variola virus in scabs of patients. Dokl Ross Akad Sci 1997; 354:832-4.

Belle Isle H., Venkatesan S., Moss B. Cell-free translation of early and late mRNAs selected by hybridization to cloned DNA fragments derived from the left 14 million to 72 million daltons of the vaccinia virus genome. Virology 1981; 112:306-17.

Benenson A.S. Immediate (so-called "immune") reaction to smallpox vaccination. J Amer Med Assoc 1950; 143:1238-49.

Bennett M., Gaskell C.J., Gaskel R.M., Baxby D., Gruffydd-Jones T.J. Poxvirus infection in the domestic cat: some clinical and epidemiological observations. Vet Rec 1986; 118:387-90.

Bennett M., Gaskell C.J., Baxby D., Gaskell R.M., Kelly D.F., Naidoo J. Feline cowpox virus infection. J Small Animal Practice 1990; 31:167-73.

Bennett M., Crouch A.J., Begon M., Duffy B., Feore S., Gaskell R.M., Kelly D.F., McCracken C.M., Vicary L., Baxby D. Cowpox in British voles and mice. J Comp Path 1997; 116:35-44.

Benzer S. 1961. On the topography of the genetic fine structure. Proc Natl Acad Sci USA 1997; 47:403-7.

Berche P. The threat of smallpox and bioterrorism. Trends Microbiol 2001; 9:15-8.

Berger, K., Heinrich, W. "Decrease of Post-Vaccinal Deaths in Austria after Introducing a Less Pathogenic Virus Strain." In *Proceedings of the 37th International Symposium on Smallpox Vaccine Organized by the International Association of Biological Standardization; 1972 October 11–13*. H. Cohen, H. Regamey, eds. Basel, New-York: S. Karger, 1973.

Berger K., Putingam F. Experimentelle Kuhpockeninfection beim Rind. Zbl Bakt 1958; 172:363-9.

Betakova T., Wolffe E.J., Moss B. Regulation of vaccinia virus morphogenesis: phosphorylation of the A14L and A17L membrane proteins and C-terminal truncation of the A17L protein are dependent on the F10L kinase. J Virol 1999; 73:3534-43.

Betakova T., Wolffe E.J., Moss B. The vaccinia virus A14.5L gene encodes a hydrophobic 53-amino-acid virion membrane protein that enhances virulence in mice and is conserved among vertebrate poxviruses. J Virol 2000; 74:4085-92.

Beutler B., van Huffel C. An evolutionary and functional approach to the TNF receptor/ligand family. Ann NY Acad Sci 1994; 730:118-33.

Birnbaum M.J., Clem R.J., Miller L.K. An apoptosis-inhibiting gene from a nuclear polyhedrosis virus encoding a polypeptide with Cys/His sequence motifs. J Virol 1994; 68:2521-8.

Björnberg R., Björnberg A. Cowpox with a secondary rash. Svenska Läkartidningen 1956; 53:655.

Black E.P., Condit R.C. Phenotypic characterization of mutants in vaccinia virus gene G2R, a putative transcription elongation factor. J Virol 1996; 7:47-54.

Black M.E., Hruby D.E. Quaternary structure of vaccinia virus thymidine kinase. Biochem Biophys Res Commun 1990; 169:1080-6.

Blackford S., Roberts D.S., Thomas P.D. Cowpox infection causing a generalized eruption in a patient with atopic dermatitis. Br J Dermatol 1993; 129:628-9.

Blasco R., Moss B. Extracellular vaccinia virus formation and cell-to-cell virus transmission are prevented by deletion of the gene encoding the 37,000-dalton outer envelope protein. J Virol 1991; 65:5910-20.

Blasco R., Moss B. Role of cell-associated enveloped vaccinia virus in cell-to-cell spread. J Virol 1992; 66:4170-9.

Blasco R., Cole N.B., Moss B. Sequence analysis, expression, and deletion of a vaccinia virus gene encoding a homolog of profilin, a eukaryotic actin-binding protein. J Virol 1991; 65;4598-608.

Blasco R., Sisler J.R., Moss B. Dissociation of progeny vaccinia virus from the cell membrane is regulated by a viral envelope glycoprotein: effect of a point mutation in the lectin homology domain of the A34R gene. J Virol 1993; 67:3319-25.

Bleyer J.G. Ueber Auftreten von Variola unter Affen der Genera *Mycetes* und *Cebus* bei Vordringen einer Pockenepidemie im Urwaldgebiete an den Nebenflüssen des Alto Urüguay in Südbrasilien. Münch Med Wchschr 1922, 69:1009-10.

Blinov V.M., Totemin A.V., Resenchuck S.M., Olenina L.V., Chizhikov V.E., Kolykhalov A.A., Frolov I.V., Gutorov V.V., Pozdnyakov S.G., Krasnykh V.N., Serpinsky O.I., Sandakhchiev L.S., Shchelkunov S.N. Study of the structure–function organization of variola virus genome. IV. Determination and analysis of nucleotide sequence of the right genomic terminus of India-1967 strain. Mol Biol (Mosk) 1995; 29:772-89.

Blomquist M.C., Hunt L.T., Barker W.C. Vaccinia virus 19-kilodalton protein: relationship to several mammalian proteins, including two growth factors. Proc Natl Acad Sci USA 1984; 81:7363-7.

Bomhard D.V., Mahnel H., Ballauf B. Zwei Fälle von Pockeninfektionen bei Katzen. Kleintierpraxis 1989; 34:157-60.

Bomhard D., Pfleghaar S., Mahnel H., Schneekloth-Dücker J. Fallbericht: Katzenpocken-infektion als Zoonose für Hund und Mensch. Kleintierpraxis 1991; 36:511-4.

Bondarev V.N., Voitinsky E.Z. *Prophylaxis and Treatment of Vaccinal Complications in Children.* Leningrad: Meditsina, 1972.

Bork P., Doolittle R.F. Drosophila kelch motif is derived from a common enzyme fold. J Mol Biol 1994; 236:1277-82.

Born T.L., Morrison L.A., Esteban D.J., Van den Bos T., Thebeau L.G., Chen N., Spriggs M.K., Sims J.E., Buller R.M.L. A poxvirus protein that binds to and inactivates IL-18, and inhibits NK cell response. J Immunol 2000; 164:3246-54.

Bossart W., Nuss D.L., Paoletti E. Effect of UV irradiation on the expression of vaccinia virus gene products synthesized in a cell-free system coupling transcription and translation. J Virol 1978; 26:673-80.

Boue A., Baltazard M. Culture du virus de la variole sur cultures de tissues. C R Acad Sci 1956; 243:1176-8.

Boulanger D., Brochier B., Crouch A., Bennett M., Gaskell R., Pastoret P. Comparison of the susceptibilities of the red fox (*Vulpes vulpes*) to a vaccinia–rabies recombinant virus and to cowpox virus. Vaccine 1995; 13:215-9.

Boulter E.A. The titration of vaccinal neutralizing antibody on chorioallantoic membranes. J Hyg 1957; 55:502-12.

Boulter E.A., Appleyard G. Differences between extracellular and intracellular forms of poxvirus and their implications. Prog Med Virol 1973; 16:86-108.

Bourke A.T.C., Dumbell K.R. An unusual poxvirus from Nigeria. Bull WHO 1972; 46:621-3.

Braginskaya V.P., Gurvich E.B., Ozeretskovsky N.A. "Specific Antismallpox Immunoglobulin in the Prevention and Treatment of Postvaccination Encephalitis in Children." In *Standards, Strains, and Methods for Control of Bacterial and Viral Preparations.* Moscow, vol. 4, 1978.

Brandt T.A., Jacobs B.L. Both carboxy- and amino-terminal domains of the vaccinia virus interferon resistance gene, E3L, are required for pathogenesis in a mouse model. J Virol 2001; 75:850-6.

Bras G. The morbid anatomy of smallpox, Documenta Med Geogr Trop 1952; 4:303-51.

Bray M., Buller M. Looking back at smallpox. Clin Infect Dis 2004; 38: 882-9.

Bray M., Martinez M., Smec D.F., Kefauver D., Thompson E., Huggins J.W. Cidofovir protects mice against lethal aerosol or intranasal cowpox virus challenge. J Infect Dis 2000; 181:10-9.

Bray M., Martinez M., Kefauver D., West M., Roy C. Treatment of aerosolized cowpox virus infection with aerosolized cidofovir. Antiviral Res 2002; 54:129-42.

Breman, Joel G. "Monkeypox: an Emerging Infection of Humans?" In *Emerging Infections,* W.M. Scheid, W.A. Craig, J.M. Hughes, eds. Washington, DC: ASM Press, 2000.

Breman J.G., Henderson D.A. Diagnosis and management of smallpox. N Eng J Med 2002; 346:1300-8.

Breman J.G., Bernadon J., Nakano J.H. Poxvirus in West African non-human primates: serological survey results. Bull WHO 1977; 55:605-12.

Brenner S., Horne R.W. A negative staining method for high resolution electron microscopy of viruses. Biochim Biophys Acta 1959; 34:103-10.

Brick D.J., Burke R.D., Minkley A.A., Upton C. Ectromelia virus virulence factor p28 acts upstream of caspase-3 in response to UV light-induced apoptosis. J Gen Virol 2000; 81:1087-97.

Brinckerhoff W.R., Tyzzer E.E. Studies upon experimental variola and vaccinia in *Quadrumana*. J Med Res 1906; 14:213-359.

Brown A., Elsner V., Officer J.E. Growth and passage of variola virus in mouse brain. Proc Soc Exp Biol Med 1960; 104:605-8.

Brown A., Bennett M., Gaskell C.J. Fatal poxvirus infection in association with FIV infection. Vet Rec 1989; 124:19-20.

Brown C.K., Turner P.C., Moyer R.W. Molecular characterization of the vaccinia virus hemagglutinin gene. J Virol 1991; 65:3598-606.

Brown J.P., Twardzik D.R., Marquardt H., Todaro G.J. Vaccinia virus encodes a polypeptide homologous to epidermal growth factor and transforming growth factor. Nature 1985; 313:491-2.

Brown S.D., Piechaczyk M. Insertion sequences and tandem repetitions as sources of variation in a dispersed repeat family. J Mol Biol 1983; 165:249-56.

Broyles S.S. A role for ATP hydrolysis in vaccinia virus early gene transcription. Dissociation of the early transcription factor–promoter complex. J Biol Chem 1991; 266:15545-8.

Broyles S.S. Vaccinia virus encodes a functional dUPTase. Virology 1993; 195:863-5.

Broyles S.S. Vaccinia virus transcription. J Gen Virol On-line 2003.

Broyles S.S., Fesler B.S. Vaccinia virus gene encoding a component of the viral early transcription factor. J Virol 1990; 64:1523-9.

Broyles S.S., Moss B. Homology between RNA polymerase of poxviruses, prokaryotes, and eukaryotes: Nucleotide sequence and transcriptional analysis of vaccinia virus genes encoding 147-kDa and 22-kDa subunits. Proc Natl Acad Sci USA 1986; 83:3141-5.

Broyles S.S., Moss B. Identification of the vaccinia virus gene encoding nucleoside triphosphate phosphohydrolase I, a DNA-dependent ATPase. J Virol 1987; 61:1738-42.

Broyles S.S., Pennington M.J. Vaccinia virus gene encoding a 30-kilodalton subunit of the viral DNA-dependent RNA polymerase. J Virol 1990; 64:5376-82.

Broyles S.S., Li J., Moss B. Promoter DNA contacts made by the vaccinia virus early transcription factor. J Biol Chem 1991; 266:15539-44.

Broyles S.S., Liu X., Zhu M., Kremer, M. Transcription factor YY1 is a vaccinia virus late promoter activator. J Biol Chem 1999; 274:35662-7.

Bryant NA, Smith VP, Saraiva M, Rivera J, Shair K, Alcami A. Characterisation of vaccinia virus strains isolated from mice. Proceedings of the XIIIth International Poxvirus and Iridovirus Symposium; 2000 September 2–6, Montpellier, France.

Buddingh G.J. Infection of the chorioallantois of the chick embryo as a diagnostic test for variola. Amer J Hyg 1938; 28:130-7.

Bugert J.J., Darai G. Poxvirus homologues of cellular genes. Virus Genes 2000; 21:111-33.

Buist J.B. The life-history of the microorganisms associated with variola and vaccinia. Proc R Soc Edinburgh 1886; 13:603.

Buller R.M.L., Palumbo G.J. Poxvirus pathogenesis. Microbiol Rev 1991; 55:80-122.

Buller R.M.L., Chakrabarti S., Cooper J.A., Twardizik D.R., Moss B. Cell proliferative response to vaccinia virus is mediated by VGF. Virology 1988a; 164:182-92.

Buller R.M.L, Chakrabarti S., Cooper J.A., Twardizik D.R., Moss B. Deletion of the vaccinia virus growth factor gene reduces virus virulence. J Virol 1988b; 62:866-74.

Burgasov, Petr N., Nikolaevsky, Georgy P. *Smallpox*. Moscow: Meditsina, 1972.

Byrd C.M., Bolken T.C., Hruby D.E. The vaccinia virus I7L gene product is the core protein proteinase. J Virol 2002; 76:8973-6.

Byrd C.M., Bolken T.C., Hruby D.E. Molecular dissection of the vaccinia virus I7L core protein proteinase. J Virol 2003; 77:11279-83.

Cacoullos N., Bablanian R. Role of polyadenylated RNA sequences (POLADS) in vaccinia virus infection: correlation between accumulation of POLADS and extent of shut-off in infected cells. Cell Mol Biol Res 1993; 39:657-64.

Cairns J. The initiation of vaccinia infection. Virology 1960; 11:603-23.

Calderara S., Xiang Y., Moss B. Orthopoxvirus IL-18 binding proteins: affinities and antagonistic activities. Virology 2001; 279:619-23.

Calmette A., Guérin J.C., Recherches sur la vaccine expérimentale. Ann Inst Pasteur 1901; 15:161-8.

Calos M.P., Galas D., Miller J.H. Genetic studies of the lac repressor. VIII. DNA sequence change results from intragenic duplication. J Mol Biol 1978; 126:865-9.

Cao J., Koop B.F., Upton C. A human homolog of the vaccinia virus *Hind*III K4L gene is a member of the phospholipase D superfamily. Virus Res 1997; 48:11-8.

Carmichael A.G., Silverstein A.M. Smallpox in Europe before the seventeenth century: Virulent killer or benign disease? J Hist Med Allied Sci 1987; 42:147-68.

Carpenter M.S., DeLange A.M. A temperature-sensitive lesion in the small subunit of the vaccinia virus-encoded mRNA capping enzyme causes a defect in viral telomere resolution. J Virol 1991; 65:4042-50.

Carra L, Dumbell K.R. Characterization of poxviruses from sporadic human infections. S Afr Med J 1987; 72:846-51.

Cassel W.A., Fater B. Vaccinia virus haemagglutinin inhibitor in ascitis fluids. Virology 1959 7:467.

Cassel W.A., Garrett R.E., Blair W.L. Haemagglutinin loss by vaccinia virus and its relationship to virus adaptation in tumor and L cells. Proc Soc Exp Biol Med 1962; 109:396-8.

Cassetti M.A., Moss B. Interaction of the 82-kDa subunit of the vaccinia virus early transcription factor heterodimer with the promoter core sequence directs downstream DNA binding of the 70-kDa subunit. Proc Natl Acad Sci USA 1996; 93:7540-5.

Cassetti M.C., Merchlinsky M., Wolffe E.J., Weisberg A.S., Moss B. DNA packaging mutant: repression of the vaccinia virus A32 gene results in noninfectious, DNA-deficient, spherical, enveloped particles. J Virol 1998; 72:5769-80.

CDC. Update: multistate outbreak of monkeypox in Illinois, Indiana, Kansas, Missouri, Ohio, and Wisconsin, 2003. MMWR Morb Mortal Wkly Rep 2003; 52:642-6.

Chakrabarti S., Brechling K., Moss B. Vaccinia virus expression vector: coexpression of beta-galactosidase provides visual screening of recombinant virus plaques. Mol Cell Biol 1985 5:3403-9.

Chang W., Lim J.G. Hellstrom J., Gentry L.E. Characterization of vaccinia virus growth factor biosynthetic pathway with an antipeptide antiserum. J Virol 1988; 62:1080-3.

Chang H.W., Watson J.C., Jacobs B.L. The E3L gene of vaccinia virus encodes an inhibitor of the interferon-induced, double-stranded RNA-dependent protein kinase. Proc Natl Acad Sci USA 1992; 89:4825-9.

Chantrey J., Meyer H., Baxby D., Begon M., Bown K.J., Hazel S.M., Jones T., Montgomery W.J., Bennett M. Cowpox: reservoir hosts and geographic range. Epidemiol Infect 1999; 122:455-60.

Chapin C.V. Variation in type of infectious disease as shown by the history of smallpox in the United States 1895–1912. J Inf Dis 1913; 13:171-96.

Chapin C.V., Smith J. Permanency of the mild type of smallpox. J Prevent Med 1932; 6:273-320.

Chen W., Drillien R., Spehner D., Buller R.M. Restricted replication of ectromelia virus in cell culture correlates with mutations in virus-encoded host range gene. Virology 1992; 187:433-42.

Chen N, Li G, Feng Z, Buller RML., Buck C, Wang C, Lefkowitz EJ, Jahrling PB, Ropper RL, Upton C. Comparison of the genomic sequences of West African and Democratic Republic of the Congo isolates of monkeypox. Proceedings of the XV International Poxvirus and Iridovirus Conference; 2004 3–8 September; Keble College, Oxford, England.

Cheng C., Kussie P., Pavletich N., Shuman S. Conservation of structure and mechanism between eukaryotic topoisomerase I and site-specific recombinases. Cell 1998; 92:841-50.

Chernos V.I., Unanov S.S., Antonova T.P., Nemtsov I.M. Tissue smallpox vaccine. Characterization of the main properties. Vopr Virusol 1977; (2):71-5.

Chertov O.Yu., Telezhinskaya I.N., Ziatseva E.V., Golubeva T.B., Zinov'ev V.V., Ovechkina L.G., Mazkova L.B., Malygin E.G. Amino acid sequence determination of vaccinia virus immunodominant protein p34 and identification of the gene. Biomed Sci 1991; 2:151-4.

Chimishkyan, Kornely L., *Comparative Study of Antigenic and Immunogenic Activities of Vaccinia Virus Strain EM-63 and Its Stability.* Moscow: Candidate of Science Dissertation, 1971.

Chiu W.L., Chang W. Vaccinia virus J1R protein: a viral membrane protein that is essential for virion morphogenesis. J Virol 2002; 76:9575-87.

Cho C.T., Wenner H.A. Monkeypox virus. Bact Reviews 1973; 37:1-18.

Cho C.T., Bolano C.R., Kamitsuka P.S., Wenner H.A. Methisazone and monkeypox virus: studies in cell cultures, chick embryos, mice and monkey. Amer J Epidemiol 1970; 92:137-44.

Cho C.T., Liu C, Wenner H.A. Monkeypox virus: responses to infection in chick embryos. Proc Soc Exp Biol Med 1972; 139:1206-12.

Christen L.M., Sanders M., Wiler C., Niles E.G. Vaccinia virus nucleotide triphosphate phosphohydrolase I is an essential viral early gene transcription termination factor. Virology 1998; 245:360-71.

Chung C.S., Hsiao J.C., Chang Y.S., Chang W. A27L protein mediates vaccinia virus interaction with cell surface heparan sulfate. J Virol 1998; 72:1577-85.

Clark A.G. Invasion and maintenance of a gene duplication. Proc Natl Acad Sci USA 1994; 91:2950-4.

Clark E., Nagler F.P.O. Haemagglutination by viruses: range of susceptible cells with special reference to agglutination by vaccinia virus. Austral J Exp Biol Med Sci 1943; 21:103-6.

Cohen J. Is an old virus up to new tricks? Science 1997; 277:312-3.

Cohen J.J., Duke R.C., Fadok V.A., Sellins K.S. Apoptosis and programmed cell death in immunity. Annu Rev Immunol 1992; 10:267-93.

Colinas R.J., Condit R.C., Paoletti E. Extrachromosomal recombination in vaccinia-infected cells requires a functional DNA polymerase participating at a level other than DNA replication. Virus Res 1990; 18:49-70.

Collier L.H. The development of a stable smallpox vaccine. J Hyg 1955; 53:76-101.

Collier W.A. Untersuchungen über Pocken-Hämagglutinine und Anti-Hämagglutimne. Zbl Bakt Abt I Orig 1951; 157:119-23.

Collier W.A., Schonfeld I.K. Studies on the serological diagnosis of smallpox. Med J Australia 1950; 2:363-6.

Comeau M.R., Johnson R., DuBose R.F., Petersen M., Gearing P., Van den Bos T., Park L., Farrah T., Buller R.M., Cohen J.I., Strockbine L.D., Rauch C., Spriggs M.K. A poxvirus-encoded semaphorin induces cytokine production from monocytes and binds to a novel cellular semaphorin receptor, VESPR. Immunity 1998; 8:473-82.

Condit R.C., Niles E.G. Orthopoxvirus genetics. Curr Top Microbiol Immunol 1990; 163:1-39.

Condit R.C., Niles E.G. Regulation of viral transcription elongation and termination during vaccinia virus infection. Biochim Biophys Acta 2002; 1577:325-36.

Condit R.C., Lewis J.I., Quinn M., Christen L.M., Niles E.G. Use of lysolecithin-permeabilized infected-cell extracts to investigate the *in vitro* biochemical phenotypes of poxvirus *ts* mutations altered in viral transcription activity. Virology 1996; 218:169-80.

Cooper J.A., Wittek R., Moss B. Hybridization selection and cell free translation of mRNA encoded within the inverted terminal repetition of the vaccinia virus genome. J Virol 1981a; 37:284-94.

Cooper J.A., Wittek R., Moss B. Extension of the transcriptional and translational map of the left end of the vaccinia virus genome to 21 kilobase pairs. J Virol 1981b; 39:733-45.

Copeman S.M. Variola and vaccinia, their manifestations and interrelations in the lower animals: a comparative study. J Path Bact 1894; 2:407-27.

Councilman, William T., *Smallpox*. Osler, McRae, eds. Modern Medicine, 1907.

Craigie J., Wishart F.O. Studies on the soluble precipitating substances of vaccinia. I. The dissociation *in vitro* of soluble precipitable substances from elementary bodies of vaccinia. J Exp Med 1936a; 64:803-18.

Craigie J., Wishart F.O. Studies on the soluble precipitating substances of vaccinia. II. The soluble precipitable substances of dermal filtrate. J Exp Med 1936b; 64:819-30.

Crook N.E., Clem R.J., Miller L.K. An apoptosis-inhibiting gene with a zinc finger-like motif. J Virol 1993; 67:2169-74.

Cross R.M., Kaplan C., McClean D. The heat resistance of dried smallpox vaccine. Lancet 1957; I:446-8.

Crouch A.C., Baxby D., McCracken C.M., Gaskell R.M., Bennett M. Serological evidence for the reservoir hosts of cowpox virus in British wildlife. Epidemiol Infect 1995; 115:185-91.

Cruickshank J.G., Bedson H.S., Watson D.H. Electron microscopy in the rapid diagnosis of smallpox. Lancet 1966; II:527-30.

Cudmore S., Reckmann I., Griffiths G., Way M. Vaccinia virus: a model system for actin–membrane interactions. J Cell Sci 1996; 109:1739-47.

Cywicki J., Michowicz S. Przypadek ospy krowiej u dojarki. Przeglad Epidemiologiczny. Rok XXII 1968; No 1:139-40.

Czerny C.P., Eis-Hübinger A.M., Mayr A., Schneweis K.E., Pfeiff B. Animal poxviruses transmitted from cat to man: current event with lethal end. J Vet Med 1991; B38:421-31.

Czerny CP, Henning K, Müler T, Kramer M, Meyer H. Orthopoxvirus in the environment: a seroepidemiological survey of cats and red foxes (*Vulpes vulpes*) in Germany. Proceedings of the 10th Poxvirus and Iridovirus International Meeting; 1994; Banff, Canada.

Da Fonseca F.G., Lanna M.C., Campos M.A, Kitajima E.W., Peres J.N., Golgher R.R., Ferreira P.C., Kroon E.G. Morphological and molecular characterization of the poxvirus BeAn 58058. Arch Virol 1998;143:1171-86.

Da Fonseca F.G., Weisberg A., Wolffe E.J., Moss B. Characterization of the vaccinia virus H3L envelope protein: topology and post-translational membrane insertion via the *C*-terminal hydrophobic tail. J Virol 2000; 74:7508-17.

Da Fonseca F.G., Trindade G.S., Silva R.L., Bonjardim C.A., Ferreira P.C., Kroon E.G. Characterization of a vaccinia-like virus isolated in a Brazilian forest. J Gen Virol 2002; 83:223-8.

Dales S., Mosbach E.H. Vaccinia as a model for membrane biogenesis. Virology 1968; 35:564-83.

Dales S., Pogo B.G. Biology of poxviruses. Virol Monogr 1981; 18:1-109.

Dales S., Siminovitch L. The development of vaccinia virus in Earle's L strain cells as examined by electron microscopy. J Biophys Biochem Cytology 1961; 10:475-503.

Dalton D.K., Pitts-Meek S., Keshav S., Figari I.S., Bradley A., Stewart T.A. Multiple defects of immune cell function in mice with disrupted interferon-gamma genes. Science 1993; 259:1739-42.

Damaso C.R., Esposito J.J., Condit R.C., Moussatche N. An emergent poxvirus from humans and cattle in Rio de Janeiro State: Cantagalo virus may derive from Brazilian smallpox vaccine. Virology 2000; 277:439-49.

Darrasse H., Bres P., Camain R. Note sur le diagnostic de la variola au laboratoire. Ann Inst Pasteur 1958; 94:577-82.

Davies M.V., Elroy-Stein O., Jagus R., Moss B., Kaufman R.J. The vaccinia virus K3L gene product potentiates translation by inhibiting double-stranded-RNA-activated protein kinase and phosphorylation of the alpha subunit of eukaryotic initiation factor 2. J Virol 1992; 66:1943-50.

Davies M.V., Chang H.W., Jacobs B.L., Kaufman R.J. The E3L and K3L vaccinia virus gene products stimulate translation through inhibition of the double-stranded RNA-dependent protein kinase by different mechanisms. J Virol 1993; 67:1688-92.

Davis R.E., Mathews C.K. Acidic C-terminus of vaccinia virus DNA-binding protein interacts with ribonucleotide reductase. Proc Natl Acad Sci USA 1993; 90:745-9.

Davison A.J., Moss B. The structure of vaccinia virus early promoters. J Mol Biol 1989a; 210:749-69.

Davison A.J., Moss B. The structure of vaccinia virus late promoters. J Mol Biol 1989b; 210:771-84.

De Jong M. The alastrim epidemic in the Hague, 1953–1954. Docum Med Geogr Trop 1956; 8:207-36.

De Korte W.E. Amaas, or kaffir milk-pox. Lancet 1904; I:1273-6.

DeCarlos A., Paez E. Isolation and characterization of mutants of vaccinia virus with a modified 94 kDa inclusion protein. Virology 1991; 185:768-78.

Defries R.D., McKinnon N.E. The laboratory diagnosis of smallpox virus utilizing the rabbit. Amer J Hyg 1928; 8:107-24.

Dekking P., Rao A.R., St. Vincent L., Kempe C.H. The weeping mother, an unusual source of variola virus. Arch ges Virusforsch 1967; 22:215-8.

Delange A.M., Reddy M., Scraba D., Upton C., McFadden G. Replication and resolution of cloned poxvirus telomeres *in vivo* generates linear minichromosomes with intact viral hairpin termini. J Virol 1986; 59:249-59.

Demasi J., Traktman P. Clustered charge-to-alanine mutagenesis of the vaccinia virus H5 gene: isolation of a dominant, temperature-sensitive mutant with a profound defect in morphogenesis. J Virol 2000; 74:2393-405.

DeMasi J., Du S., Lennon D., Traktman P. Vaccinia virus telomeres: interaction with the viral I1, I6, and K4 proteins. J Virol 2001; 75:10090-105.

Demkowicz W.E., Ennis P.A. Vaccinia virus-specific $CD8^+$ cytotoxic T lymphocytes in humans. J Virol 1993; 67:1538-44.

Deng L., Shuman S. Vaccinia NPH-I, a DExH-box ATPase, is the energy coupling factor for mRNA transcription termination. Genes Dev 1998; 12:538-46.

Derrien M., Punjabi A., Khanna M., Grubisha O., Traktman P. Tyrosine phosphorylation of A17 during vaccinia virus infection: involvement of the H1 phosphatase and the F10 kinase. J Virol 1999; 73:7287-96.

Dick, G. "Complications of Smallpox Vaccination in the United Kingdom." In *Proceedings of the 37th International Symposium on Smallpox Vaccine Organized by the International Association of Biological Standardization; 1972 October 11–13*. H. Cohen, H. Regamey, eds. Basel, New-York: S. Karger, 1973.

Dinger J.E. Difference in persistence of smallpox and alastrim virus on the chorioallantoic. Docum Med Geogr Trop 1956; 8:202-7.

Dixon, Cyril W., *Smallpox*. London: J. & A. Churchill Ltd., 1962.

Doglio L., De Marco A., Schleich S., Roos N., Krijnse Locker J. The vaccinia virus E8R gene product: a viral membrane protein that is made early in infection and packaged into the virions' core. J Virol 2002; 76:9773-86.

Domi A., Moss B. Cloning the vaccinia virus genome as a bacterial artificial chromosome in *Escherichia coli* and recovery of infectious virus in mammalian cells. Proc Natl Acad Sci USA 2002; 99:12415-20.

Douglass N., Dumbell K. Independent evolution of monkeypox and variola viruses. J Virol 1992; 66:7565-7.

Douglass N.J., Richardson M., Dumbell K.R. Evidence for recent genetic variation in monkeypox viruses. J Gen Virol 1994; 75:1303-9.

Downie A.W. A study of the lesions produced experimentally by cowpox virus. J Path Bacteriol 1939; 48:361-79.

Downie A.W. Infection and immunity in smallpox. Lancet 1951; 1:419-23.

Downie, A.W. "Poxvirus Group." In *Viral and Rickettsial Infections of Man,* F.L. Horsfall, J. Tamm, eds. Philadelphia, Pennsylvania: Lippincott, 1965

Downie, A.W. "Smallpox." Im *Infectious Agents and Host Reactions,* S. Mudd, ed. Philadelphia: Saunders, 1970.

Downie A.W., Dumbell K.R. The isolation and cultivation of variola virus on the chorioallantois of chick embryos. J Path Bact 1947; 59:189-98.

Downie A.W., Dumbell K.R. Poxviruses. Annu Rev Microbiol 1956; 10:237-52.

Downie A.W., Haddock D.W. A variant of cowpox virus. Lancet 1952; I:1049-50.

Downie A.W., McCarthy K. The antibody response in man following infection with viruses of the pox group. III. Antibody response in a smallpox. J Hyg 1958; 56:479-87.

Downie A.W., McCarthy K., Macdonald A., McCallum F.O., Macrae A.D. Virus and virus antigen in the blood of smallpox patients. Their significance in early diagnosis and prognosis. Lancet 1953; II:164-6.

Downie A.W., St. Vincent L., Meicklejohn G., Ratnakannan N.R., Rao A.R., Krishnan G.N.V., Kempe C.H. Studies on the virus content of mouth washing in the acute phase of smallpox. Bull WHO 1961; 25:49-53.

Downie A.W., Meicklejohn G., St. Vincent L., Rao A.R., Sundara Babu B.V., Kempe C.H. The recovery of smallpox virus from patients and their environment in a smallpox hospital. Bull WHO 1965; 33:615-22.

Downie A.W., Fedson D.S., St. Vincent L., Rao A.K., Kempe C.H. Haemorrhagic smallpox. J Hyg 1969; 67:619-29.

Du S., Traktman P. Vaccinia virus DNA replication: Two hundred base pairs of telomeric sequence confer optimal replication efficiency on minichromosome templates. Proc Natl Acad Sci USA 1996; 93:9693-8.

Dubochet J., Adrian M., Richter K., Garces J., Wittek R. Structure of intracellular mature vaccinia virus observed by cryoelectron microscopy. J Virol 1994; 68:1935-41.

Duffy J. Smallpox and the Indians in the American Colonies. Bull Hist Med 1951; 25:324-41.

Dulbecco R. Production of plaques in monolayer tissue cultures by single particles of animal virus. Proc Natl Acad Sci USA 1952; 38:747-52.

Dumbell K.R., Huq F. Epidemiological implications of the typing of variola isolates. Trans Roy Soc Trop Med Hyg 1975; 69:303-6.

Dumbell K.R., Huq F. The virology of variola minor. Correlation of laboratory tests with the geographic distribution and human virulence of variola isolates. Amer J Epidemiol 1986; 123:403-15.

Dumbell K.R., Kapsenberg J.G. Laboratory investigation of two "white-pox" viruses and comparison with two variola strains from southern India. Bull WHO 1982; 60:381-7.

Dumbell K., Richardson M. Virological investigations of specimens from buffaloes affected by buffalopox in Maharashtra State, India between 1985 and 1987. Arch Virol 1993; 128:257-67

Dumbell K.R., Bedson H.S., Nizamuddin M. Thermoefficient strains of variola major virus. J Gen Virol 1967; 1:379-81.

Duncan S.A., Smith G.L. Identification and characterization of an extracellular envelope glycoprotein affecting vaccinia virus egress. J Virol 1992; 66:1610-21.

Duran-Reynals M.L., Phyllis J. Infection vaccinale chez les jennes souris. Accroissement par l'extrait testiculaire du lapin du pouvoir irifectant du virus vaccinal. Biol Med 1963; 52:76-82.

Dyster L.M., Niles E.G. Genetic and biochemical characterization of vaccinia virus genes D2L and D3R which encode virion structural proteins. Virology 1991; 182:455-67.

Earl P.L., Jones E.V., Moss B. Homology between DNA polymerase of poxviruses, herpesviruses, and adenoviruses: Nucleotide sequence of the vaccinia virus DNA polymerase gene. Proc Natl Acad Sci USA 1986; 83:3659-63.

Easterbrook K.B. Controlled degradation of vaccinia virions *in vitro:* an electron microscopic study. J Ultrastruct Res 1966; 14:484-96.

Edelman G.M., Murray B.A., Mege R.-M., Cunningham B.A., Gallin W.J. Cellular expression of liver and neural cell adhesion molecules after transfection with their cDNAs results in specific cell–cell binding. Proc Natl Acad Sci USA 1987; 84:8502-6.

Efstratiadis A., Posakony J.W., Maniatis T., Lawn R.M., O'Connell C., Spritz R.A., DeRiel J.K., Forget B.G., Weissman S.M., Slightom J.L., Blechl A.E., Smithies O., Baralle F.E., Shoulders C.C., Proudfoot N.J. The structure and evolution of the human β-globin gene family. Cell 1980; 21:653-68.

Ehrengut, Wolfgang, "Die Pockenschutzimpfung." In *Impffibel.* Stuttgart: F.K. Schattauer-Verlag, 1966.

Ehrengut, Wolfgang, Ehrengut-Lange, Jutta, "Non-Infectious Smallpox Vaccine in the Prophylaxis of Postvaccinial Encephalitis." In *Proceedings of the 37ᵗʰ International Symposium on Smallpox Vaccine Organized by the International Association of Biological Standardization; 1972 October 11–13.* H. Cohen, H. Regamey, eds. Basel, New-York: S. Karger, 1973.

Ehrengut, Wolfgang, Ehrengut-Lange, Jutta, Seitz, Dieter, Weber, Gerhard, *Die postvakzinale Enzephalopathie.* Stuttgart-New York: F.K. Schattauer Verlag, 1972.

Eijl H. van, Hollinshead M., Smith G.L. The vaccinia virus A36R protein is a type Ib membrane protein present on intracellular but not extracellular enveloped virus particles. Virology 2000; 271:26-36.

Eis-Hübingen A.M., Gerritzen A., Schneweis K.E., Pfeiff B., Pullmann H., Mayr A., Czerny C.P. Fatal cowpox-like virus infection transmitted by cat. Lancet 1990; I:880.

El Dahaby H., El Sabbagh A., Nassar M., Kamell M., Iskander M. Investigations of an outbreak of cowpox with special reference to the disease in Egypt. J Arab Vet Med Assoc 1966; 26:11-24.

Engelstad M., Smith G.L. The vaccinia virus 42 kDa envelope protein is required for envelopment and egress of extracellular virus and for virulence. Virology 1993; 194:627-37.

Engelstad M., Howard S.T., Smith G.L. A constitutively expressed vaccinia gene encodes a 42-kDa glycoprotein related to complement control factors that forms part of the extracellular virus envelope. Virology 1992; 188:801-10.

Eppstein D.A., Marsh Y.V., Schreiber A.B., Newman S.R., Todaro G.J., Nestor J.J. Epidermal growth factor receptor occupancy inhibits vaccinia virus infection. Nature 1985; 318:663-5.

Ericsson M., Cudmore S., Shuman S., Condit R.C., Griffiths G., Locker J.K. Characterization of *ts*16, a temperature-sensitive mutant of vaccinia virus. J Virol 1995; 69:7072-86.

Espmark, J.A., Rabo, E., Heller, L. "Smallpox Vaccination before the Age of 3 Months." In *Proceedings of the 37ᵗʰ International Symposium on Smallpox Vaccine Organized by the International Association of Biological Standardization; 1972 October 11–13.* H. Cohen, H. Regamey, eds. Basel, New-York: S. Karger, 1973.

Esposito, Joseph J., Fenner, Frank, "Poxviruses." In *Fields Virology,* D.E. Knipe, P.M. Howley, eds. Philadelphia: Lippincott Williams & Wilkins, 2001.

Esposito J.J., Knight J.C. Orthopoxvirus DNA: A comparison of restriction profiles and maps. Virology 1985; 143:230-51.

Esposito J.J., Obijeski J.F., Nakano J.H. The virion and soluble antigen proteins of variola, monkeypox, and vaccinia viruses. J Med Virol 1977; 1:95-110.

Esposito J.J., Obijeski J.F., Nakano J.H. Orthopoxvirus DNA: strain differentiation by electrophoresis of restriction endonuclease fragmented virion DNA. Virology 1978; 89:53-66.

Esposito J.J., Cabradilla C.D., Nakano J.H., Obijeski J.F. Intragenomic sequence transposition in monkeypox virus. Virology 1981; 109:231-43.

Esposito J.J., Nakano J.H., Obijeski J.F. Can variola-like viruses be derived from monkeypox virus? An investigation based on DNA mapping. Bull WHO 1985; 63:695-703.

Espy M.J., Cockerill F.R., Meyer F.R., *et al.* Detection of smallpox virus DNA by LightCycler PCR. J Clin Microbiol 2002; 40:1985-8.

Essani K., Dales S. Biogenesis of vaccinia: Evidence for more than 100 polypeptides in the virion. Virology 1979; 95:385-94.

Evans E., Traktman P. Characterization of vaccinia virus DNA replication mutants with lesions in the D5 gene. Chromosoma 1992; 102:72-82.

Evans E., Klemperer N., Ghosh R., Traktman P. The vaccinia virus D5 protein, which is required for DNA replication, is a nucleic acid-independent nucleoside triphosphatase. J Virol 1995; 69:5353-61.

Falkner F.G., Moss B. Transient dominant selection of recombinant vaccinia viruses. J Virol 1990; 64:3108-11.

Farabaugh P.J., Miller J.H. Genetic studies of the lac repressor. VII. On the molecular nature of spontaneous hot spots in the lacI gene of *Escherichia coli.* J Mol Biol 1978; 126:847-63.

Fathi Z., Sridhar P., Pacha R.F., Condit R.C. Efficient targeted insertion of an unselected marker into the vaccinia virus genome. Virology 1986; 155:97-105.

Fathi Z., Dyster L.M., Seto J., Condit R.C., Niles E.G. Intragenic and intergenic recombination between temperature-sensitive mutants of vaccinia virus. J Gen Virol 1991; 72:2733-7.

Fedorov, Vladimir V., *Specific Features of Organization of the Program of Smallpox Eradication in Afghanistan and the Materials of Study and Use of the Method of Needle-Free Injection in Anti-Smallpox Vaccination.* Moscow: Candidate of Science Dissertation, 1977.

Fenner F. The pathogenesis of the acute exanthems. Lancet 1948; II:915-20.

Fenner F. Mouse-pox (infectious ectromelia of mice). J Immunol 1949; 63:341-73.

Fenner F. The biological characters of several strains of vaccinia, cowpox, and rabbitpox viruses. Virology 1958; 5:502-29.

Fenner F., Comben B.M. Genetic studies with mammalian poxviruses. I. Demonstration of recombination between two strains of vaccinia virus. Virology 1958 5:530-48.

Fenner, Frank, Nakano, James H. "Poxviridae." *Laboratory Diagnosis of Infectious Diseases: Principles and Practices,* E.H. Lennette, P. Halonen, F.A. Murphy, eds. New York: Springer Verlag, 1988.

Fenner F., Woodroofe G.M. The reactivation of poxviruses. II. The range of reactivating viruses. Virology 1960; 11:185-200.

Fenner, Frank, Henderson, Donald A., Arita, Isao, Jezek, Zdenek, Ladnyi, Ivan D., *Smallpox and Its Eradication.* Geneva: World Health Organization, 1988.

Fenner, Frank, Wittek, Riccardo, Dumbell, Keith R., *The Orthopoxviruses.* San Diego, New York, Berkeley, Boston, London, Sydney, Tokyo, Toronto: Academic Press, Inc., 1989.

Ferrier A, Crance JM, Garin D. A model for cowpox virus pathogenesis and immunity based on intranasal infection of BALB/C mice. Proceedings of XIV International Poxvirus and Iridovirus Symposium; 2002 September 20–24; Lake Placid, New York.

Foege, W.H. "Smallpox Eradication in West and Central Africa." In *Proceedings of the 37th International Symposium on Smallpox Vaccine Organized by the International Association of Biological Standardization; 1972 October 11–13.* H. Cohen, H. Regamey, eds. Basel, New-York: S. Karger, 1973.

Foster S.O., Brink E.W., Hutchins D.L., Pifer J.M., Lourie B., Moser C.R., Cummings E.G., Kuteyi O.E.K., Eke R.E.A., Titus J.B., Smith E.A., Hicks, J.W., Foege W.H. Human monkeypox. Bull WHO 1972; 46:569-76.

Franke C.A., Rice C.M., Strauss J.H., Hruby D.E. Neomycin resistance as a dominant selectable marker for selection and isolation of vaccinia virus recombinants. Mol Cell Biol 1985; 5:1918-24.

Freemont P.S., Hanson I.M., Trowsdale J. A novel cysteine-rich sequence motif. Cell 1991; 64:483-4.

Fuerst T.R., Fernandez M.P., Moss, B. Transfer of the inducible lac repressor/operator system from *Escherichia coli* to a vaccinia virus expression vector. Proc Natl Acad Sci USA 1989; 86:2549-53.

Funahashi S., Sato T., Shida H. Cloning and characterization of the gene encoding the major protein of the A-type inclusion body of cowpox virus. J Gen Virol 1988; 69:35-47.

Gagliardini V., Fernandez P.-A, Lee R.K., Drexler H.C.A., Rotello R.J., Fishman M.C., Yuan J. Prevention of vertebrate neuronal death by the crmA gene. Science 1994; 263:826-8.

Galasso, G.J., Mattheis, M.J., Alling, D.W., Cherry, J.D., Bairan, A., *et al.* "Clinical Evaluation of Four Smallpox Vaccines." In *Proceedings of the 37th International Symposium on Smallpox Vaccine Organized by the International Association of Biological Standardization; 1972 October 11–13.* H. Cohen, H. Regamey, eds. Basel, New-York: S. Karger, 1973.

Gamaleya, Nikolai F., *Smallpox Vaccination.* St. Petersburg: Tipographiya Br. V & I Linnik, 1913.

Gancheva Ts., Andreev N. Postvaccination encephalitis as a consequence of smallpox vaccination in the People's Republic of Bulgaria during 1960–69. Epidem Microbiol Infekts Bolesti (Sofia) 1972; 9: 141-8.

Garcia A.D., Moss B. Repression of vaccinia virus Holliday junction resolvase inhibits processing of viral DNA into unit-length genomes. J Virol 2001; 75:6460-71.

Garcia A.D., Aravind L., Koonin E.V., Moss B. Bacterial-type DNA Holliday junction resolvases in eukaryotic viruses. Proc Natl Acad Sci USA 2000; 97:8926-31.

Gates AJ, Ulaeto DO, Gopal MR. Cross-absorption of polyclonal monkeypox sera with vaccinia-infected LLCMK2 cells. Proceedings of the XIIIth International Poxvirus and Iridovirus Symposium; 2000 September 2–6; Montpellier, France.

Gavrilova E.V., Babkin I.V., Shchelkunov S.N. A multiplex PCR assay for species-specific rapid identification of orthopoxviruses. Mol Gen Mikrobiol Virusol 2003; (1):45-52.

Gehring H., Mahnel H., Mayer H. Elefantenpocken. Zbl Vet Med 1972; B19:258-61.

Gelfand H.M., Posch J. The recent outbreak of smallpox in Meschede, West Germany. Amer J Epidemiol 1971; 93:234-7.

Gendon Yu.Z., Chernos V.I. The character of plaques as a genetic trait of variola virus group. Vopr Virusol 1963; (6):676-9.

Gendon Yu.Z., Chernos V.I. Comparative study of genetic traits of several strains of variola virus group. Acta Virol 1964; 8: 359-68.

Gershon P.D. mRNA 3' end formation by vaccinia virus: mechanism of action of a heterodimeric poly (A) polymerase. Semin Virol 1998; 8:343-50.

Gershon P.D., Moss B. Early transcription factor subunits are encoded by vaccinia virus late genes. Proc Natl Acad Sci USA 1990; 87:4401-5.

Gershon P.D., Moss B. Uridylate-containing RNA sequences determine specificity for binding and polyadenylation by the catalytic subunit of vaccinia virus poly(A) polymerase. EMBO J 1993; 12:4705-14.

Gershon P.D., Ahn B.-Y., Garfield M., Moss B. Poly(A) polymerase and a dissociable polyadenylation stimulatory factor encoded by vaccinia virus. Cell 1991; 66:1269-78.

Gileva I.P., Ryazankin I.A., Maksyutov A.Z., Totmenin A.V., Lebedev L.R., Nesterov A.E., Ageenko V.A., Shchelkunov S.N., Sandakhchiev L.S. Comparative study of the properties of orthopoxvirus soluble tumor necrosis factor receptor. Dokl Ross Akad Nauk 2003; 290:688-92.

Gillard S., Spehner D., Drillien R., Kirn A. Localization and sequence of a vaccinia virus gene required for multiplication in human cells. Proc Natl Acad Sci USA 1986; 83:5573-7.

Gispen R. De herbesmetting van Indonesie met pokken. Nederl Tijdschr Geneeskd 1949; 93:3686-95.

Gispen R. Silver impregnation of smallpox elementary bodies after treatment of xylol. Antonie van Leeuwenhoek 1952; 19:157-65.

Gispen R. Analysis of pox virus antigens by means of double diffusion. A method for direct serological differentiation of cowpox. J Immunol 1955; 74:134-41.

Gispen R., Brand-Saathof B. "White" poxvirus strains from monkeys. Bull WHO 1972; 46:585-92.

Gispen R., Brand-Saathof B. Three specific antigens produced in vaccinia, variola and monkey pox infections. J Inf Dis 1974; 129:289-93.

Gispen R., Kapsenberg J.G. Monkeypox virus-infectie in cultures van apeniercellen zonder duidelijk epizootisch verband met pokken en in een kolonie van apen lijdende aan pokken. Versl meded Btrpf Volksgezond 1966; 12:140-4.

Gispen R., Verlinde J.D., Zwart P. Histopathological and virological studies on monkeypox. Arch ges Virusforsch 1967; 21:205-16.

Gispen R., Brand-Saathof B., Hekker A.C. Monkeypox-specific antibodies in human and simian sera from the Ivory Coast and Nigeria. Bull WHO 1976; 53:355-60.

Goebel S.J., Johnson G.P., Perkus M.E., Davis S.W., Winslow J.P., Paoletti E. The complete DNA sequence of vaccinia virus. Virology 1990; 179:247-66.

Golini F., Kates J.R. Transcriptional and translational analysis of a strongly expressed early region of the vaccinia virus genome. J Virol 1984; 49:459-70.

Golubchikova, K.V., Samvelova, S.A. "On the Epidemiology of Smallpox Outbreak in Moscow in January 1960." In *Smallpox*, S.S. Marennikova, ed. Moscow, 1961.

Graham K.A., Lalani A.S., Macen J.L., Ness T.L., Barry M., Liu L.-Y., Lucas A., Clark-Lewis I., Moyer R.W., McFadden G. The T1/35 kDa family of poxvirus-secreted proteins binds chemokines and modulate leukocyte influx into virus-infected tissues. Virology 1997; 229:12-24.

Greene H.S.N. A pandemic of rabbitpox. Proc Soc Exp Biol Med 1933; 30:892-4.

Greiffendorf J, Höhr D, Pilaski J. Cowpox infection in elephants and other zoo-kept mammals. Proceedings of XII International Poxvirus Symposium; 1998 June 6–10; St. Thomas, USA.

Griffiths G., Roos N., Schleich S., Locker J.K. Structure and assembly of intracellular mature vaccinia virus: thin-section analyses. J Virol 2001a; 75:11056-70.

Griffiths G., Wepf R., Wendt T., Locker J.K., Cyrklaff M., Roos, N. Structure and assembly of intracellular mature vaccinia virus: isolated-particle analysis. J Virol 2001b; 75:11034-55.

Gross C.H., Shuman S. Vaccinia virions lacking the RNA helicase nucleoside triphosphate phosphohydrolase II are defective in early transcription. J Virol 1996; 70:8549-57.

Gross C.H., Shuman S. The nucleoside triphosphatase and helicase activities of vaccinia virus NPH-II are essential for virus replication. J Virol 1998; 72:4729-36.

Groth A., Münsterer H.O. Studien über vakzination vakzinale Immunität. Zschr Immunforsch 1935; 85:139-46.

Grubisha O, Traktman P. Genetic analysis of the vaccinia virus I6 telomere-binding protein uncovers a key role in genome encapsidation. J Virol 2003; 77:10929-42.

Guan K., Broyles S.S., Dixon J.E. A Tyr/Ser protein phosphatase encoded by vaccinia virus. Nature 1991; 350:359-62.

Guarner J., Johnson B.J., Paddock C.D., Shieh W.-Z., Goldsmith C.S., Reynolds M.G., et al. Monkeypox transmission and pathogenesis in Prairie dogs. Emerg Infect Dis 2004; 10:426-31.

Guarnieri G. Ricerche sulla patogenesi ed etiologia dell' infezione vaccinica e variolosa. Arch Sci Med 1892; 16:403-24.

Gubert, V.O., Smallpox and Smallpox Vaccination. St. Petersburg: tipografiya P.P. Soikina, 1896.

Gubser C., Smith G.L. The sequence of camelpox virus shows it is most closely related to variola virus, the cause of smallpox. J Gen Virol 2002; 83:855-72.

Gubser C., Hue S., Kellam P., Smith G.L. Poxvirus genomes: a phylogenetic analysis. J Gen Virol 2004; 85:105-17.

Gudkov, Vladimir G. Experimental Vaccinia Virus Infection of Cotton Rats. Moscow: Candidate of Science Dissertation, 1980.

Gudkov VG, Shved IA, Rytik PG. Clinical, virological, and pathomorphological characterization of vaccinia virus infection of cotton rats. Proceedings of the Symposium on Laboratory Diagnostics of Viral Infections; 1976; Minsk, USSR.

Gurvich, Emma B., Differentiation of Variola Virus and the Viruses Causing Diseases Clinically Similar to Smallpox in Cell Culture. Moscow: Candidate of Science Dissertation, 1964.

Gurvich, Emma B., Pathogenetic Mechanisms and Factors Facilitating Development of Postvaccination Complications. Moscow: Doctor of Science Dissertation, 1983.

Gurvich E.B. The age-dependent risk of postvaccination complications in vaccinees with smallpox vaccine. Vaccine 1992; 10:96-7.

Gurvich E.B., Marennikova S.S. Laboratory diagnosis of smallpox and similar viral diseases by means of tissue culture methods. 3. Additional modes of differentiating viruses of the pox group in tissue culture. Acta Virol 1964; 8:435-42.

Gutman N.R. Study of the properties of variola virus. Trudy NIIVS Mechnikova 1957; 9:182-90.

Gvakharia B.O., Koonin E., Mathews C. Vaccinia virus G4L gene encodes a second glutaredoxin. Virology 1996; 226:408-11.

Hahon N. Survival of variola virus on chorioallantoic membrane preparation. J Bact 1959; 78:731-3.

Hahon N. Smallpox and related poxvirus infections in Simian host. Bacteriol Rev 1961; 25:459-76.

Hahon N., Wilson B.J. Pathogenesis of variola in Macaca irus monkeys. Amer J Hyg 1960; 71:69-80.

Hahon N., Ratner M., Kozikowski E. Factors influencing variola virus growth on the chorioallantoic membrane of embryonated eggs. J Bact 1958; 75:707-12.

Hansen H., Sandvik T., Tryland M., Olsvik O., Traavik T. Comparison of thymidine kinase and A-type inclusion protein gene sequences from Norwegian and Swedish cowpox virus isolates. APMIS 1999; 107:667-75.

Harper L., Bedson H.S., Buchan A. Identification of orthopoxviruses by polyacrylamide gel electrophoresis of intracellular polypeptides. Virology 1979; 93:435-44.

Harte M.T., Haga I.R., Maloney G., Gray P., Reading P.C., Bartlett N.W., Smith G.L., Bowie A., O'Neill L.A. The poxvirus protein A52R targets toll-like receptor signaling complexes to suppress host defense. J Exp Med 2003; 197:343-51.

Hashizume S. A new attenuated strain LC16m8 of vaccinia virus for safe smallpox vaccination. J Clin Virology 1975; 3:229-35.

Hashizume, S., Morita, T., Yoshizawa, H., Suzuki, K., Arita, M., Komatsu, T., Amano, H., Tagaya, I. "Intracerebral Inoculation of Monkeys with Several Vaccinia Strains: An Approach to the Comparison of Different Strains." In *Proceedings of the 37th International Symposium on Smallpox Vaccine Organized by the International Association of Biological Standardization; 1972 October 11–13.* H. Cohen, H. Regamey, eds. Basel, New-York: S. Karger, 1973.

Hassett D.E., Condit R.C. Targeted construction of temperature-sensitive mutations in vaccinia virus by replacing clustered charged residues with alanine. Proc Natl Acad Sci USA 1994; 91:4554-8.

Hassett D.E., Lewis J.I., Xing X., DeLange L., Condit R.C. Analysis of a temperature-sensitive vaccinia virus mutant in the viral mRNA capping enzyme isolated by clustered charge-to-alanine mutagenesis and transient dominant selection. Virology 1997; 238:391-409.

Havell E.A. Evidence that tumor necrosis factor has an important role in antibacterial resistance. J Immunol 1989; 143:2894-9.

Hay S., Kannourakis G. A time to kill: viral manipulation of the cell death program. J Gen Virol 2002; 83:1547-64.

Hazel S.M., Bennett M., Chantrey J., Bown K., Cavanagh R., Jones T.R., Baxby D., Begon M. A longitudinal study of endemic disease in its wildlife reservoir: cowpox and wild rodents. Epidemiol Infect 2000; 124:551-62.

Heberling R.L., Kalter S.S., Rodriguez A.R. Poxvirus infection of the baboon (*Papio cynocephalus*). Bull WHO 1976; 54:285-94.

Heiner G.G., Fatima N., Daniel R.W., Cole J.L., Anthony R.L., McCrumb F.R. A study of unapparent infection in smallpox. Amer J Epidemiol 1971; 94:252-68.

Hekker, A.C., Huisman, J., Polak, M.F., Sawrt-van der Hoeven, J.Th., O'Breen, M.H., Gertenbach, J., Mollema, R.M. "Field Work with a Stable Freeze-dried Smallpox Vaccine Prepared in Monolayers of Primary Rabbit Kidney Cells." In *Proceedings of the 37th International Symposium on Smallpox Vaccine Organized by the International Association of Biological Standardization; 1972 October 11–13.* H. Cohen, H. Regamey, eds. Basel, New-York: S. Karger, 1973.

Helbert D. Smallpox and alastrim. Use of the chick embryo to distinguish between the viruses of variola major and variola minor. Lancet 1957; I:1012-4.

Henderson D.A. The looming threat of bioterrorism. Science 1999; 283:1279-82.

Henning K., Czerny C.-P., Meyer H., Müller F., Kramer M. A seroepidemiological survey for orthopoxvirus in the red fox (*Vulpes vulpes*). Vet Microbiol 1995; 43:251-9.

Hentschke J. von, Meyer H., Wittstatt U., Ochs A., Burkhardt S., Aue A. Kuhpocken bei kanadischen Bibern (*Castor fiver canadensis*) und Katzenbären (*Ailurus fulgens*) Tierärztl. Umschau 1999; 54:311-7.

Herrlich A. Über die Altersdisposition bei der postvakzinalen Enzephalitis. Münch Med Wchschr 1958; 100:1567-70.

Herrlich A. Über Vakzine-Antigen Versuch einer Prophylaxe neuraler impfschäden. Münch Med Wchschr 1959; 101:12-4.

Herrlich, Albert, *Die Pocken*. Stuttgart: Georg Thieme Verlag, 1960.

Herrlich A., Mayr A. Vergleichende experimentelle Arbeiten über die Vaccine-Kuhpocken-Viren. Arch Hyg Bakt 1954; 138:479-504.

Herrlich A., Mayr A., Mahnel H. Das Antikörperbild der Variola-Vaccineinfektion. II. Mitteilung. Serologische Untersuchungen an Variolapatienten. Zbl Bakt 1959; 175:163-82.

Herrlich A., Mayr A., Mahnel, H., Munz E. Experimental studies on transformation of the variola virus into the vaccinia virus. Arch ges Virusforsch 1963; 12:579-99.

Hertig C., Coupar B.E.H., Gould A.R., Boyle D.B. Field and vaccine strains of fowlpox virus carry integrated sequences from the avian retrovirus, reticuloendotheliosis virus. Virology 1997; 235:367-76.

Higashi N., Ichimiya M. Growth of variola virus in HeLa cells. Annu Rept Inst Virus Res Kyoto Univ 1959; 132:114-31.

Higman M.A., Christen L.A., Niles E.G. The mRNA (guanine-7-)methyltransferase domain of the vaccinia virus mRNA capping enzyme. Expression in *Escherichia coli* and structural and kinetic comparison to the intact capping enzyme. J Biol Chem 1994; 269:14974-81.

Hinshaw V.S., Olsen C.W., Dybdahl-Sissoko N., Evans D. Apoptosis: a mechanism of cell killing by influenza A and B viruses. J Virol 1994; 68:3667-73.

Hirt P., Hiller G., Wittek R. Localization and fine structure of a vaccinia virus gene encoding an envelope antigen. J Virol 1986; 58:757-64.

Hochstein-Mintzel V. Smallpox vaccine, then and now. From the cow lymph to the cell-culture vaccine. Fortschr Med 1977; 95:79-84.

Hollinshead M., Vanderplasschen A., Smith G.L., Vaux D.J. Vaccinia virus intracellular mature virions contain only one lipid membrane. J Virol 1999; 73:1503-17.

Hollinshead M., Rodger G., van Eijl H., Law M., Hollinshead R., Vaux D.J., Smith G.L. Vaccinia virus utilizes microtubules for movement to the cell surface. J Cell Biol 2001; 154:389-402.

Honeychurch KM, Hedegren-Olcott M, Byrd CM, Hruby DE. Biology and biochemistry of G1L expression during the vaccinia virus replication cycle. Proceedings of the XVth International Poxvirus and Iridovirus Conference; 2004 September 3–8; Keble College, Oxford, England.

Hooda-Dhingra U., Thompson C.L., Condit R.C. Detailed phenotypic characterization of five temperature-sensitive mutants in the 22- and 147-kilodalton subunits of vaccinia virus DNA-dependent RNA polymerase. J Virol 1989; 63:714-29.

Hooda-Dhingra U., Patel D.D., Pickup D.J., Condit R.C. Fine structure mapping and phenotypic analysis of five temperature-sensitive mutations in the second largest subunit of vaccinia virus DNA-dependent RNA polymerase. Virology 1990; 174:60-9.

Hooper J.W., Schmaljohn A.L., Schmaljohn C.S. Prophylactic and therapeutic monoclonal antibodies. Patent No US 6,451,309B2, 2002.

Hopkins, Donald R. *The Greatest Killer—Smallpox in History.* Chicago, Illinois: University of Chicago Press, 2002.

Howard J., Justus D.E., Totmenin A.V., Shchelkunov S.N., Kotwal G.J. Molecular mimicry of the inflammation modulatory proteins (IMPs) of poxviruses: evasion of the inflammatory response to preserve viral habitat. J Leukoc Biol 1998; 64:68-71.

Hruby D.E., Ball L.A. Mapping and identification of the vaccinia virus thymidine kinase gene. J Virol 1982; 43:403-9.

Hsiao J.-C., Chung C.-S., Chang W. Vaccinia virus envelope D8L protein binds to cell surface chondroitin sulfate and mediates the adsorption of intracellular mature virions to cells. J Virol 1999; 73:8750-61.

Hu F., Smith C.A., Pickup D.J. Cowpox virus contains two copies of an early gene encoding a soluble secreted form of the type II TNF receptor. Virology 1994; 204:343-56.

Hu X., Wolffe E.J., Weisberg A.S., Carroll L.J., Moss B. Repression of the A8L gene, encoding the early transcription factor 82-kilodalton subunit, inhibits morphogenesis of vaccinia virions. J Virol 1998: 72:104-12.

Huang S., Hendriks W., Althage A., Hemmi S., Bluethmann H., Kamijo R., Vilcek J., Zinkernagel R.M., Aguet M. Immune response in mice that lack the interferon-gamma receptor. Science 1993; 259:1742-5.

Huggins JW, Martinez MJ, Zaucha GM, Jahrling PB, Smee D, Bray M. The DNA polymerase inhibitor cidofovir (HPMPC, VISTIDE™) is a potential antiviral therapeutic agent for the treatment of monkeypox and other orthopoxvirus infections. Proceedings of XII International Poxvirus Symposium; 1998 June 6–10; St. Thomas, USA.

Huggins JW, Baker RO, Martinez MJ, Hostetler KY, Beadle JR, Hensley LE, Jahrling PB. Status of IV cidofovir (VISTIDE™) and orally active ether lipid prodrugs for the treatment of smallpox resulting from bioterrorist or biowarfare attack. Proceedings of XIV International Poxvirus and Iridovirus Symposium; 2002 September 20–24; Lake Placid, New York.

Hughes S.J., Johnston L.H., De Carlos A., Smith G.L. Vaccinia virus encodes an active thymidylate kinase that complements a cdc8 mutant of *Saccharomyces cerevisiae*. J Biol Chem 1991; 266:20103-9.

Hutchinson H.D., Ziegler D.W., Wells D.E., Nakano J.H. Differentiation of variola, monkeypox and vaccinia antisera by radioimmunoassay. Bull WHO 1977; 55:613-23.

Hutin Y.J.F., Williams R.J., Malfati P., Pebody R., Loparev V., Ropp S., *et al.* Outbreak of human monkeypox, Democratic Republic of Congo, 1996 to 1997. Emerg Infect Dis 2001; 7:434-8.

Ibrahim M.S., Kulesh D.A., Saleh S.S., *et al.* Real-time PCR assay to detect smallpox virus. J Clin Microbiol 2003; 41:3835-9.

Ichihashi Y. Vaccinia specific hemagglutinin. Virology 1977; 76:527-38.

Ichihashi, Yasuo. "Poxviruses." In *Immunochemistry of Viruses. II. The Basis for Serodiagnosis and Vaccines*, M.H.V. Van Regenmortel, A.R. Neurath, eds. Amsterdam, New York, Oxford: Elsevier. 1990.

Ichihashi Y., Matsumoto S. The relationship between poxvirus and A-type inclusion body during double infection. Virology 1968; 36:262-70.

Ichihashi Y., Matsumoto S., Dales S. Biogenesis of poxviruses: role of A-type inclusions and host cell membranes in virus dissemination. Virology 1971; 46:507-32.

Ichihashi Y., Oie M., Tsuruhara T. Location of DNA-binding proteins and disulfide-linked proteins in vaccinia virus structural elements. J Virol 1984; 50:929-38.

Ichihashi Y., Takahashi T., Oie M. Identification of a vaccinia virus penetration protein. Virology 1994; 202:834-43.

Ikič, D., Weisz-Maleček, R., Manhalter, T. "Field Trials with a New Smallpox Vaccine Prepared in the Human Diploid Cells WI-38 In *Proceedings of the 37th International Symposium on Smallpox Vaccine Organized by the International Association of Biological Standardization; 1972 October 11–13*. H. Cohen, H. Regamey, eds. Basel, New-York: S. Karger, 1973.

Ikuta K., Miyamoto H., Kato S. Serologically cross-reactive polypeptides in vaccinia, cowpox and Shope fibroma viruses. J Gen Virol 1979; 44:557-63.

Irons J.V., Bohls S.W., Cook E.B.M., Murphy J.N. The chick membrane as a differential culture medium with suspected cases of smallpox and varicella. Amer J Hyg 1941; 33:50-5.

Isaacs S.N., Kotwal G.J., Moss B. Vaccinia virus complement-control protein prevents antigen-dependent complement-enhanced neutralization of infectivity and contributes to virulence. Proc Natl Acad Sci USA 1992a; 89:628-32.

Isaacs S.N., Wolffe E.J., Payne L.G., Moss B. Characterization of a vaccinia virus-encoded 42-kilodalton class I membrane glycoprotein component of the extracellular virus envelope. J Virol 1992b; 66:7217-24.

Ishii K., Moss B. Role of vaccinia virus A20R protein in DNA replication: construction and characterization of temperature-sensitive mutants. J Virol 2001; 75:1656-63.

Jackson R.J., Ramsay A.J., Christensen C.D., Beaton S., Hall D.F., Ramshaw I.A. Expression of mouse interleukin-4 by a recombinant ectromelia virus suppresses cytolytic lymphocyte responses and overcomes genetic resistance to mousepox. J Virol 2001; 75:1205-10.

Jansen J. Tödlich Infektionen von Kaninchen durch ein filtrierbares Virus. Zbl Bakteriol Parasitenk Infectionskr Hyg 1941; 148:65.

Jansen J. Vaccinia (runderpokken) bij rund en mensch. T Diergenesk 1949; 74:897-901.

Janson C. Versuche zur Erlangung künstlicher Immunität bei Variola Vaccina. Zbl Bakteriol Parasitenk Infectionskr Hyg 1891; 10:40-5.

Jenner, Edward, *An Inquiry into the Causes and Effects of the Variolae Vaccinae, a Disease Discovered in Some of the Western Counties of England, Particularly Gloucestershire and Known by the Name of the Cow Pox.* London: Sampson Low, 1798.

Jensen O.N., Houthaeve T., Shevchenko A., Cudmore S., Ashford T., Mann M., Griffiths G., Krijnse L.J. Identification of the major membrane and core proteins of vaccinia virus by two-dimensional electrophoresis. J Virol 1996; 70:7485-97.

Jezek, Zdenek, Fenner, Frank. "Human Monkeypox." In *Monographs in Virology,* J.L. Melnick, ed. Basel-Munchen Paris-London-New-York-New Delhi-Singapore-Tokyo-Sydney: Karger, 1988.

Jezek Z., Gromyko A.I., Szczeniowski M.V. Human Monkeypox. J Hyg Epidemiol Microbiol Immunol 1983; 27:13-23.

Jezek Z., Marennikova S.S., Mutombo M., Nakano J.H., Paluku K.M., Szczeniowski M. Human monkeypox: A study of 2510 contacts of 214 patients. J Inf Dis 1986; 154:551-5.

Jezek Z., Szczeniowski M.V., Paluku K.M., Mutombo M., Grab B. Human monkeypox: confusion with chickenpox. Acta Tropica 1988; 45:297-307.

Jin D., Li Z., Jin Q., Yuwen H., Hou Y. Vaccinia virus hemagglutinin. A novel member of the immunoglobulin superfamily. J Exp Med 1989; 170:571-6.

Joklik W.K., Moss B., Fields B.N., Bishop D.H., Sandakhchiev L.S. Why the smallpox virus stocks should not be destroyed. Science 1993; 262:1225-6.

Kalter S.S., Rodriguez A.R., Cummins L.B., Heberling R.L., Foster S.O. Experimental smallpox in chimpanzees. Bull WHO 1979; 57:637-41.

Kane E.M., Shuman S. Temperature-sensitive mutations in the vaccinia virus H4 gene encoding a component of the virion RNA polymerase. J Virol 1992; 66:5752-62.

Kane E.M., Shuman S. Vaccinia virus morphogenesis is blocked by a temperature-sensitive mutation in the 17 gene that encodes a virion component. J Virol 1993; 67:2689-98.

Kao S.-Y., Bauer W.R. Biosynthesis and phosphorylation of vaccinia virus structural protein VP11. Virology 1987; 159:339-407.

Kaplan C. The heat inactivation of vaccinia virus. J Gen Microbiol 1958; 18:58.

Kaplan C. A non-infectious smallpox vaccine. Lancet 1962; II:1027-8.

Kaplan C., Healing T.D., Evans N., Healing L., Prior A. Evidence of infection by viruses in small British field rodents. J Hyg 1980; 84:285-94.

Kaptsova, Tat'yana I., *Development of Experimental Models of Smallpox.* Moscow: Candidate of Science Dissertation, 1967.

Kates J., Beeson J. Ribonucleic acid synthesis in vaccinia virus. I. The mechanism of synthesis and release of RNA in vaccinia cores. J Mol Biol 1970a; 50:1-18.

Kates J., Beeson J. Ribonucleic acid synthesis in vaccinia virus. II. Synthesis of polyriboadenylic acid. J Mol Biol 1970b; 50:19-33.

Kates J.R., McAuslan B. Messenger RNA synthesis by a "coated" viral genome. Proc Natl Acad Sci USA 1967; 57:314-20.

Kato S., Takahashi M., Kameyama S., Kamahora J. A study on the morphological and cytoimmunological relationship between the inclusions of variola, cowpox, rabbitpox, vaccinia (variola origin) and vaccinia IHD and a consideration of the term "Guarnieri body". Biken's J 1959; 2:353-63.

Katz E., Ward B.M., Weisberg A.S., Moss B. Mutations in the vaccinia virus A33R and B5R envelope proteins that enhance release of extracellular virions and eliminate formation of actin-containing microvilli without preventing tyrosine phosphorylation of the A36R protein. J Virol 2003; 77:12266-75.

Keck J.G., Baldick C.J., Moss B. Role of DNA replication in vaccinia virus gene expression: A naked template is required for transcription of three late trans-activator genes. Cell 1990; 61:801-9.

Keck J.G., Fiegenbaum F., Moss B. Mutational analysis of a predicted zinc-binding motif in the 26-kilodalton protein encoded by the vaccinia virus A2L gene: correlation of zinc binding with late transcriptional transactivation activity. J Virol 1993a; 67:5749-53.

Keck J.G., Kovacs G.R., Moss B. Overexpression, purification, and late transcription factor activity of the 17-kilodalton protein encoded by the vaccinia virus A1L gene. J Virol 1993; 67:5740-8.

Kempe H. Studies on smallpox and complications of smallpox vaccination. Pediatrics 1960; 26:176-89.

Kempe C.H, Benenson A.S. Smallpox and vaccinia. Symposium on Unusual Infection of Childhood. Pediat Clin North America 1955; 2:19-32.

Kempe C.H., Fulg, geniti V., Minamitani M., Shinefield H. Smallpox vaccination of eczema patients with a strain of attenuated live vaccinia (CVI-78). Pediatrics 1968; 42:980-9.

Kempe C.H., Dekking F., St. Vincent L., Rao A.R., Downie A.W. Conjunctivitis and subclinical infection in smallpox. J Hyg 1969; 67:631-6.

Kerr S.M., Smith G.L. Vaccinia virus encodes a polypeptide with DNA ligase activity. Nucl Acids Res 1989; 17:9039-50.

Kettle S., Alcami A., Khanna A., Ehret R., Jassoy C., Smith G.L. Vaccinia virus serpin B13R (SPI-2) inhibits interleukin-1-β converting enzyme and protects virus-infected cells from TNF- and Fas-mediated apoptosis, but does not prevent IL-1β-induced fever. J Gen Virol 1997; 78:677-85.

Khodakevich, Lev N., *Ecological and Epidemiological Aspects of Smallpox.* Moscow: Doctor of Science Dissertation, 1990.

Khodakevich L.N., Widy-Wirski R., Arita J., Marennikova S.S., Nakano J., Meunier D. Orthopoxvirose simienne de l'homme en Republique Centralafricaine. Bull Soc Path Exot 1985; 78:311-20.

Khodakevich L., Jezek Z., Kinzanzka K. Isolation of monkeypox virus from wild squirrel infected in nature. Lancet 1986: I:98-9.

Khodakevich L.N., Szczeniowski M., Mambu-ma-Disu, Jezek Z., Marennikova S.S., Nakano J., Meier F. Monkeypox virus in relation to ecological features surrounding human settlements in Bumba zone, Zaire. Trop Geogr Med 1987; 39:56-63.

Kibler KV, Langland JO, Shots T, Zeeman C, Jacobs BL. A double-stranded RNA-binding protein is necessary to inhibit induction of apoptosis and for interferon-resistance in vaccinia virus infected cells. Proceedings of the Xth International Congress of Virology; 1996 August 11–16; Jerusalem, Israel.

Kirillova, F.M., Al'tshtein, A.D., Bykovsky, A.F. "Use of the Method of Fluorescent Antibodies for Smallpox Diagnostics." In *Smallpox,* S.S. Marennikova, ed. Moscow, 1961.

Kirkitadze M.D., Henderson C., Price N.C., Kelly S.M., Mullin N.P., Parkinson J., Dryden D.T.F., Barlow P.N. Central modules of the vaccinia virus complement control protein are not in extensive contact. Biochem J 1999; 344:167-75.

Kitamoto N., Tanimoto S., Hiroi K., Ozaki M., Miyamoto H., Wakamiya N., Ikuta I., Ueda S., Kato S. Monoclonal antibodies to cowpox virus: polypeptide analysis of several major antigens. J Gen Virol 1987; 68:239-46.

Kitamura T. Studies on the formation of hyperplastic focus by variola virus in human cell cultures. I. *In vitro* quantitation of variola virus by focus counting in HeLa and FL cell cultures. Virology 1968; 36:174-9.

Klemperer N., Ward J., Evans E., Traktman P. The vaccinia virus I1 protein is essential for the assembly of mature virions. J Virol 1997; 71:9285-94.

Klemperer N., McDonald W., Boyle K., Unger B., Traktman P. The A20R protein is a stoichiometric component of the processive form of vaccinia virus DNA polymerase. J Virol 2001; 75:12298-307.

Klietmann W.F., Ruoff K.L. Bioterrorism: implications for the clinical microbiologist. Clin Microbiol Rev 2001; 14:364-81.

Klingebiel T., Vallbracht A., Döller G., Stierhof Y.D., Gerth H.J., Glashauser E., Herzau V.A severe human cowpox infection in south Germany. Pediat Inf Dis 1988; 7:883-5.

Kluczyk A., Siemion I.Z., Szewczuk Z., Wieczorek Z. The immunosuppressive activity of peptide fragments of vaccinia virus C10L protein and a hypothesis on the role of this protein in the viral invasion. Peptides 2002; 23:823-34.

Kochneva G., Kolosova I., Maksyutova T., Ryabchikova E., Shchelkunov S. Effects of deletions of *kelch*-like genes on cowpox virus biological properties. Virus Res 2004; (in press).

Kolhapure R.M., Deolankar R.P., Tupe C.D., Raut C.G., Basu A., Dama B.M., Pawar S.D., Joshi M.V., Padbidri V.S., Goverdhan M.K., Banerje K. Investigation of buffalopox outbreaks in Maharashtra State during 1992–1996. Indian J Med Res 1997; 106:441-6.

Kolosova I.V., Seregin S.V., Kochneva G.V., Ryabchikova E.I., Bessemel'tseva E.V., Babkina I.N., Solenova T.E., Babkin I.V., Shchelkunov S.N. Orthopoxvirus genes for kelch-like proteins: II. Construction of cowpox virus variants with targeted gene deletions. Mol Biol (Mosk) 2003; 37:585-94.

Koonin E.V. A highly conserved sequence motif defining the family of MutT-related proteins from eubacteria, eukaryotes and viruses. Nucl Acids Res 1993; 21:4847.

Koonin E.V., Senkevich T.G., Chernos V.I. Gene A32 product of vaccinia virus may be an ATPase involved in viral DNA packaging as indicated by sequence comparisons with other putative viral ATPases. Virus Genes 1993; 7:89-94.

Kotwal G.J. The great escape—immune evasion by pathogens. Immunologist 1996; 4:157-64.

Kotwal G.J., Moss B. Vaccinia virus encodes a secretory polypeptide structurally related to complement control proteins. Nature 1988a; 335:176-8.

Kotwal G.J., Moss B. Analysis of a large cluster of nonessential genes deleted from a vaccinia virus terminal transposition mutant. Virology 1988b; 167:524-37.

Kotwal G.J., Moss B. Vaccinia virus encodes two proteins that are structurally related to members of the plasma serine protease inhibitor superfamily. J Virol 1989; 63:600-6.

Kotwal G.J., Hugin A.W., Moss B. Mapping and insertional mutagenesis of a vaccinia virus gene encoding a 13,800-Da secreted protein. Virology 1989; 171:579-87.

Kovacs G.R., Moss B. The vaccinia virus H5R gene encodes late gene transcription factor 4: purification, cloning and overexpression. J Virol 1996; 70:6796-802.

Kovacs G.R., Resales R., Keck J.G., Moss B. Modification of the cascade model for regulation of vaccinia virus gene expression: purification of a prereplicative, late-stage-specific transcription factor. J Virol 1994; 68:3443-7.

Kovacs G.R., Vasilakis N., Moss B. Regulation of viral intermediate gene expression by the vaccinia virus B1 protein kinase. J Virol 2001; 75:4048-55.

Krag P., Bentzon M. The international reference preparation of smallpox vaccine: an international collaboration essay. Bull WHO 1963; 29:299-309.

Kretzschmar M., van den Hof S., Wallinga J., van Wijngaarden J. Ring vaccination and smallpox control. Emerg Infect Dis 2004; 10:832-41.

Krikun, V.A. "Isolation of Pneumotropic Virus from Rats during Outbreak of a Disease with Unknown Etiology in a Reproduction Colony." In *Laboratory Animals in Medical Studies.* Moscow, 1974.

Kumar, Sudhir, Tamura, Koichiro, Nei Masatoshi. MEGA: *Molecular Evolutionary Genetics Analysis.* Pennsylvania State University, University Park, 1993.

Kyrle I., Morawetz G. Tierexperimentelle Studien über Variola. Wien klin Wchschr 1915; 28:697-701.

Laassri M., Chizhikov V., Mikheev M., Shchelkunov S., Chumakov K. Detection and discrimination of orthopoxviruses using microarrays of immobilized oligonucleotides. J Virol Methods 2003; 112:67-78.

Lackner C.A., Condit R.C. Vaccinia virus gene A18R DNA helicase is a transcript release factor. J Biol Chem 2000; 275:1485-94.

Lackner C.A., D'Costa S.M., Buck C., Condit R.C. Complementation analysis of the dales collection of vaccinia virus temperature-sensitive mutants. Virology 2003; 305:240-59.

Ladnyi, Ivan D., *Smallpox Eradication and Prevention of Its Re-emergence.* Moscow: Meditsina, 1985.

Ladnyi I.D., Ziegler P., Kima E. Human infection caused by monkeypox virus in Basankusu territory, Democratic Republic of the Congo. Bull WHO 1972; 46:593-7.

Ladnyi I.D., Ogorodnikova Z.I., Shelukhina E.M., Gerasimenko R.T., Voronin Yu.S. Study of the prevalence of variola group viruses among animals. Probl Osobo Opasnykh Infektsii (Saratov) 1975; 44:165-7.

Laidlaw SM, Skinner MA. Analysis of the poxviral gene, A19L. Proceedings of the XVth International Poxvirus and Iridovirus Conference; 2004 September 3-8; Keble College, Oxford, England.

Lal S.M., Singh J.P. Serological characterization of buffalopox virus. Arh Ges Virusforsch 1973; 43:393-6.

Lamontagne N., Marcolais G., Murois P., Assaf R. Diagnosis of rotavirus, adenovirus and herpesvirus infections by immune electron microscopy using a serum-in-agar diffusion method. Can J Microbiol 1980; 26:261-4.

Lancaster M.G., Boulter E.A., Westwood J.C.N., Randles J. Experimental respiratory infection with poxviruses. II. Pathological studies. Brit J Exp Path 1966; 47:466-71.

Lane, J.M. "Complications Following Smallpox Vaccination." In *Proceedings of the 37th International Symposium on Smallpox Vaccine Organized by the International Association of Biological Standardization; 1972 October 11-13.* H. Cohen, H. Regamey, eds. Basel, New-York: S. Karger, 1973.

Lane J.M., Ruben F.L., Heff J.M., Millar J.D. Complication of smallpox vaccination, 1968: results of ten statewide surveys. J Inf Dis 1970; 122:303-9.

Lapa S., Mikheev M., Shchelkunov S., Mikhailovich V., Sobolev A., Blinov V., Babkin I., Guskov A., Sokunova E., Zasedatelev A., Sandakhchiev L., Mirzabekov A. Species-level identification of orthopoxviruses with an oligonucleotide microchip. J Clin Microbiol 2002; 40:753-7.

Latner D.R., Xiang Y., Lewis J.I., Condit J., Condit R.C. The vaccinia virus bifunctional gene J3 (nucleoside-2'-*O*-)-methyltransferase and poly(A) polymerase stimulatory factor is implicated as a positive transcription elongation factor by two genetic approaches. Virology 2000; 269:345-55.

Latner D.R., Thompson J.M., Gershon P.D., Storrs C., Condit R.C. The positive transcription elongation factor activity of the vaccinia virus J3 protein is independent from its (nucleoside-2'-*O*-)methyltransferase and poly(A) polymerase stimulatory functions. Virology 2002; 301:64-80.

Law K.M., Smith G.L. A vaccinia serine protease inhibitor which prevents virus-induced cell fusion. J Gen Virol 1992; 73:549-57.

Law M., Hollinshead R., Smith G.L. Antibody-sensitive and antibody-resistant cell-to-cell spread by vaccinia virus: role of the A33R protein in antibody-resistant spread. J Gen Virol 2002; 83:209-22.

Lazarus A.S., Eddie A., Meyer K.F. Propagation of variola virus in the developing egg. Proc Soc Exp Biol Med 1937; 36:7-8.

Ledingham J.C.G., McClean D. The propagation of vaccinia virus in the rabbit dermis. Brit J Exp Path 1928; 9:216-24.

LeDuc J.W., Jahrling P.B. Strengthening national preparedness for smallpox: an update. Emerg Infect Dis 2001; 7:155-7.

Lee S.F., Buller R., Chansue E., Hanika W.C., Brunt E.M., Aquino T., Storch G.A., Pepose J.S. Vaccinia keratouveitis manifesting as a masquerade syndrome. Am J Ophalmol 1994; 117:480-7.

Lee-Chen G.-J., Bourgeois N., Davidson K., Condit R.C., Niles E.G. Structure of the transcription and termination sequence of seven early genes in the vaccinia virus *Hin*dIII D fragment. Virology 1988; 163:64-79.

Lepine P., Croissant O. Application de la microscopie electronique an diagnostic de la variola. Presse Med 1952; 60:1427-8.

Levaditi C., Nicolau S. Ectodermoses neurotropes. Etudes sur la vaccine. Ann Inst Pasteur 1923; 37:1-106.

Lewis-Jones M.S., Baxby D., Cefai C., Hart C.A. Cowpox can mimic anthrax. Br J Dermatol 1993; 129:625-7.

Li Khe Min', *Experimental Study of Biological Properties of Vaccinia Virus*. Candidate of Science Dissertation, 1955.

Lin S., Broyles S.S. Vaccinia protein kinase 2: A second essential serine/threonine protein kinase encoded by vaccinia virus. Proc Natl Acad Sci USA 1994; 91:7653-7.

Lin S., Chen W., Broyles S.S. The vaccinia virus B1R gene product is a serine/threonine protein kinase. J Virol 1992; 66:2717-23.

Lin C.L., Chung C.S., Heine H.G., Chang W. Vaccinia virus envelope H3L protein binds to cell surface heparan sulfate and is important for intracellular mature virion morphogenesis and virus infection *in vitro* and *in vivo*. J Virol 2000; 74:3353-65.

Liszewski M.K., Atkinson J.P. In: *The Human Complement System in Health and Disease*, J.E. Volanakis, M.M. Frank, eds. New York-Basel-Hong Kong: Marcel Dekker, Inc., 1998.

Litvinjenko, S., Arsić, B., Borjanović, S. "Epidemiološki Aspekti Variole u Jugoslaviji 1972 Godine." In *Variola u Jugoslaviji, 1972 Godine*, L. Stojkovic, ed. Ljubljana: ČGP Delo, 1973.

Liu K., Lemon B., Traktman P. The dual-specificity phosphatase encoded by vaccinia virus, VH1, is essential for viral transcription *in vivo* and *in vitro*. J Virol 1995; 69:7823-34.

Loparev V.N., Parsons J.M., Knight J.C., Panus J.F., Ray C.A., Buller R.M., Pickup D.J., Esposito J.J. A third distinct tumor necrosis factor receptor of orthopoxviruses. Proc Natl Acad Sci USA 1998; 95:3786-91.

Loparev V.N., Massung R.F., Esposito J.J., Meyer H. Detection and differentiation of Old World orthopoxviruses: restriction fragment length polymorphism of the *crmB* gene region. J Clin Microbiol 2001; 39:94-100.

Lopes O.S., Lacerda J.P.G., Fonseca I.E.M., Castro D.P., Forattini O.P., Rabelo E.X. Cotia virus, a new agent isolated from sentinel mice in Sao Paulo, Brazil. Amer J Trop Med Hyg 1965; 14:156-7.

Lopez P., Espinosa M., Greenberg B., Lacks S.A. Generation of deletions in pneumococcal *mal* genes cloned in *Bacillus subtilis.* Proc Natl Acad Sci USA 1984; 81:5189-93.

Lourie B., Bingham P.G., Evans H.H., Foster S.O., Nakano J.H., Hermann K.L. Human infection with monkeypox virus: Laboratory investigation of six cases in West Africa. Bull WHO 1972; 46:633-9.

Lourie B., Nakano J.H., Kemp G.E., Setzer H.W. Isolation of poxvirus from an African rodent. J Inf Dis 1975; 132:677-81.

Lovering R., Hanson I.M., Borden K.L.B., Martin S., O'Reilly N.J., Evan G.I., Rahman D., Pappin D.J.C., Trowsdale J., Freemont P.S. Identification and preliminary characterization of a protein motif related to the zinc finger. Proc Natl Acad Sci USA 1993; 90:2112-6.

Luftig R.B. Does the cytoskeleton play a significant role in animal virus replication? J Theor Biol 1982; 99:173-91.

Lum G.S., Soriano F., Trejos A., Llerena J. Vaccinia epidemic and epizootic in El Salvador. Amer J Trop Med Hyg 1967; 16:332-6.

Luo Y., Mao X., Deng L., Cong P., Shuman S. The D1 and D12 subunits are both essential for the transcription termination factor activity of vaccinia virus capping enzyme. J Virol 1995; 69:3852-6.

Lux S.E., John K.M., Bennet V. Analysis of cDNA for human erythrocyte ankyrin indicates a repeated structure with homology to tissue-differentiation and cell-cycle control proteins. Nature 1990; 344:36-42.

L'vov S.D., Gromashevsky V.L., Marennikova S.S., Bogoyavlenksy G.V., Bailuk F.N., Butenko A.M., Gushchina E.N., Shelukhina E.M., Zhukova O.A., Morozova T.N. Isolation of a poxvirus (Poxviridae, *Orthopoxvirus*, cowpox virus group) from the vole *Microtus (M.) oeconomus* Pall. 1778 in the forest–tundra of the Cola Peninsular. Vopr Virus 1988; (1):92-4.

Maa, J.-S., Esteban, M. 1987. Structural and functional studies of a 39000-Mr immunodominant protein of vaccinia virus. J. Virol. 61. P.3910-3919.

Maa J.-S., Rodriguez J.F., Esteban M. Structural and functional characterization of a cell surface binding protein of vaccinia virus. J Biol Chem 1990; 265:1569-77.

Macaulay C., Upton C., McFadden G. Tumorigenic poxviruses: transcriptional mapping of the terminal inverted repeats of Shope fibroma virus. Virology 1987; 158:381-93.

Mackett, M., *Restriction Endonuclease Analysis of Orthopoxvirus DNA.* Univ. of London: Ph.D. Thesis, 1981.

Mackett M., Archard L.C. Conservation and variation in orthopoxvirus genome structure. J Gen Virol 1979; 45:683-701.

Magnus P. von, Andersen E.K., Petersen K.B., Birch-Andersen A. A pox-like disease in cynomolgus monkeys. Acta Path Microbiol Scand 1959; 46:156-76.

Magrath G.B., Brinckerhoff W.R. On experimental variola in the monkey. J Med Res 1904a; 11:230-46.

Magrath G.B., Brinckerhoff W.R. On the occurrence of cytoryctes variolae (Guarnieri) in the skin of the monkey inoculated with variola virus. J Med Res 1904b; 11:173-80.

Mahnel H., Czerny C.P., Mayr A. Nachweis und Identifizierung von Pockenvirus bei Hauskatzen. J Vet Med 1989; 36:231-6.

Mahr A., Roberts B.E. Arrangement of late RNAs transcribed from a 7.1-kilobase *Eco*RI vaccinia virus DNA fragment. J Virol 1984; 49:510-20.

Mahy B.W., Almond J.W., Berns K.I., Chanock, R.M., Lvov D.K. The remaining stocks of smallpox virus should be destroyed. Science 1993; 262:1223-4.

Maiboroda A.D. Experimental infection of Norwegian rats (*Rattus norvegicus*) with ratpox virus. Acta Virol 1982; 26:288-91.

Maiboroda A.D., Lobanova Z.I. Asymptomatic form of pox in laboratory rats. Zh Mikrobiol 1980; (2):106-9.

Mal'tseva, Nelli N., *Comparative Study of the Properties of Various Vaccinia Virus Strains.* Moscow: Candidate of Science Dissertation, 1965.

Mal'tseva, Nelli N., *Rapid Diagnostics of Disease Caused by Orthopoxvirus and Several Herpes Viruses.* Moscow: Doctor of Science Dissertation, 1980.

Maltseva N.N., Marennikova S.S. A method for serological differentiation of closely related poxviruses. Acta Virol 1976; 20:250-2.

Mal'tseva N.N., Akatova-Shelukhina E.M., Yumasheva M.A., Marennikova S.S. Etiology of several outbreaks of pox-like cattle diseases and methods for differentiation of vaccinia, cowpox, and swinepox viruses. Zh Gig Epidemiol Mikrobiol Immunol 1966; (10):193-9.

Mal'tseva N.N., Marennikova S.S., Nakano J., Matsevich G.R., Khabakhpasheva N.A., Shelukhina E.M., Arita I., Gromyko A.I., Stepanova L.G. Data of comparative study of population of Republic of the Congo for the presence of antibodies to orthopoxviruses. II. Species-level identification of antibodies using solid-phase variant of immunoenzyme method. Zh Mikrobiol 1984; (4):64-7.

Mal'tseva N.N., Marennikova S.S., Matsevich G.R., Stepanova L.G., Khabakhpasheva N.A. Standardization of ELISA for detection of antibodies to orthopoxviruses. Acta Virol 1985; 29:294-9.

Mao X., Shuman S. Intrinsic RNA (guanine-7) methyltransferase activity of the vaccinia virus capping enzyme D1 subunit is stimulated by the D12 subunit. Identification of amino acid residues in the D1 protein required for subunit association and methyl group transfer. J Biol Chem 1994; 269:24472-9.

Marchal J. Infectious ectromelia. A hitherto undescribed virus disease of mice. J Path Bacteriol 1930; 33:713-28.

Marennikova, Svetlana S., *Smallpox Ovovaccine.* Moscow, 1958.

Marennikova S.S. Isolation of variola virus in chick embryos. Zh Mikrobiol 1960; (6):102-5.

Marennikova S.S. Isolation and study of the properties of variola virus. II. Susceptibility of laboratory animals to variola virus. Vopr Virusol 1961; (1):73-8.

Marennikova, Svetlana S. *Materials on Study of the Agent, Laboratory Diagnostics, and Emergency Prophylaxis of Smallpox.* Moscow: Doctor of Science Dissertation, 1962.

Marennikova SS. Virus Strains for Smallpox Vaccine Production. International Symposium über Fragen des Pockenschutzes; 1967; Berlin, Leipzig: Johann Ambrosius Barth, 1968.

Marennikova, Svetlana S. "Evaluation of Vaccine Strains by Their Behaviour in Vaccinated Animals and Possible Implications of the Revealed Features for Smallpox Vaccination Practice." In *Proceedings of the 37ᵗʰ International Symposium on Smallpox Vaccine Organized by the International Association of Biological Standardization; 1972 October 11–13.* H. Cohen, H. Regamey, eds. Basel, New-York: S. Karger, 1973.

Marennikova S.S. Vaccine prevention of smallpox. Vestn Akad Med Nauk SSSR 1975; (4):33-7.

Marennikova S.S., Akatova E.M. On hemagglutination activity of variola virus. Byull Exp Biol Med 1958; (4):88-91.

Marennikova S.S., Kaptsova T.I. Age-dependence of susceptibility of white mice to variola virus. Acta Virol 1965; 9:230-4.

Marennikova S.S., Mal'tseva N.N. Use of microprecipitation reaction on slide for laboratory diagnostics of smallpox. Vopr Virusol 1961; (2):204-6.

Marennikova S.S., Mal'tseva N.N. Comparative study of several vaccinia virus strains. I. Specific features of behavior in chick embryos, hemagglutinating activity, and temperature resistance. Vopr Virusol 1964a; (3):280-6.

Marennikova S.S., Mal'tseva N.N. Comparative study of several vaccinia virus strains. II. Pathogenicity for laboratory animals. Vopr Virusol 1964b; (3):287-91.

Marennikova S.S., Matsevich G.R. Postvakzinale Komplikationen seitens des Zentralnervensystems und Behandlung mit Gammaglobulin mit erhöhtem Gehalt von Antikörpern gegen das Vakzinevirus. Mitt Österr Sanitätsverwaltung 1968; 69:1-4.

Marennikova S.S., Matsevich G.R. On neurological complications after vaccination against smallpox. Zh Mikrobiol 1971; (9):3-7.

Marennikova S.S., Matsevich G.R. The state and certain aspects of the struggle against complications after smallpox vaccination. Vestn Akad Med Nauk SSSR 1974; (12):42-7.

Marennikova S.S., Macevič G.R. Experimental study of the role of inactivated vaccine in two-step vaccination against smallpox. Bull WHO 1975; 52:51-6.

Marennikova S.S., Shafikova R.A. Comparative studies on the properties of variola virus strains. I. Characteristics of chorioallantoic membrane lesions and pathogenicity for chick embryos after different methods of inoculation. Acta Virol 1969; 13:538-43.

Marennikova S.S., Shafikova R.A. On the effect of incubation temperature on development of the specific lesions caused by variola virus. Vopr Virusol 1970; (2):239-40.

Marennikova, Svetlana S., Shchelkunov, Sergei N., *Orthopoxviruses Pathogenic for Humans.* Moscow: KMK Press Ltd., 1998.

Marennikova S.S., Stepanova L.G. New methods for determination of the virulence of vaccinia and variola viruses. Proceedings of the Conference of the Tarasevich State Institute of Control of Medical and Biological Preparations; 1958; Moscow.

Marennikova S.S., Shelukhina E.M. White rats as a source of pox infection in Carnivora of the family Felidae. Acta Virol 1976a; 20:442.

Marennikova S.S., Šeluhina E.M. Susceptibility of some rodent species to monkeypox virus, and course of the infection. Bull WHO 1976b; 53:13-20.

Marennikova S.S., Tashpulatov G.M. Comparative study of the vaccination efficiencies, reactogenicities, and antigenic activities of smallpox vaccines produced from various strains. Vopr Virusol 1966; (4):266-72.

Marennikova S.S., Mastyukova Yu.N., Ogorodnikova Z.I. On cultivation of variola and vaccinia viruses in developing chick embryos. Vopr Virusol 1956: (6):36-40.

Marennikova S.S., Gurvich E.B., Yumasheva M.A. Isolation and study of variola virus. I. A direct isolation of variola virus in cell culture. Vopr Virusol 1959; (6):703-10.

Marennikova, S.S., Akatova, E.M., Gurvich, E.B., Zuev, V.A., Ogorodnikova, Z.I., Yumasheva, M.A. "Methods of Laboratory Diagnostics of Smallpox and Their Comparative Evaluation Using the Materials of Outbreak in 1960." In *Smallpox,* Marennikova S.S., ed. Moscow, 1961a.

Marennikova, S.S., Eremyan, A.V., Ogorodnikova, Z.I. "Use of Antismallpox Gamma-Globulin in Smallpox Seroprophylaxis and Serotherapy." In *Smallpox,* Marennikova S.S., ed. Moscow, 1961b.

Marennikova S.S., Gurvich E.B., Ogorodnikova Z.I. The experience of comparative study of domestic and foreign smallpox vaccines. Vopr Virusol 1962; (1):62-8.

Marennikova S.S., Gurvich E.B., Yumasheva M.A. Laboratory diagnosis of smallpox and similar viral diseases by means of tissue culture methods. I. On sensitivity of tissue culture method for indication of variola virus. Acta Virol 1963; 7:124-30.

Marennikova S.S., Gurvich E.B., Yumasheva M.A. Laboratory diagnosis of smallpox and similar viral diseases by means of tissue culture methods. II. Differentiation of smallpox virus from varicella, vaccinia, cowpox and herpes viruses. Acta Virol 1964; 8:135-42.

Marennikova S.S., Akatova-Shelukhina E.M., Gurvich E.B. On the properties of alastrim virus. Vopr Virusol 1965; (4):439-47.

Marennikova S.S., Matsevich G.R., Svet-Moldavskaya I.A. On the efficiency of γ-globulin with an increased content of anti-smallpox antibodies in treatment of postvaccination encephalitides. Vopr Virusol 1968; (1):9-13.

Marennikova SS, Chimishkyan KL, Maltseva NN, Shelukhina EM, Fedorov VV. Characteristics of virus strains for production of smallpox vaccine. Proceedings of the Symposium on Smallpox; 1969 September 2–3; Symposium on Acute Respiratory Disease; 1969 October 1–2; Zagreb.

Marennikova S.S., Ladny I.D., Shelukhina E.M., Mal'tseva N.N., Matsevich G.R., Chimishkyan K.L. Chimpanzee pox under natural conditions (?). Vopr Virusol 1971a; (4):469-70.

Marennikova S.S., Shelukhina E.M., Mal'tseva N.N., Ladny I.D. Monkeypox—casual agent of smallpox-like human disease. Vopr Virusol 1971b; (4):463-9.

Marennikova S.S., Gurvich E.B., Shelukhina E.M. Comparison of the properties of five pox virus strains isolated from monkeys. Arch ges Virusforsch 1971c; 33:201-10.

Marennikova S.S., Seluhina E.M., Malceva N.N., Cimiskjan K.L., Macevic G.R. Isolation and properties of the causal agent of a new variola-like disease (monkeypox) in man. Bull WHO 1972a; 46:599-611.

Marennikova S.S., Šeluhina E.M., Mal'ceva N.N., Ladny I.D. Poxviruses from clinically ill and asymptomatically infected monkeys and a chimpanzee. Bull WHO 1972b; 46:613-20.

Marennikova S.S., Cimiškjan K.L., Šenkman L.S., Macevič G.R. Some factors determining differences in the antigenicity of vaccinia virus strains. Bull WHO 1972c; 46:159-63.

Marennikova S.S., Shelukhina E.M., Shenkman L.S. Role of the temperature of incubation of infected chick embryos in the differentiation of certain poxviruses according to pock morphology. Acta Virol 1973; 17:362.

Marennikova S.S., Maltseva N.N., Korneeva V.I., Garanina V.M. Pox infection in Carnivora of the family Felidae. Acta Virol 1975a; 18:260.

Marennikova S.S., Matsevich G.R., Yanchaitis B., et al. Materials to the assessment of vaccination against smallpox under protection of gamma-globulin as a method for prevention of neurological complications. Pediatriya 1975b; (1):10-3.

Marennikova S.S., Shelukhina, E.M., Shenkman L.S., Mal'tseva N.N., Matsevich G.R. Results of examination of wild monkeys for the presence of antibodies to variola virus and viruses of variola group. Vopr Virusol 1975c; (3):321-6.

Marennikova S.S., Shelukhina E.M., Matsevich G.R., Habahpasheva N.A. An antigenically atypical strain of variola virus. J Gen Virol 1976a; 33:513-5.

Marennikova S.S., Maltseva N.N., Korneeva V.I. Pox in giant anteaters due to agent similar to cowpox virus. Brit Vet J 1976b; 132:182-6.

Marennikova S.S., Shelukhina E.M., Shenkman L.S. "White-wild" (variola-like) poxvirus strains from rodents in equatorial Africa. Acta Virol 1976c; 20:80-2.

Marennikova S.S., Matsevich G.R., Sokolova A.F., Shul'ga L.G., et al. Immunological characterization of a two-step method of anti-smallpox vaccination. Zh Mikrobiol 1977a; (5):105-9.

Marennikova S.S., Maltseva N.N., Korneeva V.I., Garanina N.M. Outbreak of pox disease among Carnivora (Felidae) and Edentata. J Inf Dis 1977b; 135:358-66.

Marennikova S.S., Ladnyi I.D., Ogorodnikova Z.I., Shelukhina E.M., Maltzeva N.N. Identification and study of a poxvirus isolated from wild rodents in Turkmenia. Arch Virol 1978a; 56:7-14.

Marennikova S.S., Shelukhina E.M., Fimina V.A. Pox infection in white rats. Lab Animals 1978b; 12:33-6.

Marennikova S.S., Gancheva Ts., Donchev D., Matsevich G.R., Petrov V., Rumenova I., Matova E. Reactogenicity and immunological efficiency of a two-stage vaccination against variola virus. Epidem Mikrobiol Infekts Bolesti (Sofia) 1978c; 15:229-36.

Marennikova S.S., Maltseva N.N., Habahpaševa N.A. ELISA—a simple test for detecting and differentiating antibodies to closely related orthopoxviruses. Bull WHO 1981; 59:365-9.

Marennikova S.S., Voinarovska I., Bokhenek V., Dziok A.F., Shelukhina E.M., Mal'tseva N.N., Matsevich G.R., Efremova E.V. Cowpox in humans. Zh Mikrobiol 1984a; (8):64-9.

Marennikova S.S., Shelukhina E.M., Efremova E.V. New outlook on the biology of cowpox virus. Acta Virol 1984b; 28:437-44.

Marennikova S.S., Shelukhina E.M., Maltseva N.N., Efremova E.V., et al. Data of serological examination of the population of Republic of the Congo for the presence of antibodies to orthopoxviruses. I. Comparative estimation of various examination methods and the total results. Zh Mikrobiol 1984c; (3):95-100.

Marennikova S.S., Ezek Z., Szczeniowski M., Paluku Malenga Mbudi, Vernett M. On contagiousness of monkeypox for humans: results of examination of two infection outbreaks in Zaire. Zh Mikrobiol 1985; (8):38-43.

Marennikova S.S., Shelukhina E.M., Khodakevich L.N., Yanova N.N. Isolation of monkeypox virus from a wild African squirrel. Vopr Virusol 1986; (2):238-41.

Marennikova S.S., Nagieva F.G., Matsevich G.R., Shelukhina E.M., Khabahpasheva N.A., Platonova G.M. Monoclonal antibodies to monkeypox virus: preparation and application. Acta Virol 1988a; 32:19-26.

Marennikova S.S., Zhukova O.I., Manenkova G.M., Yanova N.N. A laboratory confirmed case of human infection with ratpox (cowpox). Zh Mikrobiol 1988b; (6):30-2.

Marennikova S.S., Shelukhina E.M., Matsevich G.R., Ezek Z., Khodakevich L.N., Zhukova O.A., Janova N.N., Chekunova E.V. Monkeypox in humans: current results. Acta Virol 1989; 33:246-53.

Marennikova S.S., Yanova N.N., Zhukova O.A. Electron microscopy as a method for diagnostics in epidemiological surveillance over poxvirus infections. Zh Mikrobiol 1990; (8):57-62.

Marennikova S.S., Zhukova O.A., Nagieva F.G., Shelukhina E.M., Matsevich G.R. Development and application of EIA test kit for detection and species-level identification of monkeypox virus. Zh Mikrobiol 1991; (2):60-3.

Marennikova SS, Zhukova OA, Panina I. An unusual form of pathology caused by vaccinia virus. Proceedings of the IXth International Congress of Virology; 1993 August 8–13; Glasgow.

Marennikova S, Zhukova O, Shelukhina E, Tsanava Sh, Gashnikov P, Shchelkunov S, Yanova N. Rats in the ecological chain of cowpox virus. Proceedings of the 3^d Congress of European Society of Veterinarian Virology on Immunobiology of Viral Infections; 1994; Interlaken, Switzerland.

Marennikova S.S., Gashnikov P.V., Zhukova O.A., Ryabchikova E.I., Strel'tsov V.V., Ryazankina O.I., Chekunova E.V., Yanova N.N., Shchelkunov S.N. Biotype and genetic characterization of a cowpox virus isolate that caused infection of a child. Zh Mikrobiol 1996; (4):6-10.

Marennikova S.S., Onischenko G.G., Matsevich G.R., Sandakhchiev L.S. Smallpox vaccination in the light of past experiences. Epidemiologiya 2003; (6):73-78.

Marotta C.A., Wilson J.T., Forget B.G., Weissman S.M. Human β-globin messenger RNA. III. Nucleotide sequences derived from complementary DNA. J Biol Chem 1977; 252:5040-53.

Marsden J.P., Greenfield C.R.M. Inherited smallpox. Arch Dis Childhood 1934; 9:309-14.

Martin K.H., Grosenbach D.W., Franke C.A., Hruby D.E. Identification and analysis of three myristylated vaccinia virus late proteins. J Virol 1997; 71:5218-26.

Massague J., Pandiella A. Membrane-anchored growth factors. Annu Rev Biochem 1993; 62:515-41.

Massung R.F., Liu L.-I., Qi J., Knight J.C., Yuran T.E., Kerlavage A.R., Parsons J.M., Venter J.C., Esposito J.J. Analysis of the complete genome of smallpox variola major virus strain Bangladesh-1975. Virology 1994; 201:215-40.

Massung R.F., Knight J.C., Esposito J.J. Topography of variola smallpox virus inverted terminal repeats. Virology 1995; 211:350-5.

Massung R.F., Loparev V.N., Knight J.C., Totmenin A.V., Chizhikov V.E., Parsons J.M., Safronov P.F., Gutorov V.V., Shchelkunov S.N., Esposito J.J. Terminal region sequence variations in variola virus DNA. Virology 1996; 221:291-300.

Mathew T. Virus study of pock disease among buffaloes. Indian J Pathol Bacteriol 1967; 10:101-2.

Mathew T., Mathew Z. Isolation, cultivation and haemadsorption of buffalo pox virus on BHK-21 cell line from Dhule epidemic (Western India). Int J Zoonoses 1986; 13:45-8.

Matsevich, Gennady R., *Vaccinal Encephalitis after Immunization against Smallpox (Prevalence, Serotherapy, and Prospects of Prophylaxis.* Moscow: Candidate of Science Dissertation, 1970.

Matsevich, Gennady R., *Specific Prophylaxis of Smallpox Infection upon Cancellation of Smallpox Vaccination.* Moscow: Doctor of Science Dissertation, 1983.

Matsevich G.R., Frolova M.A., Gudkova R.G. Experimental study of cell-mediated immunity upon introduction of inactivated and live vaccinia virus. Zh Mikrobiol 1978; (4):24-6.

Maxam A., Gilbert W. Sequencing end-labeled DNA with base-specific chemical cleavages. Methods Enzymol 1980; 65:499-560.

Mayer H. Eine generalisierte Vaccinia-Pockeninfektion eines Zirkusele fanten als Ursache menschlicher Erkrankungen. Zbl Bakt I Orig 1973; A224:448-52.

Mayer H., Gehring H., Mahnel H. Elefantenpocken in einem Wenderzirkus. Verhandlungsber 1972; 14: Int Symp Erkrank Zootiere (Wroclaw) 14:211-6.

Mayr A. Ein Beitrag zum Problem der qualitativen Differenzierung einzelner Vaccinevirusstamme. Zbl Bakt 1958; 171:7.

Mayr A., Herrlich A. Züchtung des Variolavirus in der infantilen Maus. Arch ges Virusforsch. 1960; 10:226-35.

Mayr A., Stickl H., Müller H.K., Danner K., Singer H. Der Pockenimpfstamm MVA: Marker genetische Struktur, Erfahrungen mit der parenteralen Schutzimplung und Verhalten im abwehrgeschwächten Organismus. Zbl Bakt Parasitenk Infectionskr Hyg 1978; 167:375-90.

Mayr A., Lauer J., Czerny C.P. Neue Fakten über die Verbreitung von Orthopockenvirusinfektionen. Der praktische Tierärtzl 1995; 11:961-7.

McCallum F.O., McDonald J.K. 1957. Survival of variola virus in raw cotton. Bull WHO 1995; 16:247-54.

McCarthy K., Downie A.W. An investigation of immunological relationships between the viruses of variola, vaccinia, cowpox and ectromelia by neutralization tests on the chorioallantois of chick embryos. Brit J Exp Path 1948; 29:501-10.

McCarthy K., Helbert D. A comparison of haemagglutinins of variola, alastrim, vaccinia, cowpox and ectromelia viruses. J Path Bacteriol 1960; 79:416-20.

McCarthy K., Downie A.W., Bradley W.H. The antibody response in man following infection with viruses of the Pox group. II. Antibody response following vaccination. J Hyg 1958; 56:466-78.

McConnell S.J., Herman Y.F., Mattson D.E., Erickson L. Monkeypox disease in irradiated cynomolgus monkeys. Nature 1962; 195:1128-9.

McConnell S.J., Hickman R.L., Wooding W.J., Huxsoll D.L. Monkeypox: experimental infection in chimpanzee (*Pan satyrus*) and immunization with vaccinia virus. Amer J Vet Res 1968; 29:1675-80.

McDonald W.F., Traktman P. Vaccinia virus DNA polymerase. *In vitro* analysis of parameters affecting processivity. J Biol Chem 1994; 269:31190-7.

McDonald W.F., Klemperer N., Traktman P. Characterization of a processive form of the vaccinia virus DNA polymerase. Virology 1997; 234:168-75.

McFadden G., Stuart D., Upton C., *et al.* Replication and resolution of poxvirus telomeres. Cancer Cells 2003; 6:77-85.

McGeoch D.J. Protein sequence comparisons show that 'pseudoproteases' encoded by the poxviruses and certain retroviruses belong to the deoxyuridine triphosphatase family. Nucl Acids Res 1990; 18:4105-10.

McIntosh A.A.G., Smith G.L. Vaccinia virus glycoprotein A34R is required for infectivity of extracellular enveloped virus. J Virol 1996; 70:272-81.

McKelvey T, Andrews SC, Miller SE, Pickup DJ. A major non-essential component of the virus particle is required for the formation of A-type inclusion bodies that contain virus particles. Proceedings of the 9th International Conference on Poxviruses and Iridoviruses; 1992 September 1–6; Les Diablerets, Switzerland.

McKelvey T.A. Andrews S.C., Miller S.E., Ray C.A., Pickup D.J. Identification of the orthopoxvirus p4c gene, which encodes a structural protein that directs intracellular mature virus particles into A-type inclusions. J Virol 2002; 76:11216-25.

Meis R.J., Condit R.C. Genetic and molecular biological characterization of vaccinia virus gene which renders the virus dependent on isatin-β-thiosemicarbazone (IBT). Virology 1991; 182:442-54.

Meltzer M.I., Damon I., LeDuc J.W., Millar J.D. Modeling potential responses to smallpox as a bioterrorist weapon. Emerg Infect Dis 2001; 7:959-69.

Merchlinsky M. Intramolecular homologous recombination in cells infected with temperature-sensitive mutants of vaccinia virus. J Virol 1989; 63:2030-5.

Merchlinsky M. Resolution of poxvirus telomeres: processing of vaccinia virus concatemer junctions by conservative strand exchange. J Virol 1990; 64:3437-46.

Merchlinsky M., Moss B. Resolution of linear minichromosomes with hairpin ends from circular plasmids containing vaccinia virus concatemer junctions. Cell 1986; 45:879-84.

Merchlinsky M., Moss B. Nucleotide sequence required for resolution of the concatemer junction of vaccinia virus DNA. J Virol 1989; 63:4354-61.

Merchlinsky M., Moss B. Introduction of foreign DNA into the vaccinia virus genome by *in vitro* ligation: recombination-independent selectable cloning vectors. Virology 1992; 190:522-6.

Meyer H., Pfeffer M., Rziha H.-J. Sequence alteration within and downstream of the A-type inclusion protein genes allow differentiation of *Orthopoxvirus* species by polymerase chain reaction. J Gen Virol 1994; 75:1975-81.

Meyer H., Ropp S.L., Esposito J.J. Gene for A-type inclusion body protein is useful for a polymerase chain reaction assay to differentiate orthopoxviruses. J Virol Methods 1997; 64:217-21.

Meyer H., Schay C., Mahnel H., Pfeffer M. Characterization of orthopoxviruses isolated from man and animals in Germany. Arch Virol 1999; 144:491-501.

Meyer H., Neubauer H., Pfeffer M. Amplification of 'variola virus-specific' sequences in German cowpox virus isolates. J Vet Med 2002a; 49:17-9.

Meyer H., Perrichot M., Stemmler M., Emmerich P., Schmitz H., Varaine F., Shungu R., Tshioko F., Formenty P. Outbreaks of disease suspected of being due to human monkeypox virus infection in the Democratic Republic of Congo in 2001. J Clin Microbiol 2002b; 40:2919-21.

Mika L., Pirsch J. Differentiation of variola virus from other members of the poxvirus group by plaques technique. J Bact 1960; 80:861-3.

Mikheev M.V., Lapa S.A., Shchelkunov S.N, Chikova A.K., Mikhailovich V.M., Sobolev A.Iu., Babkin I.V, Griadunov D.A., Bulavkina M.A., Gus'kov A.A., Sokunova E.B., Kochneva G.V., Blinov V.M., Sandakhchiev L.S., Zasedatelev A.S.,

Mirzabekov A.D. Identification of orthopoxvirus species using oligonucleotide microchips. Vopr Virusol 2003; 48:4-9.

Mikheev M.V., Feshchenko M.V., Shchelkunov S.N. Phylogenetic analysis of chemokine-binding protein gene from orthopoxviruses Mol Gen Mikrobiol Virusol 2004; (1):29-36.

Milhaud C., Klein M., Virat J. Analyse d'un cas de variole du singe (monkeypox) chez le chimpanzee (*Pan troglodytes*). Exp Anim 1969; 2:121-35.

Miller C.G., Shchelkunov S.N., Kotwal G.J. The cowpox virus-encoded homolog of the vaccinia virus complement control protein is an inflammation modulatory protein. Virology 1997; 229:126-33.

Miller N.F. *Smallpox Inoculation (Vaccination)*. Moscow: A.A. Kartsev, 1887.

Millns A.K., Carpenter M.S., DeLange A.M. The vaccinia virus-encoded uracil DNA glycosylase has an essential role in viral DNA replication. Virology 1994; 198:504-13.

Mims C.A. The response of mice to the intravenous injection of cowpox virus. Brit J Exp Path 1968; 49:24-32.

Mineeva R.M. Comparative data on dynamics of antibodies in the blood of smallpox vaccinees and convalescents. Trudy Kirg IEMG (Frunze) 1961; 5:47-57.

Miranda M.P., Reading P.C., Tscharke D.C., Murphy B.J., Smith G.L. The vaccinia virus kelch-like protein C2L affects calcium-independent adhesion to the extracellular matrix and inflammation in a murine intradermal model. J Gen Virol 2003; 84:2459-71.

Mitra A.C., Chatterjee S.N., Sarkar J.K., Manji P., Das A.K. Viraemia in haemorrhagic and other forms of smallpox. J Indian Med Assoc 1966; 47:112-4.

Mohamed M.R., Niles E.G. Interaction between nucleoside triphosphate phosphohydrolase I and the H4L subunit of the viral RNA polymerase is required for vaccinia virus early gene transcript release. J Biol Chem 2000; 275:25798-804.

Mohamed M.R., Niles E.G. The viral RNA polymerase H4L subunit is required for vaccinia virus early gene transcription termination. J Biol Chem 2001; 276:20758-65.

Moore J.B., Smith G.L. Steroid hormone synthesis by a vaccinia enzyme: a new type of virus virulence factor. EMBO J 1992; 11:1973-80.

Morgan C. The insertion of DNA into vaccinia virus. Science 1976; 193:591-2.

Morgan J.R., Roberts B.E. Organization of RNA transcripts from a vaccinia virus early gene cluster. J Virol 1984; 51:283-97.

Morgan J.R., Cohen L.K., Roberts B.E. Identification of the DNA sequence encoding the large subunit of the mRNA-capping enzyme of vaccinia virus. J Virol 1984; 52:206-14.

Morozov M.A. On Paschen bodies. Vrach Delo 1924; 24–26:1449-50.

Morozov M.A. Staining of Paschen bodies by silvering. Labor Praktika 1926; 5:6-7.

Morozov, Mikhail A., Solov'ev, Vladimir C. *Smallpox*. Moscow: Medgiz, 1948.

Morozov M.A., Korol'kova M.I., Kasatkevich S.S., Dolinov K.E. Dry smallpox vaccine. Zh Mikrobiol 1943; (6):76-80.

Mortimer P.P. The new cell culture smallpox vaccine should not be offered to the general population. Rev Med Virol 2003; 13:17-20.

Moss B. "Poxviridae: the Viruses and Their Replication." *Fields Virology.* B.N. Fields, D.M. Knipe, P.M. Howley, *et al.,* eds. Philadelphia: Lippincott–Raven Publishers 1996; 2637-71.

Moss B., Shisler J.L. Immunology 101 at poxvirus U: immune evasion genes. Semin Immunol 2001; 13:59-66.

Moss B., Shisler J.L., Xiang Y., Senkevich T.G. Immune-defense molecules of molluscum contagiosum virus, a human poxvirus. Trends Microbiol 2000; 8:473-7.

Mossman K., Upton C., Buller R.M.L., McFadden G. Species specificity of ectromelia virus and vaccinia virus interferon-γ binding proteins. Virology 1995; 208:762-9.

Mossman K., Nation P., Macen J., Garbutt M., Lucas A., McFadden G. Myxoma virus M-T7, a secreted homolog of the interferon-γ receptor, is a critical virulence factor for the development of myxomatosis in European rabbits. Virology 1996; 215:17-30.

Moyer R.W., Graves R.L. The mechanism of cytoplasmic orthopoxvirus DNA replication. Cell 1981; 27:391-401.

Moyer R.W., Graves R.L. The late white pock (mu) host range (hr) mutants of rabbit poxvirus are blocked in morphogenesis. Virology 1982; 119:332-46.

Moyer R.W., Brown G.D., Graves R.L. The white pock mutants of rabbit poxvirus. II. The early white pock (mu) host range (hr) mutants of rabbit poxvirus uncouple transcription and translation in nonpermissive cells. Virology 1980a; 106:234-49.

Moyer R.W., Graves R.L., Rothe C.T. The white pock (mu) mutants of rabbit poxvirus. III. Terminal DNA sequence duplication and transposition in rabbit poxvirus. Cell 1980b; 22:545-53.

Mukinda V.B.K., Mweta G., Kilundu M., Heymann D.L., Khan A.S., Esposito J.J., *et al.* Re-emergence of human monkeypox in Zaire in 1996. Lancet 1997; 349:1449-50.

Murphy P.M. Molecular mimicry and generation of host defense protein diversity. Cell 1993; 72:823-6.

Murti B.R., Shrivastav J.B. A study of biological behaviour of variola virus. I. Biological behaviour on the chorio-allantios. Indian J Med Sci 1957a; 11:574-9.

Murti B.R., Shrivastav J.B. A study of biological behaviour of variola virus. II. Experimental inoculation of laboratory animals. Indian J Med Sci 1957b; 11:580-7.

Mutombo M.W., Arita J., Jezek Z. Human monkeypox transmitted by a chimpanzee in a tropical rain-forest area of Zaire. Lancet 1983; I:735-7.

Müller T., Henning K., Kreimer M., Czerny C.P., Meyer H., Ziedler K. Seroprevalence of orthopoxvirus specific antibodies in red foxes (*Vulpes vulpes*) in the Federal state Brandenburg, Germany. J Wildlife Dis 1996; 32:348-53.

Nagafuchi A.Y., Shirayoshi Y., Okazaki K., Yasuda K., Takeichi M. Transformation of cell adhesion properties by exogenously introduced E-cadherin cDNA. Nature 1987; 329:341-3.

Nagler F.P. Application of Hirst's phenomenon to the titration of vaccinia virus and vaccinia immune serum. Med J Australia 1942; 1:281-3.

Nagler F.P.O., Rake G. The use of electron microscope in diagnosis of variola, vaccinia and varicella. J Bacteriol 1948; 55:45-51.

Naidoo J., Baxby D., Bennett M., Gaskell R.M., Gaskell C.J. Characterization of orthopoxviruses isolated from feline infections in Britain. Arch Virol 1992; 125:261-72.

Najarro P., Traktman P., Lewis J.A. Vaccinia virus blocks gamma interferon signal transduction: Viral VH1 phosphatase reverses Stat1 activation. J Virol 2001; 75:3185-96.

Nakano, James H., "Comparative Diagnosis of Poxvirus Diseases." In *Comparative Diagnosis of Viral Diseases: Human and Related Viruses,* E. Kurstak, C. Kurstak, eds. New York: Acad. Press, 1978.

Nakano E., Panicali D., Paoletti E. Molecular genetics of vaccinia virus: demonstration of marker rescue. Proc Natl Acad Sci USA 1982; 79:1593-6.

Nanning W. Prophylactic effect of antivaccinia gammaglobulin against post-vaccinal encephalitis. Bull WHO 1962; 27:317-24.

Neff, J.M. "Vaccination of Healthy Children with CVI-78 and Calf-Lymph Smallpox Vaccine." In *Proceedings of the 37th International Symposium on Smallpox Vaccine Organized by the International Association of Biological Standardization; 1972 October 11–13.* H. Cohen, H. Regamey, eds. Basel, New-York: S. Karger, 1973.

Nelson J.B. The behavior of pox viruses in the respiratory tract. II. The response of mice to the nasal instillation of variola virus. J Exp Med 1939; 70:107-15.

Nelson J.B. The stability of variola virus propagated in embryonated eggs. J Exp Med 1943; 78:231-9.

Neubauer H., Reischl U., Ropp S., *et al.* Specific detection of monkeypox virus by polymerase chain reaction. J Virol Methods 1998; 74:201-7.

Nevins J.R., Joklik W.K. Isolation and properties of the vaccinia virus DNA-dependent RNA polymerase. J Biol Chem 1977; 252:6930-8.

Ng A., Tscharke D.C., Reading P.C., Smith G.L. The vaccinia virus A41L protein is a soluble 30 kDa glycoprotein that affects virus virulence. J Gen Virol 2001; 82:2095-105.

Niles E.G., Seto J. Vaccinia virus gene D8 encodes a virion transmembrane protein. J Virol 1988; 62:3772-8.

Niles E.G., Lee-Chen G.-J., Shuman S., Moss B., Broyles S.S. Vaccinia virus gene D12L encodes the small subunit of the viral mRNA capping enzyme. Virology 1989; 172:513-22.

Nizamuddin M., Dumbell K.R. A simple laboratory test to distinguish the virus of smallpox from that of alastrim. Lancet 1961; I:68-9.

Noble J.Jr. A study of New and Old World monkeys to determine the likelihood of simian reservoir of smallpox. Bull WHO 1970; 42:509-14.

Noble J.Jr., Rich J.A. Transmission of smallpox by contact and by aerosol routes in *Macaca irus*. Bull WHO 1969; 40:279-86.

Noordaa, Jan, van der., *Primary Vaccination of Adults with an Attenuated Strain of Vaccinia Virus*. N.V. Uitgeverij W.P. van Stockum and Zoon-Gravenhage, 1964.

North E.A., Broben J.A., Mengoni A.H. The use of the chorio-allantois of the developing chick embryo in the diagnosis of smallpox. Med J Australia 1944; 1:437-8.

Novick D., Kim S.-H., Fantuzzi G., Reznikov L.L., Dinarello C.A., Rubinstein M. Interleukin-18 binding protein: a novel modulator of the Th1 cytokine response. Immunity 1999; 10:127-36.

Nowotny N. The domestic cat: a transmitter of viruses from rodents to man? A serological study of orthopox-, hanta- and encephalomyocarditis virus. Proceedings of the 3^d Congress of European Society of Veterinarian Virology on Immunobiology of Viral Infections; 1994a; Interlaken, Switzerland.

Nowotny N. The domestic cat: a possible transmitter of viruses from rodents to man. Lancet 1994b; 343:921.

Nowotny N. von, Fischer O.W., Schilcher F., Schwendenwein I., Loupal G., Schwarzmann Th., Meyer J., Hermanns W. Pockenvirusinfectionen bei Hauskatzen: klinische parho-histologische, virologische und epizootiologische Untersuchungen. Wien Tierärtzl Monatsschr 1994; 81:362-9.

Noyes W.F. A simple technical for demonstrating plaque formation with virus of vaccinia. Proc Soc Exp Biol Med 1953; 83:426-9.

Oda K., Joklik W.K. Hybridization and sedimentation studies on "early" and "late" vaccinia messenger RNA. J Mol Biol 1967; 27:395-419.

Ogorodnikova, Zlata I., *Several Issues of Vaccination Immunity to Variola Virus*. Moscow: Candidate of Science Dissertation, 1969.

Oie M., Shida H., Ichihashi Y. The function of the vaccinia hemagglutinin in the proteolytic activation of infectivity. Virology 1990; 176:494-504.

Okamura H., Tsutsui H., Kashiwamura S., Yoshimoto T., Nakanishi K. Interleukin-18: a novel cytokine that augments both innate and acquired immunity. Adv Immunol 1998; 70:281-9.

Olson V.A., Laue T., Laker M.T., Babkin I.V., Drosten C., Shchelkunov S.N., Niedrig M., Damon I.K., Meyer H. Real-time PCR system for detection of orthopoxviruses and simultaneous identification of smallpox virus. J Clin Microbiol 2004; (in press).

Onishchenko G.G., Maekov V.I., Ustyushin V.N., Borisevich S.N., Kuznetsova G.I., Loginova S.N., Berezhnoi A.M., Vasil'ev N.T., Maksimov V.A., Makhlai A.A. Isolation and identification of smallpox vaccine virus that caused iatrogenic vaccinia in children in Vladivostok. Zh. Mikrobiol 2003; 2:40-5

Opgenorth A., Strayer D., Upton C., McFadden G. Deletion of the growth factor gene related to EGF and TGFα reduces virulence of malignant rabbit fibroma virus. Virology 1992; 186:175-91.

Opgenorth A., Nation N., Graham K., McFadden G. Transforming growth factor alpha, Shope fibroma growth factor, and vaccinia growth factor can replace myxoma growth factor in the induction of myxomatosis in rabbits. Virology 1993; 192:701-9.

Ortiz M.A., Paez E. Identification of viral membrane proteins required for cell fusion and viral dissemination that are modified during vaccinia virus persistence. Virology 1994; 198:155-68.

Ouchterlony O. Antigen–antibody reactions in gels. Acta Path Microbiol Scand 1949; 26:507-15.

Owen J.E., Schultz D.W., Taylor A., Smith G.R. Nucleotide sequence of the lysozyme gene of bacteriophage T4: analysis of mutations involving repeated sequences. J Mol Biol 1983; 165:229-48.

Palumbo G.J., Pickup D.J., Fredrickson T.N., McIntyre L.J., Buller R.M.L. Inhibition of an inflammatory response is mediated by a 38-kDa protein of cowpox virus. Virology 1989; 171:262-73.

Palumbo GJ, Glasgow W, Pickup D, Buller RML. Multigenic evasion of an inflammatory response by poxviruses; Proceedings of the 9th International Conference on Poxviruses and Iridoviruses; 1992 September 1–6; Les Diablerets, Switzerland.

Palumbo G.J., Glasgow W.C., Buller R.M.L. Poxvirus-induced alteration of arachidonate metabolism. Proc Natl Acad Sci USA 1993; 90:2020-4.

Palumbo G.J., Buller R.M., Glasgow W.C. Multigenic evasion of inflammation by poxviruses. J Virol 1994; 68:1737-49.

Panicali D., Paoletti E. Construction of poxviruses as cloning vectors: Insertion of the thymidine kinase gene from herpes simplex into the DNA of infectious vaccinia virus. Proc Natl Acad Sci USA 1982; 79:4927-31.

Panus J.F., Smith C.A., Ray C.A., Smith T.D., Patel D.D., Pickup D.J. Cowpox virus encodes a fifth member of the tumor necrosis factor receptor family: a soluble, secreted CD30 homologue. Proc Natl Acad Sci USA 2002; 99:8348-53.

Parkinson J.E., Smith G.L. Vaccinia virus gene A36R encodes a Mr 43-50K protein on the surface of extracellular enveloped virus. Virology 1994; 204:376-90.

Parkinson J.E., Sanderson C.M., Smith G.L. The vaccinia virus A38L gene product is a 33-kDa integral membrane glycoprotein. Virology 1995; 214:177-88.

Paschen E. Was wissen wir über den Vakzineerreger? Münch med Wchschr 1906; 53:2391-3.

Paschen, E. Technick der Darstellung der Elementarkoerperchen (Paschen'sche Koerperchen in der Variola-pustel). Dtsch med Wchschr 1917; 43:1036.

Paschen, Enrique. "Technick der microskopischen Untersuchung der Pockenvirus." In *Handbuch der biologischen Arbeitsmethoden*. E. Abderhalden, ed. 1924; 13:567-94.

Patel D.D., Pickup D.J. Messenger RNAs of a strongly expressed late gene of cowpox virus contain 5'-terminal poly(A) sequences. EMBO J 1987; 6:3787-94.

Patel D.D., Pickup D.J. The second-largest subunit of the poxvirus RNA polymerase is similar to the corresponding subunits of procaryotic and eucaryotic RNA polymerases. J Virol 1989; 63:1076-86.

Patel D.D., Pickup D.J., Joklik W.K. Isolation of cowpox virus A-type inclusions and characterization of their major protein component. Virology 1986; 149:174-89.

Patel A.H., Gaffney D.F., Subak-Sharpe J.H., Stow N.D. DNA sequence of the gene encoding a major secreted protein of vaccinia virus, strain Lister. J Gen Virol 1990; 71:2013-21.

Paul G. Zur Differentialdiagnose der Variola und den Varicellen. Die Erscheinungen an der variolierten Hornhaut des Kaninchens und ihre frühzeitige Erkennung. Centralbl Bakt 1915; 75:518-24.

Payne L.G. Polypeptide composition of extracellular enveloped vaccinia virus. J Virol 1978; 27:28-37.

Payne L.G. Identification of the vaccinia hemagglutinin polypeptide from a cell system yielding large amounts of extracellular enveloped virus. J Virol 1979; 31:147-55.

Payne L.G. Significance of extracellular enveloped virus in the *in vitro* and *in vivo* dissemination of vaccinia. J Gen Virol 1980; 50:89-100.

Payne L.G. Characterization of vaccinia virus glycoproteins by monoclonal antibody preparations. Virology 1992; 187:251-60.

Payne L.G., Kristensson K. Extracellular release of enveloped vaccinia virus from mouse nasal epithelial cells *in vivo*. J Gen Virol 1985; 66:643-6.

Payne L.G., Norrby E. Presence of haemagglutinin in the envelope of extracellular vaccinia virus particles. J Gen Virol 1976; 32:63-72.

Pearson W.R., Lipman D.J. Improved tools for biological sequence comparison. Proc Natl Acad Sci USA 1988; 85:2444-8.

Pelham H.R., Sykes J.M., Hunt T. Characteristics of a coupled cell-free transcription and translation system directed by vaccinia cores. Eur J Biochem 1978; 82:199-209.

Pelkonen P.M., Tarvainen K., Hynninen A., Kallio E.R.K., Henttonen H., Palva A., Vaheri A., Vapalahti O. Cowpox with severe generalized eruption, Finland. Emerg Infect Dis 2003; 9:1458-61.

Pepose J.S., Margolis T.P., LaRussa P., Pavan-Langston D. Ocular complications of smallpox vaccination. Am J Ophthalmol 2003; 136:343-52.

Perkus M.E., Goebel S.J., Davis S.W., Johnson G.P., Norton E.K., Paoletti E. Deletion of 55 open reading frames from the termini of vaccinia virus. Virology 1991; 180:406-10.

Peters J.C. A monkeypox enzooty in the "Blijdorp" Zoo. Tijdschr Diergeneeskd 1966; 91:387-91.

Pfeffer K.T., Matsuyama T., Kunig T.M., Wakeham A., Kishihara K., Shahinian A., Wiegmann K., Ohashi P.S., Kronke M., Mak T.W. Mice deficient for the 55 kd tumor necrosis factor receptor are resistant to endotoxic shock, yet succumb to *L. monocytogenes* infection. Cell 1993; 73:457-67.

Pfeffer M., Burck G., Meyer H. Kuhpockenviren in Deutschland: Eine Analyse von 5 Fällen aus dem Jahr 1998. Berl Münch Tierärztl Wschr 1999; 112:334-8.

Pfeiff B., Pullmann H., Eis-Hübinger A.M., Gerritzen A., Schneweis K.E., Mayr A. Letale Tierpockeninfektion bei einem Atopiker unter dem Bild einer Variola vera. Hautarzt 1991; 42:293-7.

Pickup D.J., Ink B.S., Parsons B.L., Hu W., Joklik W.K. Spontaneous deletions and duplications of sequences in the genome of cowpox virus. Proc Natl Acad Sci USA 1984; 81:6817-21.

Pickup D.J., Ink B.S., Hu W., Ray C.A., Joklik W.K. Hemorrhage in lesions caused by cowpox virus is induced by a viral protein that is related to plasma protein inhibitors of serine proteases. Proc Natl Acad Sci USA 1986; 83:7698-702.

Pilaski J, Jacoby F. Are wild-living rodents the primary reservoir of cowpox virus in Europe? Proceedings of the IX Int. Congress of Virology; 1993 August 8–13 Glasgow.

Pilaski J., Rösen A., Dara, G. Comparative analysis of the genomes of orthopoxviruses isolated from elephant, rhinoceros and okapi by restriction enzymes. Arch Virol 1986; 88:135-42.

Pogo B.G., Dales S. Two deoxyribonuclease activities within purified vaccinia virus. Proc Natl Acad Sci USA 1969; 63:820-7.

Polak, M.F. 1973. "Complications of Smallpox Vaccination in the Netherlands, 1959-1970." In *Proceedings of the 37th International Symposium on Smallpox Vaccine Organized by the International Association of Biological Standardization; 1972 October 11–13.* H. Cohen, H. Regamey, eds. Basel, New-York: S. Karger, 1973.

Polak M.F., Beunders B.J.W., Werff A.R. van der, Sanders W., Klaveren J.N. van, Brans L.M. A comparative study of clinical reactions observed after application of several smallpox vaccines in primary vaccination of young adults. Bull WHO 1963; 29:311-22.

Porterfield J.S., Allison A.S. Studies with poxviruses by an improved plaque technique. Virology 1960; 10:233-44.

Post L.E., Arfsten A.E., Davis G.R., Nomura M. DNA sequence of the promoter region for the alpha ribosomal protein operon in *Escherichia coli*. J Biol Chem 1980; 255:4653-9.

Prier J.E., Sauer R.M. A pox disease of monkeys. Ann NY Acad Sci 1960; 85:951-9.

Prier J.E., Sauer R.M., Malsberger R.G., Sillaman J.M. Studies on a pox disease of monkeys. II. Isolation of the etiological agent. Amer J Vet Res 1960; 21:381-4.

Puckett C., Moss B. Selective transcription of vaccinia virus genes in template dependent soluble extracts of infected cells. Cell 1983; 35:441-8.

Pulford D.J., Meyer H., Ulaeto D. Orthologs of the vaccinia A13L and A36R virion membrane protein genes display diversity in species of the genus *Orthopoxvirus*. Arch Virol 2002; 147:995-1015.

Punjabi A., Boyle K., DeMasi J., Grubisha O., Unger B., Khanna M., Traktman P. Clustered charge-to-alanine mutagenesis of the vaccinia virus A20 gene: temperature-sensitive mutants have a DNA-minus phenotype and are defective in the production of processive DNA polymerase activity. J Virol 2001; 75:12308-18.

Rajagopal I., Ahn B.Y., Moss B., Mathews C.K. Roles of vaccinia virus ribonucleotide reductase and glutaredoxin in DNA precursor biosynthesis. J Biol Chem 1995; 270:27415-8.

Ramirez J.C., Tapia E., Esteban M. Administration to mice of a monoclonal antibody that neutralizes the intracellular mature virus form of vaccinia virus limits virus replication efficiently under prophylactic and therapeutic conditions. J Gen Virol 2002; 83:1059-67.

Rao, Ayyagari R. 1972. *Smallpox*. Bombay: The Kothari Book Depot, 1972.

Rao P.S. Immunizing potency against variola of strains of vaccinia virus seeds used in India for the preparations of vaccine lymph. Indian J Med Res 1952; 40:341-51.

Ravanello M.P., Hruby D.E. Conditional lethal expression of the vaccinia virus L1R myristylated protein reveals a role in virion assembly. J Virol 1994; 68:6401-10.

Ray C.A., Black R.A., Kronheim S.R., Greenstreet T.A., Sleath P.R, Salvesen G.S., Pickup D.J. Viral inhibition of inflammation: cowpox virus encodes an inhibitor of the interleukin-1β converting enzyme. Cell 1992; 69:597-604.

Reed K.D., Melski J.W., Graham M.B., Regnery R.L., Sotir M.J., Wegner M.V., Kazmierczak J.J., Stratman E.J., Li Y., Fairley J.A., Swain G.R., Olson V.A., Sargent E.K., Kehl S.C., Frace M.A., Kline R., Foldy S.L., Davis J.P., Damon I.K. The detection of monkeypox in humans in the Western Hemisphere. N Engl J Med 2004; 350:342-50.

Regamey R.H. Propos sur la vaccination antivariolique. Med Hyg (Geneva) 1965; 710:1117-9.

Regnery D.C. Isolation and partial characterization of an orthopoxvirus from a California vole (*Microtus californicus*). Arch Virol 1987; 94:159-62.

Reid R.A., Hemperly J.J. Human *N*-cadherin: nucleotide and deduced amino acid sequence. Nucl Acids Res 1990; 18:5896.

Reith RW, Williamson JD. Pathogenesis of vaccinia and cowpox infections, proceedings of the XI Poxvirus and Iridovirus Meeting; 1996 May 4–8; Spain.

Rempel R.E., Traktman P. Vaccinia virus B1 kinase: phenotypic analysis of temperature-sensitive mutants and enzymatic characterization of recombinant proteins. J Virol 1992; 66:4413-26.

Resch W, Weisberg AS, Moss B. Vaccinia virus gene A11R is required for the formation of viral membrane structures. Proceedings of the XVth International Poxvirus and Iridovirus Conference; 2004 September 3–8; Keble College, Oxford, England.

Ribas E. Alastrim, amaas or milk-pox. Trans Roy Soc Trop Med Hyg 1910; 4:224-32.

Rice A.P., Roberts B.E. Vaccinia virus induces cellular mRNA degradation. J Virol 1983; 47:529-39.

Richter C., Schweizer M., Cossarizza A., Franceschi C. Control of apoptosis by the cellular ATP level. FEBS Lett 1996; 378:107-10.

Ringwald M., Schuh R., Vestweber P., Eistetter H., Lottspeich F., Engel J., Dolz R., Jahnig F., Epplen J., Mayer S., Muller C., Kemler R. The structure of cell adhesion molecule uvomorulin. Insights into the molecular mechanism of Ca^{2+}-dependent cell adhesion. EMBO J 1987; 6:3647-53.

Risco C., Rodriguez J.R., Lopez-Iglesias C., Carrascosa J.L., Esteban M., Rodriguez D. Endoplasmic reticulum–Golgi intermediate compartment membranes and vimentin filaments participate in vaccinia virus assembly. J Virol 2002; 76:1839-55.

Rivas C., Gil J., Melkova Z., Esteban M., Diaz-Guerra M. Vaccinia virus E3L protein is an inhibitor of the interferon (IFN)-induced 2-5A synthetase enzyme. Virology 1998; 243:406-14.

Rivers T.M., Ward S.M., Baird R.D. Amount and duration of immunity induced by intradermal inoculation of cultured vaccine virus. J Exp Med 1939; 69L857-66.

Roberts J.A. Histopathogenesis of mousepox. I. Respiratory infection. Brit J Exp Path 1962; 43:451-61.

Robinson D.N., Cooley L. Drosophila kelch is an oligomeric ring canal actin organizer. J Cell Biol 1997; 138:799-810.

Rochester S.C., Traktman P. Characterization of the single-stranded DNA binding protein encoded by the vaccinia virus I3 gene. J Virol 1998; 72:2917-26.

Rodriguez J.F., Esteban M. Mapping and nucleotide sequence of the vaccinia virus gene that encodes a 14-kilodalton fusion protein. J Virol 1987; 61:3550-4.

Rodriguez J.F., Smith G.L. Inducible gene expression from vaccinia virus vectors. Virology 1990; 177:239-50.

Rodriguez J.F., Kahn J.S., Esteban M. Molecular cloning, encoding sequence, and expression of vaccinia virus nucleic acid-dependent nucleoside triphosphatase gene. Proc Natl Acad Sci USA 1986; 83:9566-70.

Rodriguez D., Esteban M., Rodriguez J.R. Vaccinia virus A17L gene product is essential for an early step in virion morphogenesis. J Virol 1995; 69:4640-8.

Rodriguez J.R., Risco C., Carrascosa J.L., Esteban M., Rodriguez D. Characterization of early stages in vaccinia virus membrane biogenesis: implications of the 21-kilodalton protein and a newly identified 15-kilodalton envelope protein. J Virol 1997; 71:1821-33.

Rodriguez J.R., Risco C., Carrascosa J.L., Esteban M., Rodriguez D. Vaccinia virus 15-kilodalton (A14L) protein is essential for assembly and attachment of viral crescents to virosomes. J Virol 1998; 72:1287-96.

Roger H., Weil E. Inoculation su singe. C R Soc Biol 1902; p. 1271.

Rohde, Walter. "Pocken." In W. Rohde, U. Schneeweiss, F.M.G. Otto, *Grundriβ der Impfpraxis*. Leipzig: J.A. Barth, 1968.

Roizman B., Joklik W., Fields B., Moss B. The destruction of smallpox virus stocks in national repositories: a grave mistake and a bad precedent. Infect Agents Dis 1994; 3:215-7.

Rollins B.J. Chemokines. Blood 1997; 90:909-28.

Romano P.R., Zhang F., Tan S.L., Garcia-Barrio M.T., Katze M.G., Dever T.E., Hinnebusch A.G. Inhibition of double-stranded RNA-dependent protein kinase PKR by vaccinia virus E3: role of complex formation and the E3 *N*-terminal domain. Mol Cell Biol 1998; 18:7304-16.

Rondle C.J.M., Dumbell K.R. Antigens of cowpox virus. J Hyg 1962; 60.:41-9.

Rondle C.J.M., Sayeed K.A.R. Studies on monkeypox virus. Bull WHO 1972; 46:577-83.

Roper R.L., Moss B. Envelope formation is blocked by mutation of a sequence related to the HKD phospholipid metabolism motif in the vaccinia virus F13L protein. J Virol 1999; 73:1108-17.

Roper R, Moss B. Vaccinia A35R is important in virulence. Proceedings of the XIVth International Poxvirus and Iridovirus Symposium; 2002 September 20–24; Lake Placid, New York.

Roper R.L., Payne L.G., Moss B. Extracellular vaccinia virus envelope glycoprotein encoded by the A33R gene. J Virol 1996; 70:3753-62.

Roper R.L., Wolffe E.J., Weisberg A., Moss B. The envelope protein encoded by the A33R gene is required for formation of actin-containing microvilli and efficient cell-to-cell spread of vaccinia virus. J Virol 1998; 72:4129-204.

Ropp S.L., Jin Q., Knight J.C., Massung R.F., Esposito J.J. Polymerase chain reaction strategy for identification and differentiation of smallpox and other orthopoxviruses. J Clin Microbiol 1995; 33:2069-76.

Rosahn P.D., Hu C.K. Rabbitpox. Report of an epidemic. J Exp Med 1935; 62:331.

Rosales R., Harris N., Ahn B.-Y., Moss B. Purification and identification of a vaccinia virus-encoded intermediate stage promoter-specific transcription factor that has homology to eukaryotic transcription factor SII (TFIIS) and an additional role as a viral RNA polymerase subunit. J Biol Chem 1994a; 269:14260-7.

Rosales R., Sutter G., Moss B. A cellular factor is required for transcription of vaccinia viral intermediate-stage genes. Proc Natl Acad Sci USA 1994b; 91:3794-8.

Rosel J.L., Moss B. Transcriptional and translational mapping and nucleotide sequence analysis of a vaccinia virus gene encoding the precursor of the major core polypeptide 4b. J Virol 1985; 56:830-8.

Rosel J.L., Earl P.L., Weir J.P., Moss B. Conserved TAAATG sequence at the transcriptional and translational initiation sites of vaccinia virus late genes deduced by structural and functional analysis of the *Hin*dIII-H genome fragment. J Virol 1986; 60:436-49.

Rosengard A.M., Liu Y., Nie Z.P., Jimenez R. Variola virus immune evasion design: Expression of a highly efficient inhibitor of human complement. Proc Natl Acad Sci USA 2002; 99:8808-13.

Rotz L.D., Khan A.S., Lillibridge S.R., Ostroff S.M., Hughes J.M. Public health assessment of potential biological terrorism agents. Emerg Infect Dis 2002; 8:225-30.

Ruby J, Senkevich T, Buller M, Cuff S. A poxvirus protein inhibits apoptosis mediated by CD40 and the p75 TNF receptor. Proceedings of the XIth Poxvirus and Iridovirus Meeting; 1996 May 4–9; Toledo, Spain.

Ruffer, Marc Armand, *Studies in the Palaeopathology of Egypt,* R.L. Moodie, ed. Chicago: University of Chicago Press, 1921.

Ruffer M.A., Ferguson A.R. Note on an eruption resembling that of variola in the skin of a mummy of the twentieth dynasty (1200–1100 B.C.). J Path Bact 1910; 15:4.

Ruiz-Arguello B, Alejo A, Smith VP, Saraiva M, Ho Y, Alcami A. Novel chemokine binding properties of the tumor necrosis factor receptor encoded by ectromelia and variola viruses. Proceedings of the XVth International Poxvirus and Iridovirus Conference; 2004 September 3–8; Keble College, Oxford, England.

Rutherfurd K.J., Chen S., Shively J.E. Isolation and amino acid sequence analysis of bovine adrenal 3β-hydroxysteroid dehydrogenase/steroid isomerase. Biochemistry 1991; 30:8108-16.

Ryazankina O.I., Muravlev A.I., Gutorov V.V., Mikrjukov N.N., Cheshenko I.O., Shchelkunov S.N. Comparative analysis of the conserved region of the orthopoxvirus genome encoding the 36K and 12K proteins. Virus Res 1993; 29:281-303.

Rytik P.G., Boiko V.I., Votyakov V.I., Pinchuk S.I., Gudkov V.G. A new experimental model of vaccinia infection. Vopr Virusol 1976; (4):465-8.

Safronov P.F., Petrov N.A., Ryazankina O.I., Totmenin A.V., Shchelkunov S.N., Sandak-hchiev L.S. Host range genes of cowpox virus. Dokl Ross Akad Nauk 1996; 249:829-33.

Saitou N., Nei M. The neighbor-joining method: a new method for reconstructing phylogenetic trees. Mol Biol Evol 1987; 4:406-25.

Salles-Gomes L.F. de, Angulo J.J., Menezes E., Zamith V.A. Clinical and subclinical *variola minor* in a ward outbreak. J Hyg 1965; 63:49-58.

Salmons T., Kuhn A., Wylie F., Schleich S., Rodriguez J.R., Rodriguez D., Esteban M., Griffiths G., Locker J.K. Vaccinia virus membrane proteins p8 and p16 are cotranslationally inserted into the rough endoplasmic reticulum and retained in the intermediate compartment. J Virol 1997; 71:7404-20.

Samuel C.E. Antiviral actions of interferon. Interferon regulated cellular proteins and their surprisingly selective antiviral activities. Virology 1991; 183:1-11.

Sanderson C.M., Parkinson J.E., Hollinshead M., Smith G.L. Overexpression of the vaccinia virus A38L integral membrane protein promotes Ca^{2+} influx into infected cells. J Virol 1996; 70:905-14.

Sanderson C.M., Hollinshead M., Smith G.L. The vaccinia virus A27L protein is needed for the microtubule-dependent transport of intracellular mature virus particles. J Gen Virol 2000; 81:47-58.

Sandvik T., Tryland M., Hansen H., Mehl R., Moens U., Olsvik Ø., Traavik T. Naturally occurring orthopoxviruses: potential recombination with vaccine vectors. J Clin Microbiol 1998; 36:2542-7.

Sanger F., Nicklen S., Coulson A.R. DNA sequencing with chain-terminating inhibitors. Proc Natl Acad Sci USA 1977; 74:5463-7.

Sanz P., Moss B. Identification of a transcription factor, encoded by two vaccinia virus early genes, that regulates the intermediate stage of viral gene expression. Proc Natl Acad Sci USA 1999; 96:2692-7.

Saravia M., Alcami A. *CrmE*, a novel soluble tumor necrosis factor receptor encoded by poxviruses. J Virol 2001; 75:226-33.

Sarkar J.K., Mitra A.C. Virulence of variola virus isolated from smallpox cases of varying severity. Indian J Med Res 1967; 55:13-20.

Sarkar J.K., Mitra A.C., Mukherjee M.K., De S.K., Guha Mazumdar D. Virus excretion in smallpox. I. Excretion in the throat, urine and conjunctive of patients. Bull WHO 1973a; 48:517-22.

Sarkar J.K., Mitra A.C., Mukherjee M.K., De S.K. Virus excretion in smallpox. II. Excretion in the throats of household contacts. Bull WHO 1973b; 48:523-7.

Sarov I., Joklik W.K. Studies on the nature and location of the capsid polypeptides of vaccinia virions. Virology 1972; 50:579-92.

Schay C, Meyer H, Mahnel H. Classification of orthopoxviruses isolated in Germany during 1985 to 1991. Proceeding of the 9th International Conference on Poxviruses and Iridoviruses; 1992 September 1–6; Les Diablerets, Switzerland.

Scheiflinger F., Dorner F., Falkner F.G. Construction of chimeric vaccinia viruses by molecular cloning and packaging. Proc Natl Acad Sci USA 1992; 89:9977-81.

Schmelz M., Sodeik B., Ericsson M., Wolffe E.J., Shida H., Hiller G., Griffiths G. Assembly of vaccinia virus: the second wrapping cisterna is derived from the trans Golgi network. J Virol 1994; 68:130-47.

Schmidt, M., *Zoologische Klinik. Die Krankheiten der Affen*. Berlin: A. Hirschwald, 1870.

Schmitt J.F.C., Stunnenberg H.G. Sequence and transcriptional analysis of the vaccinia virus *Hind*HI I fragment. J Virol 1988; 62:1889-97.

Schnierle B.S., Gershon P.D., Moss B. Cap-specific mRNA(nucleotide-O2'-)-methyltransferase and poly(A) polymerase stimulatory activities of vaccinia virus are mediated by a single protein. Proc Natl Acad Sci USA 1992; 89:2897-901.

Schönbauer M., Schönbauer-Langle A., Kölbl S. Pockeninfection bei einer Hauskatze. Zbl Vet Med 1982; B29:434-40.

Schultz G.S., White M., Mitchell R., Brown G., Lynch J., Twardzik D.R., Todaro G.J. Epithelial wound healing enhanced by transforming growth factor-α and vaccinia growth factor. Science 1987; 235:350-1.

Schüppel K.-F., Menger S., Eulenberger K., Bernhard A., Pilaski J. Kuhpockeninfektion bei Alpakas (*Lama glama pacos*). Verhandlungsbericht über die Erkrankungen der Zootiere 1997; 38:259-65.

Schupp P., Pfeffer M., Meyer H., Rurck G., Kölmel K., Neumann C. Cowpox virus in a 12-year-old boy: rapid identification by an orthopoxvirus-specific polymerase chain reaction. Br J Dermatol 2001; 145:146-50.

Schwarz D.A., Katayama C.D., Hedrick S.M. *Schlafen*, a new family of growth regulatory genes that affect thymocyte development. Immunity 1998; 9:657-68.

Seet B.T., Johnston J.B., Brunetti C.R., Barrett J.W., Everett H., Cameron C., Sypula J., Nazarian S.H., Lucas A., McFadden G. Poxviruses and immune evasion. Ann Rev Immunol 2003; 21:377-423.

Sehgal C.L., Ray S.N. Survey of rhesus monkeys (*Macaca mulatta*) for haemaglutination-inhibition antibody against vaccinia/variola and monkeypox viruses. J Commun Dis 1974; 6:233-5.

Seki M., Oie M., Ichihashi Y., Shida H. Hemadsorption and fusion inhibition activities of hemagglutinin analyzed by vaccinia virus mutants. Virology 1990; 175:372-84.

Senkevich TG, Moss B. Vaccinia virus H2 protein, like the A28 protein, is an essential component of the virion membrane specifically required for virus entry into cells. Proceedings of the XVth International Poxvirus and Iridovirus Conference; 2004 September 3–8; Keble College, Oxford, England.

Senkevich T.G., Muravnik G.L., Pozdnyakov S.G., Chizhikov V.E., Ryazankina O.I., Shchelkunov S.N., Koonin E.V., Chernos V.I. Nucleotide sequence of *Xho*I fragment of ectromelia virus DNA reveals significant differences from vaccinia virus. Virus Res 1993; 30:73-88.

Senkevich T.G., Koonin E.V., Buller R.M.L. A poxvirus protein with a RING zinc finger motif is of crucial importance for virulence. Virology 1994; 198:118-28.

Senkevich T.G., White C.L., Koonin E.V., Moss B. A viral member of the ERV1/ALR protein family participates in a cytoplasmic pathway of disulfide bond formation. Proc Natl Acad Sci USA 2000; 97:12068-73.

Senkevich T.G., White C.L., Weisberg A., Granek J.A., Wolffe E.J., Koonin E.V., Moss B. Expression of the vaccinia virus A2.5L redox protein is required for virion morphogenesis. Virology 2002a; 300:296-303.

Senkevich T.G., White C.L., Koonin E.V., Moss B. Complete pathway for protein disulfide bond formation encoded by poxviruses. Proc Natl Acad Sci USA 2002b; 99:6667-72.

Senkevich T.G., Ward B.M., Moss B. Vaccinia virus entry into cells is dependent on a virion surface protein encoded by the A28L gene. J Virol 2004; 78:2357-66.

Seregin S.V., Babkina I.N., Nesterov A.E., Sinyakov A.N., Shchelkunov S.N. Comparative studies of gamma-interferon receptor-like proteins of variola major and variola minor viruses. FEBS Lett 1996; 382:79-83.

Sergiev P.G., Svet-Moldavsky G.Ya. Cultivation of viruses in developing eggs of reptiles. I. Cultivation of vaccinia virus on the chorioallantois of developing eggs of turtles and grass snakes. Bull Eksperim Biol Med 1951; 31:141-4.

Shafikova, Risalya A, *Properties of the Freshly Isolated Variola Virus Strains and Study of Variation of This Virus under Natural and Experimental Conditions.* Moscow: Candidate of Science Dissertation, 1970.

Shafikova R.A., Marennikova S.S., Characterization of properties of the variola virus strains isolated during various forms of the disease. Vopr Virusol 1970; (6):699-702.

Shchelkunov S.N. Functional organization of variola major and vaccinia virus genomes. Virus Genes 1995; 10:53-71.

Shchelkunov S.N. Species-specific differences in organization of molecular virulence factors of variola, cowpox, and vaccinia virus. Infec Dis Rev Suppl 2001; 3:16-25.

Shchelkunov S.N. Immunomodulatory proteins of orthopoxviruses. Mol Biol (Mosk) 2003; 37:41-53.

Shchelkunov S.N., Totmenin A.V. Two types of deletions in orthopoxvirus genomes. Virus Genes 1995; 9:231-45.

Shchelkunov S.N., Marennikova S.S., Totmenin A.V., Blinov V.M., Chizhikov V.E., Gutorov V.V., Safronov P.F., Pozdnyakov S.G., Shelukhina E.M., Gashnikov P.V., Andzhaparidze O.G., Sandakhchiev L.S. Construction of libraries of fragments of smallpox virus DNA and structure–function analysis of viral host range genes. Dokl Ross Akad Nauk 1991; 321:402-6.

Shchelkunov S.N., Blinov V.M., Totmenin A.V., Marennikova S.S., Kolykhalov A.A., Frolov I.V., Chizhikov V.E., Gutorov V.V., Gashnikov P.V., Belanov E.F., Belavin P.A., Resenchuk S.M., Shelukhina E.M., Netesov S.V., Andzhaparidze O.G., Sandakhchiev L.S. Study of the structure–function organization of variola virus genome. I. Cloning of *Hind*III and *Xho*I fragments of the viral DNA and sequencing of *Hind*III-M, -L, and –I fragments. Mol Biol (Mosk) 1992a; 26:1099-115.

Shchelkunov SN, Marennikova SS, Blinov VM, Totmenin AV, Chizhikov VE, Netesov SV, Andzhaparidze OG, Sandakhchiev LS. The nucleotide sequence of genome of variola major virus strain India-1967. Proceedings of the 9th International Conference on Poxviruses and Iridoviruses; 1992b September 1–6; Les Diablerets, Switzerland.

Shchelkunov S.N., Blinov V.M., Sandakhchiev L.S. Genes of variola and vaccinia viruses necessary to overcome the host protective mechanisms. FEBS Lett 1993a; 319:80-3.

Shchelkunov S.N., Blinov V.M., Sandakhchiev L.S. Ankyrin-like proteins of variola and vaccinia viruses. FEBS Lett 1993b; 319:163-5.

Shchelkunov S.N., Blinov V.M., Resenchuk S.M., Totmenin A.V., Sandakhchiev L.S. Analysis of the nucleotide sequence of 43 kbp segment of the genome of variola virus India-1967 strain. Virus Res 1993c; 30:239-58.

Shchelkunov, Sergei N., Blinov, Vladimir M., Totmenin, Aleksei V., Chizhikov, Vladimir E., Olenina, L.V., Gutorov, V.V., Safronov, P.F., Sandakhchiev, L.S., Marennikova, S.S., Andzhaparidze, O.G., Li-Ing L., Utterbach, T., Kerlavage, A., Selivanov, N., Venter, J.C., Esposito, J.J., Massung, R.F., Jin, Q., Knight, J.C., Loparev, V., Cavallaro, K.F., Yuran, T.E., Parsons, J.M., Mahy, B.W.J. "Sequencing of the Variola Virus Genome." In *Concepts in Virology: From Ivanovsky to the Present*, B.W.J. Mahy, D.K. Lvov, eds. Singapore: Harwood Academic Publishers, 1993d.

Shchelkunov S.N., Blinov V.M., Totmenin A.V., Marennikova S.S., Kolykhalov A.A., Frolov I.V., Chizhikov V.E., Gutorov V.V., Gashnikov P.V., Belanov E.F., Belavin P.A., Resenchuk S.M., Andzhaparidze O.G., Sandakhchiev L.S. Nucleotide sequence analysis of variola virus *Hind*III M, L, I genome fragments Virus Res 1993e; 27:25-35.

Shchelkunov S.N., Resenchuk S.M., Totmenin A.V., Blinov V.M., Marennikova S.S., Sandakhchiev L.S. Comparison of the genetic maps of variola and vaccinia viruses. FEBS Lett 1993f; 327:321-4.

Shchelkunov S.N., Blinov V.M., Resenchuk S.M., Totmenin A.V., Olenina L.V., Chirikova G.B., Sandakhchiev L.S. Analysis of the nucleotide sequence of 53 kbp from the right terminus of the genome of variola major virus strain India-1967. Virus Res 1994a; 34:207-36.

Shchelkunov S.N., Resenchuk S.M., Totmenin A.V., Blinov V.M., Sandakhchiev L.S. Analysis of the nucleotide sequence of 48 kbp of the variola major virus strain India-1967 located on the right terminus of the conservative genome region. Virus Res 1994b; 32:1-19.

Shchelkunov S.N., Massung R.F., Esposito J.J. Comparison of the genome DNA sequences of Bangladesh-1975 and India-1967 variola viruses. Virus Res 1995; 36:107-18.

Shchelkunov S.N., Totmenin A.V., Sandakhchiev L.S. Analysis of the nucleotide sequence of 23.8 kbp from the left terminus of the genome of variola major virus strain India-1967. Virus Res 1996; 40:169-183.

Shchelkunov S.N., Safronov P.F., Totmenin A.V., Petrov N.A., Ryazankina O.I., Gutorov V.V., Kotwal G.J. The genomic sequence analysis of the left and right species-specific terminal region of a cowpox virus strain reveals unique sequences and a cluster of intact ORFs for immunomodulatory and host range proteins. Virology 1998; 243:432-60.

Shchelkunov S.N., Totmenin A.V., Loparev V.N., Safronov P.F., Gutorov V.V., Chizhikov V.E., Knight J.C., Parsons J.M., Massung R.F., Esposito J.J. Alastrim, smallpox, variola minor virus genome DNA sequences. Virology 2000; 266:361-86.

Shchelkunov S.N., Totmenin A.V., Babkin I.V., Safronov P.F., Ryazankina O.I., Petrov N.A., Gutorov V.V., Uvarova E.A., Mikheev M.V., Sisler J.R., Esposito J.J., Jahrling P.B., Moss B., Sandakhchiev L.S. Human monkeypox and smallpox viruses: genomic comparison. FEBS Lett 2001; 509:66-70.

Shchelkunov S.N., Totmenin A.V., Safronov P.F., Mikheev M.V., Gutorov V.V., Ryazankina O.I., Petrov N.A., Babkin I.V., Uvarova E.A., Sandakhchiev L.S., Sisler J.R., Esposito J.J., Damon I.K., Jahrling P.B., Moss B. Analysis of the monkeypox virus genome. Virology 2002a; 297:172-94.

Shchelkunov S., Totmenin A., Kolosova I. Species-specific differences in organization of orthopoxvirus kelch-like proteins. Virus Genes 2002b; 24:157-62.

Shedlovsky T., Smadel J.E. The LS-antigen of vaccinia. II. Isolation of a single substance combining both L- and S-activity. J Exp Med 1942; 75:165-78.

Shelukhina, Emma M., *Biology and Ecology of Orthopoxviruses Pathogenic for Humans*. Moscow: Doctor of Science Dissertation, 1980.

Shelukhina E.M., Marennikova S.S. Generalized monkeypox in baby rabbits and white mice upon oral infection. Vopr Virusol 1975; (6):703-5.

Shelukhina E.M., Marennikova S.S., Mal'tseva N.N., Matsevich G.R., Hasmi A.A. Results of a virological study of smallpox convalescents and contacts. Zh Gig Epidemiol Mikrobiol Immunol 1973; 17:266-71.

Shelukhina E.M., Maltseva N.N., Shenkman L.S., Marennikova S.S. Properties of two isolates (MK-7-73 and MK-10-73) from wild monkeys Brit Vet J 1975; 131:746-8.

Shelukhina E.M., Marennikova S.S., Shenkman L.S., Froltsova A.E. Variola virus strains of 1960–1975: the range of intraspecies variability and relationship between properties and geographic origin. Acta Virol 1979a; 23:360-6.

Shelukhina E.M., Shenkman L.S., Rozina E.E., Marennikova S.S. A possible mechanism for preservation of certain orthopoxviruses in nature. Vopr Virusol 1979b; (24):368-71.

Shelukhina E.M., Andronnikov V.A., Tabakov V.A., *et al.* Group infections of dairymaids with pseudocowpox. Zh Gig Epidemiol Mikrobiol Immunol 1986a; (4):431-8.

Shenkman, Lyudmila S., *Experimental Study of the Vaccination Process upon Administration of Various Vaccinia Virus Strains.* Moscow: Candidate of Science Dissertation, 1972.

Shida H. Nucleotide sequence of the vaccinia virus hemagglutinin gene. Virology 1986; 150:451-62.

Shida H., Dales S. Biogenesis of vaccinia: Carbohydrates of the hemagglutinin molecule. Virology 1981; 111:56-72.

Shida H., Tanabe, K., Matsumoto S. Mechanism of virus occlusion into A-type inclusion during poxvirus infection. Virology 1977; 76:217-33.

Shimoyama Y., Yoshida T., Terada M., Shimosato Y., Abe O., Hirohashi S. Molecular cloning of a human Ca^{2+}-dependent cell–cell adhesion molecule homologous to mouse placental cadherin: its low expression in human placental tissues. J Cell Biol 1989; 109:1789-94.

Shisler J.L., Jin X.L. The vaccinia virus K1L gene product inhibits host NF-kappaB activation by preventing IkappaBalpha degradation. J Virol 2004; 78:3553-60.

Shisler J.L., Moss B. Immunology 102 at poxvirus U: avoiding apoptosis. Semin Immunol 2001; 13:67-72.

Shneiderman, Anna E., *The Factors Determining Reactogenicity and Antigenic Activity of Smallpox Vaccines.* Moscow: Candidate of Science Dissertation, 1978.

Shooter R.A. (Chairman). *Report of the Investigation into the Cause of the 1978 Birmingham Smallpox Occurrence.* London: H.M. Stationery Office, 1980.

Shors T., Keck J.G., Moss B. Down regulation of gene expression by the vaccinia virus D10 protein. J Virol 1999; 73:791-6.

Shuman S. Vaccinia virus RNA helicase: An essential enzyme related to the DE-II family of RNA-dependent NTPases. Proc Natl Acad Sci USA 1992; 89:10935-9.

Shuman S., Morham S.G. Domain structure of vaccinia virus mRNA capping enzyme. Activity of the Mr 95,000 subunit expressed in *Escherichia coli.* J Biol Chem 1990; 265:11967-72.

Shuman S., Moss B. Identification of a vaccinia virus gene encoding a type I DNA topoisomerase. Proc Natl Acad Sci USA 1987; 84:7478-82.

Shuman S., Moss B. Factor-dependent transcription termination by vaccinia virus RNA polymerase. Evidence that the *cis*-acting termination signal is in nascent RNA. J Biol Chem 1988; 263:6220-5.

Shuman S., Moss B. Purification and use of vaccinia virus messenger RNA capping enzyme. Methods Enzymol 1990; 181:170-80.

Shuman S., Broyles S.S., Moss B. Purification and characterization of a transcription termination factor from vaccinia virions. J Biol Chem 1987; 262:12372-80.

Shuman S., Golder M., Moss B. Insertional mutagenesis of the vaccinia virus gene encoding a type I DNA topoisomerase: evidence that the gene is essential for virus growth. Virology 1989; 170:302-6.

Simpson D.A., Condit R.C. Vaccinia virus gene A18R encodes an essential DNA helicase. J Virol 1995; 69:6131-9.

Singh J.P., Singh S.B. Isolation and characterization of the aetiologic agent of buffalopox. J Res Ludhiana 1967; 4:440-8.

Skinner M.A., Moore J.B., Binns M.M., Smith G.L., Boursnell M.E.G. Deletion of fowlpox virus homologues of vaccinia virus genes between the 3β-hydroxysteroid dehydrogenase (A44L) and DNA ligase (A50R) genes. J Gen Virol 1994; 75:2495-8.

Slabaugh M.B., Mathews C.K. Hydroxyurea-resistant vaccinia virus: overproduction of ribonucleotide reductase. J Virol 1986; 60:506-14.

Slabaugh M.B., Roseman N., Davis R., Matthews C. Vaccinia virus encoded ribonucleotide reductase: Sequence conservation of the gene for the small subunit and its amplification in hydroxyurea-resistant mutants. J Virol 1988; 62:519-27.

Smadel J.E., Rivers T.M., Hoagland C.L. Nucleoprotein antigen of vaccine virus. I. A new antigen from elementary bodies of vaccinia. Arch Path 1942; 34:275-85.

Smee D.F., Bailey K.W., Sidwell R.W. Treatment of cowpox virus respiratory infection in mice with ribavirin as a single agent or followed sequentially by cidofovir. Antivir Chem Chemother 2000a; 11:303-9.

Smee D.F., Bailey K.W., Wong M., Sidwell R.W. Intranasal treatment of cowpox virus respiratory infection in mice with cidofovir. Antivir Res 2000b; 47:171-7.

Smith, C.A., Goodwin, R.G. "Tumor Necrosis Factor Receptors in the Poxvirus Family: Biological and Genetic Implications." In *Viroceptors, Virokines and Related Immune Modulators Encoded by DNA Viruses*, G. McFadden, ed. Austin: R.G. Landes Company, 1994.

Smith C.A., Farrah T., Goodwin R.G. The TNF receptor superfamily of cellular and viral proteins: Activation, costimulation, and death. Cell 1994; 76:959-62.

Smith C.A., Hu F.Q., Smith T.D., Richards C.L., Smolak P., Goodwin R.G., Pickup D.J. Cowpox virus genome encodes a second soluble homologue of cellular TNF receptors, distinct from CrmB, that binds TNF but not LT alpha. Virology 1996; 223:132-47.

Smith C.A., Smith T.D., Smolak P.J., Friend D., Hagen H., Gerhart M., Park L., Pickup D.J., Torrance D., Mohler K., Schooley K., Goodwin R.G. Poxvirus genomes encode a secreted, soluble protein that preferentially inhibits β chemokine activity yet lacks sequence homology to known chemokine receptors. Virology 1997; 236:316-27.

Smith G.L. Vaccinia virus glycoproteins and immune evasion. J Gen Virol 1993; 74:1725-40.

Smith G.L., Chan Y.S. Two vaccinia virus proteins structurally related to the interleukin-1 receptor and the immunoglobulin superfamily. J Gen Virol 1991; 72:511-8.

Smith G.L., Carlos A. de, Chan Y.S. Vaccinia virus encodes a thymidylate kinase gene: Sequence and transcriptional mapping. Nucl Acids Res 1989a; 17:7581-90.

Smith G.L., Chan Y.S., Kerr S.M. Transcriptional mapping and nucleotide sequence of a vaccinia virus gene encoding a polypeptide with extensive homology to DNA ligases. Nucl Acids Res 1989b; 17:9051-62.

Smith G.L., Howard S.T., Chan Y.S. Vaccinia virus encodes a family of genes with homology to serine proteinase inhibitors. J Gen Virol 1989c; 70:2333-43.

Smith G.L., Chan Y.S., Howard S.T. Nucleotide sequence of 42 kbp of vaccinia virus strain WR from near the right inverted terminal repeat. J Gen Virol 1991; 72:1349-76.

Smith G.L., Vanderplasschen A., Law M. The formation and function of extracellular enveloped vaccinia virus. J Gen Virol 2002; 83:2915-31.

Smith S.A., Mullin N.P., Parkinson J., Shchelkunov S.N., Totmenin A.V., Loparev V.N., Srisatjaluk R., Reynolds D.N., Keeling K.L., Justus D.E., Barlow P.N., Kotwal G.J. Conserved surface-exposed K/R-X-K/R motifs and net positive charge on poxvirus complement control proteins serve as putative heparin binding sites and contribute to inhibition of molecular interactions with human endothelial cells: a novel mechanism for evasion of host defense. J Virol 2000; 74:5659-66.

Sodeik B., Krijnse-Locker J. Assembly of vaccinia virus revisited: *de novo* membrane synthesis or acquisition from the host? Trends Microbiol 2002; 10:15-24.

Sodeik B., Doms R.W., Ericsson M., Hiller G., Machamer C.E., Hof W., Meer G. van, Moss B., Griffiths G. Assembly of vaccinia virus: role of the intermediate compartment between the endoplasmic reticulum and the Golgi stacks. J Cell Biol 1993; 121:521-41.

Solov'ev V.D., Bektemirov T.A. On the question of differentiation of vaccinia and variola viruses in cell culture. Vopr Virusol 1962; (4):24-7.

Solov'ev V.D., Mastyukova Yu.N. On the methods for titration of vaccinia and variola viruses and neutralizing antibodies. Vopr Virusol 1958; (6):342-6.

Solov'ev, Valentin D., Mastyukova, Yuliya N., *Vaccinia Virus and Problems of Smallpox Vaccination.* Moscow: Medgiz, 1961.

Solov'ev V.D., Bektemirov V.A., Marchenko A.T., Nikolaevsky G.P. Study of cross-immunity of monkeys to vaccinia and variola viruses. Vopr Virusol 1962; (6):701-5.

Spaander, J. "Opening Address." In *Proceedings of the 37th International Symposium on Smallpox Vaccine Organized by the International Association of Biological Standardization; 1972 October 11–13.* H. Cohen, H. Regamey, eds. Basel, New-York: S. Karger, 1973.

Spehner D., Gillard S., Drillien R., Kirn A. A cowpox virus gene required for multiplication in Chinese hamster ovary cells. J Virol 1988; 62:1297-304.

Spriggs M.K. One step ahead of the game: Viral immunomodulatory molecules. Annu Rev Immunol 1996; 14:101-30.

Spriggs M.K. Shared resources between the neural and immune systems: semaphorins join the ranks. Curr Oppin Immunol 1999; 11:387-91.

Spriggs M.K., Hruby D.E., Maliszewski C.R., Pickup D.J., Sims I.E., Buller R.M.L., VanSlyke J. Vaccinia and cowpox viruses encode a novel secreted interleukin-1-binding protein. Cell 1992; 71:145-52.

Sprössig M, Macmerth R, Miksch R. Isolierung eines Virus der Pocken Gruppe bei einem erkrankten Elefanten. Proceedings of the International Symposium über Fragen der Pockenschutzes; 1968; Leipzig: J.A. Barth, 1968.

Stearn, Esther W., Stearn, Allen E., *The Effect of Smallpox on the Destiny of the Amerindian.* Boston, Massachusetts: Humphries, 1945.

Steinert P.M., Roop D.R. Molecular and cellular biology of intermediate filaments. Ann Rev Biochem 1988; 57:593-625.

Stern W., Dales S. Biogenesis of vaccinia: Isolation and characterization of a surface component that elicits antibody suppressing infectivity and cell–cell fusion. Virology 1976; 75:232-41.

Stickl, H., Hochstein-Mintzel, V., Huber, H.Ch. "Preliminary Results with the Highly Attenuated Vaccinia Virus MVA." In *Proceedings of the 37ᵗʰ International Symposium on Smallpox Vaccine Organized by the International Association of Biological Standardization; 1972 October 11–13.* H. Cohen, H. Regamey, eds. Basel, New-York: S. Karger, 1973.

Stokes G.V. High-voltage electron microscope study of the release of vaccinia virus from whole cells. J Virol 1976; 18:636-43.

Stolz W., Gotz A., Thomas P. Characteristic but unfamiliar—the cowpox infection, transmitted by a domestic cat. Dermatology 1996; 193:140-3.

Stone J.D., Burnet F.M. The production of vaccinia haemagglutinin in rabbit skin. Austral J Exp Biol Med Sci 1946; 24:9-13.

Stuart D.T., Upton C., Higman M.A., Niles E.G., McFadden G. A poxvirus-encoded uracil DNA glycosylase is essential for virus viability. J Virol 1993; 67:2503-12.

Studier F.W., Rosenberg A.H., Simon M.N., Dunn J.J. Genetic and physical mapping in the early region of the bacteriophage T7 DNA. J Mol Biol 1979; 135:917-37.

Subrahmanyan T.P. A study of the possible basis of age-dependent resistance of mice to poxvirus diseases. Austral J Exp Biol Med Sci 1968; 46:251-65.

Sugai T. Ueber den Komplementbindungsversuch bei Variola vera. Zbl Bakt 1909; 49:650-3.

Sung T.-C., Roper R.L., Zhang Y., Rudge S.A., Temel R., Hammond S.M., Morris A.J., Moss B., Engebrecht J., Frohman M.A. Mutagenesis of phospholipase D defines a superfamily including a *trans*-Golgi viral protein required for poxvirus pathogenicity. EMBO J 1997; 16:4519-30.

Suvaković V., Kecmanović M., Serhati A., Gazideda K. "Flat Type of Smallpox—Presentation of 12 Cases." *Variola u Jugoslaviji 1972 Godine,* L. Stojković, ed. Ljubljana: ČGP Delo, 1973.

Svet-Moldavsky G.Ya. Viral infection and body temperature. Vaccinia virus-induced keratitis in grass snakes. Byull Eksperim Biol Med 1954; 37:64-7.

Svet-Moldavskaya I.A. Vaccinia in white rats after total gamma irradiation. Acta Virol 1968; 12:271-4.

Svet-Moldavskaya, Inna A., *Regularities of Immunogenesis Induced by Vaccinia Virus and Experimental Approaches to Development of New Tests for Evaluation of Reactogenicity of Vaccine Strains and Efficiency of Antismallpox Immunity.* Moscow: Doctor of Science Dissertation, 1970.

Symons J.A., Alcami A., Smith G.L. Vaccinia virus encodes a soluble type I interferon receptor of novel structure and broad species specificity. Cell 1995; 81:551-60.

Szajner P., Weisberg A.S., Moss B. Physical and functional interactions between vaccinia virus F10 protein kinase and virion assembly proteins A30 and G7. J Virol 2004; 78:266-74.

Szathmary J. Studies on the non-specific inhibition by some body fluids against vaccinia haemagglutination. Arch ges Virusforsch 1961; 10:540.

Takahashi T., Oie M., Ichihashi Y. *N*-terminal amino acid sequences of vaccinia virus structural proteins. Virology 1994; 202:844-52.

Tamin A., Villarreal E.G., Weinrich S.L., Hruby D.E. Nucleotide sequence and molecular genetic analysis of the vaccinia virus *Hin*dIII N/M region encoding the genes responsible for resistance to α-amanitin. Virology 1988; 165:141-50.

Tantawi H.H., Fayed A.A., Shalaby M.A., Skalinsky E.I. Isolation, cultivation and characterization of poxvirus from Egyptian water buffaloes. J Egypt Vet Med Assoc 1977; 37:15-20.

Teissier P., Marie P.-L. Essais de serotherapie variolique. Compt Rend Acad Sci 1921; 155:1536-9.

Teissier P., Duvoir M., Stevenin H. Experiences de variolisation sur les singes (*M. rhesus* et *M. nemestrinus*). C R Soc Biol 1911; 70:654-6.

Tengelsen L.A., Slabaugh M.B., Bibler J.K., Hruby D.E. Nucleotide sequence and molecular genetic analysis of the large subunit of ribonucleotide reductase encoded by vaccinia virus. Virology 1988; 164:121-31.

Thompson C.L., Condit R.C. Marker rescue mapping of vaccinia virus temperature-sensitive mutants using overlapping cosmid clones representing the entire virus genome. Virology 1986; 150:10-20.

Thomsett L.R., Baxby D., Denham E.M. Cowpox in the domestic cat. Vet Rec 1978; 108:567.

Thorpe L.E., Mostashari F., Karpati A.M., Schwartz S.P., Manning S.E., Marx M.A., Frieden Th.R. Smallpox vaccination and adverse cardiac events. Emerg Infect Dis 2004; 10:962.

Tint, H. "The Rational for Elective Pre-Vaccination with Attenuated Vaccinia (CVI-78) in Preventing Some Vaccination Complications." In *Proceedings of the 37th International Symposium on Smallpox Vaccine Organized by the International Association of Biological Standardization; 1972 October 11–13*. H. Cohen, H. Regamey, eds. Basel, New-York: S. Karger, 1973.

Tongeren H.A.E. van. Spontaneous mutation of cowpox virus by means of egg passage. Arch ges Virusforsch 1952; 5:35-52.

Torres C.M. Further studies on the pathology of alastrim and their significance in the variola alastrim problem. Proc Roy Soc Med 1935–1936; 29:1525-39.

Tracey K.J., Cerami A. Tumor necrosis factor: a pleiotropic cytokine and therapeutic target. Annu Rev Med 1994; 45:491-503.

Traktman P., Sridhar P., Condid R.C., Roberts B.E. Transcriptional mapping of the DNA polymerase gene of vaccinia virus. J Virol 1984; 49:125-31.

Traktman P., Caligiuri A., Jesty S.A., Sankar U. Temperature-sensitive mutants with lesions in the vaccinia virus F10 kinase undergo arrest at the earliest stage of morphogenesis. J Virol 1995; 69:6581-7.

Traktman P., Liu K., Demasi J., Rollins R., Jesty S., Unger B. Elucidating the essential role of the A14 phosphoprotein in vaccinia virus morphogenesis: construction and characterization of a tetracycline-inducible recombinant. J Virol 2000; 74:3682-95.

Tryland M., Myrmel H., Holtet L., Haukenes G., Traavik T. Clinical cowpox cases in Norway. Scand J Infect Dis 1998a; 30:301-3.

Tryland M., Sandvik T., Arnemo J.M., Gudbrand S., Olsvik Ø, Traavik T. Antibodies against orthopoxviruses in wild carnivores from Fennoscandia. J Wildlife Dis 1998b; 34:443-50.

Tryland M., Sandvik T., Hansen H., Haukenes G., Holtet L., Bennett M., Mehl R., Moens U., Olsvik Ø., Traavik T. Characteristics of four cowpox virus isolates from Norway and Sweden. APMIS 1998c; 106:623-35.

Tryland M., Sandvik T., Holtet L., Nilsen H., Olsvik Ø., Traavik T. Antibodies to orthopoxvirus in domestic cats in Norway. Vet Rec 1998d; 143:105-9.

Tryland M., Sandvik T., Mehl R., Bennet M., Traavik T., Olsvik Ø. Serosurvey for orthopoxviruses in rodents and shrews from Norway. J Wildlife Dis 1998e 34:240-50.

Tsanava, Shota A., *Search for Natural Reservoir of Orthopoxviruses on the Territory of Georgian SSR.* Moscow: Candidate of Science Dissertation, 1990.

Tsanava Sh., Sakvarelidze L., Shelukhina E. Serological survey of wild rodents in Georgia for antibodies to orthopoxviruses. Acta Virol 1989; 33:91.

Tseng M., Palaniyar N., Zhang W.D., Evans D.H. DNA binding and aggregation properties of the vaccinia virus I3L gene product. J Biol Chem 1999; 274:21637-44.

Tsuchiya Y., Tagaya I. Isolation of a multinucleated giant cell-forming and hemagglutinin-negative variant of variola virus. Arch ges Virusforsch 1972; 39:292-5.

Tulman E, Alfonso CL, Lu Z, Zsak L, Sandybaev NT, Kerembekova UZ, Zaitsev VL, Kutish GF, Rock DL. The genome of horsepox virus. Proceedings of the XIVth International Poxvirus and Iridovirus Symposium; 2002 September 20–25; Lake Placid, New York.

Turner G.S., Squires E.J. Inactivated smallpox vaccine: immunogenicity of inactivated intracellular and extracellular vaccinia virus. J Gen Virol 1971; 13:19-25.

Turner G.S., Squires E.J., Murray H.G.S. Inactivated smallpox vaccine. A comparison of inactivation methods. J Hyg 1970; 68:197-210.

Turner, P.C., Musy, P.Y., Moyer, R.W. "Poxvirus Serpins." In *Viroceptors, Virokines and Related Immune Modulators Encoded by DNA Viruses.* J. McFadden, ed. Austin: R.G. Landes Company, 1995.

Turner P.C., Baquero M.T., Yuan S., Thoennes S.R., Moyer R.W. The cowpox virus serpin SPI-3 complexes with and inhibits urokinase-type and tissue-type plasminogen activators and plasmin. Virology 2000; 272:267-80.

Ueda Y., Morikawa S., Matsuura Y. Identification and nucleotide sequence of the gene encoding a surface antigen induced by vaccinia virus. Virology 1990; 177:588-94.

Ulaeto D., Grosenbach D., Hruby D. E. The vaccinia virus 4c and A-type inclusion proteins are specific markers for the intracellular mature virus particle. J Virol 1996; 70:3372-7.

Unanov S.S., Marennikova S.S., Akatova E.M., Gurvich E.B., Kaptsova T.I., Mal'tseva N.N., Milushin V.N., Plutenko K.N., Shutov A.V. Comparative estimation of smallpox vaccines. Vopr Virusol 1967; (5):519-25.

Upton C., McFadden G. Tumorigenic poxviruses: analysis of viral DNA sequences implicated in the tumorigenicity of Shope fibroma virus and malignant rabbit virus. Virology 1986; 152:308-21.

Upton C., Macen J.L., McFadden G. Mapping and sequencing of a gene from myxoma virus that is related to those encoding epidermal growth factor and transforming growth factor alpha. J Virol 1987; 61:1271-5.

Upton C., Macen J.L., Maranchuk R.A., DeLange A.M., McFadden G. Tumoriogenic poxviruses: fine analysis of the recombination junctions in malignant rabbit fibroma virus, a recombinant between Shope fibroma virus and myxoma virus. Virology 1988; 166:229-39.

Upton C., Macen J.L., Wishart D.S., McFadden G. Myxoma virus and malignant rabbit fibroma virus encode a serpin-like protein important for virus virulence. Virology 1990; 179:618-31.

Upton C., Macen J.L., Schreiber M., McFadden G. Myxoma virus expresses secreted protein with homology to the tumor necrosis factor receptor gene family that contributes to viral virulence. Virology 1991; 184:370-82.

Upton C., Mossman K., McFadden G. Encoding of a homolog of IFN-γ receptor by myxoma virus. Science 1992; 258:1369-72.

Upton C., Stuart D., McFadden G. Identification of a poxvirus gene encoding a uracil DNA glycosylase. Proc Natl Acad Sci USA 1993; 90:4518-22.

Upton C., Schiff L. Rice S.A., Dowdeswell T., Yang X., McFadden G. A poxvirus protein with a RING finger motif binds zinc and localizes in virus factories. J Virol 1994; 68:4186-95.

Upton C., Slack S., Hunter A.L., Ehlers A., Roper R.L. Poxvirus orthologous clusters: toward defining the minimum essential poxvirus genome. J Virol 2003; 77:7590-600.

Uvarova E.A., Shchelkunov S.N. Species-specific differences in the structure of orthopoxvirus complement-binding protein. Virus Res 2001; 81:39-45.

Uvarova EA, Shchelkunov SN. Western African monkeypox virus strains lack the gene of complement-binding protein. Proceedings of the XVth International Poxvirus and Iridovirus Conference; 2004 September 3–8; Keble College, Oxford, England.

Van Meir E., Wittek R. Fine structure of the vaccinia virus gene encoding the precursor of the major core protein 4a. Arch Virol 1988; 102:19-27.

Varich N.L., Sychova I.V., Kaverin N.V., Antonova T.P., Chernos V.T. Transcription of both DNA strands of vaccinia virus genome *in vivo*. Virology 1979; 96:412-30.

Vaux D.L., Haecker G., Strasser A. An evolutionary perspective on apoptosis. Cell 1994; 76:777-9.

Venkatesan S., Gershowitz A., Moss B. Complete nucleotide sequences of two adjacent early vaccinia virus genes located within the inverted terminal repetition. J Virol 1982; 44:637-46.

Verlinde J.D. Koepokken bij de mens. Tschr Diergeneesk 1951; 76:334-42.

Verlinde J.D., Tongeren H.A.E. van. Isolation of smallpox virus from the nasopharynx of patients with variola sine eruptione. Antonie van Leeuwenhoek 1952; 18:109-12.

Vieuchange J., Brion G., Gruest J. Virus variolique en cultures de cellules. Ann Inst Pasteur 1958; 95:681-93.

Virus Taxonomy: The Classification and Nomenclature of Viruses. The Seventh Report of the International Committee on Taxonomy of Viruses. M.H.V. van Regenmortel, C.M. Fauquet, D.H.L. Bishop, E.B. Carstens, M.K. Estes, S.M. Lemon, J. Maniloff, M.A. Mayo, D.J. McGeoch, C.R. Pringle, R.B. Wickner, eds. VII Report of the ICTV. SanDiego: Academic Press, 2000.

Vito P., Lacano E., D'Adamio L. Interfering with apoptosis: Ca^{2+}-binding protein ALG-2 and Alzheimer's disease gene ALG-3. Science 1996; 271:521-5.

Voegeli, C.F. "Production and Use of Smallpox Egg Vaccine." In *Proceedings of the 37th International Symposium on Smallpox Vaccine Organized by the International Association of Biological Standardization; 1972 October 11–13.* H. Cohen, H. Regamey, eds. Basel, New-York: S. Karger, 1973.

Vorob'ev, Anatoly A., Lebedinsky, Vladimir A., *Mass Immunization Methods.* Moscow: Meditsina, 1977.

Vorob'ev A.A., Podkuiko V.N., Mikhailov V.V., Makhlai A.A. Non-parenteral immunization methods. Zh Mikrobiol 1996; (5): 117-21.

Vos J.C., Stunnenberg H.G. Derepression of a novel class of vaccinia virus genes upon DNA replication. EMBO J 1988; 7:3487-92.

Vos J.C., Sasker M., Stunnenberg H.G. Vaccinia virus capping enzyme is a transcription initiation factor. EMBO J 1991; 10:2553-8.

Walls H.H., Ziegler D.W., Nakano J.H. A study of the specificities of sequential antisera to variola and monkeypox viruses by radioimmunoassay. Bull WHO 1980; 58:131-8.

Walls H.H., Ziegler D.W., Nakano J.H. Characterization of antibodies to orthopoxviruses in human sera by radioimmunoassay. Bull WHO 1981; 59:253-62.

Wang S., Shuman S. Vaccinia virus morphogenesis is blocked by temperature-sensitive mutations in the F10 gene, which encodes protein kinase 2. J Virol 1995; 69:6376-88.

Wang S.P., Shuman S. A temperature-sensitive mutation of the vaccinia virus E11 gene encoding a 15-kDa virion component. Virology 1996; 216:252-7.

Ward B.M., Moss B. Vaccinia virus A36R membrane protein provides a direct link between intracellular enveloped virions and the microtubule motor kinesin. J Virol 2004; 78:2486-93.

Ward G.A., Stover C.K., Moss B., Fuerst T.R. Stringent chemical and thermal regulation of recombinant gene expression by vaccinia virus vectors in mammalian cells. Proc Natl Acad Sci USA 1995; 92:6773-7.

Wasilenko S.T., Stewart T.L., Meyers A.F., Barry M. Vaccinia virus encodes a previously uncharacterized mitochondrial-associated inhibitor of apoptosis. Proc Natl Acad Sci USA 2003; 100:14345-50.

Watson J.C., Chang H.-W., Jacobs B. Characterization of a vaccinia virus-encoded double-stranded RNA-binding protein that may be involved in inhibition of the double-stranded RNA-dependent protein kinase. Virology 1991; 185:206-16.

Weekly Epidemiological Record. Human monkeypox in Kasai Oriental, Zaire (1996–1997). 1997a; 72:101-4.

Weekly Epidemiological Record. Monkeypox in the Democratic Republic of the Congo (former Zaire). 1997b; 72:258.

Weekly Epidemiological Record. Human monkeypox in Kasai Oriental, Democratic Republic of the Congo (former Zaire). 1997c; 72:369-72.

Wehrle P.F., Posch J., Richter K.H., Henderson D.A. An airborne outbreak of smallpox in a German hospital and its significance with respect to other recent outbreaks in Europe. Bull WHO 1970; 43:669-79.

Wei C.M., Moss B. Methylated nucleotides block 5'-terminus of vaccinia virus messenger RNA. Proc Natl Acad Sci USA 1975; 72:318-22.

Weinrich S.L., Hruby D.E. A tandemly-oriented late gene cluster within the vaccinia virus genome. Nucl Acids Res 1986; 14:3003-16.

Weintraub S., Dales S. Biogenesis of poxviruses: genetically controlled modifications of structural and functional components of the plasma membrane. Virology 1974; 60:96-127.

Weir J.P., Moss B. Nucleotide sequence of the vaccinia virus thymidine kinase gene and the nature of spontaneous frameshift mutations. J Virol 1983; 46:530-7.

Weltzin R., Liu J., Pugachev K.V., Myers G.A., Coughlin B.C., Cruz J., Kennedy J.S., Ennis F.A., Monath T.P. Clonal vaccinia virus grown in cell culture as a new smallpox vaccine. Nat Med 2003; 9:1115-6.

Wenner H.A., Macosaet F.D., Kamitsuka P.S., Kidd P. Monkeypox. I. Clinical, virological, and immunologic studies. Amer J Epidemiol 1968; 87:551-66.

Wenner H.A., Bolano C.R., Cho, C.T., Kamitsuka P.S. Studies on the pathogenesis of monkeypox. III. Histopathological lesions and sites of immunofluorescence. Arch ges Virusforsch 1969a; 27:179-97.

Wenner H.A., Cho C.T., Bolano C.R., Kamitsuka P.S. Studies on the pathogenesis of monkeypox. II. Dose response and virus dispersion. Arch ges Virusforsch 1969b; 27:166-78.

Westwood J.C.N., Boulter E.A., Bowen E.T.W., Maber H.B. Experimental respiratory infection with poxviruses. I. Clinical, virological and epidemiological studies. Brit J Exp Path 1966; 47:453-65.

Whitaker-Dowling P., Younger J.S. Characterization of a specific kinase inhibitory factor produced by vaccinia virus which inhibits the interferon-induced protein kinase. Virology 1984; 137:171-81.

White C.L., Weisberg A.S., Moss B. A glutaredoxin, encoded by the G4L gene of vaccinia virus, is essential for virion morphogenesis. J Virol 2000; 74:9175-83.

Whitehead S.S., Hruby D.E. Differential utilization of a conserved motif for the proteolytic maturation of vaccinia virus proteins. Virology 1994a; 200:154-61.

Whitehead S.S., Hruby D.E. A transcriptionally controlled trans-processing assay: putative identification of a vaccinia virus-encoded proteinase which cleaves precursor protein P25K. J Virol 1994b; 68:7603-8.

WHO Advisory Committee on Variola Virus Research. *Report of the 5th Meeting.* Geneva, Switzerland: World Health Organization, 2003.

WHO. Requirements for Biological Substances. WHO Technical Report Series. 1959; No. 180.

WHO. Requirements for Biological Substances. WHO Technical Report Series. 1966; No. 323.

Wienecke R., Wolffe H., Schaller M., Meyer H., Plewig G. Cowpox virus infection in an 11-year-old girl. J Am Acad Dermatol 2000; 42:892-4.

Wildy, P. "Classification and Nomenclature of Viruses." In *Monographs in Virology.* Basel, München, Paris, London, New-York, Sydney: S. Karger, 1971.

Willemse A., Egberink H.F. Transmission of cowpox virus infection from domestic cat to man. Lancet 1985; I:1515.

Williams G.T., Smith C.A. Molecular regulation of apoptosis: genetic controls on cell death. Cell 1993; 74:777-9.

Williams O., Wolffe E.I., Weisberg A.S., Merchlinsky M. Vaccinia virus WR gene A5L is required for morphogenesis of mature virions. J Virol 1999; 73:4590-9.

Wilson, Graham S. *The Hazards of Immunization.* University of London: The Athlone Press, 1967.

Wilton S., Mohandas A.R., Dales S. Organization of the vaccinia envelope and relationship to the structure of intracellular mature virions. Virology 1995; 214:503-11.

Wisser J., Pilaski J., Strauss G., Meyer H., Burck-Truyen U., Rudolph M., Frölich K. Cowpox virus infection causing stillbirth in an Asian elephant (*Elephas maximus*). Vet Rec 2001; 149:244-6.

Wittek R., Cooper J.A., Moss B. Transcriptional and translational mapping of a 6.6-kilobase-pair DNA fragment containing the junction of the terminal repetition and unique sequences at the left end of the vaccinia virus genome. J Virol 1980; 39:722-32.

Wittek R., Hänggi, M., Hiller G. Mapping of a gene coding for a major late structural polypeptide on the vaccinia virus genome. J Virol 1984a; 49:371-8.

Wittek R., Richner B., Hiller G. Mapping of the genes coding for the two major vaccinia virus core polypeptides. Nucl Acids Res 1984b; 12:4835-48.

Wolff H.L., Croon J.J.A.B. The survival of smallpox virus (variola minor) in natural circumstances. Bull WHO 1968; 38:492-3.

Wolffe E.J., Isaacs S.N., Moss B. Deletion of the vaccinia virus B5R gene encoding a 42 kilodalton membrane glycoprotein inhibits extracellular virus envelope formation and dissemination. J Virol 1993; 67:4732-41.

Wolffe E.J., Moore D.M., Peters P.J., Moss B. Vaccinia virus A17L open reading frame encodes an essential component of nascent viral membranes that is required to initiate morphogenesis. J Virol 1996; 70:2797-808.

Wolffe E.J., Katz E., Weisberg A., Moss B. The A34R glycoprotein gene is required for induction of specialized actin-containing microvilli and efficient cell-to-cell transmission of vaccinia virus. J Virol 1997; 71:3904-15.

Wolffe E.J., Weisberg A.S., Moss B. Role for the vaccinia virus A36R outer envelope protein in the formation of virus-tipped actin-containing microvilli and cell-to-cell virus spread. Virology 1998; 244:20-6.

Wolfs T.F.W., Wagenaar J.A., Niesters H.G.M., Osterhaus A.D.M.E. Rat-to-human transmission of cowpox infection. Emerg Infect Dis. 2002; 9:1495-6.

Woodroofe G.M., Fenner F. Serological relationship within the poxvirus group: an antigen common to all members of the group. Virology 1962; 16:334-41.

World Health Organization. Guide to the Laboratory Diagnosis of Smallpox for Smallpox Eradication Programmes. Geneva: WHO, 1969.

World Health Organization. "The global Eradication of Smallpox. Final Report of the Global Commission for the Certification of Smallpox Eradication." In *History of International Public Health, Geneva.* No. 4, 1980.

World Health Organization. Smallpox: Yugoslavia. Wkly Epidemiol Rec 2004; 47:161-2.

Wright C.F., Coroneos A.M. The H4 subunit of vaccinia virus RNA polymerase is not required for transcription initiation at a viral late promoter. J Virol 1995; 69:2602-4.

Wright C.F., Oswald B.W., Dellis S. Vaccinia virus late transcription is activated *in vitro* by cellular heterogeneous nuclear ribonucleoproteins. J Biol Chem 2001; 276:40680-6.

Wu A.M., Chapman A.B., Platt T., Guarente L., Beckwith J. Deletions of distal sequence affect termination of transcription at the end of the tryptophan operon in *E. coli.* Cell 1980; 19:829-36.

Xiang Y., Simpson D.A., Spiegel J., Zhou A., Silverman R.H., Condit R.C. The vaccinia virus A18R DNA helicase is a postreplicative negative transcription elongation factor. J Virol 1998; 72:7012-23.

Xiang Y., Latner D.R., Niles E.G., Condit R.C. Transcription elongation activity of the vaccinia virus J3 protein *in vivo* is independent of poly(A) polymerase stimulation. Virology 2000; 269:356-69.

Xue F., Cooley L. Kelch encodes a component of intercellular bridges in *Drosophila* egg chambers. Cell 1993; 72:681-93.

Yamaguchi N., Macdonald D.W., Passanisi W.C., Harbour D.A., Hopper C.D. Parasite prevalence in free-ranging farm cats, *Felis silvestris catus.* Epidemiol Infect 1996; 116:217-23.

Yang W.-P., Kao S.-Y., Bauer W.R. Biosynthesis and post-translational cleavage of vaccinia virus structural protein VP8. Virology 1988; 167:585-90.

Yanova, Natal'ya N., *Electron Microscopy in Diagnostics and Investigation of Several poxviruses.* Moscow: Candidate of Science Dissertation, 1996.

Yeh W.W., Moss B., Wolffe E.J. The vaccinia virus A9L gene encodes a membrane protein required for an early step in virion morphogenesis. J Virol 2000; 74:9701-11.

Yu L., Shuman S. Mutational analysis of the RNA triphosphatase component of vaccinia virus mRNA capping enzyme. J Virol 1996; 70:6162-8.

Yuen L., Moss B. Oligonucleotide sequence signaling transcriptional termination of vaccinia virus early genes. Proc Natl Acad Sci USA 1987; 84:6417-21.

Yumasheva M.A. Cytological changes in the culture of skin–muscle tissue infected with variola and vaccinia viruses. Vopr Baktriol Virusol Immunol 1959; (14):191.

Yumasheva, Marina A., *Comparative Study of Cyto- and Histological Changes Caused by Several Viruses of Variola Group in Susceptible Tissues and Cells.* Moscow: Candidate of Science Dissertation, 1968.

Zajac P., Spehner D., Drillien R. The vaccinia virus J5L open reading frame encodes a polypeptide expressed late during infection and required for viral multiplication. Virus Res 1995; 37:163-73.

Zhang W., Evans D.H. DNA strand exchange catalyzed by proteins from vaccinia virus-infected cells. Virology 1993; 67:204-12.

Zhang Y., Moss B. Immature viral envelope formation is interrupted same stage by lac operator-mediated repression of the vaccinia virus D13L gene and by the drug rifampicin. Virology 1992; 187:643-53.

Zhang Y., Keck J.G., Moss B. Transcription of viral late genes is dependent on expression of the viral intermediate gene G8R in cells infected with an inducible conditional-lethal mutant vaccinia virus. J Virol 1992; 66:6470-9.

Zhang Y., Ahn B.-Y., Moss B. Targeting of a multicomponent transcription apparatus into assembling vaccinia virus particles requires RAP94, an RNA polymerase-associated protein. J Virol 1994; 68:1360-70.

Zhukova, Ol'ga A., *Modern Human Orthopoxvirus Infections: Diagnostics and Ecology of the Causative Agents.* Moscow: Candidate of Science Dissertation, 1993.

Zhukova O.A., Tsanava S.A., Marennikova S.S. Experimental infection of domestic cats by cowpox virus. Acta Virol 1992; 36:329-31.

Ziegler D.W., Hutchinson H.D., Koplan J.P., Nakano J.H. Detection by radioimmunoassay of antibodies in human smallpox patients and vaccinees. J Clin Microbiol 1975; 1:311-7.

Zimmer K., Bogantes J.C., Herbst W., Rather W. Pockenvirusinfektion bei einer Katze und deren Besitzerin. Tierärztl Prax 1991; 19:423-7.

Zollman S., Godt D., Prive G.G., Couderc J.-L., Laski F.A. The BTB domain found primarily in zing finger proteins, defines an evolutionarily conserved family that includes several developmentally regulated genes in *Drosophila*. Proc Natl Acad Sci USA 1994; 91:10717-21.

Zuelzer W. Zur Aetiologie der Variola. Centr med Wissensch 1874; 12:82-3.

Index